Perfumes, Soaps, Detergents & Cosmetics

Volume I

(Soaps & Detergents)

Perfumes, Soaps, Detergents & Cosmetics

Volume I
(Soaps & Detergents)

S.C.Bhatia
B.E.(Chemical), M.B.A.

CBS

CBS PUBLISHERS & DISTRIBUTORS
4596/1A, 11 Darya Ganj, New Delhi - 110 002 (India)

ISBN : 81-239-0662-5

First Edition : 2001
_Reprint : 2009

Published by :
Satish Kumar Jain for CBS Publishers and Distributors,
4596/1-A, 11 Darya Ganj, New Delhi - 110 002 (India)

Printed at :
Asia Printograph,
Shahdara, Delhi - 110 032

Preface

This volume, dealing with soaps and detergents, is the first part of a comprehensive textbook on the perfume, soap, detergent and cosmetic industries. It is concerned primarily with the manufacturing processes, raw materials and equipment related to these industries but also considers ancillary topics needed for a full understanding of the subject. This reference textbook is intended for chemists, postgraduates in chemistry, specialists in oil, fats, soap and detergents, chemical engineers and research students.

Part I focuses on soaps and detergents. It covers all the operations that a practical manufacturing chemist will require and use, taking care to provide information that is accurate and up-to-date for readers. The purpose is to bridge the gap between general and specialised information on surfactants and detergency, to set forth the fundamentals and then to apply them to a practical process.

Written in a simple and readable manner, the text includes dimensions of the theory of detergency and surfactants, the properties and roles of the various ingredients used, principles of formulation and the interpretation and use of analytical data. Also dealt with at length are environmental and safety factors.

Chapter 1 elaborates on raw materials used in soaps and detergents, keeping in mind the fact that in terms of tonnage, the Rs. 9,000 crore soaps and detergents industry in India is the biggest in the world, ahead of even the U.S. and China. Soap technology has come a long way from earlier methods of kettle boiling and frame cooling and new raw materials are rapidly replacing the old. The emphasis in soap research is, therefore, increasingly on the interdependence of raw materials used, the process technology, material science aspects and the soap quality. After examining this important area, the next topic discussed is surfactants, which include detergents as cleaning agents. Surfactants, the performance chemicals used for human cleanliness and for industrial surface active applications, occupy a vital place in today's chemical industry. The main type of chemical surfactant products anionic, cationic and non-ionic find application in household products as well as personal care and industrial uses such as detergents, paints and coatings. Nowadays, thanks to the idiot box revolution, cosmetic and toiletry products are becoming increasingly popular as people become more hygienic and appearance conscious. Therefore, products like shampoos and bath products have generated a demand for lighter and milder surfactants. The latest surfactants are detergent sanitisers, made by germicidally treated equipment, and now popular in the food industry for the same reason.

Chapter 3 describes the principle groups of synthetic detergents, often preferred over common washing soaps on the ground of economics and efficiency. They also remain unaffected by the presence of naturally occurring salts and perform well even in hard water. These detergents have been developed to maintain their cleansing potentiality with practically no wastage, even in acidic water, unlike soaps which get partially decomposed in acidic water. Synthetic detergents on the other hand, improve and accelerate the cleansing property. The Chapter explains the groupings as materials based on molecular structure and their solubility or insolubility. Chapter 4 goes on to soapmaking oils and fats which are essentially glycerides and need pre-treatment before they can be manufactured into soap. The grouping of fatty acids and pre-treatment processes are also explained.

Soaps are covered in the next three Chapters. Every effort is made to establish the principles which control the elements of manufacturing processes. Chapter 5 explains the principles of the manufacture of soap base by direct saponification while Chapter 6 deals with fatty acids, chemical that have been produced, purified and separated into fractions for very many years. After discussing their manufacture and purification as well as their use in the production of soap base, the topic moves to the manufacture of solid household and toilet soaps from soap base in Chapter 7. Chapter 8 then goes on the recovery and refining of glycerine which is normally purified by distillation, sometimes followed by bleaching and after treatments. Alternative processes which use ion exclusion or ion exchange are also discussed.

There has been an emphasis recently on the development of detergents with improved washing power. In the detergent industry, distinctions are drawn between cleaning compositions on the basis of their functional utility. Generally, cleaning compositions for laundering purposes employ high foaming organic surfactants as the main cleaning agents and must wash both natural and synthetic fibres. Chapter 9 describes the industrial processes for the synthesis of detergents and the raw materials used. Chapter 10 goes on to the manufacture of finished detergents according to their classification, based on their mode of chemical action. An attempt has been made in Chapter 11 to incorporate a wide range of formulations of detergents, giving various process details and emphasising the techno-economical and functional aspects of the ingredients used. It also highlights the various industrial and household applications of detergents.

Chapter 12 explores the area of perfumery in soaps and detergents. A perfume is used to impart a 'pleasant' and 'suitable' odour to a detergent or soap to mask the mild but detectable odour which is derived from the raw materials in the formulation and can detract from the aesthetic character of the product. The perfumery raw materials and processes are explained, linking the topic to Part II of this textbook. In recent years, the alarming increase in pollution has underscored the need for biodegradable and non-polluting detergents and soap. Chapter 13 discusses environmental and safety factors, stressing on the control of foam and syndets in water systems as a first step in reducing the effects of chemicals used in the soap and detergent industries on the ecology. Ways of handling wastes discharged from laundry etc., are also highlighted. Safety evaluation is particularly important since detergents can cause skin allergies or irritation, giving hardship to consumers. Chapter 14 discusses ways of testing materials and current practices followed to improve their precision and accuracy. It explains how performance tests are ideal yardsticks for a practical and realistic evaluation of soap and detergents; with a statistical presentation of data. Tests discussed are radioactive tracer techniques, screening tests, cotton wash test methods, wool washing, detergency comparator test methods, washing procedures for other fibres, hard surface cleaning and miscellaneous tests.

This reference textbook also includes a glossary of technical terms that will help processors, consumers and others connected with this trade in correct interpretation of the terms. BIS specifications related to soap and detergents, plant machinery and raw material manufacturers are listed for quick reference.

The text is throughout supplemented by Flow diagrams, Figures and Tables wherever needed.

The author, S. C. Bhatia, is a chemical engineer with management qualifications who has written several books on chemical and allied subjects. At present, he is a renowned consultant in the field of environment, waste heat recovery/energy conservation and petrochemicals.

Acknowledgements are due to Mr. Keshav Kumar, the Computer Operator who worked long hours to complete the task of bringing out the book on time. Also to Ms. Dipali Singh (M. A. Eng., Miranda College, Delhi University) for editing and proof-reading and Mr. Maduresh Kumar for proof-reading. The author also wishes to express his appreciation for the work of Mr. Prakash and Mr. Santosh Kumar Shrivastava who drew and labelled the Flow diagrams.

S. C. Bhatia

CONTENTS AT A GLANCE

Volume I: Soaps and Detergents

x Contents at a Glance

CONTENTS IN DETAIL

Volume I: Soaps and Detergents

CHAPTER 2

CHAPTER 3

CHAPTER 4
OILS, FATS AND THEIR PRETREATMENT 175-204

CHAPTER 5

CHAPTER 6

CHAPTER 7

MANUFACTURE OF SOLID, HOUSEHOLD AND

CHAPTER 8

CHAPTER 9

CHAPTER 11

CHAPTER 12

CHAPTER 13

CHAPTER 14

Chapter 1

Raw Materials Used in Soap Industry

INTRODUCTION

Although some cleaning operations can be carried out with organic solvents and other materials, most are done with water; and the detergents (cleaning agents) discussed in this book are intended to improve the cleaning performance of water.

Fats from domestic animals which had been killed for food, and olive oil (which could be obtained more readily than other vegetable oils), were used for the manufacture of soap. Until caustic alkalis became available, saponification must have been difficult and probably incomplete. The fats may have been allowed to deteriorate with the development of free fatty acids which react readily with alkali carbonates; and the soap so formed would have helped the reaction. The process has been said to have been one in which the fats and alkali carbonate solution were boiled to dryness in a pot over an open fire. More water was added, and the mixture boiled down again; this process being repeated as many times as necessary. The first reaction may have been slow hydrolysis of the fat to give free fatty acid which was then neutralised by the alkali carbonate to give soap. If sodium carbonate was used, a 'hard soap' was produced, but with potassium carbonate the product would have been some sort of 'soft soap', unless a further stage of salting out with common salt (sodium chloride) was employed to convert the potassium soap to sodium soap.

As the chemical industry developed, alkalis (sodium carbonate and sodium hydroxide) made from salt have largely replaced the sources mentioned earlier. Soap cannot be used in acid solution (because it is decomposed with the liberation of fatty acids); and it forms a precipitate

1

with the calcium and magnesium in hard water. Hence, although soap is a good all-round detergent, it has limitations, particularly for some industrial purposes in the textile field.

RAW MATERIALS

Surfactants, Builders and Fillers

Soaps, and the alternative non-soapy detergents, have a number of important properties and one such property is the reduction of the surface energy, or surface tension (more strictly interfacial tension), of solutions which contain them, so that they are called surface active agents. This is commonly contracted to *surfactants*. Surfactants contain a hydrophobic (water hating, or repelling) and a hydrophilic (water loving, or attracting) end. Alternatively, this can be expressed as a lipophilic (fat loving, equals hydrophobic) and a lipophobic (fat hating, equals hydrophilic) end. Such chemical compounds are having amphipathic properties. Non-soapy surfactants are also sometimes called *syndets*.

Solutions of acids, bases (alkalis) and salts contain charged particles called ions. Ions which carry positive electrical charges are called cations and others which carry negative charges are called anions. Ions can carry one unit of charge (called monovalent or univalent), two units (divalent), and so on. Examples are :

Hydrochloric acid which ionizes $HCl = H^+ + Cl^-$

Sulphuric acid which ionizes $H_2SO_4 = H^+ + HSO_4^-$ $HSO_4^- = H^+ + SO_4^{2-}$

Sodium hydroxide which ionizes $NaOH = Na^+ + OH^-$

Ammonium hydroxide which ionizes $NH_4OH = NH_4^+ + OH^-$

Sodium carbonate which ionizes $Na_2CO_3 = 2Na^+ + CO_3^{2-}$

Magnesium sulphate which ionizes $MgSO_4 = Mg^{2+} + SO_4^{2-}$

Trisodium orthophosphate which ionizes $Na_3PO_4 = 3Na^+ + PO_4^{3-}$

All normal chemical compounds are electrically neutral, so that the charges on the positive and negative ions which they contain, or produce in aqueous solution, balance. That is sodium chloride and magnesium sulphate are Na^+Cl^- and $Mg^{2+}SO_4^{2-}$ respectively, but sodium sulphate is $Na_2^+SO_4^{2-}$.

It should be noted that chloride ions (Cl^-) have completely different properties from those of chlorine gas (Cl_2). Similarly, the sodium ion (Na^+) differs from metallic sodium (Na).

Many surfactants ionise; and those with the hydrophobic-hydrophilic portion in the anion are called anionic surfactants. Examples are :

(i) Soap, which is a mixture of salts (usually sodium) of long chain fatty acids, $R.COO^-Na^+$, where R represents a long hydrocarbon chain.

(ii) Alkyl benzene sulphonates, $R - C_6H_4 - SO_3^-N_a^+$

Similarly, surfactants with the hydrophobic-hydrophilic portion in the cation are described as cationic. Examples are :

(i) Quaternary ammonium salts, such as alkyl trimethyl ammonium chloride, $R—N(CH_3)_3^+.Cl^-$.

(ii) Alkyl pyridinium salts, such as cetyl pyridinium bromide, $C_{16}H_{33}—C_5H_4N^+Br^-$.

The balancing ions, such as Na^+ or Cl^-, in anionic and cationic surfactants respectively, are sometimes called *gengenions*. Unlike the anionics, which are the most important group of surfactants used in detergent products, the cationics are used mainly in fabric softeners and as germicides.

A third group of surfactants, which is becoming increasingly important in the detergents field is the *nonionic* type. By definition, these compounds do not ionise, but they contain a hydrophobic and a hydrophilic end to make them surfactants. There are naturally occurring nonionic surfactants, such as saponin, but those of importance in the detergents industry are synthetic. The hydrophobic portion is normally an alkyl chain, or an alkyl substituted aromatic hydrocarbon. The hydrophilic portion is usually a polyglycol chain produced by the condensation of ethylene oxide with itself and with an active hydrogen atom in the hydrophobe molecule. In the most important sub-group, the active hydrogen atom is in an hydroxyl group of a long chain alcohol, or an alkyl phenol. The reactions can be illustrated by reference to a long chain alcohol, R—OH.

$$R—OH + CH_2\overset{\diagdown}{\underset{\diagup}{}}CH_2 \rightarrow R—O—CH_2—CH_2—OH$$
$$O$$

This compound, in which one molecule of ethylene oxide has been attached, is itself an alcohol; and more ethylene oxides can be added similarly to form a polyglycol chain containing n—O—CH_2—CH_2—groups.

$$R—O—CH_2—CH_2—OH + n(CH_2)_2O \rightarrow R—O—CH_2—CH_2—$$
$$(O—CH_2—CH_2)_{n-1}—O—CH_2—CH_2—OH$$

Other compounds to which ethylene oxide can be added to form nonionic surfactants are fatty acids (in which the carboxyl group, –COOH, contains an –OH), long chain amines and long chain amides, in which there is an active hydrogen atom in the $–NH_2$ group.

The ethylene oxide condensates just considered end with an alcohol group; and this can be sulphated to produce another type of anionic surfactant.

Certain nonionic surfactants are made with propylene oxide in conjunction with ethylene oxide. There is also a fourth group of surfactants known as the *amphoterics*. These can ionise as either anionic or cationic substances, depending upon the alkalinity, or acidity, of the solution in which they are present. Amphoteric surfactants are generally expensive and used only in rather specialised products. A very large number of surfactants have been developed and the more important types, with their properties and uses, are discussed in Chapter 2.

It will be appreciated that soap is a detergent, but the word 'detergent' is frequently confined to the non-soapy surfactants and products based upon them. In this book the term 'detergent' will be taken to include soap; and non-soapy types will be called non-soapy detergents (abbreviated to NSDs). NSDs are also called soapless detergents, or synthetic detergents. The latter term is not very appropriate, as soaps can also be made from fatty acids produced by the oxidation of petroleum-based hydrocarbons.

Soaps and non-soapy surfactants may require the addition of other substances to make them perform well. This is particularly so with the NDSs, which need additives to make them effective heavy duty fabric washing products capable of dealing with really dirty clothes and other articles. Such additives are called *builders*. The overall cleaning effect of a detergent solution is the net result of soil removal and soil redeposition; and some authors say that only substances which perform both functions should be called builders. *Fillers* are substances added to a detergent to cheapen the product. In some cases the distinction between fillers and builders is blurred. For example, sodium silicate and sodium carbonate which are, or were, added to household soaps to reduce cost also have some beneficial effects.

TYPES OF SOAPS AND SOAPMAKING TERMS

The oils and fats used in the manufacture of soap are essential triglycerides, that is esters of the trihydric alcohol, glycerol. Here the common, but not universal, convention which reserves the word *'glycerol'* for the pure compound, $CH_2OH–CHOH–CH_2OH$ (1,2,3 trihydroxypropane or 1,2,3 propanetriol), and calls the actual materials *'glycerines'*, is followed. For example, a soap lye crude glycerine may be found to contain 83.5% glycerol and a distilled glycerine 99.1% glycerol. The fatty acids which are combined with the glycerol in a fat are almost exclusively straight chain monocarboxylic acids with an even number of carbon atoms, including the one in the carboxylic group. For example :

Stearic acid	$C_{17}H_{35}.COOH$
Oleic acid	$C_{17}H_{33}.COOH$
Lauric acid	$C_{11}H_{23}.COOH$

The triglycerides, which comprise the bulk of an oil or fat which is fresh, have the formula $CH_2-O-OCR_1-CH-O-OCR_2-CH_2-O-OC-R_3$ where R_1, R_2 and R_3 are long chain alkyl groups which may be the same, or different. Here it is only necessary to add that an oil is a fat which is liquid under the prevailing conditions, but the distinction is somewhat arbitrary and a matter of convention. Coconut oil is normally solid in temperate areas, but it is called an oil because it is liquid in its tropical countries of origin. Tallow, which is a typical fat, is now commonly transported as an oil in heated road and ships' tanks.

The soaps used as detergents are the soluble salts of mixtures of fatty acids. The cation is usually sodium, but potassium, or a mixture of sodium and potassium, is used in soft and shaving soaps. Ammonia (giving ammonium soaps) and various organic nitrogenous bases are sometimes used to neutralise the fatty acids in specialist products. The fatty acid salts of other metals are substantially insoluble. Those of calcium and magnesium are formed as undesirable precipitates when soap is used in hard water, but they, and some other insoluble soaps, are used for various industrial purposes.

Household sodium soaps are called '*hard soaps*' to distinguish them from jelly-like '*soft soaps*', which were at one time used extensively in the textile industry and the home. As already mentioned, *soft soaps* are commonly made with potassium as the cation; and from liquid oils. The term hard soap can be confusing in that these soaps are normally softer when made than the other main group of *solid soaps* and the *toilet soaps*.

Sodium soaps can be made by :

(i) *Direct saponification,* almost always with caustic soda :

$$CH_2.O.OCR.CH.O.OCR.CH_2.O.OCR + 3NaOH \rightarrow$$

fat   caustic soda

$$CH_2OH-CHOH-CH_2OH + 3R.COO\ Na$$

glycerol   soap

(ii) *Splitting*, or *hydrolysis*, of the fat with water to fatty acids plus glycerol; followed by separation of the glycerol, and neutralisation of the fatty acids with caustic soda, or sodium carbonate, to form soap.

(iii) *Neutralisation of fatty acids produced in other ways*, particularly by the oxidation of a petroleum hydrocarbon.

A major distinction is between soaps made by an '*all-in*' process and soaps made by the '*fully-boiled*' process. All-in soaps are so-called because the fat (or fatty acid) and the alkali are just mixed, so that everything put into the pot goes into the product. Glycerol is not separated; and it is important to avoid any undue excess of alkali. All-in processes are divided into '*semi-boiled*', in which external heat is applied, and '*cold-process*' in which it is not, the only heat coming from the heat of the saponification reaction. All-in processes can be carried out very simply; and they are still used by some small local manufacturers in some of the less-developed countries, where these manufacturers are sometimes known as traditional soapmakers. Soaps which include potassium are made by an all-in process, because graining with salt, which is an inherent part of the fully-boiled process, would displace potassium by sodium.

Most soap made by direct saponification of fats is produced by some version of the fully-boiled process. This process includes three main stages, although other operations are sometimes added :

(i) *Saponification*, the chemical reaction which forms soap and liberates glycerol. This is sometimes called *killing* the fat.

(ii) *Washing*, in which the soap mass is *grained*, or *salted-out*, with salt (or other electrolyte) and, on standing, separates into a lower aqueous layer called *lye*, and an upper layer of *soap curd*, also called *kettle wax*. The soap curd contains a proportion of enmeshed lye. A washing operation transfers part of the glycerol to the lye; and also removes dirt from the soap.

(iii) *Fitting* or *finishing*. In this operation the electrolyte content of the soap mass is adjusted so that on prolonged standing (or passage through a centrifuge) an upper layer of *neat soap*, or *soap base*, and a lower layer of *nigre*, or *nigre lye*, is obtained. The nigre is reprocessed. Fitting achieves a further purification of the soap and produces the neat soap in a desirable physico-chemical state.

Removal of the neat soap from a vessel in which it is settled is called cleansing. Soapmakers use the term *closing* to describe a reduction in electrolyte content (with consequent reduction in tendency to separate) and *opening* for the converse operation. Those terms, however, have rather electrolyte content (with consequent reduction in tendency to separate) and *opening* for the converse operation. Those terms, however, have rather different connotations in relation to the washing and the fitting stages.

The large vessels in which soap is made in the traditional form of the process are often called *pans* or *kettles*.

Bar soaps, known as *curd soaps*, were made at one time by an elaboration of the washing stage just described; but neat soap is the normal starting material for the production of most soap products (hence the name 'soap base'). If the fitting stage is carried out properly, and the soap is well-settled, the neat soap has a fairly constant composition, although this can be varied somewhat. Usually neat soap contains about 61–64% *TFM* (*total fatty matter*), equivalent to about 67–69% anhydrous soap and 30–32% water.

When neat soap is cooled with no more than minor additions (if any), a *genuine soap* is produced. It is also possible to reduce the soap content of the product by addition of suitable alkaline solutions (or dilute brine) to give what are called *filled*, or *run*, soaps. A special type of filled soap is described as *glued-up*.

Genuine soaps made by the older processes using frames, or water coolers, have substantially the same composition as the base from which they are produced. With the modern spray-chilling and extrusion processes, the distinction between genuine and filled soaps is obscured by the fact that evaporation occurs during the process and dilute electrolyte has to be added if it is required to bring the TFM back to 62–64%. Genuine and filled soaps can be sold as *bars*, or cut into *billets* and stamped into *tablets*. These soaps are known by various names: *laundry*, *common*, or *household*, as well as *hard* or *bar*.

To a soapmaker a *lye* is an aqueous liquor which normally contains only a negligible concentration of soap, the exception being a *thick*, or *nigre lye* which may carry about 1–3% soap. The term lye is used in several ways. It came originally from the alkali industry, where it means a solution of caustic soda (or possibly other alkali); and it is sometimes used in this sense in the soap industry, particularly by older soapmakers. The lye separated from soap curd in the washing stage of soapmaking is called *spent lye* provided it contains little caustic soda. If it contains a substantial concentration of caustic soda (and hence requires to be processed further in the soapmaking department), it is called *half-spent lye*. The glycerol concentration in the spent lye produced is very important; but when a soapmaker speaks of the strength of a lye (or a soap mass), the reference is normally to caustic soda rather than glycerol content.

When fats are hydrolysed with water, with or without a catalyst, the process is often called *splitting*, and the aqueous glycerine separated *sweet water*. In some countries, soap lyes are called sweet water, and some glycerine distillers use the term sweet water for the dilute liquor produced in some types of still, which is otherwise called *weaks*.

The term *lye* is used for any liquor separated after soap has been grained, and *sweet water* for the glycerol bearing liquor from a splitting operation.

ACIDS, ALKALIS, BUFFERS, pH AND INDICATORS

Acids and bases are now defined as proton donors and proton receivers respectively, but for present purposes, the older and narrower definition of an acid as a substance which yields hydrogen ions in solution and a base (alkali) as one which yields hydroxyl ions is adequate. Actually, all aqueous solutions contain both hydrogen and hydroxyl ions, and the definitions should be in terms of an excess of either hydrogen or hydroxyl ions.

Pure water itself ionises reversibly to a very small extent, $H_2O \rightleftharpoons H^+ + OH$, and the equilibrium constant is equal to $\dfrac{[H^+] \times [OH^-]}{[H_2O]}$,

where $[H^+]$ = concentration of hydrogen ions in gram equivalents (moles) per litre.

$[OH^-]$ = concentration of hydroxyl ions in gram equivalents (moles) per litre.

and $[H_2O]$ = concentration of molecular water in gram equivalents (moles) per litre.

As the extent of the ionisation is very slight, the concentration of molecular water is constant, and $[H^+] \times [OH^-] = K$, the ionic product for water. At 25°C, $K = 10^{-14}$, so that $[H^+] \times [OH^-] = 10^{-14}$. To avoid the need to consider concentrations in terms of 10 raised to various negative powers, the pH and pOH scales have been devised. pH is defined as $-\log_{10}[H^+]$ and pOH as $-\log_{10}[OH^-]$. The ionic product equation can then be rewritten pH + pOH = 14. From the definitions of pH and pOH it follows :

(i) The lower the pH the higher the concentration of hydrogen ions, that is the more acidic the solution; and similarly for pOH.

(ii) Because the scales are logarithmic, a solution with a pH of 3 contains ten times the hydrogen ion concentration of one with a pH of 4.

From the relationship pH + pOH = 14, pOH = 14 − pH, and for most purposes the pOH scale is unnecessary, as the whole range of acidity and alkalinity can be expressed in terms of pH.

In solutions which are neutral (neither acidic nor alkaline) pH = pOH = 7; and those with a pH below 7 are acidic and those with a pH above 7 are alkaline, or basic.

A decinormal (N/10) solution of any strong acid, assumed to be fully ionised, contains 0.1 gram equivalents, or moles, of hydrogen ions per litre, that is $[H^+] = 0.1$; and its pH = $-\log_{10}[0.1] = 1$. More concentrated solutions of strong acids can have a negative pH, as well as one between 1 and 0. Similarly, an N/10 solution of a strong base, say caustic soda, has $[OH^-] = 0.1$ and a pOH of 1. Its pH is, therefore, 14 − 1 = 13. Again, more concentrated solutions of caustic soda have a pH above 13.

To set N/10 caustic soda in context with other alkalis to be discussed later, it contains 4 g/l of NaOH, or 0.4% m/m as its specific gravity is approximately 1.

Strong lavatory cleaners are commonly acids, but most other cleaning with aqueous solutions requires alkaline conditions. Both pH (alkaline potential) and quantity of alkali present are important, but should be distinguished. From the previous discussion it will be noted that an N/10 solution of caustic soda has a high alkaline potential, but only a small alkaline capacity (that is, it will only neutralise a small quantity of acid).

Classes of Salts

Four classes of salts, those from strong acid-strong base, weak acid-strong base, strong acid-weak base, and weak acid-weak base are now considered.

Salts of a strong acid and a strong base

Solutions of salts of strong acids and strong bases, such as sodium chloride and sodium sulphate, are substantially neutral. The alkylbenzene sulphonic acids are strong, so that the sodium alkylbenzene sulphonates are neutral.

Salts of a weak acid and a strong base

The alkalis used in detergent products are nearly always salts of this type, so that their properties are important to the detergents technologist.

Consider the addition to pure water of sodium acetate, $CH_3.COONa$, a substance similar in many ways to a soap, although acetic acid is soluble in water and stronger than the fatty acids in soap (but still weak compared with hydrochloric acid). The anion of the weak acid, the acetate ion in this example, reacts with water to form molecular acetic acid and hydroxyl ions

$$CH_3.COO^- + H_2O \rightarrow CH_3.COOH + OH^-$$

so that the solution becomes alkaline. A more complete way to consider the situation is to say that the solution contains sodium ions (Na^+) and that all the following reactions must reach and maintain equilibrium :

$$CH_3.COO^- + H_2O \rightleftharpoons CH_3.COOH + OH^- \tag{a}$$

$$H_2O \rightleftharpoons H^+ + OH^- \tag{b}$$

$$CH_3.COOH \rightleftharpoons H^+ + CH_3COO^- \tag{c}$$

The sequence of events can be indicated as follows :

(1) Reaction (a) forward increases $[OH^-]$.

(2) The increase in $[OH^-]$ from (1) causes a decrease in $[H^+]$ from (b) in reverse.

(3) The decrease in $[H^+]$ from (2) leads to an increase in $[CH_3.COO^-]$ from (c) forward.

(4) The increase in $[CH_3.COO^-]$ from (3) causes further increase in $[OH^-]$ from (1) above, and so on until all the equilibria are established.

The sodium ion in the sodium acetate solution does not react significantly with water, $Na^+ + H_2O \rightarrow NaOH + H^+$ because NaOH is a strong base which ionises $NaOH \rightarrow Na^+ + OH^-$, followed by $H^+ + OH^- \rightarrow H_2O$, which returns the situation to $Na^+ + H_2O$.

The weaker the acid, the higher the concentration of unionised acid and the lower that of its ions; hence the higher the concentration of hydroxyl ions produced and the more alkaline the solution. A solution of a pure soap develops a pH around 9, depending upon the type of soap and its concentration. Carbonic acid and the silicic acids are very weak, so that their sodium salts give more alkaline solutions than soap. Sodium carbonate is strongly alkaline and sodium bicarbonate (which can be regarded as carbonic acid half neutralised with caustic soda) is mildly alkaline.

The hydrolysis reaction for the bicarbonate ion is $HCO_3^- + H_2O \rightleftharpoons H_2CO_3 + OH^-$. That for the carbonate ion takes place in two stages

$$CO_3^{2-} + H_2O \rightleftharpoons HCO_3^- + OH^-$$

$$HCO_3^- + H_2O \rightleftharpoons H_2CO_3 + OH^-$$

Salts of a strong acid and a weak base

In this case the cation of the weak base reacts with water leaving the solution acidic. For example with ammonium chloride $NH_4^+ + H_2O \rightleftharpoons NH_4OH + H^+$.

Salts of a weak acid and a weak base

In a solution of a salt of a weak acid and a weak base both cation and anion react with water. For ammonium acetate

$$NH_4^+ + CH_3.COO^- + H_2O \rightleftharpoons NH_4OH + CH_3.COOH$$

The pH of the solution then depends upon the relative strengths of the acid and base.

The different strengths of sulphuric and carbonic acids cause sodium bisulphate ($NaHSO_4$) to be a strong acid whereas sodium bicarbonate ($NaHCO_3$) is a weak alkali, although both are their respective acids half neutralised with caustic soda.

Since the pH of some important alkalis varies with concentration. It will be noted that the pH of solutions of the less strong alkalis tends to level off as concentration increases and never reaches the levels produced by caustic soda and the very alkaline silicates. In some cases it even falls with an increase in concentration. Table 1.1 gives information concerning the pH of 1% m/m solutions of various alkalis.

A buffer, or buffer solution, is one in which the pH changes relatively little when an acid or alkali is added to it. Many 'soils' (i.e. the matter to be removed by the washing process) (particularly those which contain sweat) include acidic constituents, and it is important that detergent solutions should remain at near the selected pH throughout the washing process; this requires a buffering effect. Buffers are also used to produce solutions of known pH, for example, in the preparation of a set of comparator tubes which show the colours developed by an indicator at a series of known pH values.

Table 1.1. Approximate pH values for 1% m/m solutions of various alkalis.

Alkali	pH 1% m/m solution
Caustic soda	13.2
Sodium orthosilicate, anhydrous	12.9
Sodium sesquisilicate, anhydrous	12.6
Sodium metasilicate, anhydrous	12.3
Trisodium orthophosphate, anhydrous	12.0
Ammonium hydroxide	11.4
Sodium carbonate, anhydrous	11.4
Alkaline sodium silicate, anhydrous	11.2
Sodium percarbonate	10.6
Neutral sodium silicate, anhydrous	10.5
Sodium pyrophosphate, anhydrous	10.5
Sodium perborate, $NaBO_2.H_2O_2.3H_2O$	10.2
Sodium sesquicarbonate, $Na_2CO_3.NaHCO_3.2H_2O$	10.0
Sodium tripolyphosphate, anhydrous	9.5
Borax, $Na_2B_4O_7.10H_2O$	9.2
Disodium hydrogen phosphate, anhydrous	8.8
Sodium bicarbonate	8.4
Graham's salt (Calgon), sometimes called sodium hexametaphosphate	6.9

Unless there are other differences in composition, as may occur in commercial materials, a hydrate should show the same pH as its anhydrous salt after adjustment of concentration to the same level of anhydrous salt in the two solutions.

Effective buffers are produced by a combination of a weak acid and its salt, or a weak base and its salt. Examples are :

Boric acid (HBO_2) and borax ($Na_4B_2O_7.10H_2O$),

Sodium bicarbonate ($NaHCO_3$) and sodium carbonate (Na_2CO_3), and

sodium dihydrogen orthophosphate (NaH_2PO_4), and

disodium hydrogen phosphate (Na_2HPO_4).

In the second and third examples, the sodium bicarbonate and the sodium dihydrogen phosphate act as weak acids. The sodium silicates work in a similar way and are very effective buffers.

The buffering action of a weak acid (HA) and its sodium salt (NaA) can be explained as follows.

The solution contains Na^+ and A^- ions from the salt, together with the unionised acid, its ions and water. The equilibria

$$HA \rightleftharpoons H^+ + A^-$$ (a)

$$\text{and} \quad H_2O \rightleftharpoons H^+ + OH^-$$ (b)

have to remain satisfied.

If acid is added to the system $[H^+]$ is increased, but it is then reduced nearly to its original value by reaction (a) in reverse. If alkali is added, OH⁻ is temporarily increased, but is reduced by reaction (b) in reverse, leaving less $[H^+]$. HA then ionises according to (a) to restore equilibrium at near its original value.

The changes in pH as an alkali is added to an acid, as in an analytical titration, depend upon the strengths of the acid and the alkali. The end point in a titration can be determined using an electrical pH meter, or from conductivity measurements, but indicators which change colour at certain values of pH are commonly used, particularly in factory practice. Many indicators are available and their colour changes occur at different pH ranges; this makes the proper selection of indicator important in many cases. Some common indicators with the pH range over which the colour change occurs are shown in Table 1.2. In order to obtain the correct end point in a volumetric analysis it is necessary to select an indicator which changes colour at near the equivalence point. Suitable indicators can be selected as shown in Table 1.3 by combining the information in Table 1.2.

Table 1.2. Some common indicators.

Indicator	Colour change	pH range
Methyl orange	red - yellow orange	3.1 – 4.5
Bromcresol green	yellow - blue	3.8 – 5.4
Methyl red	red - yellow	4.7 – 6.4
Litmus	red - blue	5.0 – 7.0
Cresol red	yellow - red	7.2 – 8.8
Thymol blue	yellow - blue	8.0 – 9.6
Phenolphthalein	colourless - purple	8.5 – 10.5
Thymolphthalein	colourless - blue	9.4 – 10.6

Table 1.3.   Indicator selection.

Titration	Approx. pH range at equivalence point	Suitable indicator
Strong acid - strong base	4 - 10	Almost any
Weak acid - strong base	7 - 10	Phenolphthalein
Strong acid - weak base	4 - 7	Methyl orange, methyl red
Weak acid - weak base	very narrow	No indicator satisfactory

Some important applications in the detergents industry are :

(i)   The estimation of *free fatty acid (FFA)* in a fat is carried out in alcoholic solution (FFA is soluble in alcohol), by titration with standard caustic soda, or caustic potash, to the phenolphthalein end point (estimation of a weak acid with a strong base).

(ii)   To estimate *free caustic alkali in a soap*, the sample is dissolved in hot alcohol (neutralised if necessary), barium chloride solution is added to precipitate any sodium carbonate present, and the caustic soda is titrated with standard sulphuric acid to the phenolphthalein end point. Phenolphthalein is chosen because it is required to estimate only the strong alkali, caustic soda, and not to include mild alkalis such as soap. It is necessary to remove sodium carbonate from solution because it is alkaline to phenolphthalein (in the titration of a sodium carbonate solution the phenolphthalein end point is at the bicarbonate stage).

Soaps are often checked for caustic soda content by 'spotting' a freshly cut surface with phenolphthalein solution, noting whether a purple colour develops, and, if so, how quickly. With experience it is possible to judge roughly the free caustic soda content of the sample. Soapmakers now use this test, instead of the traditional testing on the tongue, at relatively high levels of caustic soda. For the routine examination of production samples of toilet soap, which should contain little, or no, free caustic soda, spotting is useful to decide whether, or not, a quantitative estimation is necessary.

(iii)   If it is required to estimate the *total free alkali* (hydroxide plus carbonate) in a soap, an excess of standard sulphuric acid is added to the sample in alcoholic solution, the mixture is boiled to remove carbon dioxide liberated from the carbonate, and the excess acid (actually fatty acid liberated from the soap by the stronger sulphuric acid) is back titrated with standard caustic soda to the phenolphthalein end point.

(iv)   An important sequence of tests is used to determine *total fatty matter (TFM), total alkali* and *'soda as soap'*. The sample of soap is decomposed by the addition of a known quantity of standard acid, the fatty matter liberated is separated (by ether extraction) and weighed. The excess acid is back titrated with standard alkali to the methyl orange end point. Methyl orange is selected for this estimation, because it is required to include alkali combined with weak acids, such as silicic and boric. Hence weak acids left in the aqueous liquid from which the fatty matter has been separated should not be allowed to act as acids in the analysis. The alkali which is combined with the fatty acids to form soap, known as 'soda as soap', is determined by dissolving the extracted fatty matter in neutral alcohol and titrating to the phenolphthalein end point with standard caustic soda. In many routine tests, it is only required to estimate TFM. It is then not necessary to use standard acid, and it is possible in many cases to add a weighed piece of beeswax and to separate the fatty matter which has been liberated from the soap in the form of a wax cake. The weight of TFM is found by the deduction of the weight of the added beeswax from that of the cake. Some sample calculations using the results of tests such as those just outlined are given later, after various materials have been discussed.

INORGANIC MATERIALS IMPORTANT IN THE DETERGENTS INDUSTRY

Common Salt, or Sodium Chloride (NaCl)

Salt is used mainly to grain soap. It can be purchased in crystalline form, but it is now often bought as a saturated brine containing about 25% NaCl. When crude glycerine is being recovered from soap lyes, a major part of the salt in the lyes is recovered in the evaporation process and is returned to soapmaking as either solid salt, or brine. Salt is now often recovered directly as brine. In any case some new salt has to be purchased to make up for losses.

Solid salt is used in some soapmaking procedures, but modern processes usually require saturated brine, made by dissolving salt in water, or in a liquor containing a small concentration of glycerol. A convenient system for the preparation of brine on the large scale is shown diagrammatically in Fig. 1.1. The salt is fed in a stream (usually by a worm conveyor) to the top of a small tower to the base of which water is introduced at the required rate. The slurry overflows to one of a group of tanks in which dissolution is completed by compressed air introduced through a perforated pipe. With experience, the flows of water and salt can be adjusted so that saturated brine is produced, leaving only a small excess of salt in the tanks. This has to be dissolved occasionally by addition of water to the tank and blowing with air. If the salt is slightly dirty, or if it contains calcium or magnesium, it is often useful to make the brine slightly alkaline and to filter it before it is sent to the soapmaking department. This also reduces the content of heavy metals, such as copper and iron.

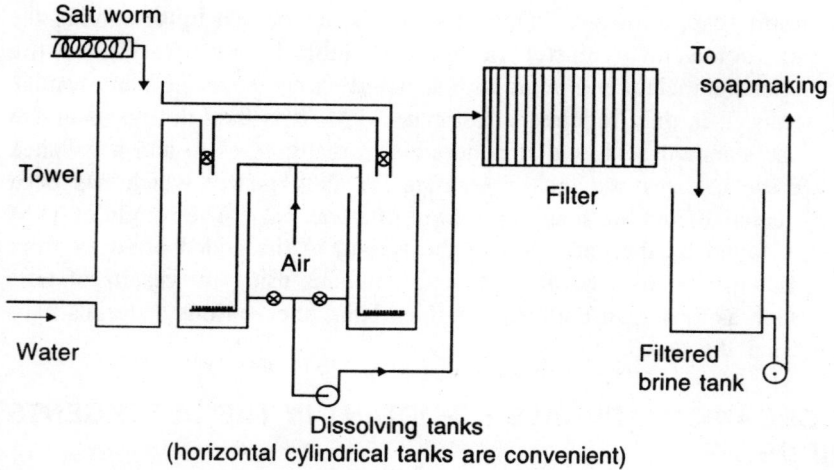

Fig. 1.1. Brine preparation.

Caustic Soda or Sodium Hydroxide (NaOH)

Caustic soda is sold as powder, flake, solid blocks produced by running molten, anhydrous, material into lightweight steel drums, and also as a concentrated solution containing 45–50% NaOH. Detergent manufacturers normally need to dissolve solid to produce a concentrated solution before use and, when possible, there are considerable advantages in buying solution, provided transport costs are not excessive. Caustic soda is a dangerous chemical and adequate precautions, including the wearing of goggles and other protective clothing, are important when it is being handled. Efficient soapmaking normally requires the use of a concentrated solution containing 45–50% NaOH which is subsequently diluted with lyes. The heat evolved when solid caustic soda is dissolved to produce these concentrated solutions is sufficient to raise the temperature to near boiling point. Solid blocks in steel drums are normally the cheapest form of solid, and they can be dissolved by cutting off the ends of the drums with hammer and chisel (and sometimes perforating the cylinders with a pick) before placing the material in the dissolving tank with suitable lifting tackle. The tank should be sized so that the rather small quantity of water needed to produce a solution of the required concentration covers the drum, or drums. Some operators use a tank with a perforated plate, or grid, near the base. The rise in temperature and density gradients provide circulation and reasonably rapid solution. The empty drum shells must be rinsed with a little water from a hose to prevent loss of alkali and to make the metal safe to handle.

It is often convenient to express quantities and concentrations of alkaline materials in terms of Na_2O. The conversion factors can be calculated by the usual methods of elementary chemistry. Using a table of atomic masses, and considering caustic soda as an example,

$$2 \ NaOH = Na_2O + H_2O$$

$$2 \times (23 + 16 + 1) = (46 + 16) + (2 + 16)$$

$$\text{or } 2 \times 40 = 80 \qquad 62 \qquad 18$$

The equivalent mass of $Na_2O = 62/2 = 31$; and $x\%$ of NaOH is equivalent to $x \times 31/40 = 0.775x\% \ Na_2O$.

The concentration of a caustic soda solution is often expressed, and determined approximately, from its density measured at a known temperature around ambient. Density is mass per unit volume (that is g/ml, or lb/ft^3), but it is often expressed in terms of specific gravity, or relative density, which is density relative to that of water. Table 1.4 gives approximate relationships between concentration and density for some caustic soda solutions.

Caustic Potash or Potassium Hydroxide (KOH)

Considerations similar to those for caustic soda apply to caustic potash. In most factories much less caustic potash than caustic soda is used, and somewhat less concentrated solutions are usually adequate. The heat of solution is even greater than that for caustic soda and solutions are not normally produced at more than about 40% KOH. The equivalent mass of caustic potash is $39.1 + 16 + 1 = 56.1$ and that of K_2O is $[(2 \times 39.1) + 16]/2 = 47.1$, so that $y\%$ KOH is equivalent to $y \times (47.1/56.1) = 0.84y\% \ K_2O$, $y \times (40/56.1) = 0.713y\% \ NaOH$, or $y \times (31/56.1) = 0.553y\% \ Na_2O$.

Table 1.4.   Approximate concentrations and densities of caustic soda solutions at 15.6°C (60°F).

% NaOH	% Na$_2$O	S.G.	°Tw	°Be
1	0.78	1.01	2	1.4
2	1.55	1.02	4	2.8
4	3.10	1.04	8	5.6
8	6.20	1.09	18	12.0
10	7.75	1.11	22	14.4
20	15.5	1.22	44	26.1
25	19.4	1.28	56	31.7
30	23.3	1.33	66	36.0
35	27.1	1.38	76	39.9
40	31.0	1.43	86	43.6
45	34.9	1.48	96	47.0
50	38.8	1.53	106	50.2

The concentrated solution containing 45–50% NaOH is commonly called 100°Tw liquor.

Sodium Carbonate (Na$_2$CO$_3$)

Anhydrous sodium carbonate, commonly known as soda ash, is available in light and heavy forms, which essentially differ only in bulk density. Most sodium carbonate is produced from salt. Care is necessary in selecting natural sodium carbonate for use in detergent products because it is liable to contain iron and organic dirt.

Sodium carbonate forms hydrates, the decahydrate, Na$_2$CO$_3$.10H$_2$O, being traditional washing soda once widely found in households. Other hydrates are Na$_2$CO$_3$.H$_2$O and Na$_2$CO$_3$.7H$_2$O. Detergent manufacturers seldom need to buy hydrated sodium carbonate, but the formation of these hydrates is very important in the manufacture of certain types of soap and NSD powders.

Sodium Bicarbonate (NaHCO$_3$)

Sodium bicarbonate is a fine white powder which is only mildly alkaline, so that it is little used in detergent formulations. Mixtures of pure soap powder and sodium bicarbonate have been made for special purposes, and the buffering action of sodium bicarbonate may be useful in reducing the pH of certain high carbonate formulations.

Sodium Sesquicarbonate ($Na_2CO_3.NaHCO_3.2H_2O$)

Sodium sesquicarbonate, also known as 'modified alkali', exists in the form of needle-shaped crystals. It is intermediate in alkalinity between sodium carbonate and sodium bicarbonate, and has a number of uses. In particular :

(i)   As a mild alkali in commercial laundries, although this use is declining.

(ii) As a major constituent of certain floor and other hard surface cleaners.

The Sodium Silicates

A number of different types of sodium silicate are available and are important in many industries including detergents. They can be regarded as combinations of SiO_2 (silicon dioxide), Na_2O and, in some cases, H_2O. There are two major types :

(i)   *Crystalline silicates* which are definite chemical compounds.

sodium orthosilicate, $SiO_2.2Na_2O$ (Na_4SiO_4),

sodium sesquisilicate, $2SiO_2.3Na_2O$ ($Na_6Si_2O_7$),

and, sodium metasilicate, $SiO_2.Na_2O$ (Na_2SiO_3).

(ii) *Colloidal silicates* which are not definite compounds and which can have any ratio of SiO_2/Na_2O between about 1.6 and 3.85. The ratio is normally expressed by mass, but as the molecular mass of SiO_2 (60) is nearly the same as that of Na_2O (62), the composition by mass is nearly the same as that on a molar basis (mass ratio SiO_2/Na_2O × (62/60) = molar ratio). In the detergents industry, the usual ratios, by mass, are about 3.2 known as 'neutral silicate', and 2.0, known as 'alkaline silicate'. Despite these names, all the sodium silicates are alkaline, as shown in Table 1.1, but the higher the ratio the less alkaline the material. The alkalinity of a sodium silicate solution is directly related to its Na_2O content, the SiO_2 having little, if any, effect.

Orthosilicate and sesquisilicate are very alkaline and are used for specially difficult jobs in some commercial laundries. Metasilicate, sometimes known by the proprietary *brand* name 'Metso', is widely used in laundries and as an ingredient in some detergent formulations. It is available anhydrous and as the pentahydrate $Na_2O.SiO_2.5H_2O$. Both are white solids, and various granularities are commonly sold. Both materials are delivered in paper sacks.

Sodium silicates are normally made by fusing together pure quartz sand (more than 99% SiO_2) and soda ash which provides the Na_2O; carbon dioxide is evolved. The colloidal types leave the cooler which follows the furnace as a 'soluble glass'. This glass is not readily soluble and has to be dissolved in a digester under pressure. The solution is settled to remove impurities and concentrated by evaporation. This is normally done by the silicate manufacturer, but some large users may find it economical to instal the equipment needed to handle soluble glass, particularly if the silicate has to be transported long distances. The silicate solutions produced from soluble glass are sometimes called 'water glass', an old name for sodium silicate solutions.

Concentrated sodium silicate solutions are viscous, and the viscosity is greater for those with a high SiO_2/Na_2O ratio than for those with a lower ratio (above about 1.7 where a minimum is found). For this reason, alkaline silicate (ratio 2) is commonly purchased at 100°Tw or 120°Tw (120°Tw alkaline silicate contains 32.15% SiO_2 plus 16.07% Na_2O giving a total solids content of 48.2%), whereas neutral silicate (ratio 3.2) is usually bought at around 80°Tw (28.9% SiO_2 plus 9.0% Na_2O giving 37.9% total solids).

Sodium silicate solutions can be handled in mild steel, or cast iron, equipment. Tanks should be covered to avoid formation of hard skins by evaporation of water from the surface. Suitable heating facilities are necessary in tanks subject to cold, or temperate, climates to avoid excessive viscosity at low temperatures.

Sodium silicate solutions can be drum dried to produce soluble powders containing 15–20% of water. Crosfields sell 'C' soluble powder (an alkaline silicate of ratio 2.0, containing 56.0% SiO_2, 28.0% Na_2O and 16.0% H_2O) and 'M' soluble powder (a neutral silicate of ratio 3.3, containing 64.5% SiO_2, 19.5% Na_2O and 16.0% H_2O). Other silicate manufacturers produce similar products. 'C' soluble powder will dissolve readily in cold water, but 'M' soluble requires water at 75–80°C for easy dissolution.

Potassium Silicates

Potassium silicates are sometimes used in soft soaps, liquid soaps and heavy duty liquid NSD products.

Borax

Borax is available as the decahydrate, $Na_2B_4O_7.10H_2O$, and as the pentahydrate, $Na_2B_4O_7.5H_2O$. It is a white powder which is a weak alkali used in some mild soap and NSD products.

Sodium Phosphates

Sodium phosphates are very important constituents of detergent products, although now sometimes restricted for environmental reasons. A number of different types of sodium phosphates are, or have been, used in detergent products, and it is important to distinguish between them.

All the sodium salts of the chain polyphosphoric acids are sequestering, complexing, or chelating, agents, that is they take alkaline earth and heavy metal ions into complexes which leave only a very low concentration of the ions of the complexed metal free in the solution. For example, sodium pyrophosphate complexes calcium from a solution of calcium chloride by the reaction.

$$4Na^+ + P_2O_7^{4-} + Ca^{2+} + 2\ Cl^- \rightarrow 4\ Na^+ + [CaP_2O_7]^{2-} + 2Cl^-$$

There are also a number of ring sodium phosphates with the general formula $(NaPO_3)n$, of which the only one significant in the detergent industry is sodium trimetaphosphate, $Na_3P_3O_9$. Unlike the chain polyphosphates, this compound is not a chelating agent, but, it can be used as precursor of sodium tripolyphosphate.

The various condensed phosphates can be produced by elimination of H_2O from a phosphate containing a higher ratio of H_2O (or Na_2O) to P_2O_5, or by addition of H_2O (or a base such as NaOH) to one with a lower ratio. The most important condensed phosphates are tetrasodium pyrophosphate and penta-sodium tripolyphosphate. The pyrophosphate is formed by heating disodium hydrogen orthophosphate

$$2\ Na_2HPO_4 = Na_4P_2O_7 + H_2O,$$

and the tripolyphosphate by heating a mixture of disodium hydrogen orthophosphate and sodium dihydrogen orthophosphate

$$2Na_2HPO_4 + NaH_2PO_4 = Na_5P_3O_{10} + 2H_2O.$$

In both cases small proportions of other sodium phosphates are found in the product.

The very important pentasodium tripolyphosphate, often called polyphosphate or TPP, exists in two forms or phases. Form 1 is the form stable at high temperatures and it hydrates rapidly to the hexahydrate, $Na_5P_3O_{10}.6H_2O$. Form 2, the low temperature form, hydrates more slowly. In practice, the sodium tripolyphosphate of commerce is a mixture of the two forms, but manufacturers can produce grades with various Form 1 contents. The approximate Form 1 content is commonly measured by the Temperature Rise Test; this is carried out by measuring the rise in temperature when water is added to the sample of tripolyphosphate in glycerine under standard conditions. Under the test conditions about 90% of the potentially available heat of hydration is released by Form 1 and 10% by Form 2. From this, Form 1 content = $(TR - 6) \times 4\%$. Three grades produced are as under :

T.R. °C	Form 1 from formula %	Description
16.4	42	Fast hydrating
12.8	27	Fast hydrating
6.5	2	Slow hydrating

Particle size also influences rate of hydration; fine powders hydrating more rapidly than coarser powders of the same Form 1 content. The material with TR = 16.4°C just mentioned is a granular grade, whereas the other two are relatively fine powders. Some practical implications of rate of hydration are considered later.

The condensed phosphates hydrolyse slowly in solution, and tripolyphosphate changes to ortho and pyrophosphate. The rate of hydrolysis increases with temperature and under acidic conditions. If the hydrate $Na_5P_3O_{10}.6H_2O$ is heated sufficiently to remove water, decomposition to ortho and pyrophosphate occurs and little, or no, tripolyphosphate remains.

The various sodium phosphates are white powders which can be handled in bags, or in bulk.

Potassium Phosphates

Potassium phosphates corresponding to the sodium salts can be produced; as in the case of the silicates, potassium pyrophosphate is sometimes selected for use in liquid detergents because of its superior solubility.

BLEACHES, INORGANIC AND ORGANIC

General

Chemical bleaching (as distinct from adsorption bleaching with activated earths or activated carbon) is used to improve the colour of ingredients in the manufacture of detergent products, and chemical bleaches are also used in formulations to improve performance by the removal of organic colour from fabrics or other articles. Chemical bleaching is usually carried out with oxidising agents, but reducing agents have sometimes been employed.

Sodium and other Hypochlorites

These so-called chlorine bleaches are made by reacting chlorine gas with an alkali, or alkaline earth. At one time slaked lime was used to give bleaching powder (also called 'chloride of lime', a misnomer); but the alkali is now usually caustic soda which gives sodium hypochlorite, or 'bleach'.

$$2 \text{ NaOH} + \text{Cl}_2 = \text{NaOCl} + \text{NaCl} + \text{H}_2\text{O}$$

To achieve adequate stability it is necessary to leave an excess of 1 or 2% of caustic soda, so that bleach liquor is strongly alkaline (hypochlorous acid is also weak, so that its sodium salt is itself alkaline).

It will be noted that sodium chloride equivalent to the hypochlorite is formed and is left in the bleach liquor. This is significant in that the chloride ion, Cl^-, as well as the hypochlorite ion, OCl^-, is necessary for the production of chlorine gas when bleach liquor is made acidic. Hydrochloric acid provides its own chloride ion, but with other acids, such as sulphuric, the sodium chloride plays an essential part in the reaction

$$\text{NaOCl} + 2 \text{ HCl} = \text{NaCl} + \text{Cl}_2 + \text{H}_2\text{O}$$

$$\text{NaOCl} + \text{NaCl} + \text{H}_2\text{SO}_4 = \text{Na}_2\text{SO}_4 + \text{Cl}_2 + \text{H}_2\text{O}$$

$$\text{or, ionically, } \text{OCl}^- + \text{Cl}^- + 2\text{H}^+ = \text{Cl}_2 + \text{H}_2\text{O}$$

Because of the toxic nature of chlorine gas it is important to avoid its evolution into the atmosphere, particularly in buildings. Accidents have occurred where acid has been used to adjust the pH when sodium hypochlorite is employed to sterilise water in swimming baths, etc., and too much has been added. Similarly, when sodium hypochlorite and acidic lavatory cleaners have been mixed in error. The same type of situation could occur in the detergents factory unless proper precautions are taken. A subtle variation of the problem can arise with lavatory cleaners which include trichlorcyanuric acid, and the acid, sodium bisulphate. Hypochlorous acid is generated from the trichlorcyanuric acid, but no significant evolution of chlorine occurs when normal supplies of water are used to flush the lavatory. If, however, the products are used with sea, or other salt, water an unpleasant amount of chlorine can be produced.

In the treatment of fabrics (and in the factory), hypochlorite bleach is normally used in alkaline solution. The active agent is then really active, or nascent, oxygen (O). If oxidisable materials, such as most organic coloured compounds, are present, the active oxygen will react; if not, the atoms may combine slowly to produce gaseous oxygen (O_2), which is gradually released. Although the effective bleaching agent is active oxygen, it is usual to determine the concentration of hypochlorite present in terms of available chlorine, that is the quantity of chlorine liberated when the solution is made acidic. One method is to add potassium iodide, acidify, and titrate the iodine liberated with standard sodium thiosulphate.

$$NaOCl + NaCl + H_2SO_4 \rightarrow Na_2SO_4 + Cl_2 + H_2O$$
$$\downarrow$$
$$NaCl + O$$

Hence one atom of active oxygen, atomic mass 16, is equivalent to two atoms of chlorine, atomic mass $2 \times 35.5 = 71$ (or to one molecule of chlorine, molecular mass 71). The molecular mass of sodium hypochlorite, NaOCl, is 74.5. Therefore, 74.5 parts of NaOCl are equivalent to 71 parts of available chlorine; so that a solution containing 10.0% of NaOCl contains $10.0 \times (71/74.5) = 9.53\%$ available chlorine. This close numerical similarity may seem surprising.

When used to bleach fabrics in a washing process, sodium hypochlorite is normally added to a rinse in a separate operation after the main wash, otherwise part of the available chlorine would be mopped up by oxidisable matter in the soil. Another very important use of sodium hypochlorite is as a disinfectant. It is effective agent all types of micro-organisms, but some types, such as bacterial spores, require longer times and higher concentrations than more delicate types of organism. The active chemical species is understood to be hypochlorous acid and disinfectant performance increases with fall in pH. The main limitation of sodium, and other, hypochlorites is again that they can be used up by other oxidisable matter often present with the organisms; this means that sufficient must be used to maintain the necessary concentration of active bacteriocide for the required time.

Sodium hypochlorite solution is produced with about 13–14% available chlorine content, and it is commonly received in factories at this strength. Even with strongly alkaline it tends to decompose slowly on storage. Decomposition can be with evolution of oxygen gas as explained earlier, but it also occurs by the formation of sodium chlorate,

$$3NaOCl = NaClO_3 + 2NaCl$$

Loss of available chlorine is greater when the concentration of sodium hypochlorite is high than when it is low.

Substances which produce available Chlorine when dissolved

When it is desired to add a substance to a dry detergent product which will produce available chlorine when water is added for use, there are two major types of compound available :

(i) Chlorinated trisodium orthophosphate, or sodium phosphate hypochlorite (SPH), described as $(Na_3PO_4.12H_2O)_5$ NaOCl or $(Na_3PO_4.11H_2O)_4$ NaOCl approximately. This contains about 3.5% available chlorine.

(ii) An extensive range of organic compounds which contain N – Cl groups in which the chlorine is 'available'. The most important are probably trichlorcyanuric acid (TCCA), with about 90% available chlorine, and potassium dichlorcyanurate, with 59% available chlorine.

An important practical difference between the two groups of compounds is their behaviour in relation to the water content of the product. With SPH, the water shown in the formula is necessary for stability, and if the product dries out on storage available chlorine is likely to be lost. On the other hand, the chlorcyanurates and similar compounds are readily decomposed with loss of available chlorine if the product picks up water. Scouring powders containing TCCA, etc. usually need an ingredient, such as anhydrous trisodium orthophosphate, which will mop up water vapour which leaks into the pack, and so protects the TCCA. These products also commonly include a stabilising olefine, usually as a perfume ingredient.

Sodium Chloride, NaClO$_2$

Commercial sodium chloride containing about 80% of $NaClO_2.2H_2O$, with sodium chloride, sodium hydroxide and sodium carbonate (of which the alkalis are stabilisers) is available. A material is also sold which contains about 48% $NaClO_2.2H_2O$, plus 48% sodium nitrate which is added as a corrosion inhibitor. Sodium chlorite is used in some processes for the chemical bleaching of fats.

Hydrogen Peroxide and the per-salts

These compounds liberate active, or nascent, oxygen which is the bleaching agent. They are often considered milder on dyes and fabrics than sodium hypochlorite, but at concentrations which give equal stain removal this is not necessarily true.

Hydrogen peroxide solution is sometimes used in industry, but it cannot be included in detergent products, as can certain 'per-salts' which are widely included in detergent powders sold in many countries. The most important per-salt is sodium perborate, $NaBO_3.4H_2O$, or $NaBO_2.H_2O_2.3H_2O$, containing about 10% available oxygen. The monohydrate, containing about 16% available oxygen is also sometimes used. As shown in Table 1.1, sodium perborate is also an alkali. Sodium perborate is generally considered ineffective at temperatures below 60°C and it works best at higher temperatures of 80–100°C, although some benefits from the inclusion of sodium perborate at temperatures as low as 35°C have been claimed. A number of compounds described as *bleach activators* have been found which react with perborate in the wash liquor to form peracetic acid (or its salt) which bleaches at lower temperatures than perborate. The most important of these activators is TAED, tetra-acetyl ethylene diamine. Others are TAGU, tetra-acetyl glycol uril, tetra-acetyl methylene diamine, and p-acetoxy-benzene sulphonate.

The principal other per-salt used in detergent formulations is sodium percarbonate, $2Na_2CO_3.3H_2O_2$. This compound is less stable than sodium perborate in detergent powders, but it may have some advantage in products to be used at relatively low temperatures, and it avoids the inclusion of boron to which there are some environmental objections when the waste water containing it may be used for irrigation.

Reducing Bleaches

Reducing bleaches are of little importance in the detergent industry, but sodium hyposulphite, $Na_2S_2O_4$ (also known as sodium hydrosulphite or blankite), is sometimes used to improve the colour of soap during the boiling process.

ANALYSIS OF SOAP PRODUCTS

General

Some analytical tests used for soap products are outlined. Calculations using the results of such tests, and other information presented in subsequent sections, will now be given for two typical examples, a household soap and a soap powder. First some additional tests will be outlined.

Chlorides (Salt)

Chlorides (salt) can be determined as follows. A 10 g sample of soap is dissolved in hot distilled water and transferred to a 250 ml narrow necked graduated flask. Sufficient calcium nitrate is added to precipitate the soap (about 15 ml of a 20% solution), and the flask is filled to the 250 ml mark with distilled water. The mixture is filtered and 100 ml taken in a porcelain bowl where it is titrated with standard silver nitrate solution using a little potassium chromate as indicator. The reactions are :

$$NaCl + AgNO_3 \rightarrow AgCl + NaNO_3$$
Silver chloride
(insoluble white)

$$K_2CrO_4 + 2AgNO_3 \rightarrow Ag_2CrO_4 + 2KNO_3$$
Silver chromate
(insoluble brick red)

Because silver chloride is less soluble than silver chromate, there is no permanent change of colour from the yellow due to potassium chromate to brick red until all the chloride has been precipitated. A blank estimation, without soap, should be carried out and the figure deducted from the test titration. If 0.1 N silver nitrate solution is used, 1 ml is equivalent to (58.5/10) $\div$ 1000 = 0.00585 g NaCl, or 0.00355 g chloride expressed as Cl. With the quantities just specified, the salt found came from 10 × (100/250) = 4 g soap, and, if the titration after deduction of the blank is y ml, the salt content of the soap is y × 0.00585 × (100/4) = 0.146 y% as NaCl.

Silica, SiO$_2$

To separate silica, a sample of the soap, or soap powder, is ashed in a platinum dish with fusion mixture (potassium carbonate plus sodium carbonate) and the mix is heated to melting point. When cool, the melt is dissolved in distilled water, transferred to a porcelain basin and carefully acidified with hydrochloric acid. It is then evaporated to dryness, moistened with hydrochloric acid and again evaporated to dryness after any large lumps have been broken with a glass rod. The basin and contents are baked in an oven at 135°C, or gently on a sand bath, cooled, and then a little hydrochloric acid and water added. This process renders the silica insoluble, so that it can be filtered and weighed after the filter paper has been ashed. If insolubles other than silica may be present, the silica can be vaporised as silicon tetrafluride, SiF$_4$, by evaporating two or three times after additions of hydrofluoric acid and a few drops of sulphuric acid. This second residue is not silica and should be deducted from the crude residue first obtained.

Available Oxygen

Available oxygen can be estimated by acidification of a solution of a weighed sample with sulphuric acid and titration with standard potassium permanganate.

Household Soap

The main constituents of a genuine household soap are :

(i) Soap, comprising fatty matter (mainly fatty acids) and the alkali combined with it.

(ii) Free caustic soda.

(iii) Sodium carbonate.

(iv) Sodium chloride.

(v) Glycerol.

(vi) Water, possibly plus colour and perfume.

Soap Powder

If a powder contains phosphates, the P_2O_5 content can be determined by a gravimetric method, and various more complicated procedures are available for the estimation of different types of phosphate. If only a figure for P_2O_5% is available, it is necessary to make assumptions as to the type of phosphate, and/or to use trial and error methods to make the figures balance in a way which seems reasonable. There are also complications with regard to the determination of the total alkali.

WATER HARDNESS AND WATER SOFTENING

Water used for washing, and other purposes, in home and factory contains impurities to a degree which varies with its source. In the detergent field, the most important impurities are the calcium and magnesium salts which constitute water hardness. Traces of iron can lead to iron stains (sometimes known as iron mould), but this is more common in laundries than in the home.

The effect of water hardness is obvious with soap in that precipitates are formed; and no lather is produced until the calcium and magnesium in the water have been removed as insoluble soaps. Although it does not prevent lather, hardness, particularly calcium hardness, does have an adverse effect upon the performance of NSD products, at least in the washing of textiles, unless builders are also present in sufficient quantity. Unbuilt non-soapy surfactants are used successfully for dishwashing in hard as well as soft water.

Water Hardness

Water hardness is divided into :

(i) Temporary, or alkaline, hardness due to bicarbonates. This hardness was called temporary because it is largely removed by boiling (or heating to near boiling point), when the bicarbonates are converted to relatively insoluble carbonates which precipitate on kettles and other heating surfaces. Actually, the alkaline earth carbonates are slightly soluble, so that the hardness originally due to bicarbonates is not completely removed by boiling. The modern term is alkaline hardness, so called because this hardness can be determined by titration with standard acid to the methyl orange end point.

(ii) *Permanent hardness*, due to other salts, such as sulphate, chloride and nitrate.

(iii) *Total hardness* can be estimated by :

(a) Titration with Wanklyn's (standard soap) solution to a permanent lather (that is one which persists after the bottle in which the estimation is carried out has rested on its side for five minutes). Particularly with water high in magnesium, a false end point can occur when the calcium, but not the magnesium, has been precipitated. This lather does not last for five minutes.

(b) Titration with standard EDTA (ethylene diamine tetra-acetic acid) solution, using as indicator a black dye, known variously as Eriochrome, Solochrome, etc. EDTA is a sequestering agent which, like the condensed phosphates, complexes the calcium and magnesium in the water.

Permanent hardness is not estimated directly, but is found as the difference between total and alkaline hardness.

Hardness, calcium plus magnesium, is usually expressed in terms of $CaCO_3$, although CaO has been used in some countries.

The demarcations between waters classed as soft, moderately hard, etc., are necessarily arbitrary. Laundries usually soften the water to be used in washing operations.

Methods of Softening Water

There are three main methods to soften water :

(i) The *lime-soda process* in which calcium hydroxide (slaked lime) and sodium carbonate are added to the raw water. The main reactions are:

$$Ca(HCO_3)_2 + Ca(OH)_2 \rightarrow 2CaCO_3 \downarrow + 2H_2O$$

$$Mg(HCO_3)_2 + 2Ca(OH)_2 \rightarrow Mg(OH)_2 \downarrow + 2CaCO_3 \downarrow + 2H_2O$$

(magnesium hydroxide is considerably less soluble than magnesium carbonate).

$$CaSO_4 + Na_2CO_3 \rightarrow CaCO_3 \downarrow + Na_2SO_4$$

$CaCO_3$ and $Mg(OH)_2$ are relatively insoluble, and are separated by settling and decantation, followed, if necessary, by filtration. Because of the slight solubility of $CaCO_3$ and $Mg(OH)_2$, the water retains a little hardness.

(ii) *Base exchange.* Various materials, including natural zeolites (types of silicate minerals), synthetic zeolites, and certain synthetic resins, are capable of softening water by an exchange of sodium for calcium and magnesium. The bed of base exchange granules is then regenerated by the action of a sodium chloride (salt) solution which reverses the softening reaction. Calling any base exchange material NaZ, the reactions are roughly:

Softening $2NaZ + Ca$ salt $\rightarrow CaZ_2 + 2Na$ salt

Regeneration $CaZ_2 + 2NaCl \rightarrow 2NaZ + CaCl_2$

The excess brine (or other sodium salt used for regeneration), carrying with it the soluble calcium and magnesium salts (usually chlorides), is run to waste, and the regenerated bed is rinsed before being used for the next cycle. Alkaline hardness is converted to sodium bicarbonate in the softened water. This is a disadvantage in some situations, for example, it can reduce the pH of a wash system comprising soap and sodium sesquicarbonate. A suitable addition of caustic soda converts the bicarbonate to sodium carbonate. Water softened by base exchange may not be suitable for boiler feed, particularly if the raw water is high in alkaline hardness. One procedure is to mix a little raw water with the softened water. A base exchange bed filters out any particulate dirt from the water. It also removes any soluble iron, but the treatment of raw water high in iron content can lead to problems in regeneration.

Chapter 2

Surfactants

INTRODUCTION

Amongst chemical specialities, which are wide and varied, 'surfactants' are a group of materials that have evolved with the growing and ever-changing needs of society. They pervade the whole spectrum of the industrial horizon, and, as such, deserve a close look as regards their present global status and future potential. Substances containing in the same molecule groups which are strongly hydrophobic (water repellent) and hydrophilic (water loving) orient themselves in different ways when dissolved or dispersed in water or nonaqueous solvents. Such orientation, which could also be in the form of aggregates of oriented molecules, termed micelles, is known as 'Surface Activity' and materials exhibiting this are termed 'Surface Active Agents', often abbreviated as 'Surfactants'. Fig. 2.1 indicates the orientation of surfactants in several systems.

Practically almost every product in commercial use today has come in contact, at some stage or other of it's production process, with surfactants, which are classified by the nature of their hydrophilic moiety. The surfactant mosaic is made up of anionics, cationics, nonionics and amphoterics in broad classification. However, surfactants grouped together as polymeric surfactants, biosurfactants etc., depending on their end-use importance or functional aspects are gaining prominence.

The hydrophilic moiety carries a negative ionic charge in anionics, a positive ionic charge in cationics and is neutral in nonionics. The amphoterics contain, or have the potential to form, both positive and negative functional groups under specified conditions. The surfactants, in general, are governed by several criteria such as :

(a) Solubility in water, non-aqueous media and two-phase systems.

(b) Hydrophile-lipophile balance (HLB).

(c) Surface tension.

(d) Foaming characteristics.

(e) Cloud point.

(f) Orientation at interfaces.

(g) Substantivity such as adsorption on solid surfaces.

(h) Micelle formation.

(i) Functional properties such as detergency, wetting, emulsifying, solubilising and dispersing.

(j) Reactivity such as complex forming ability and antimicrobial properties.

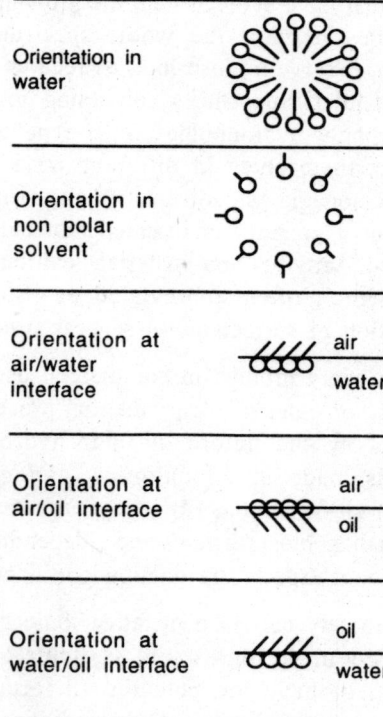

Fig. 2.1 Orientation of surfactant in several systems.

Nature has used the science of surface activity in many simple and complex ways. Surfactants have been used by man for a long time, a soap making process being recorded by Plinius as early as 70 AD. Synthetic surfactants emerged on a large scale only with the use of petroleum derived fractions in the 1950s. Over the last 50 years there has been an enormous number of new surfactants produced and many new applications. Apart from the natural and synthetic surfactants, biosurfactants have emerged on the horizon. These are a collective class of biosurface-active materials, which are basically obtained by manipulating fermentation processes with very specific microorganisms. Several biosurfactants have surface-active and emulsification properties far superior to either synthetic or natural surfactants. In addition, they have the green attributes of bio-compatibility, complete bio-degradability and superior eco-toxic characteristics. The usual picture of a surfactant as a simple rod-like hydrophobe with a hydrophilic head has changed in recent times. Many new surfactants are now being produced of more complex structure with more than one hydrophilic head. Such products form micro-emulsions more readily.

Polymeric materials, especially high molecular weight ones, have been used to give colloidal stability since almost the dawn of civilisation. In particular, naturally occurring polymeric materials, such as egg white, gum arabic and animal glues have been used to make water-based inks and paints pigmented with carbon black or metal oxide pigments. In recent years polymeric surfactants have been developed for wide applications in surface coatings. Surfactant technology thus is continuing to grow aided by a large well-established base industrial potential.

The number of surfactants available commercially today are several thousands. As a result many chemists, engineers, biologists and others, having the need to employ surfactants, feel daunted by the prospect of ploughing through the literature to find the best material for their purpose. Even if inclined to do so, they may face the investment of inordinate amounts of time and effort without guarantee of success. The breadth and complexity of the science and technology of surfactants makes it difficult to compress all the information available in a small volume. However in this chapter attempt has been made to give sufficient insight into the various classes of surfactants, major raw materials, key physico-chemical aspects, applications, present global status, market participants in the field, typical formulations, opportunities for growth and other allied matters of relevance to the surfactants industry. In analysing the surfactant markets, emphasis is laid on the household sector consuming various detergents and cleaners, industrial and institutional sectors covering important industrial segments, personal care products sector involving hair care, skin care and oral-hygience products segments and the agrosector with great demand for crop protection chemicals.

TYPES OF SURFACTANTS

Anionic Surfactants

By far the largest class of surfactants in general use today are the anionics (including soaps). For thousands of years in the history of the application of surface activity to the needs of mankind, the soaps, derived from natural animal and vegetable fats and oils, dominated the field. The major historical and economical advantage of the fatty acid soaps has always been the ready availability of the raw materials from natural, renewable sources. The weakness of the classical soaps, due to the sensitivity of the carboxylate salts which mainly constitute them, in many processes requiring surfactant action, led to the development of synthetic surfactants. Classically, anionic surfactants are divided into four basic categories, namely :

(a) Carboxylates, of which common soap is the most important.

(b) Sulphonates, of which linear alkyl benzene sulphonate (LABS) is the most significant and is the largest used synthetic surfactant worldwide.

(c) Sulphuric acid esters such as alcohol sulphates or sulphated oils, the earliest of synthetic surfactants.

(d) Other category includes the nitrogen-bearing taurates and sarcosinates, phosphorus-bearing chemicals such as the phosphate esters and others, and industrial surfactants of the alkylnaphthalene-formaldehyde condensate type.

On a global basis, the anionic surfactants usage is roughly as given in Table 2.1.

Table 2.1. Global split of anionic surfactants.

Anionic surfactant type	Percentage of the global anionics market
Alkylbenzene sulphonates	45–47
Lignin sulphonate	23–25
Alcohol ether sulphates	13–15
Alcohol sulphates	8–10
Others	6–8

The four basic categories of anionic surfactants fall under two large classes of products—'true surfactants' and 'quasi-surfactants'. The 'true surfactants' are those which exhibit the important properties of lowering the surface tension of water and the interfacial tension between water and oils, of wetting and foaming actions, of emulsification etc. The most important of these surfactants are soap, alkylbenzene sulphonates, alcohol ether sulphates and alcohol sulphates. The 'quasi-surfactants' have feeble powers sulphates and alcohol sulphates. The 'quasi-surfactants' have feeble powers in terms of reducing the surface tension or wetting action, but are vastly superior in properties such as those of dispersion, thickening, spreading, penetrating, defoaming, demulsifying etc. Typical of these are lignin sulphonates and alkylnaphthanlene sulphonates. Most of these quasi-surfactants are used in the production of industrial commodities.

Surfactants are seldom used directly in the application sectors and are always components of formulations containing both surface-active and non-surface active chemicals. The market segments for anionic surfactants can be categorised as :

(a) The household products sector.

(b) The personal care products sector.

(c) Industrial and institutional cleaners.

(d) Other industrial applications.

In the household sector there are three important segments :

(a) Laundry detergents.

(b) Dishwashing detergents.

(c) Household cleaners.

'True' anionics are consumed to a large extent in these applications. The usage is dominated by Linear alkylbenzene sulphonate (LAS), Alcohol ethoxy sulphate (AES), Sodium alpha sulphomethyl laurate (SAS), Alkyl sulphates (AS), Alpha olefin sulphonate (AOS) new anionics such as methyl ester sulphonates and phosphate esters. In all the segments of the household sector the anionics are used alongwith other surfactant types, mainly nonionics.

The laundry detergent segment is by far the most important in terms of the volume of surfactants used and the value contribution.

Whereas in the global context the laundry detergent segment is the predominant one, in the Industrial Triad the other segments are significant contributors. The three principle types of end products in the household laundry products segment are :

(a) 'Conventional' or 'universal' powders.

(b) Liquid detergents.

(c) Superconcentrates.

Momentous changes have taken place in recent years in these products and multifunctional detergents incorporating bleach, fabric softeners etc. have landed in the market.

The personal care products sector also comprises of three segments, namely :

(a) Hair care.

(b) Skin care.

(c) Oral hygiene products.

Anionic surfactants are mainly used in these segments, but they have to compete with other types such as amphoterics and cationics. At the global level, the skin care segment dominates with consumption of large volumes of soaps. In the Industrial Triad, however, both soap and other surfactants usage in this segment is limited. The hair-care sector is dominated by shampoos, which consume large volumes of anionics such as AES, AS and soap. The oral-hygiene segment is a small user of surfactants, with the anionics as the major type.

Industrial and Institutional Cleaners are products used in :

(a) Janitorial services.

(b) Food processing and commercial dishwashing.

(c) Transportation.

(d) Commercial laundry and dry cleaning.

(e) Metals industry.

Anionics figure in large volume in these applications, LAS being the major anionic. Nonionic alcohol ethoxylates are also used in this sector.

Amongst industrial uses, anionics have substantial roles in textiles, petroleum processing, emulsion polymerisation of plastics and lattices, leather and pesticidal formulations. Other application areas include cement and concrete, paints and inks, froth flotation of ores, and laying of asphalt in road making.

Cationic Surfactants

Cationic surfactants first became important when the commercial potential of their bacteriostatic properties was recognised in early fifties. Since than a wide class of cationics with literally hundreds of molecular structures have been introduced as commercial products, although they do not approach the importance of the anionic materials in sheer quantity or value. In most of these products advantage is taken of the 'substantivity' of the cation and it's semipermanent adsorption on surfaces which are normally further makes the substrate surface oleophilic.

Until the availability of straight-chain petroleum-based surfactants, the sole sources of raw materials for cationic surfactants were vegetable derivatives of fatty amines of 1, 2 or 3 alkyl chains bonded directly or indirectly to a cationic nitrogen group. The most important classes of these cationics are the amine salts, quaternary ammonium compounds and the amine oxides. The cationic nature of the fatty amines and derivatives has given rise to a wide variety of functions in many applications. These functions may be recognised as being derived from four basic functional properties, namely, (a) surface active, (b) substantive, (c) reactive and (d) antimicrobial as represented in Fig. 2.2. These properties are exploited for the following surfactant functions in various applications as indicated in Table 2.2.

Amongst the cationic surfactants there are two important categories which differ mainly in the nature of the nitrogen containing group. The first consists of the alkyl nitrogen compounds such as simple ammonium salts containing at least one long-chain alkyl group and one or more amine hydrogens, and quaternary ammonium compounds in which all amine hydrogens have been replaced by organic radical substitutions. The amine substitutions may be either long or short chain substitutions. The counterion may be a halide, sulphate, acetate etc. The second category contains heterocyclic materials typified by the pyridinium, morpholinium and imidazolinium derivatives.

Table 2.2. Basic properties of fatty amines and their derivatives and their surfactant functions in various applications.

Basic property	Surfactant functions
Surface active	Emulsifying, wetting, thickening, foaming
Substantivity	Softening, antistatic properties, lubrication, corrosion inhibition, adhesion, hydrophobation
Reactive	Flocculation, decolourisation, ion exchange
Antimicrobial	Algicidal, fungicidal, bactericidal

The economic importance of the cationic surfactants has increased greatly because of their unique properties. Their substantive property is exploited in fabric softeners, bitumen emulsions, oil-muds used in the petroleum industry, hair-care products, corrosion inhibition in oilfield and water treatment segments. The cationics are also important in the textile industry as waterproofing and dye-fixing agents. Other application areas are in flotation processing of mineral ores, lubrication, as surface modifiers in electrostatic charge control. Their biological activity seems to be characteristic of almost all-cationic surfactant species, irrespective of minor structural changes, and is utilised in killing or inhibiting the growth of many microorganisms.

Nonionic surfactants, which fit into a broader spectrum, do not carry any discrete charge when dissolved in water, unlike anionics or cationics. Some of the important advantages of this electrical neutrality are :

(a) Lower sensitivity to the presence of electrolytes in the system.

(b) A lessened effect of solution pH.

(c) The synthetic flexibility of being able to design the required degree of solubility into the molecule by the careful control of the size of the hydrophilic group.

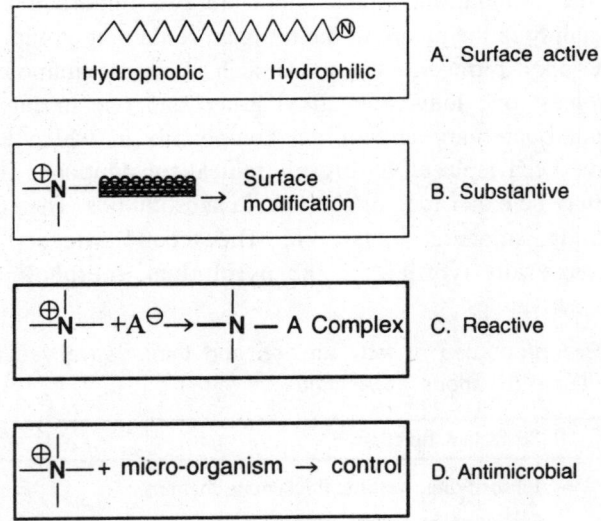

Fig. 2.2. Four basic functional properties of cationic surfactants.

Several nonionics are currently in commercial use. The most numerous and most important technically amongst these are the polyoxyethylenes (POE) of the general formula :

$$RX (CH_2CH_2O)_nH$$

where R is usually a typical surfactant hydrophobic group, but may also be a polyether such as polyoxypropylene and X is O, N or another functionality capable of linking the polyoxyethylene chain to the hydrophobe. In most capable of linking the polyoxyethylene chain to the hydrophobe. In most cases the value of 'n', the average number of oxyethylene units in the hydrophilic group must be greater than 5 or 6 to impart sufficient water solubility to make the materials useful. The most common pathway to the POE surfactants is by the reaction of ethylene oxide (EO) with alcohols, acids, amides, mercaptans etc. If primary amines are used double chain surfactants of the form :

$$RN\ [(CH_2CH_2O)_nH]_2$$

can be formed with different molecular complexities. Hydrophilicity, in the case of nonionics is, provided by hydrogen bonding with water molecules. Oxygen atoms and hydroxyl groups readily form strong hydrogen bonds whereas ester and amide groups do so less readily. Hydrogen bonding provides solubilisation in neutral and alkaline media. Under highly acidic conditions oxygen atoms are protonated giving a quasi-cationic character. As each atom of oxygen can make only small contribution to water solubility, dissolution of nonionics in aqueous media necessitates the presence of several oxygen atoms in the molecule. Polyoxyethylene solubilisation has thus been the key to the substantial and growing potential of the nonionics. The compatibility of the nonionics with both ionic and amphoteric surfactants gives a boost to this class as it gives rise to synergistic systems. Nonionics like alcohol ethoxylates and alkylphenol ethoxylates serve as building blocks for deriving anionic surfactants by sulphation. A new dimension to the nonionics resulted on account of the polyoxyethylene-polyoxypropylene block copolymers, wherein the hydrophobic tendency of the propylene oxide units is taken advantage of in tailoring the products.

Apart from ethoxylated products, alkanolamides, carboxylic amides, glycerol esters, sorbitan esters, amine oxides, 'Surfynols' based or acetylenic diols, 'Silwet' surfactants comprising of siloxane back-bone with organic polyalkylene pendants, alkyl polyglucosides, fluorosurfactants, triesters of phosphoric acid and carbohydrate derived products figure in the nonionics market.

Amongst the varied applications of nonionic surfactants, emulsification, dispersion, detergency and foaming are important phenomena. Emulsions are used in a large number of applications such as foods, cosmetics, medicinals, agriculture, paints and numerous industrial processes. Their ability to control the stability of small particles is important in detergents. An important aspects of this is that the surfactant amount necessary in a composition is relatively low as compared with anionics. The nonionics serve as both foam boosters and foam inhibitors. Anionic derivatives of nonionics are particularly good high foaming detergents, whereas other derivatives serve as low foamers. Large volume uses of nonionics associated with their foaming properties is in various types of cleaners, latex paints, insecticidal and herbicidal wettable powders and emulsifiable concentrate, aerosol foam products, spray metal cleaning, foam drilling of oil and gas wells, and emulsion and suspension polymerisation. Other applications are as demulsifiers of crude oils, lubricants, antistatic agents, thickeners and emollients, dyeing assistants and pigment dispersants.

Amphoteric Surfactants

The true amphoteric surfactant carries both cationic and anionic groupings within the molecule, acting as anionic materials in alkaline pH and cationic at acid pH. This structure makes it not only possible to combine them with the other categories of surfactants without the problem of property inhibition, but also with the choice of selecting the ionic character. Although the amphoteric surfactants represent the lowest volume in the total global surfactants market, the unique properties they can impart to a formulation are improving their market position. Amphoteric surfactants have been available for many years. Early amphoterics were glycinates based on imidazoline chemistry and were the basis of many toiletry products such as mild shampoos. Betaines were also developed and used increasingly in personal care products due to their skin friendly, foaming and free rinsing properties. Although a large group of organic functionalities hold the potential for producing amphoteric surfactants, the materials mostly encountered in this category are :

(a) Imidazoline derivatives.

(b) Betaines and sulpho-betaines.

(c) Amino acid derivatives.

(d) Lecithin and related phosphatides

(e) Protein based products.

(f) Mono and dipropionates.

Most commercially important imidazoline-derived amphoteric surfactants can be described as fatty acid—amino-ethylethanolamine condensates of the general structure :

$$RCONHCH_2NR'R''$$

where R is the fatty acid residue, and R' as well as R" are different functionalities. Their mildness and toxicity led to their use in shampoos and body care products. Their amphoteric nature also makes them useful in a wide range of water types, ranging from hard to soft water and high to low pH values. Such flexibility makes them useful in cleaning formulations under a variety of conditions.

Betaine surfactants can formally be considered to be special members of the ring-opened, imidazoline-derived surfactants. However, they do not exhibit many of the characteristics of other amphoterics with regard to their solubility and electrical nature in alkaline solution. They are compatible with anionics at all pH values and do not have any problems of complex formation. The carboxyl-containing betaines have been found to form external salts in very strong acids (such as hydrochlorides) while the sulphobetaines do not do so. This class is generally insensitive to the presence of electrolytes and usually perform well in hard water. The carboxyl betaines have found a number of commercial applications as levelling and wetting agents in textile processing, detergents, scrubbing compounds, antistatic agents, as fabric softeners, as lime soap dispersants and in personal care products. The sulpho-betaines are useful in special areas such as the control of static charge in photographic films.

NC_{10}–C_{12} and N C_{12}–C_{18} fatty acid amidoethyl-N-(2-hydroxy ethyl) glycinates are the types of amino acid derivatives used for the production of strongly alkaline industrial cleaning agents, baby shampoos and other mild surfactant shampoos.

Lecithins are commercially popular phosphatidyl surfactants and are high volume products. Because of their natural raw material sources, the lecithin surfactants are normally found as rather complex mixtures represented by :

$$CH_2OCOR$$
$$|$$
$$CH_2OCOR'$$
$$|\quad OH$$
$$|\quad |$$
$$CH_2OPOCH_2CH_2NH_2$$
$$\|$$
$$O$$

The natural lecithins have very limited solubility in water. Having good oil solubility, they have found extensive commercial application as nonaqueous emulsifiers, dispersants and wetting agents in diverse areas such as marine paints, inks, foods and cosmetics. The largest use of lecithins is in the processing of food and animal feeds and pet foods as emulsifiers or wetting agents. In the food industry they are used in baking, cake mixes, chocolates and ice creams. In cosmetics, they are used in creams, oils, soaps and hair care products. They simulate the phospholipids that are natural moisturisers in the skin and render the skin elastic. In hair care products they prevent the extraction of lipids. In soap their amphoteric nature represses the alkalinity and improves the stability of lather. Other applications of lecithins are as dispersants for dyes and pigments in the paint, ink and textile industries, in formulations of oil penetrants, in fat liquoring of leather, and as buffering agents in lubricating and other oils. Natural sources of lecithin range from egg yolks to many seed oils such as cottonseed, sunflower and soyabean. The most common source however is soyabean due to its relatively lower cost and greater availability.

Protein hydrolysed by acids and bases leads to the formation of polypeptides and amino acid residues, the proportion determined by the temperature and duration of hydrolysis. Protein sources can be vegetable protein byproducts, leather scraps and fish scales. Protein-based surfactants are produced from these hydrolysates by reaction with fatty acid chlorides, condensation with fatty acids and amines and reaction with polyamines. They find application in detergents soap and toilet bar formulations.

The need for salt-free amphoterics by industries to overcome the problems associated with high electrolyte levels and corrosion resulted in the growth of amphoteric mono- and dipropionates. Some of the many other advantages of these products are excellent dirt lifting properties, excellent hydrotropic properties, excellent free rinsing properties they impart to detergent systems, stability over a wide pH range, low toxicity, biostatic properties and lubricity. Currently, these propionate amphoterics are recommended for formulating general cleaning products, environmentally friendly vehicle cleaning systems, personal care products such as hand cleaners, shower gels, hair shampoos and foam baths, microemulsions of terpene based products and aqueous slat/chain lubricants.

Polymeric Surfactants

Colloidal stabilisation by polymeric materials in both aqueous and nonaqueous systems is described as 'Steric stabilisation'. A limited range of polymers fit this role. The polyisobutylene chain is frequently used for stabilisation in very nonpolar organic solvents like aliphatic hydrocarbons. In solvents of higher polarities such as aromatic hydrocarbons, esters and ketones, various polyesters are often used. These polyesters may be based on hydroxycarboxylic acids, with the molecular weight of the hydroxy-carboxylic acid governing the range of solvents in which the surfactant is most effective. Alternatively, a condensation polyester from a mixture of diacid, diol and a chain terminating acid or alcohol may be used. A polyethylene oxide chain may be incorporated into polyesters of this type to increase compatibility with more polar solvents. Polymethylmethacrylate is also used to provide steric stabilisation to solvents of higher polarities. The polyalkylene oxide chains can be used to give steric stabilisation in the most polar organic solvents like alcohols, glycols and glycol ethers.

Polymeric dispersants have come into prominence to solve the complex rheological problems associated with spreading, levelling, relatively capital and energy expensive dispersion processes in the ink and paint industries.

Although surface coatings are the main outlets for sophisticated polymeric surfactants, a variety of other applications have started to develop in the last few years. These include their use in oil-based drilling muds, lubricating oil additives such as ashless dispersants, processing of electroceramics, magnetic tapes and dises and inverse emulsion polymerisation processes.

Biosurfactants

Biosurfactants, derived from fermentation processes with very specific micro-organisms, depending on their nature or state typically display strong features in some respects while at the same time lacking in the versatility of conventional surfactants. Several biosurfactants have surface-active and emulsification properties far superior to either synthetic or natural surfactants. Based on their technical competence and social dictates they have the capability to penetrate all current fields of surfactant use. In particular, they have high potential for use in household detergents and cleaners and personal care products.

'Emulsan', a fully commercialised biosurface-active material is one of the best discovered emulsifiers and is classified as a 'Bioemulsifier'. 'Surfactin', another biosurfactant, has the ability to lower the surface tension of 0.1 N sodium bicarbonate solution from 72 dynes/cm to 27.9 by the addition of as little as 0.005%. Other products such as rhamnose-lipids and sophorose-lipids obtained with Torulopsis bombicola KSM35 also show excellent surfactant properties. Biological products of genetic- and protein-engineering such as Lipolase or Cellulase impart restoration and repair characteristics to fabrics and are viewed as bio-detergents. In the personal care sector, biosurfactants termed biocosmetics are expected to take long strides. Sophorose- lipids are already in use in this sector as high value skin moisturisers. In the oilfield industry, microbial enhanced oil recovery (EOR), cleaning up of offshore spills, cleaning of oil tankers, pumping of crudes using bioemulsifiers, demulsification of crudes from the enriched oil recovery, extraction of bitumen from tar sands and viscosity reduction of heavy crudes are potential application areas for these surfactants. Lecithins, described earlier under amphotenics, are also bio-surface active materials used largely as food emulsifiers. Recent projections by some agencies show significant penetration of the biosurfactants into the conventional surfactant markets in the next two decades.

Widely used in consumer and industrial applications such as soaps and cleaners, laundry detergents, cosmetics and toiletries, non-consumer applications for surfactants can also be found in almost every industry. Except for cleaning uses, most of them consume minute quantities of the surfactants as processing aids. Industries where they are used include rubber, plastics, textiles, construction, petroleum-drilling, ore processing, food processing, pharmaceuticals, herbicides, machinery and metals. As new regulations and changing consumer demands arise, the surfactants industry is evolving rapidly. However, many of the surfactants initially alleged as being potential pollutants due to lot of pressure from the environmentalists are now being cleared by environmental agencies. Simultaneous with the development of new products for new or changing industrial applications, surfactants are beginning to attract attention in energy and environmental protection applications, as they are being tried out in pollution-control systems, paper processing and ultra-thin films. Main driving forces influencing changes in the surfactant industry are :

(a) The relative state of the economy.

(b) Consumer demand for 'green' products that are milder, safer and cheaper.

(c) High expenses due to regulatory compliance.

(d) Insurance and environmental costs.

(e) Technical obsolescence.

(f) Erosion of commodity/specialty segment differentiation.

(g) General public perception of the chemical industry.

RAW MATERIALS OF SURFACTANTS

The surfactant industry has undergone major changes over the last few years. Surfactant companies generally have been rationalising their product ranges in order to concentrate on their core surfactant business. Surfactant consumers are now looking for ingredients with increasing number of surfactant industry is also turning towards raw materials from renewable sources and is trying to be less dependent on petrochemical inputs. New molecules like the alkyl polyglycosides (APG) and N-alkyl glucosamide (AGA) are now being increasingly produced. Amongst the various aspects related to the surfactants industry, key factors are :

(a) Surfactant raw materials.

(b) Major surfactants.

(c) Important physico-chemical characteristics.

(d) Technologies.

There are two major materials from which the organic component of surfactant chemicals is predominantly obtained, namely :

(a) Hydrocarbons, from petroleum and gas drilling operations.

(b) Fats and oils from vegetative, animal and marine sources.

The paper industry is an ancillary source and supplies two rather important products—tall oils from which fatty acids can be obtained and lignosulphonates.

Petroleum is a unique source for benzenoid and naphthenoid surfactant chemicals such as alkylbenzene sulphonates, alkylphenols and alkylnaphthalene sulphonates. It is also the source of the most important petrochemical intermediate, ethylene and many straight chain chemical intermediates such as normal paraffins and alpha olefins from which several surfactants are derived.

Fats and oils are unique sources for straight chain, saturated and unsaturated compounds that serve as alternatives to petroleum intermediates for surfactants. An important example is that of detergent alcohols. Other intermediates derived from fats and oils are methyl esters, soaps and fatty acids. The oleochemical industry has become an important segment of the global chemical industry and particularly of the surfactants industry because of the promise of the products being natural, very biodegradable, toxicologically safe and renewable. Movements in the production and supply of fats and oils affects the economics of the surfactants industry. Traditionally petrochemical feedstocks have been cheaper alternatives but the environmental drive and causes which have led to price decline in the key oils for the surfactants industry—palm kernel oil and coconut oil—have made them preferred materials. Fig. 2.3 gives surfactants that are derived from fats and oils.

The paper industry provides two key surfactant raw materials, namely, tall oil and lignin sulphonates. Tall oil is a coproduct in the kraft pulping process for paper pulp using pine and similar woods. About 90% of all tall oil (a mixture of unesterified oleic and linoleic acids with small amounts of rosin acids and unsaponifiables) produced is fractionally distilled to get tall oil fatty acids (TOFA) used as raw material for surfactants. Tall oil production is dependent on the growth of the paper industry.

Technically either of the two basic raw materials—hydrocarbons or fats—can be used to obtain majority of synthetic surfactants. The choice depends on the raw material availability and prices. Key surfactant intermediates worthy of attention are :

(a) Linear alkyl benzene (LAB).

(b) Detergent alcohols.

(c) Alphaolefins.

(d) Ethylene oxide/propylene oxide.

(e) Fatty amines.

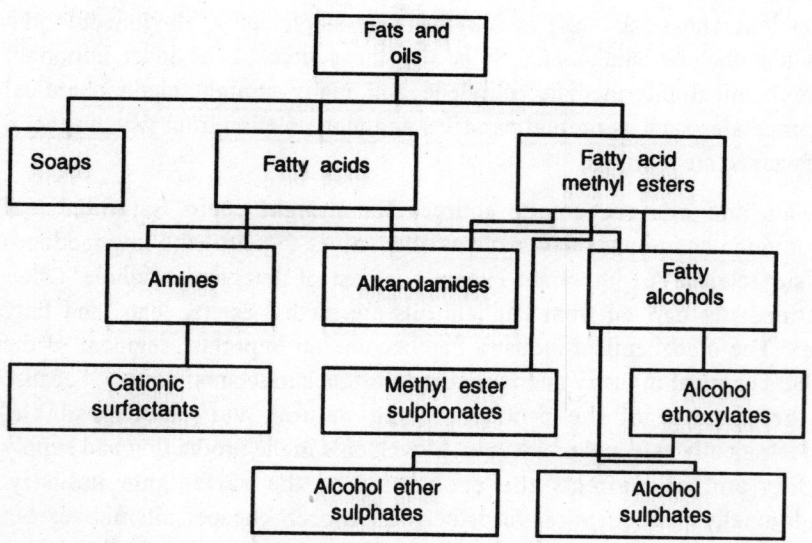

Fig. 2.3. Surfactants derived from fats and oils.

Linear Alkyl Benzene (LAB)

Linear alkylbenzene is the most important industrial raw material for surfactants derived from petrochemical sources and is even otherwise the most dominant raw material of the detergent industry. Its level of consumption in any global region is an indicator of the scope for detergent products—laundry detergents, dishwashing detergents and materials used in the maintenance of buildings, offices, hospitals, schools etc. LAB is the raw material for the production of alkylbenzene sulphonic acid and alkylbenzene sulphonate (LAS). Detergent formulators use both types of these intermediates. In many formulations, a salt of LAS such as Na-LAS (sodium salt) is added directly as a formulant. In other cases, the LAS salt is formed during the formulation process through the neutralisation of the acid with the desired base—caustic soda or triethanolamine.

The structure of LAB depends on two major factors :

(a) The petroleum feedstocks used—normal paraffins and benzene in terms of range and purities.

(b) The method of manufacture of LAB.

Because of the complexities involved in the selection of LAS for a particular formulation and end use, there have been immense improvements in the methods of manufacturing LAB, the parent material for LAS. The specifications for LAB in terms of the distribution of homologues and phenyl-position isomers makes it almost a 'designer molecule'.

LAB Manufacturing Processes

Two basic routes employed for the manufacture of LAB are :

(1) The alkylation of benzene with long-chain mono-olefins using HF and other catalysts.

(2) The alkylation of monochlorinated benzene with n-paraffins using aluminium chloride catalysts.

The mono-olefins alkylation process is predominant and based on UOP technology. In the overall UOP process, internal mono-olefins obtained through the catalytic dehydrogenation of n-paraffins are used for the alkylation of benzene with HF catalyst to give LAB. UOP employs the PACOL process for producing the olefins from n-paraffins. In this process linear paraffins are dehydrogenated in the presence of selective hydrogenation catalysts, believed to be platinum, arsenic and lithium deposited on alumina. Operating temperatures in the PACOL reactor are in the range 400–600°C with dehydrogenation occurring in the presence of hydrogen. Some by-product light ends and hydrogen are separated from the reactor effluent and a part of this hydrogen-rich gas is recycled to the reactor. An equilibrium mixture of linear olefins together with unconverted n-paraffins and side reaction products like diolefins and aromatics is produced this way. If left unchanged, the diolefins will react with two moles of benzene to yield heavy diphenylalkane products or heavier polymer stocks, some of which are removed as by-product 'heavy alkylate' in the LAB refining column.

There have been several variations of the UOP process over the years. The most recent one is the DETAL process, a joint development of UOP and Petresa of Spain. The DETAL process employs a heterogeneous bed reactor with liquid feeds and utilises a solid acidic catalyst. This innovation reduces the investment of the overall plant and improves yields. Its greatest advantage is that it eliminates the use of HF and associated problems. In another of UOP's developments, the DEFINE process has the objective of increasing LAB yield and cutting down heavy alkylate production. The equilibrium mixture from the PACOL process, termed as separator liquid, is charged to the DEFINE reactor alongwith make-up hydrogen for selective conversion of diolefins to mono-olefins. A stripping column separates dissolved light hydrocarbons, and the bottoms comprising a mixture of olefins and unconverted n-paraffins are charged together with benzene to the alkylation reactor, where LAB is produced. Benzene and unconverted paraffins are fractionated from the alkylation reactor effluent and recycled. The final column fractionates LAB overhead from heavy alkylate bottoms. Reduction of dialkylates content improves the solubility of the ultimate sulphonated and neutralised product, although the presence of dialkylates contributes to improved detergency of the sodium salt.

High quality LAB with minimum isomers and side products has been obtained by varying temperature, concentration, reaction ratios, recycle points in the LAB manufacturing processes. Fig. 2.4 shows schematic flow in the UOP process.

Typical specifications for LAB from the UOP PACOL-DEFINE process are given in Table 2.3.

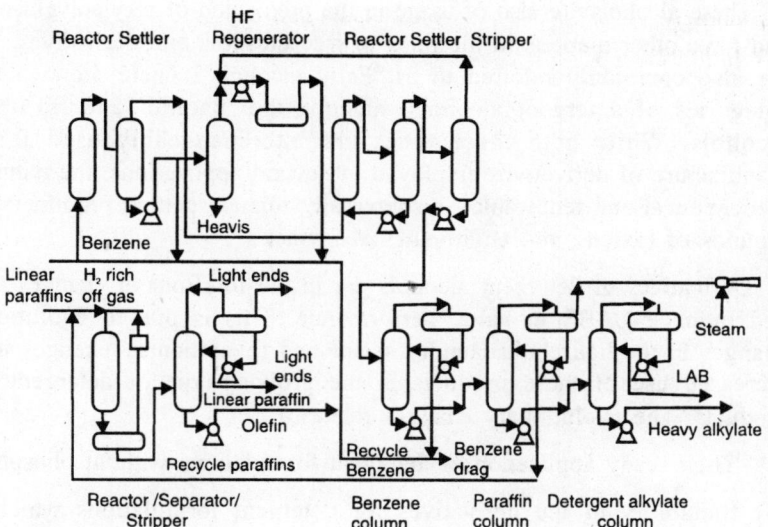

Fig. 2.4. Production of linear alkyl benzene schematic flow (UOP process).

Table 2.3. Typical specifications for LAB from UOP PACOL define process.

Bromine No.	0.02
Saybolt colour	+30
Doctor test	Negative
Water wt. %	0.01
Sulphonation wt. %	98.5
Biodegradability %	95.0
Paraffins wt. %	0.3
Indans and Tetralins	2.0
Normal LAB %	93.0
Appearance	Clear and bright

Detergent Alcohols

Detergent alcohols, as the name implies, are alcohols primarily produced for the manufacture of detergent intermediates, the most important of which are the surfactants such as alcohol ethoxylates, alcohol ether sulphates, alcohol sulphates and phosphate esters.

These alcohols are also of usage in the production of alkylpolyglucosides and have other markets in the form of free alcohols and derivatives. They are also commonly referred to as 'Fatty alcohols'. There are two basic categories of detergent alcohols, namely, the 'natural' and 'synthetic' alcohols. While both these types are interchangeably used for the manufacture of derivatives employed in several applications, the natural— oleochemical and renewable—resource alcohols are getting prominence on grounds of toxicity and environmental impact.

Derivatives of detergent alcohols are in the forefront of changes today and outpace LABS in value performance. This is due to evolutionary changes in the household laundry sector and revolutionary changes in the increased use of these products in the growing-type of detergent end-products. The evolutionary changes relate to :

(a) Their ready application in detergent formulations without phosphates.

(b) Suitability of the derivatives for detergent formulations which are effective at low washing temperatures—an aspect of importance as the washing loads are containing more and more sensitive, coloured synthetic fabrics.

(c) Their superior hardness tolerance (calcium and magnesium ions present in water) in comparison with that of LABS.

The revolutionary changes are related to four growing types of detergent end-products, namely :

(a) Low temperature compact powders.

(b) High concentrated liquids.

(c) 'One packet per wash load' products combining detergent, bleach and softener.

(d) Growing concern for skin-safe low irritant products for dishwashing as well as personal care products.

These changes which vary from region to region as in United Sates, Western Europe and Japan are benefitting the derivatives manufacturers and also formulators.

Detergent alcohols are defined as alcohols containing 12 or more carbon atoms per molecule with a carbon backbone of high linearity mainly for the manufacture of surfactant intermediates. Anionics like alcohol sulphates, alcohol ether sulphates, phosphate esters, sulphosuccinates, nonionics like alcohol ethoxylates and alkylpolyglucosides and even amines for cationic surfactants can be made from these alcohols. Highly branched alcohols, of long chain lengths, are used mainly in the manufacture of plasticisers. Alcohol structures which are slightly branched are also used in surfactants manufacture. Detergent alcohols are commercially produced from natural raw materials, principally fatty acids and methyl esters and petrochemical sources, namely, ethylene and n-paraffins. The two routes divide the alcohols into the aforementioned 'natural' and 'synthetic' alcohols.

Natural alcohols

Natural detergent alcohols, both saturated and unsaturated, are manufactured through the hydrogen reduction of fatty acids and their esters, typically by the Lurgi process, as well as the catalytic hydrolysis of methyl esters. The latter route is preferred nowadays for a variety of reasons including economy of investment. The methyl esters are, in turn, manufactured from triglyceride fats such as coconut oil, palm kernel oil, and tallow through transesterification processes. A continuous system for methyl ester production is given in Fig. 2.5 and a batch system for transesterification is shown in Fig. 2.6.

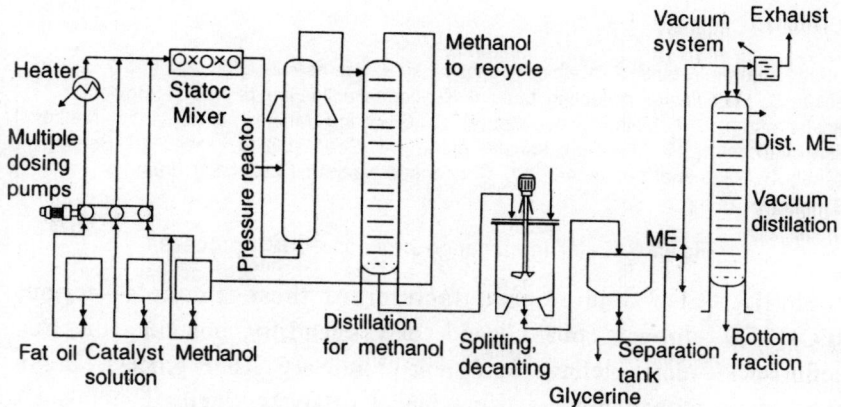

Fig. 2.5. Methyl ester production—continuous system.

Synthetic detergent alcohols

Synthetic detergent alcohols are substantially obtained through modifications of Ziegler chemistry, often referred to as 'ethylene growth' processes. In these processes, ethylene is added to triethylaluminium to produce a mixture of alkyls which is then oxidised by air to the alkoxides and subsequently hydrolysed to the alcohols. The alcohols from Ziegler chemistry are always products with even number of carbon atoms, thus imitating the structure of the natural alcohols. Vista chemicals and ethyl corporation control most of such capacity. ALFOLS are Vista products.

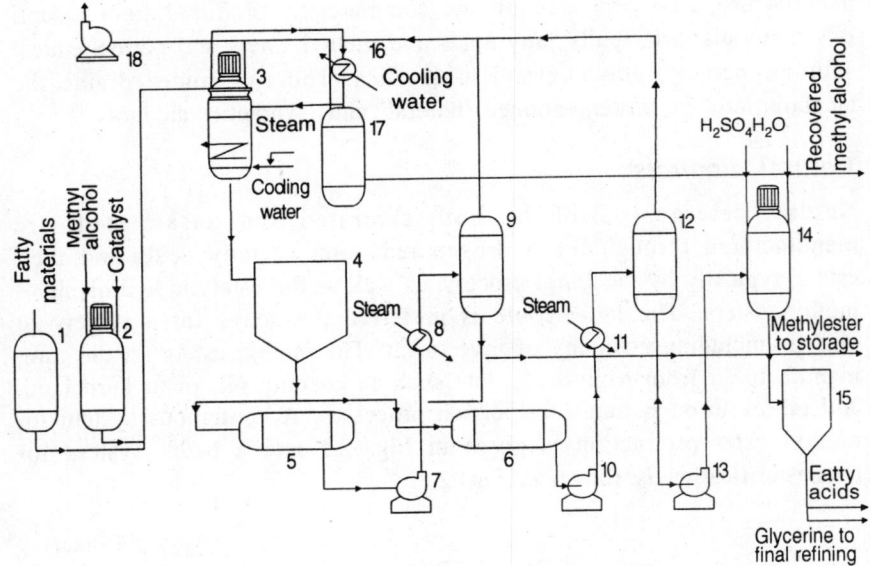

1. Fatty materials tank; 2. Methyl alcohol tank; 3. Interesterification reactor; 4. Settling vessel; 5. Methylester collecting tank; 6. Glycerine collecting tank; 7. Methylester transfer pump; 8.,9. Methanol evaporator; 10. Glycerine transfer pump; 11.,12. Methanol flash evaporator; 13. Glycerine transfer pump; 14. Soap spliting vessel; 15. Settling vessel; 16. Methanol condenser; 17. Condensed methanol collecting drum; 18. Vacuum pump

Fig. 2.6. Ester interchange process—batch process.

Shell is also a large manufacturer of these alcohols—termed NEODOLS—through cobalt ligand 'oxo' chemistry, obtaining the base olefin stock—alpha olefins—from their proprietary SHOP process. This is an ethylene growth process using nickel catalysts. Shell's alcohols are slightly branched alcohols with odd and even-numbered alcohols. Detergents from these alcohols have properties and costs comparable with those of even numbered alcohols.

Ruhrchemie and Rhone Poulenc have developed a version of oxo chemistry with water-soluble rhodium complexes. This process is reported to yield alcohols of higher linearity than the Shell process.

Internal olefins manufactured through normal paraffin dehydrogenation with special catalysts are also reported to yield slightly branched alcohols through oxo-chemistry. Secondary alcohols are made in Japan through an n-paraffin oxidation process.

Exxon Chemical also is a source for linear detergent-range alcohols, although they have been for long in the business for branched chain alcohols to the plasticisers industry and also Guerbet alcohols. It is reported that these alcohols are highly branched when compared to the modified Ziegler alcohols.

Alphaolefins (LAO)

Alpha olefins, often abbreviated as LAO, are linear mono-olefins with the double bond located at the terminal of a linear long-chain hydrocarbon. They differ from the 'internal mono-olefins' wherein the double bond is randomly located inside the hydrocarbon molecule. For detergent applications the carbon range of the raw materials used is C_8 to C_{18}. Apart from detergents, the alphaolefins figure in the production of oxo-alcohols for the plasticisers industry, in the manufacture of HDPE and LLDPE as co-monomers, in polybutene production, and as raw materials for polyalphaolefins (PAO) used as synthetic lubricants.

Alphaolefins production

The four main routes for the production of alphaolefins are :

(a) Through wax-cracking.

(b) Paraffin dehydrogenation.

(c) Ethylene oligomerisation.

(d) Shell's SHOP process.

Shell is the sole operator in Western Europe using the wax-cracking process, that too partially, to produce alphaolefins. Chevron, which operated this process in USA, abandoned it in the mid-eighties. The cracking of wax occurs thermally and a wide band of alphaolefins are produced, substantial volumes of which have little application. The process requires a special type of wax with highly linear hydrocarbon waxes which are reported to originate only from Libyan and Indonesian crude oils. Shell has also operated a paraffin chlorination-dehydrochlorination plant in USA to produce odd and even number C_{11} to C_{14} linear internal olefins of which a segment has been alphaolefins. They have been used for the manufacture of LAB and oxo-alcohols, but not for alpha-olefin based products.

By far, the most important of the processes to make alphaolefins are those based on Ziegler chemistry. In the classical process, an aluminium triethylene growth promoter (TEA)—catalyst—is used to build up long chains of low molecular weight ethylene polymers on the aluminium, the ethylene being inserted between the aluminium atom and the carbon. These are then displaced by more ethylene to yield a mix of alpha-olefins and TEA, the latter being recycled. The displacement involves dehydrometallisation followed by ethylene hydroalumination. Modifications of this process are the Gulf Process and the Ethyl Process. In the Gulf Process, the catalytic process is a one-step reaction producing a geometric distribution of alphaolefins. In the Ethyl Process, the one-step Gulf Process is changed to a two-step process carried out at lower and higher temperatures. By this method, using appropriate reaction conditions a stoichiometric distribution of alphaolefins of desired chain lengths can be produced.

Shell's SHOP process employed in plants in U.K. produces internal olefins for captive manufacture of oxo alcohols or LAB in Shell's plants. The alpha-olefins produced are for merchant sales. The SHOP Process is a variation of the Gulf Process with replacement of the aluminium catalyst by a proprietary nickel catalyst in a solvent like benzene or 1,4-butane-diol.

In India, Gujarat Godrej Innovative Chemicals whose plant is engineered by Lurgi, adopts a new route for alphaolefins. In this process, linear alcohols made from imported palm stearin are catalytically dehydrated and the alphaolefins produced are used as raw materials for alpha olefin sulphonates.

Ethylene Oxide (EO)/Propylene Oxide (PO)

Ethylene oxide is an important building block of the petrochemical industry and is used to manufacture ethylene glycol (for polyester fibre and bottles, antifreeze compounds), polyethylene glycols (brake fluids, cosmetic intermediates, epoxides), glycol ethers (solvents) and surfactants. The EO based nonionic surfactants as well as other types of importance are alcohol ethoxylates (AE), alcohol ether sulphates (AES), alkylphenol ethoxylates (APE), ethoxylated fatty esters and acids and ethoxylated amines.

The glycols have been considered over the years as major outlets for ethylene oxide. However, currently nonglycol applications of EO are in the increase.

Manufacture

EO manufacturing processes are : (a) chlorohydrin process, and (b) direct oxidation of ethylene. The chlorohydrin process, in which ethylene is reacted with hypochlorous (chlorine and water) to give ethylene chlorohydrin which is treated with slaked lime or caustic soda to form EO, has been the traditional route to EO, but now this has been largely replaced by the direct oxidation process.

The direct oxidation of ethylene may be carried out over a silver catalyst in the vapour phase at about 245°C and 21 atmospheres of pressure. Either oxygen or air enriched in oxygen may be used for direct oxidation. The reaction may be represented as follows :

$$CH_2 = CH_2 + \tfrac{1}{2}O_2 \longrightarrow \underset{\displaystyle O}{CH_2{-}CH_2}$$

Ethylene   Oxygen   Ethylene oxide

Propylene Oxide (PO) is produced either by the chlorohydrin or hydroperoxide process. ARCO Chemical USA is the largest producer of PO world over. Various modifications of these processes are used for the production of PO.

Fatty Amines

Nitrogen-based surfactants dominate the cationic types. Fatty amines comprising of a basic group of alkylamines—primary, secondary and tertiary, are important raw materials for these surfactants.

Primary, secondary and tertiary alkylamines—taking their names from the number of hydrogen atoms of the ammonia molecule substituted by one, two or three alkyl groups—are obtained by the reduction of nitriles which in turn are produced from fatty acids and ammonia in the presence of zinc oxide or manganese acetate catalysts. Alkyl group chain lengths may vary from C_8/C_{10} to higher. Whereas the primary amines are only obtained from nitriles, the secondary amines are obtained both from nitriles and from primary amines through further alkylation. The tertiary amines, the most important of this group in terms of volume and diversity of down-stream products, are produced through several routes such as :

(a) Reductive alkylation of a primary amine with formaldehyde and hydrogen.

(b) From alkyl chlorides and dimethylamine.

(c) From primary alkyl bromide and dimethylamine.

Tertiary amines are supplementally obtained from alpha-olefins.

Apart from the above types of oxygen-free amines, the fatty amines also comprise of oxygen-containing amines like ethoxylated amines. The production of nitrile and alcohol based amines in the Industrial Triad is of the order of 350–375,000 metric tons per annum.

The three groups of amines by themselves do not have any marked surfactant properties. About 20–25% of the amines and their salts are used directly and the balance converted to derivatives. Primary amines are predominantly used for the production of tertiary amines. Other uses of these amines are in metal extraction, as froth flotation collectors, as anti-caking agents for fertilisers, as filming amines in steam evaporators etc. Ethoxylated primary amines are used as cationic collectors in the flotation of zinc ore, as emulsification agents for pesticidal emulsions and in the petroleum industry. Secondary amines are the basic raw materials for the important range of dialkyl dimethyl quaternary ammonium salts used very largely as fabric softeners in home laundries and as textile conditioners in the textile industry. The major direct use of tertiary amines is in solvent extraction of metals. They are the basic raw materials for either classes of important nitrogen compounds, namely :

(a) Alkanolamides

(b) Quaternary ammonium compounds

(c) Polyamides

(d) Betaines

(e) Imidazolinium compounds

(f) Amphoteric surfactants

(g) Amine oxides

(h) Ethoxylated amines

All of these products are surface-active and are cationic in nature. The selection criteria for fatty amines and derivatives involve.

(a) Solubility in water, non-aqueous media and two-phase systems.

(b) Hydrophile—lipophile balance (HLB).

(c) Surface tension.

(d) Other surfactant properties like Forming, Krafts point, cloud point.

(e) Antimicrobial properties.

(f) Various physical and chemical properties.

Amines with bridging amids groups termed 'Amidoamines' are also building block amines. They are derived from fatty acids, i.e., the fatty acid is the source of the alkyl radical rather than the nitrile. The important economic advantage of the amids bridging is that these amines can be prepared at far lower levels of investment than are the basic amines which require the use of high pressures and expensive equipment. Also the amids group improves the multifunctionality of these amines. Many of these materials and their derivatives perform as effectively as the basic amines. In direct use they find application as flotation agents, asphalt emulsifiers and corrosion inhibitors and compete with the basic amines and their derivatives.

KEY FEATURES OF SURFACTANTS

Major Surfactants

Amongst the various types of surfactants major ones can be identified either by their large-scale use in industry or by their special chemical structures and characteristics. The products highlighted in this section are :

(a) Sulphonates—lignin sulphonates, petroleum sulphonates, alkylbenzene sulphonates, alkylnaphthalene sulphonates.

(b) Sulphates—alcohol sulphates, alcohol and alkylphenol ethersulphates.

(c) Polyhydric alcohol esters—sorbitan esters, glycerol esters, polyglycerol esters, glycols and glycol esters.

(d) Ethoxylates—Ethylene oxide/propylene oxide block copolymers, alcohol ethoxylates, alkylphenol ethoxylates, polyhydric alcohol ethoxylates, lanolin ethoxylates, amine ethoxylates, polysiloxane ethoxylates.

(e) Alkanolamides.

(f) Fatty amine oxides.

(g) Acetylenic alcohol derivatives.

(h) Silicones.

(i) Fluorosurfactants.

(j) Alkyl polyglycosides.

(k) Sulphosuccinates and sulphosuccinamates.

(l) Phosphate esters.

(m) Amine based surfactants—imidazolines, quaternary ammonium compounds, betaines.

(n) Sarcosinates and taurates.

Lignin sulphonates

Although lignin sulphonates have low surface activity, they are used in many areas where a conventional surfactant can be used. They are obtained as by-products in the pulping of wood in the paper industry.

Lignin is one of the three major constituents of wood—the celluloses, pentagens and lignin. It represents 15–30% of the wood depending on the species. There is no clear understanding of the structure of lignin, but in general, propane units are bound through ether linkages to phenols with the molecular weight of the order of tens of thousands.

When wood chips are treated with sulphite liquor the cellulose and hemicellulose are not affected, whereas the lignin is partially sulphonated on the aromatic ring, with some degradation to pentoses and hexoses (sugars). The sulphonates and the sugars are all water-soluble and thus are easily separated from cellulose pulp. The sugars in the sulphite liquor are destroyed by fermentation and the residue, together with surplus sodium or calcium sulphite (depending on the pulping method used) is spray dried to the commercial product. Unlike other products of recovery processes, the lignosulphonates do not consume sulphonation chemicals. Various refinements on the raw, dried product yield products of different molecular weights, cation and chemical composition.

Petroleum sulphonates

The anionic petroleum sulphonates are obtained either as by-products of oil processing or by synthetic methods. Both materials are used in commerce competitively.

When high-grade petroleum oils (white oils) are prepared, reactive components and sulphonatable coloured bodies in the base oil have to be removed from the non-reactive portion, which is the desired material. One of the best ways of doing this is to sulphonate the oils with oleum or sulphur tri-oxide or both. The reactive components react with the acid to form sulphonates. The non-reactive components are naphthenic-paraffinic hydrocarbons. The sludge obtained in the sulphonation of base oils, termed 'green acids' contains the water-soluble sulphonates, excess acid and occluded oil. The products are largely disulphonated. After separation, the upper layer contains all unreacted oil and oil-soluble sulphonates termed 'mahogany' acids.

The oil is now given an alkaline wash and extracted with a low-boiling alcohol. The paraffinic-naphthenic fraction is not soluble in the alcohol whereas the sodium salts of mahogany acids are. When the alcohol is stripped from the lower layer, a high molecular weight sodium petroleum sulphonate results with some 30% unreacted oil. The sulphonates by and large are monosulpho-nated products.

Synthetic petroleum sulphonates are byproducts obtained in the Friedel Craft reaction to get detergent alkylates from benzene such as :

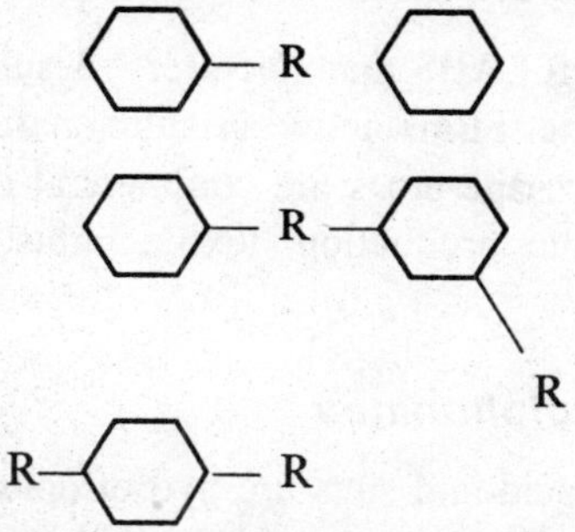

whose mixture is sulphonated with oleum. Both the natural and synthetic sulphonates are marketed as such or as overbased materials. The products marketed in largest volume are the alkaline earth salts produced by the double decomposition of the sodium sulphonate with an aqueous solution of barium, calcium or magnesium chlorides.

The petroleum sulphonates find application as emulsifiers for textile oils, dry cleaning solvents, degreases, soluble oils, metal working fluids, mold release agents and froth collectors in ore flotation. They are also used as dispersants (pigment dispersion in paints or sludge dispersion in fuel oil) and as rust inhibitors.

Alkylbenzene sulphonates

The largest used surfactant across the world are alkylbenzene sulphonates. There are two broad types of these products, namely :

(a) Linear alkylbenzene sulphonates (LABS), and

(b) Branched chain alkylbenzene sulphonates.

The branched chain product has poor biodegradability and as such does not figure in the household sector. The linear product is predominantly used in the household detergents and scourers, but the usage extends to I&I cleaners and industrial sectors. The length of the alkyl chain has profound influence on the detergency and solubility of the product.

The most common method of sulphonating alkylbenzenes has been through oleum. The disadvantage of the process is the need for the disposal of the spent acid. Air-SO_3, used predominantly for the sulphonation of long-chain alcohols is increasingly being applied to alkylbenzenes. The detergency properties of LABS (dirt removal power) are superior to those of the alcohol ethoxylates and moreover it is a cheaper product. However, it's high foaming tendency and higher efficiency only at higher temperatures are disadvantages for its use in the household sector.

There are a number of LABS manufacturers. A substantial portion of the product is for captive use. Firms active in sulphonation also undertake toll manufacture. The main usage areas are commercial dishwashing, emulsion polymerisation, petroleum production, textile industries and in pesticidal sprays.

Alkylnaphthalene sulphonates

Naphthalene sulphonic acid and salts are hydrotropes, i.e., materials which assist in solubilisation of associate actives without other impact on the system. In this respect they are similar to cumene, xylene and toluene sulphonic acids. They do not have any meaningful surfactant properties. With the introduction of alkyl side chains such as propyl, butyl and dipropyl surfactancy develops.

The sulphonates are prepared by condensing naphthalene sulphonic acid with formaldehyde. In this case the disulphonates are formed. If the homosulphonates are desired naphthalene, oleum and formaldehyde are reacted together.

Alkyl naphthalene sulphonates and the alkyl naphthalene sulphonates—formaldehyde condensates are used in applications where lignosulphonates can also be used. Most importantly they are used as dispersing agents and stabilisers wherein particularly the condensates are superior performers. The areas of application are :

(a) As water reducers in the preparation of concrete (the formaldehyde and melamine formaldehyde condensates are the 'superplasticisers').

(b) As solubilisers in the production of SBR lattices.

(c) As pigment wetting and dispersion agents in the mixing of paints (sodium diisopropyl naphthalene sulphonic acid and condensates are used).

(d) As low-foam surfactants and dye dispersion aids in the textile industry.

(e) As pitch dispersants in paper industry.

(f) As syntans—leather tanning agents.

(g) As suspension and dispersion agents in the formulation of pesticide suspension concentrates and in the formulation of wettable powders.

(h) As dispersing agents in cure pastes in the plastics and elastomer industries.

(i) As compact dispersing agents in oil muds.

(j) As free flow additives in fertiliser mixtures and other dry materials.

(k) In carpet steam cleaners.

(l) In tank soak metal cleaners, and.

(m) In high foaming alkaline and acid cleaners.

Alcohol sulphates

Alkyl sulphates or alcohol sulphates are the first of the world's synthetic detergents, pioneered in Germany during the Second World War. They are currently produced by the sulphation of linear primary alcohols with air-SO_3 or chlorosulphonic acids, followed by neutralisation with a variety of cations. Most of the alcohol sulphates are made from natural alcohols because of their large use in personal care products. The alcohol's used range from C_8 to C_{16}. In the United States, they are derived from animal tallow base alcohols and as such are the least expensive anionics. The most-used product is lauryl alcohol sulphate. Secondary alcohol sulphates are also used as surfactants in industrial applications.

In general the alcohol sulphates are more hydrophilic than the sulphonates. Sodium lauryl alcohol sulphate has strong wetting, detergent and emulsifying properties. They are good foamers but are sensitive to hard water, necessitating effective builder support in consumer products. The major outlets for alcohol sulphates are in laundry powders, personal shampoos, in emulsion polymerisation of rubber lattices for textile and paper coatings, in wool scouring, kier boiling of cotton, as dye levelling agent for anionic dyes, as wetters for fruit processing, as pigment dispersion aids in paints and inks, in foaming of gypsum board, in soaking and degreasing of leather and in binding compositions for non-wovens.

Alcohol and alkylphenol ether sulphates (AES/APES)

The hydrophilicity of ethoxylate structures can be markedly enhanced by the sulphation/sulphonation of the—OH group. In the process the ethoxylates lose their nonionic nature and become anionic materials.

When the aromatic nucleus of a compound is sulphonated we have a 'sulphonated' product. If the reacting group in a hydroxyl group is an aliphatic compound, then the process is termed 'sulphation'.

In the sulphonation of an alkylphenol, while the primary reaction is sought at the active hydrogen site, some sulphonation of the aromatic nucleus also occurs. The ether sulphates are diversely known as ethoxysulphates or sulphated polyethers. AES and APES are used interchangeably in most technical applications.

The manufacture of the ether sulphates is based on sulphonation routes akin to those used for the manufacture of LABS. The heat of reaction is however higher and special types of reactors have to be employed. Common sulphonating agents used are oleum, chlorosulphonic acid and air-sulphur trioxide.

The alcohol ethoxylates, while soluble in water, produce turbidity at higher temperatures with likelihood of layer separation. AES overcome this problem. They are also high foaming products compared to the low foaming ethoxylates and are therefore of advantage where foaming action is required. In comparison with other anionic products they are excellently soluble, have good hard water tolerance and are more compatible with enzymes. They are extensively used in personal care products due to their mildness to the skin. In household detergents they are synergistic with LABS.

Virtually any sulphonator can make AES and APES. The basic markets for AES are in light-duty liquids, bath additives and personal shampoos. A sizeable quality is also used in household liquid surface cleaners. APES on the other hand is almost entirely consumed in the industrial sector. In most industrial operations the markets for AES and APES overlap. Major ones are in emulsion polymerisation of polyvinyl acetate and acrylic polymers, as foaming agents in the production of foam concrete and plasterboard and as foam booster in commercial dishwashing applications. The ether sulphates are also well known as dispersants for coating fillers in the paper industry, as foamers in drilling mud compositions, in the surfactant mixes for enhanced oil recovery and as wetting agents for primary backing compounds in the resins and elastomers industries.

Sorbitan esters

The surfactant field comprises of several nonionics belonging to the ester class. A wide variety of esters derived from alkanoic acids by reaction with polyhydric alcohols—sorbitan, glycerol and polyethylene/polypropylene glycols have commercial significance.

The sorbitan esters with a wide range of hydrophilic and lyophilic properties are commercially produced. Sorbitol is a hexahydric alcohol derived from high purity dextrose. Sorbitan which is a mixture of anhydrosorbitols mainly consists of

(1,4 Sorbitan)   (Isosorbide)

The sorbitan esters are not water-soluble but soluble in a wide range of mineral and vegetable oils, as sorbitan is not a strongly hydrophilic group. The lipophilic character of these esters aids tapping of water in oil to form water-in-oil (w/o) emulsions. Sorbitan mono, di and tri esters are derived by the reaction of sorbitan with fatty acids. The physical form and HLB are tailor-made by varying the type of fatty acid used and the degree of esterification. The products are approved for human ingestion and are used widely in foods, beverages and pharmaceuticals. Their lipophilic, emulsifying, solubilising, softening and lubricating properties are made use of in synthetic fibre manufacture, textile processing and cosmetics.

Glycerol esters

Glycerol esters form another group of polyhydric alcohol based nonionic surfactants derived by the esterification of glycerol with saturated and unsaturated fatty acids. Mono- and di-glycerides are made both through the esterification of glycerol with the appropriate fatty acid and through 'glycerolysis', the latter gaining prominence. In its modern batch version, glycerolysis involves reacting a suitable animal or vegetable fat with excess glycerol catalysed by NaOH or KOH at around 250°C. The reaction proceeds as follows :

$$
\begin{array}{ccc}
\text{CH}_2\text{OOCR} & \text{CH}_2\text{OH} & \text{CH}_2\text{OOCR} \\
| & | & | \\
\text{CHOOCR} \quad + \quad 2\text{CHOH} & \rightleftharpoons & 3\ \text{CHOH} \\
| & | & | \\
\text{CH}_2\text{OOCR} & \text{CH}_2\text{OH} & \text{CH}_2\text{OH} \\
\text{(Fat)} & \text{(Glycerol)} & \text{(Monoglyceride)}
\end{array}
$$

Excess glycerol is used to drive the reaction to the right and to slow down the reverse reaction. The position of the acyl radical—*alpha, beta, gamma*—depends mainly on the temperature of glycerolysis, but the *beta* product reverts to the *alpha* one on cooling. In place of fats, methyl esters are also used for glycerolysis. Fats used are soya bean oil, cottonseed oil, edible tallow and food grade triglycerides. Typically the commercial product analyses 40–48% monoester, 30–40% diester and 5–10% triester with low percentages of free fatty acid and glycerol. Product concentrations can be changed through molecular distillation of the mixed ester product or a mixture can be enriched with a pure or enriched monoester.

Both the solid and liquid type glycerol esters are hydrophobic materials which can be used as w/o emulsifiers, co-emulsifiers, lubricants, rust preventive additives, mould release agents and as solvents for dyes and pigments. They are used as components in lubricants and softeners for the leather industry, as anti-icing additives in fuels and as rust preventive additives in compounded oils. Other application areas of glycerol esters are in textiles, metals and plastics.

Glycerol monostearate is used as an emulsifier and pacifier in cosmetics. The glycerol esters are effective in building up viscosity in cosmetic systems. The non-dispersible grades are used as bases for creams and lotions. By adding other non-ionics such as sorbitan esters or ethoxylated polyol esters, anionics such as soaps or cationics such as quaternary ammonium compounds, the glycerol esters can be readily dispersed. The self-emulsifying grades are used as primary emulsifiers and in anti-perspirant formulations.

An important use of monoglycerides and diglycerides is in foods. They are widely used in breads, cakes, other bakery products, candies, ice creams, yeast, butter, whipped toppings and icings. Table 2.4 gives some typical commercial glycerol esters and suppliers of monoglycerides. Other than the stearates, glycerol monocaprylate made from short chain coconut fatty acids is rather unique. Due to its structure, it is an efficient solubiliser dissolving water upto 20% by weight. It has been used very effectively in the pharmaceutical industry as a solubiliser for vitamins, flavours and medicaments.

Table 2.4 Some typical commercial glycerol esters and their functions.

Ester type	Form	Mono %	Free glyce- rine	Acid value	Iodine value	HLB
Monocaprylate	Liquid	75	15	3	1	8.3
Monooleate	-do-	40	4	4	95	3.4
Monostearate	Flake	40	6	6	1	3.8
Distearate	-do-	–	–	5	–	2.4
Monostearate SE	Bead	30	4	4	3	8.4
Monolaurate	Solid	–	–	4	–	4.9
Dilaurate	-do-	–	–	5	–	4.0
Monococoate	-do-	50	5	4	10	3.8

Functions	Application areas
Lubricant	Textile staple fibres
Lubricant/Antistatic agent	In formulations for high speed processing of polyamide, polyester filaments.
Fabric softener	In textile industry
Emulsifiers/lubricants	Drilling and cutting oils, grinding and polishing pastes for metals.
Internal lubricant/thermal stabiliser/wetting agent for additives	PVC processing
Antistatic agents	Polyethylene, polypropylene, foamed polystyrene.

Polyglycerol esters

Compounds derived from the condensation products of hydrophobic groups containing an active hydrogen and glycidol constitute an important family of nonionics. They have the general formula :

$$RX(C_3H_5O)_nOH$$

where R is a typical surfactant hydrophobic group and X is O, N or another functionality capable of linking the glycidol to the hydrophobe. The maximum value of 'n' in the case of polyglycerol is ten as further reaction with the hydrophobic group becomes difficult. The existence of three possible isomers of diglycerol and increase in isomeric possibilities with each subsequent addition of glycidyl units leads to complexity of the products. Ranging from hydrophilic monoesters lipophilic diesters, the polyglycerol esters can be represented as :

$$
\begin{array}{ccc}
\overset{\displaystyle H\ H\ H}{\underset{\displaystyle O-OH\ H}{H-\ C-C-C-O}} &
\left[\overset{\displaystyle H\ H}{\underset{\displaystyle O\ H}{C-C-C-O}}\right]_n &
\overset{\displaystyle H\ H\ H}{\underset{\displaystyle H\ OH\ O}{C-C-C-H}} \\
\overset{\displaystyle |}{C=O} & \overset{\displaystyle |}{C=O} & \overset{\displaystyle |}{C=O} \\
R & R & R
\end{array}
$$

They are prepared by the esterification of polyglycerols with specific fatty acids. Polyglycerol can also be prepared by the alkaline dehydration of glycerol and then esterified with a fatty acid.

The polyglycerol esters are effective nonionic emulsifiers for both o/w and w/o emulsions. They are used in emollient creams and lotions, in pharmaceutical vehicles and due to their nontoxicity in food emulsions such as margarine and moisturising additives for bread and other baked goods. Typical polyglycerol esters are given in Table 2.5.

Glycols and glycol esters

Polyethylene glycols are condensation polymers of ethylene oxide represented by the formula :

$$HO(CH_2CH_2O)_nH$$

where n represents the average number of ethylene oxide units. Ranging from molecular weight 200 to 3000, represented as PEG 200, PEG 400 etc. they exhibit good water solubility. Water is miscible in all proportions with PEG 200 through PEG 600. The polyethylene glycols are chemical intermediates for other nonionics. Being themselves nonionics they find applications as solubilisers, in coatings, adhesives, lubricants, metal working, paper manufacture, petroleum production, ceramic making, printing, electronics, solvents, cleaning formulations, humectants and viscosity modifiers.

Surfactants of the glycol ester group are monoesters of either ethylene or propylene glycol. The extreme hydrophobicity of the diesters render them unfit for use as surfactants. Since the simple glycol esters possess one unesterified hydroxyl group, they can be ethoxylated to increase their hydrophillic characteristics. Commercial polyethylene glycol esters contain varying proportions of monoesters, diesters and polyglycol. The physical properties of these esters vary with the type of fatty acid and the degree of ethoxylation.

Table 2.5. Typical polyglycerol esters.

Ester	Form at 25°C	Hydroxyl value	Saponification value	Iodine value	HLB (+ 1.0)
Decaglycerol tetraoleate (Drewmulse 10-4-O)	Liquid	200–250	125–145	60	6.0
Decaglycerol octaolete (Drewmulse 10-8-O)	Liquid	25–50	150–180	80	4.0
Decaglycerol decaoleate (Drewmulse 10-10-O)	Liquid	20–60	160–190	90	3.0
Decaglycerol decastearate (Drewmulse 10-10-S)	Solid	30–60	150–190	3	3.0
Hexaglycerol distearate (Drewmulse 6-2-S)	Solid	270–310	100–130	5	8.0
Triglycerol monostearate (Drewmulse 3-1-S)	Solid	315–345	120–140	3	7.0
Triglycerol monooleate (Drewmulse 3-1-O)	Liquid	230–290	120–160	90	7.0

'Drewmulse' is the trade name of PVO International Inc. and corresponding Stepan Company products are designated 'Drewpol'.

The ethoxylates contain nearly 60% by weight of ethylene oxide to enable the solubilisation of the fatty acid at room temperature. Their principle use is as emulsifiers in the textile industry for processing oils, antistatic agents and softeners. Other applications are as fibre lubricants, as detergents in scouring operations and as emulsifiers for pesticides and cosmetic preparations.

Amongst the propylene glycol esters the most important is propylene glycol monostearate (PGME) which is almost wholly utilised for manufacturing derivatised products which act as emulsifiers in the food industry. Table 2.6 gives some commercial polyethylene glycols and Table 2.7 some commercial fatty acid polyethylene glycol/polypropylene glycol esters.

Ethylene Oxide/Propylene Oxide Block Copolymers

Amongst the ethoxylates ethylene oxide - propylene oxide (EO/PO) block copolymers stand out as a unique class. In this class of materials, the careful control of monomer feed and reaction conditions allows the preparation of a series of surfactants in which the HLB, solubility, wetting and foaming properties can be closely and reproducibly controlled. The classification of these products is primarily based on the nature of the initiator employed in the formation of the initial polymer block, with subclasses determined by the compositions of the various blocks. The hydrophobicity is derived from one of the two polymeric blocks and typical initiators are monohydric alcohols such as butanol, dihydric materials such as glycol, glycerol, higher polyols or ethylene diamine.

Table 2.6. Some commercial polyethylene glycols.

(Union Carbide Carbowax brand products)

Carbowax	Molecular weight	Specific gravity	Viscosity (Cst at 100°C)	Surface tension (dynes/cm)
PEG 200	190–210	1.1239	4.3	44.5
PEG 300	285–315	1.1250	5.8	44.5
PEG 400	380–420	1.1254	7.3	44.5
PEG 600	570–630	1.1257	10.8	44.5
PEG 900	855–900	1.0927 (60°C)	15.3	–
PEG 1000	950–1050	1.0926 (60°C)	17.2	–
PEG 1450	1300–1600	1.0919 (60°C)	26.5	–
PEG 3350	3000–3700	1.0926 (60°C)	90.8	–
PEG 4600	4400–4800	1.0926 (60°C)	184.0	–
PEG 8000	7000–9000	1.0845 (70°C)	822.0	–

Table 2.7 Some commercial fatty acid polyethylene glycol/polypropylene glycol esters.

Glycol ester	Applications
PEG esters of lauric, oleic and stearic acids	Cosmetics, food, pharmaceuticals, plastics, agriculture and others
PEG 600 distearate	Fabric softener
PEG 600 dioleate	-do-
Glycol stearate	Pearl shine agent for shampoos, shower and bath preparations, emulsions
Glycol distearate	-do-
Glycol stearate	-do-
Glycol distearate	-do-
Propylene glycol stearate	-do-
PEG 300 oleate	Emulsifier, lubricant, textile softener
PEG 400 dilaurate	Emulsifier, lubricant, softeners, release agent for paper, spin finishes.
PEG 300 monopelargonate	Coemulsifier
PEG 200 monolaurate	Defoamer in water based systems
PEG 600 monostearate	Emulsifier and viscosity modifier
PEG 400 monooleate	Emulsifier
PEG 6000 monooleate	Emulsifier, stabiliser, lubricant
PEG 400 sesquioleate	Emulsifier for kerosene in agricultural sprasy

Depending on the nature of the polymer blocks in the molecule this class of nonionics is made up of three subclasses, namely, 'all-block', 'heteric-block', and 'all-heteric' nonionics. Structurally the simplest form of the surfactants of this group is :

$$H—(EO)_m (PO)_n (EO)_m—H$$

In all-block surfactants each block is homogeneous—meaning that a single alkylene oxide is used in the monomer feed during each step in preparation. In heteric-block surfactants one portion of the molecule is composed of a single alkylene oxide, while the other is a mixture of two or more such materials, one of which may be the same as that of the homogeneous block portion of the molecule. In the preparation of such materials, the hetero portion of the molecule will be totally random. In the all-heteric block the monomer feed for the alkylene oxide in each step is composed of a mixture of two or more materials. These materials have a potential manufacturing advantage in that it is possible to vary the monomer feed composition during the reaction to continuously change the composition of the polymer as it grows on the initiator. The EO-PO block copolymers can belong to either of the classes :

$$I—A_m—B_n \text{ or } [B_nA_m]_x \text{ } I' \text{ } [A_m B_n]_y$$

where I and I' are the initiators. A is an alkylene oxide in which at least one hydrogen has been replaced by an alkyl or aryl group and B is an aqueous solubilising group such as oxyethylene. 'm' and 'n' are the degrees of polymerisation, usually greater than six. A wide variety of surfactant properties can be achieved by controlling 'm' and 'n'.

A method of manufacturing the EO-PO block copolymers is to form the polymerised PO—substrate and copolymerise the hydrophilic substrate onto it. The inverse type in which PO is condensed onto a polyethylene glycol substrate has also interesting properties. Other versatile group of products are made using ethylene diamine and synthetic C13 to C15 alcohols as starting materials.

EO-PO block copolymers encompass almost the entire spectrum of applications by the surfactants. Essentially they are defoamers, wetters, rinse aids, detergents, emulsifiers, emulsion stabilisers, dispersants, emulsifiers, antistats, antidusts and thickeners. Industrially they are prominent in three groups, namely :

(a) defoamers—in bottle washing, sugar production, fermentation of antibiotics, paper manufacture, in metal/phosphate cleaning, drilling muds, dishwashing, commercial laundry and inks.

(b) detergents—in hard surface cleaners, alkaline metal degreasing, wool scouring, cotton desizing, hand dishwashing, food and drink industry.

(c) Wetters—in agricultural chemicals, paper, leather and paints industries and commercial dishwashing.

The efforts to effect energy savings during the last decade frequently led to employing very high agitation rates to compensate for lower temperatures in many industrial operations. This led to greater use of low foaming surfactants. Apart from EO/PO block copolymers, ethoxylated and propoxylated linear alcohols (AEOPO) and alkylphenols (APEOPO) have been inducted for this purpose. Global production of EO/PO copolymers is around 140000 metric tons per annum.

Alcohol Ethoxylates (AE)

Alcohol ethoxylates (AE), also known as alcohol ethers, ethoxylated alcohols, polyglycol ethers, alkoxy alcohols, alcohol—EO adducts are ethers and are made through the catalytic reaction of one mole of a long chain alcohol with one or more moles of ethylene oxide (EO). The reaction is exothermic and solid or liquid products result depending on both the molecular weight of the products and their distribution. The substrated alcohols used are popular detergent alcohols (C12–C18), obtained either from natural sources or petrochemical feedstock. Alcohol ethoxylates with a low level of EO are usually sulphated to the anionic ether sulphates. For the ethoxylation, predominantly alcohols with mixed lengths comprising both linear and slightly-branched types are used.

Secondary reaction products, mainly polyethylene glycols (PEG) of molecular weight exceeding 200, are also formed in the ethoxylation of alcohols. The PEG content of a commercial product is of the order of 2–3%. Alcohol ethoxylates are true surfactants with clear hydrophobic and hydrophilic moieties. They do not dissociate on dissolution in water and thus are nonionic surfactants. Alcohol ethoxylates with 1-4 moles EO are oil-soluble, and water solubility begins to increase after 5–6 moles EO. The solubility decreases with increase in temperature and as such the cloud point of an alcohol ethoxylate is an important parameter in its utility for a particular application. The cloud point also is a good indicator of the foaming tendency of the alcohol ethoxylate. Commercial products have between 10–50 moles EO.

Replacement of EO by propylene oxide (PO) results in propoxylates. Just as the alcohol ethoxylates with low level of EO can be sulphated, they undergo phosphation to yield surfactants that perform as emulsifiers, detergents and dispersants. In general, the larger the EO chain, the higher is the HLB or hydrophilic character, pour point, viscosity, density, cloud point and flash point of the product. In the case of solid products the pour points can be reduced by adding a small amount of water to facilitate handling of the alcohol ethoxylate. Secondary alcohol ethoxylates, a subclass of alcohol ethoxylates and containing 3–40 moles EO per mole of alcohol, remain fluid over a wider temperature range than primary alcohol ethoxylates and thus eliminate the expensive heating during storage and handling. Furthermore, they possess higher solubility in both water and hydrocarbon solvents than the primary alcohol ethoxylates.

The three most important industrial applications of alcohol ethoxylates are in industrial institutional cleaners, paper and textiles. In the cleaner sector they are used extensively in hard surface cleaners and in commercial laundry compositions. Other usage areas are as low-foam wetters for commercial dishwashing powders and rinse aids, as components of sanitiser formulations, in hand cleaning gels and in wax as well as silicone polishes. They are also excellent emulsifiers for degreasing of metals with solvents. In the paper industry, they are used for deinking of waste paper, as wax emulsifiers in chipboard manufacture, fluff pulp bulking agents and pitch dispersants. In textile industry they are used in the scouring of synthetic fibres and in ship finish formulations. Other applications are in agricultural sprays, as wetting agents for leather, as surfactants in emulsion polymerisation and for modification of PVC plastisol viscosities.

Alkyl Phenol Ethoxylates (APE)

Resembling alcohol ethoxylates in physical and performance properties. APE are chemically stable and highly versatile surfactants used in a large number of industrial products such as acid and alkaline metal cleaning formulations, detergents, wetting agents and efficient emulsifiers or co-emulsifiers. They are also used as dispersants, solubilisers, coupling agents and rewetting agents. Industrially their most important applications are in industrial institutional cleaners, emulsion polymerisation of paint emulsions, as wetting, scouring agents and detergents in many areas of textile processing, and as wetting agents and emulsifiers in the formulation of pesticidal concentrates. They are also used in recycling of paper (resin removal, deinking and wetting), concrete (plasticiser and foaming agent), gypsum board (foaming agent), leather tanning (wetting agent), formulation of hand cleaning gels, in coal dust suppression, filter cake dewatering, surface treatment in zinc and aluminium production, drilling mud chemicals and as raw materials for the anionic alkylphenol ethoxysulphates or ether sulphates.

The alkylphenol ethoxylates are represented by

$$RC_6H_4 (OCH_2CH_2)_nOH$$

where R is the alkyl group attached to the aryl group and n is the average number of EO moles, generally 5–40, in these products. Most important of the alkyl phenols is nonylphenol followed by octylphenol, dodecyl and other alkylphenols including those with more than one side chain. Nonylphenol with 4 moles of EO is insoluble in water, but soluble in hydrocarbons. With 6–7 moles EO the product becomes soluble in cold water and yet remains oil soluble. With 8–12 moles EO good water solubility results. Higher ethoxylates with 20 and more moles EO per mole of alkylphenol are used in specialty applications such as emulsion polymerisation.

The cloud point is a significant parameter in identifying various alkylphenol ethoxylates and their formulations. The larger the number of moles of EO in the molecule, the more soluble it is and the higher the cloud point. At 15 moles EO, the product will not cloud even in boiling water. The foaming tendency increases upto 18 moles EO and thereafter it begins to decline. The solubility of the alkylphenol ethoxylates is influenced by the level of calcium and magnesium ions. An important aspect of these products is their tolerance to water hardness. The surfactant properties are governed by the length of the alkyl chains. With an alkyl chain of 7–10 carbon atoms, good wetting and detergent action are favoured and with more than 12 carbon atoms these properties are lowered and the emulsifying properties predominate.

Polyhydric alcohol ethoxylates

Ethoxylated sorbitan esters (Polysorbates) and ethoxylated glycerides are two important categories in this class.

Ethoxylation of sorbitan esters leads to a series of more hydrophilic surfactants. All hydroxyl groups of sorbitan can react with ethylene oxide. A typical POE (20) sorbitan ester is represented by the structure :

$$
\begin{array}{l}
\text{H} \\
| \\
\text{H—C} \\
| \\
\text{H—C—O(C}_2\text{H}_4\text{O)}_x\text{H} \\
| \\
\text{H(OC}_2\text{H}_4)_y\text{O—C—H} \\
| \\
\text{H—C} \\
| \\
\text{H—C—O(C}_2\text{H}_4\text{O)}_z\text{H} \\
| \\
\text{H—C—O(C}_2\text{H}_4\text{O)}_w\text{OCR} \\
| \\
\text{H}
\end{array}
$$

wherein $w + x + y + z$ is equal to 20. The polysorbates either alone or with their counterparts, the sorbitan fatty acid esters, have a number of applications. Their properties are exploited in their uses as emulsifiers for w/o hydraulic fluid emulsions, in quenching emulsions for heat treated metal parts, in emulsion polymerisation, as dispersants for pigment colourants, in shampoos and other anti-irritant products in the cosmetics industry.

The ethoxylated glycerides comprise three chemically diverse groups of substances, all of which are derived from various acyl glycerides. One group is obtained by ethoxylation of monoglyceride. Thus POE 20 glyceryl oleate is obtained by ethoxylation of glyceryl oleate wherein both the OH groups present react with EO until the total level of polyoxyethylene (POE) equals 20. A second group is derived from OH group containing glycerides, which includes glyceryl ricinoleate, castor oil and trihydroxy stearin. The EO addition can occur on any available hydroxyl grouping. In addition some reactions can occur resembling those given below for the third type. When a natural glyceride is allowed to react with EO, a third type of ethoxylated glyceride results. Water solubility and HLB of products in the above referred three groups depend on the starting glycerides and the level of ethoxylation.

The ethoxylated glycerides, solids or liquids, are useful emulsifying agents, suspending and solubilising agents, skin conditioners and emollients. Typical members of this class are POE 4 castor oil. POE 120 glyceryl stearate and POE 6 triolein. The low HLB value products act as defoamers and w/o emulsifiers. The high HLB value products serve as o/w emulsifiers. Other applications are in metal working fluids, floor polishes, cutting oils, mold release agents, universal colourant vehicles in aqueous and non-aqueous systems for paints and inks, oilfield drilling fluids, hydraulic fluids and metal cleaners.

Lanolin Ethoxylates

Ethoxylated lanolin derivatives are intended to provide the benefits of lanolin in finished cosmetic products without imparting the unctuous characteristics of lanolin. They are formed by ethoxylation of the hydroxyl groups present in lanolin or its hydrogenation product—giving rise to complex mixtures of ethoxylated sterols, lanolin alcohols, lanolin fatty acids and unaltered lanolin. This complexity arises due to the presence of water in the lanolin and the transesterification reactions taking place during the ethoxylation. Most members of this group are waxy solids, water soluble due to the high levels of ethoxylation. Specific examples are POE 20 lanolin and PPG 12 POE 65 lanolin oil.

Ethoxylated Polysiloxanes

The introduction of the first organosilican block copolymers made it possible to lower the surface tension of many organic liquids that are not influenced by hydrocarbon surfactants. Organosilicon block copolymers emerged as commercial products at a time when new polymer chemistry and its applications were taking form and the technology of polyurethane had just begun to unfold itself. Nonionic silicone surfactants are the dominant stabilisers of polyurethane foam. Foam ingredient solubility is imparted to the silicone by such hydrogen bonding groups as the ether oxygens and hydroxyls of polyoxyethylene and polyoxypropylene glycols.

Silicone—oxyalkylene block copolymers typified by Union Carbide 'Silwet' surfactants are of two distinct structural types. The major class is a polydimethylsiloxane to which polyethers have been grafted through a hydrosilylation reaction. This process results in an alkyl pendant type (AP type) copolymer in which the polyoxyalkylene groups are attached along the siloxane backbone through a series of hydrolytically stable Si-C bonds. These products have the general formula :

$$Me_3SiO \; (Me_2SiO)_x \; (Me \; SiO)y \; SiMe_3$$
$$\mid$$
$$PE$$

wherein Me is CH_3 group and PE is $—CH_2CH_2O(EO)_m(PO)_nZ$. EO and PO are ethylene and propylene oxides and Z is either hydrogen or a lower alkyl group. The second class is a branched polydimethylsiloxane to which polyethers have been attached through condensation chemistry. This creates an alkoxy-end-blocked (AEB) type copolymer in which the polyoxyalkylene groups are attached to the ends of the silicone backbone through Si–O–C bonds. This linkage offers limited resistance to hydrolysis under neutral or slightly alkaline conditions, but breaks down quickly in acidic environments. These products have the general formula

$$(MeSi)_{y-z} \; [(O \; Si \; Me_2)_{x/y} \; O—PE]_y$$

wherein Me is CH_3, PE is $—(EO)_m \; (PO)_n \; R$, R being a lower alkyl group.

A broad range of this class of surfactants has been developed by varying x, y, m and n values in the above formula. These products offer unique properties and performance that are not achievable with conventional organic surfactants. The ethoxylated polysiloxanes are especially used in the cosmetic industry as emulsifiers for silicones. The low surface tension, high wetting, superior dispersing, emulsifying, lubrication, sheen, gloss enhancing, static suppressing, moderate prefoaming, broad range of aqueous cloud points and solubility, good thermal stability and good compatibility with other surfactants and system components, are the various parameters taken advantage of in the applications of these products. Their low toxicity, environmental acceptability and their cost effectiveness due to their performance at low concentrations are added advantages. Other than cosmetics, end uses are in adhesives and sealants, agriculture, automotive specialties, coatings, fibres, hydrocarbon antifoams, mineral processing, paper, petroleum extraction and processing, pigments and dyes, plastics and rubber, pharmaceuticals, printing inks and textiles. A significant application is in their role as emulsifying agent in the formation of flexible urethane foams smoothening the reaction between polyol, isocyanate and water. In rigid urethane foams also they contribute to structural soundness.

Commercially available Union Carbide 'Silwet' brand products are indicated in Fig. 2.7.

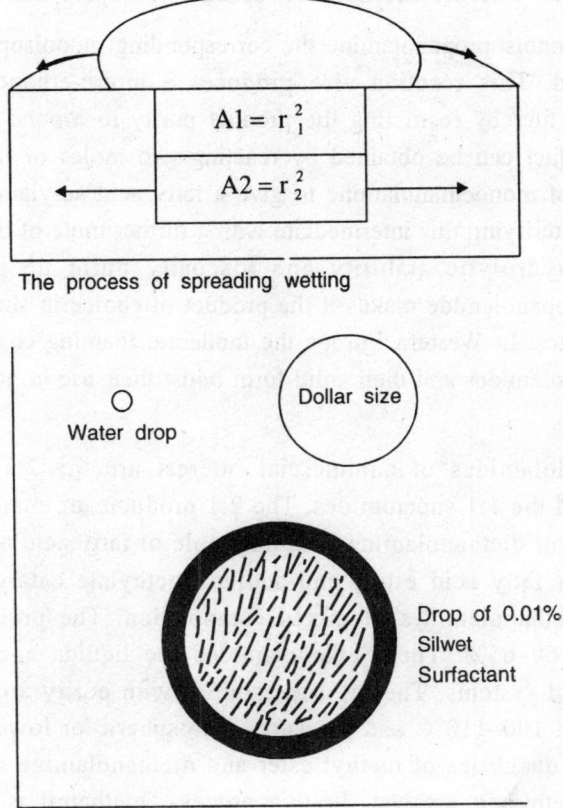

$$A1 = r_1^2$$
$$A2 = r_2^2$$

The process of spreading wetting

Water drop

Dollar size

Drop of 0.01% Silwet Surfactant

Fig. 2.7. Wetting and spreading of silwet surfactants.

Alkanolamides

The alkanolamides are another class of major nonionic surfactants comprising of:

(a) Monoalkanolamides

(b) 2:1 knitchevsky diethanolamides

(c) 1:1 superamides, and

(d) The mixed ratio amides.

Monoalkanolamides are made by condensing one mole of a monoalkanolamide with one mole of fatty acid or ester at 180°C according to the equation :

$$RCOOH + NH_2CH_2CH_2OH \longrightarrow R\ CO\ NHCH_2CH_2OH + H_2O$$

With monoisopropanolamine the corresponding monoisopropanolamide is obtained. This reaction also produces a mono-ethanolamide ester byproduct, thereby restricting the product purity to around 95%. A 99% purity product can be obtained by reacting two moles of fatty acid with one mole of monoethanolamine to give a fatty acid acrylaminoethyl ester and transesterifying this intermediate with a further mole of the amine. The superior hydrolytic stability and viscosity build up properties of monoisopropanolamide makes it the product of choice in shampoos in the United States. In Western Europe the moderate foaming characteristics of monoethanolamides and their solid form boost their use in powder laundry detergents.

Diethanolamides of commercial interest are the 2:1 knitchevsky amides and the 1:1 superamides. The 2:1 products are made by reacting two moles of diethanolamine with one mole of fatty acid at 140–170°C. The use of fatty acid esters and sodium methylate catalyst also gives the 2:1 diethanolamides of mixed composition. The product purity is thus only 60–65%. These alkanolamides are liquids and are used in water-based systems. The 1:1 superamides with purity around 90% are obtained at 100–110°C and pressures atmospheric or lower by reacting equimolar quantities of methyl ester and diethanolamine using alkaline sodium methylate catalyst. In this process, methanol is continuously removed using explosion-proof equipment. Typical compositions of the 2:1 and 1:1 amides are as follows :

Composition	Percent by weight	
	2:1 knitchevskyn amide	1:1 superamide
Diethanolamide	65	90
Free diethanolamine	22	5
Amine salt	10	Trace
Amide ester	1	4
Water	2	Trace
Methanol	–	0.2
Methyl ester	–	0.8

The superamides are gels or soft, low melting solids with slight solubility in water. The viscosity building properties of these amides are directly related to their water solubilities and in this respect the lauric isopropanolamide is superior to the lauric diethanolamide.

The alkanolamides figure in a number of applications. In detergent and cosmetic formulations they basically perform as foam quality improvers, foam stabilisers, viscosity builders and detergency synergists. They are used in light duty washing liquids, heavy-duty laundry detergents, and home and institutional cleaners, including carpet cleaners. In the formulation of shampoos, bubble baths, synthetic toilet bars and institutional scrub soaps they are used directly. Ethoxylated alkanolamides and sulphosuccinates derived from their reaction with maleic anhydride are also used in cosmetics. Their asset in shampoos is due to their effectiveness in the presence of high levels of grease. The alkanolamides are compatible with other nonionic and anionic surfactants. Thus a 5% addition of alkanolamide in a formulation containing 20% anionics and 75% builders enhances the detergency of the product. The superamides are excellent foam boosters and valuable constituents in handwash detergents and shampoos. The addition of fatty alkanolamides to surfactant formulations reduces the 'Zein number' of the formulation, rendering it thereby skin non-irritant. Other applications are as bases for deodorant sticks, as textile greasing agents, and as rust inhibitors in drilling and cutting oils. Lauric acid based alkanolamides dominate the market.

Fatty Amine Oxides (FAO)

The substitution of the traditional alkanolamides as foam boosters in dishwashing liquids was the specific instance, which brought recognition to the nonionic fatty amine oxides. These are produced by reaction of tertiary amines and hydrogen peroxide or peroxyacids. Their general structure is given by :

$$R\ R'\ R''\ N^+\ O^-$$

wherein R, R' and R'' may be alkyl, aryl or any or several other structures, with at least one being a fatty acid residue. Two typical reactions showing their formations are :

$$C_{12}H_{25}(CH_3)_2\ N + H_2O_2 \longrightarrow C_{12}H_{25}\ (CH_3)_2N \longrightarrow O + H_2O$$

These materials while possessing a formal charge separation on the nitrogen and oxygen atoms behave as electrolytes. In aqueous solution FAO exhibit nonionic or cationic properties depending on the pH. At lower pH the weakly cationic form precipitates metathetically with anionic surfactants such as sodium dodecyl benzene sulphonate. Under neutral or alkaline conditions they exist in solution as non-ionised hydrates.

The most common amine used for the production of FAO is the alkyldimethylamine where the alkyl group contains 12 to 16 carbon atoms. In turn, the alkyldimethylamines are derived from either linear alpha-olefins or detergent alcohols. The synthesis of FAO involves the conversion of the alpha-olefin to the primary alkyl bromide which when reacted with dimethylamine gives the corresponding alkyldimethylamine.

FAO are good foam stabilisers, thickeners and emollients, emulsifiers and conditioning agents exhibiting mildness towards the skin. In dishwashing liquids they show superior foam stabilisation properties, when used along with alcohol ether sulphates in presence of grease, compared with the more commonly used fatty alkanolamides. Other applications are as rinse additives in laundering, shampoo formulations, waterless hand cleaners and hair conditioners.

Acetylenic Alcohol Derivatives

Surfactants based on acetylenic diols are a series of products under the trade name 'Surfynols' marketed by Air Products Co. The core member of this group, Surfynol 104, is a nonionic molecule containing a hydrophilic portion in the middle of two symmetric hydrophobes. Chemically, it can be represented as :

$$CH_3-CH-CH_2-\underset{\underset{OH}{|}}{C}-C \equiv C-\underset{\underset{OH}{|}}{C}-CH_2-CH-CH_3$$

with CH_3 groups on the CH, C, C, and CH positions

(2,4,7,9-tetramethyl-5-decyn-4,7-diol)

This unique chemical structure allows Surfynol 104 to provide multifunctional properties such as surface tension and foam control. Other members of the Surfynol group are other acetylenic alcohols and diols and ethoxylated products. Surfynol 104 effectively reduces the surface tension of water to 33.1 dynes/cm under dynamic conditions at a concentration of 0.1%. In this respect it is superior to the performance of Triton X. Surfynol 104 does not have a cloud point and as such functions as a defoamer over a very broad temperature range. Table 2.8 gives the application areas in which the wetting, defoaming and dispersant properties of some Surfynols are exploited.

Table 2.8. Application areas of 'Surfynol' surfactants.

Applications		Surfynol types				
	61	82S	104A	440	465	485
Coatings						
(a) (Acrylic, alkyl, epoxy SBR, polyurethane)	WD	–	WD	W	W	W
(b) Powder	–	WG	–	–	–	–
(c) Radiation curable	–	–	W	W	W	W
(d) Paper (SBR, PVA)	–	–	WD	WD	–	–
Graphic arts						
(a) Flexographic & gravure inks	WD	–	WD	W	W	W
(b) Overprint varnish	W	–	WD	W	–	–
Dyes/pigments						
(a) Pigment dispersion	WD	–	WD	–	–	–
(b) Dye milling	–	–	WD	–	–	–
(c) Crystallisation & filtration	–	–	WDC	–	–	–

(*Contd...*)

Applications	Surfynol types					
	61	82S	104A	440	465	485
Adhesives (acrylic, PVA, phenolics, styrene butaeidne	–	–	WD	W	W	W
Foundry	–	–	WD	–	–	–
Emulsion poly-merisation	–	–	WD	W	E	E
PVC production/ plastisols	–	–	WD	–	–	–
Agricultural chemicals	W	W	WD	W	W	W
Oilfield chemicals	W	–	WD	WD	–	–
Hard surface cleaners	W	W	WD	W	W	W
Metalworking fluids	–	–	WD	W	W	–
Textiles	W	–	WD	W	W	–
Cement/Concrete	–	–	WD	WD	–	–

W– low or nonfoaming wetting agents; D–defoaming, anti-foaming and/or deair entrainment agents; G–pigment grinding aid or dispersant; C–crystallisation enhancer; E–emulsifier; 61–3, 5 dimethyl-1-hexyn3-ol; 82-S – 3,6-dimethyl-4-octyne-3,6-diol; 104A–2, 4, 7, 9 tetramethyl-5-decyne-4,7 diol; 440–104 + 3.5 moles EO; 465–104 + 10 moles EO; 485–104 + 30 moles EO.

Silicones

Like all true surfactants the silicone surfactants possess two moieties termed 'lipophilic' and 'lipophobic', rather than 'hydrophilic' and 'hydrophobic' as is the case with hydrocarbon surfactants. The variety of solubilising groups with which the silicones can be reacted permit tailoring a silicone surfactant for specific systems. The generalised structure of a silicone surfactant is

$$CH_3-\underset{\underset{CH_3}{|}}{\overset{\overset{CH_3}{|}}{Si}}-O\left[-\underset{\underset{X}{|}}{\overset{\overset{CH_3}{|}}{Si}}-O\right]_n-\left[\underset{\underset{CH_3}{|}}{\overset{\overset{CH_3}{|}}{Si}}-O\right]_m-\underset{\underset{CH_3}{|}}{\overset{\overset{CH_3}{|}}{Si}}-CH_3$$

wherein X is any of the hydrophilic groups of anionic, cationic or nonionic character or organophilic groups such as esters, amides and polyethers. Unlike most common organic surfactants, silicone surfactants are multifunctional. The many possible combinations of siloxane molecular weight, functionality and polyether composition control the HLB and molecular geometry, the primary influences of surfactant behaviour. The silicone backbone can be considered, as a 'Superhydrocarbon' because of the density of methyl groups present, and is very stable and very flexible. Most silicone surfactants thus are liquids and remain so even at high molecular weights. Major applications of silicone surfactants are as wetters, emulsifiers for silicone fluids and concrete mould release agents.

Other silicone based surfactants are fluorosurfactants and ethoxylated polysiloxanes, referred to in other sections.

Fluorosurfactants

A fluorosurfactant consists of a hydrophobic and oleophobic perfluorinated tail and a hydrophilic or organophilic head group, structurally represented as

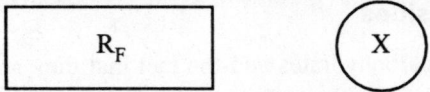

The fluorochemical tail, R_F consists of the stable perfluorocarbon group. This group can be modified in length and structure to meet end use needs, and provides exceptional resistance to thermal and chemical attack. This fluorochemical portion is basically responsible for its capacity to dramatically reduce surface tension as well as being the major difference between fluorosurfactants, conventional hydrocarbon surfactants and silicone surfactants. The head group X is commonly a water- soluble group but can also be designed to be oil-soluble for use in non-aqueous systems. The nature of X can be varied to give anionic, cationic, nonionic or amphoteric surfactants such as:

$$R_F SO_3^- K^+; \quad R_F NH_3^+ Cl^-; \quad R_F (CH_2 CH_2 O)_n H;$$
$$R_F N^+ H(CH_3) \, CH_2 \, COO^-$$

The fluorosurfactants can be made commercially by the following processes

(a) *Simons electrochemical fluorination process (ECF)*

$$HF + C_8 H_{17} SO_2 F_{18} \xrightarrow{\text{ECF}} C_8 Fl_8 SO_2 F + H_2$$

followed by conversion of $C_8 F_{17} SO_2 F$ to various types of surfactants

(b) *Telomerisation process*

$$C_2 F_5 I + n \, (CF_2 - CF_2) \longrightarrow C_2 F_5 \, (CF_2 CF_2)_n \, I$$

followed by conversion of the telomer iodides to surfactants

(c) *Anionic oligomerisation of fluorine monomers*

$$n \, (CF_2 = CF_2) \longrightarrow (C_2 F_4)n$$

followed by reaction of the oligomers with alcohols, phenols, oleum or polyethylene glycols leading to various surfactants.

The advantages of fluorosurfactants are lower surface tension in aqueous systems, better effectiveness at very low concentrations, excellent stability in hostile environments and good surface activity in organic media. Areas of application include floor polish emulsions, photographic processes, production of fluoropolymers, electroplating bath additives, electronic coatings, as additives for oil/gas well stimulation, in battery electrolytes, in fire-fighting foams and as additives in paints, coatings, inks and adhesives.

Alkyl Polyglycosides

The abundance of carbohydrates and the fact that they are both natural and renewable has evoked fresh interest in the last few years in carbohydrate derived surfactants. Nonionic alkyl glucosides, commonly known as alkyl polyglycosides (APG), have attracted attention as highly efficient and mild surfactants. They are obtained by the reaction of glucose with fatty alcohol, starch providing the source for glucose. APG has a generic structure :

$$C_nH_{2n+1}O \ (C_6H_{10}O_5)_x \ H$$

where x ranges from 1 to 4.

APG is reportedly more soluble than a typical alcohol ethoxylate or alkylphenol ethoxylate designed for household laundry applications. Detergent formulations based on APG are also more stable than the above-referred ethoxylates over a wider range of temperatures.

Consequently its liquid detergent formulations require less solubilising agent and can be formulated with higher levels of soluble builders. Liquid dishwashing formulations incorporating it reportedly require no foam booster. Its mildness to the skin promotes its use in hand dishwashing personal care applications. APG can also be used effectively in combination with anionic surfactants. Henkel is the major producer of this type of nonionic surfactant.

Sulphosuccinates, Sulphosuccinamates

The sulphosuccinates are made by forming a maleic ester and then sulphonating it. The starting raw materials for these anionics are maleic anhydride and a fatty alcohol, alcohol ethoxylate or an alkanolamide as the second reactant. If amines are used in the reaction replacing the alcohols, then sulphosuccinamate surfactants are produced. The products obtained in these reactions can be half esters or diesters. For the half ester an aqueous solution of sodium sulphite is used as the sulphonating agent and for the diester, sodium bisulphite. Typical half ester and diester sulphosuccinates are illustrated below :

Diester sulphosuccinate   Half ester sulphosuccinate

The sulphosuccinamates are typically manufactured by the reaction :

Primary Maleic anhydride
amine

involving a primary amine and maleic anhydride, and then adding sodium bisulphite across the double bond.

The sulphosuccinates are good to excellent foamers and wetting agents. The half esters have application in household detergents, but the sodium dialkylesters are industrially used in the following areas :

(a) Emulsion polymerisation of PVA emulsions and PVC dispersion resins.

(b) Foam stabilisers in rug shampoos.

(c) As components of solvent degreasers for metals.

(d) As solubilisers in industrial dry cleaning formulations.

(e) Dewatering of mineral slurries.

(f) Mineral flotation.

(g) Oil-spill chemicals.

(h) Pigment dispersion and wetting agents in paints.

(i) Foaming agents for secondary foam backings.

(j) Coatings of textiles, leather and leather-cloth.

(k) Modification of plastisol viscosity.

Phosphate Esters

Analogous to the sulphate esters, the phosphate esters are reaction products of phosphoric acid with alcohols or ethoxylates resulting in an anionic product. As it is difficult to phosphate an alcohol with orthophosphoric acid, pyrophosphoric acid is used. Likewise, oleum replaces sulphuric acid. It is possible this way to generate mono, di and polyphosphoric esters. Aromatic phosphate esters are also feasible and these can be further ethoxylated. The phosphated and ethoxylated esters form an important group of industrial surfactants.

The phosphate esters find application primarily in their exploitation as :

(a) Detergents

(b) Dispersants

(c) Emulsifiers

As detergents they are used in industrial and institutional cleaners. As dispersants they are used in paper coatings, paints and inks, textile dyeing and pesticide/herbicide suspension concentrates. As emulsifiers they are used in emulsion polymerisation of lattices, cutting oils, drilling aids, crop protection chemical concentrates. They also are used in dry cleaning, as corrosion inhibitors, and as antistats in textile cleaning formulations.

Amine Based Surfactants

Amines are the building blocks for the entire train of cationic surfactants as illustrated below :

The amines find direct application as surface-active agents in ore flotation, filming amines for corrosion inhibition and as anti-caking agents for granular and blended fertilisers.

Amine ethoxylates

These are prepared by ethoxylation and the products contain 2–50 EO units. Their uses are as :

(a) Cationic collectors in ore flotation and frothing agents.

(b) Part of filming amine formulations.

(c) Additives for lubricating oils and gasoline.

(d) Emulsifiers for agricultural sprays.

(e) Emulsifiers for textile spinning oils.

(f) Emulsion breakers.

(g) Dye-levelers and textile finishing agents.

(h) Corrosion inhibitors.

(i) Antistats.

Quaternary Ammonium Compounds (QAC)

The quaternary ammonium compounds are commercially the most important of the group of cationic and amphoteric surfactants accounting for 45–50% of the total market for this class. There are over 200 QAC in commercial use. The QAC are made by 'quaternising' primary, secondary and tertiary amines with four different types of quaternising agents:

(1) Methyl chloride or bromide.

(2) Dimethyl or diethyl sulphate.

(3) Benzyl chloride.

(4) Ethylene oxide or diethanolamine.

Products obtained by the use of different quaternising agents have different properties and serve widely different markets. The alkyl, benzyl or ethoxy groups attach themselves to the nitrogen atom of the amine.

The counterion is the halide or the methosulphate group. Alkyl chain lengths are determined by fatty acids used in preparing the amine and profoundly affect the properties of the product. If a symmetrical tertiary amine is the starting material and methyl chloride is used as the quaternising agent, a symmetrical QAC :

$$[R-\underset{\underset{R}{|}}{\overset{\overset{R}{|}}{N}}(CH_3)]^+ \ Cl^-$$

is obtained. Distearyldimethyl ammonium chloride (DSDMAC) is the most important QAC and is used as fabric softener. It is also used in the preparation of organic-modified clays used in petroleum drilling, as thixotropic agents in paints and inks, and also for its bacteriostatic properties.

When monoalkyl type QAC are quaternised with benzyl chloride, compounds known as 'benzyl quats' are obtained. They have the structure :

$$[R—N\ (CH_3)_2.CH_2.C_6H_5]^+\ Cl^-$$

These quaternaries form a special class of biocides, sanitisers etc., used in hospitals, restaurants, food processing centres and swimming pools.

The other groups of QAC, namely,

$$R—\overset{+}{C}H_2N[CH_2CH_2OH]_2\ Cl^-$$
$$|$$
$$Me$$

and

$$\text{and } R—\overset{+}{N}\overset{(CH_2CH_2O)_x}{\underset{(CH_2CH_2O)_y}{\diagup}}\ Cl^-$$
$$\diagup$$
$$Me$$

are obtained by quaternising with diethanolamine and ethylene oxide. They have a wide range of hydrophilicity. They are important in the formulation of asphalt/bitumens, in vehicle cleaning formulations and as emulsifiers in pesticide formulations.

Imidazolines and Polyamines

The cationic properties of amines are greatly improved by the incorporation of additional nitrogen atoms into their structures, resulting in polyamines of the type :

$$H_2N\ CH_2\ CH_2\ NH_2\ CH_2\ CH_2\ NH_2$$
diethylenetriamine

They are made by reducing the product obtained by the cyanoethylation of primary and other amines with acrylonitrile. Their difunctional character makes them very effective corrosion inhibitors.

Polyamines react with fatty acids and dimer/trimer acids to form the imidazolines such as :

$$RC\overset{N—CH_2}{\underset{NH—CH_2}{\diagdown}}$$

These, with various substituents, are used in fabric softeners, corrosion inhibitors in oil-field applications and other areas.

Betaines

As already pointed out, the amphoteric betaines are special members of the ring opened imidazoline derived surfactants. They differ from other amphoterics with regard to their solubility and electrical nature in alkaline solution. They are compatible with anionics at all pH values. The carboxyl containing betaines are insensitive to water hardness. They are used as levelling and wetting agents in textile processing, detergents, scrubbing compounds, antistatic agents, as fabric softeners, as lime soap dispersants and in personal care products. Coconut fatty acid amidopropylbetaine is particularly used in hair shampoos, foam baths, shower foams and liquid soaps where a high, creamy foam is required. It has a good skin tolerance.

The most common route to carboxyl-betaines is through the quaternisation of long-chain alkyldimethyl tertiary amines with chloroacetic acid :

$$R(CH_3)_2 \ N + ClCH_2COOH \longrightarrow R(CH_3)_2N^+CH_2COO^-$$

There are several commercial producers of cocoamido-propylbetaine and other betaines. The betaines play an important role in the amphoterics market which in United States alone is currently around 12000 metric tons per annum.

Sarcosinates and Taurates

The sarcosinates and taurates are two fairly important groups of surfactant products with amino acid moieties in their structure. They were initially introduced by IG Farben.

The sarcosinates are condensation products of fatty acids with N-methyl glycine (sarcosine). They are manufactured from fatty acid chlorides and sodium sarcosine :

$$\underset{\substack{\| \\ R-C-Cl}}{\overset{O}{}} \overset{CH_3}{\underset{\substack{| \\ NH-CH_2COONa}}{}} \xrightarrow{-NaCl} \underset{\substack{\| \\ R-C-N-CH_2COONa}}{\overset{O \quad CH_3}{}}$$

The alkyl groups in the sarcosinates are usually lauryl or oleyl, the former dominating. These products are anionic foam boosters and are used in hair conditioners and shampoos, rug shampoos, tooth paste, hand soaps and pharmaceuticals. They are also used in emulsion polymerisation and in window cleaning formulations. They are important as emulsifying agents for the removal of certain anti-corrosive agents on metal prior to their further processing.

The taurates, also referred to as taurides and N-acylmethyl taurates, are made in a manner similar to that of the sarcosinates with the reaction occurring between N-methyl taurine and fatty acid chlorides under alkaline conditions :

$$\underset{\text{R–C–Cl}}{\overset{\displaystyle O}{\overset{\|}{}}} + \underset{\text{NH–CH}_2\text{–CH}_2\text{–SO}_3\text{Na}}{\overset{\displaystyle CH_3}{\overset{|}{}}} \longrightarrow \underset{\text{R–C–N–CH}_2\text{–CH}_2\text{–SO}_3\text{Na}}{\overset{\displaystyle O\;\;CH_3}{\overset{\|\;\;\;|}{}}}$$

Coconut oil fatty acids and oleic acid, again, are the preferred fatty feedstocks but in this case oleic acid is the preferred fatty acid. Both sodium and potassium salts prevail. The taurates combine the properties of soap (foaming power and emulsifying property) with those of the alkylaryl sulphonates (greater insensitivity of hard water). The taurates are particularly water-soluble. They are used as additives in all areas of cleaning materials and in cosmetics and pharmaceuticals.

The global consumption of sarcosinates is of the order of 5500–6000 metric tons per annum with the United States accounting for 60% of this volume. The global consumption of taurates is in the range 3500–4000 metric tons per annum, Japan being the major user.

TECHNOLOGIES FOR MANUFACTURE OF SURFACTANTS

Important Physico-chemical Characteristics of Surfactants

When one considers the impact of surface science in general, and emulsions, dispersions, foaming agents, wetting agents etc., in particular, on our day-to-day existence, one begins to realise the overwhelming extent to which these areas of chemistry and chemical technology permeate our lives. Although surfactant science and technology is a wide field, some specific aspects of these materials related to their application are highlighted in this section. These are:

(a) The HLB system

(b) Surfactant solubility and the Kraft temperature

(c) Emulsion stabilisation

(d) Foam stabilisation

(e) Wetting and related phenomena

(f) Dispersion.

The HLB system

Quantitative correlation of the chemical structure of surfactant molecules with surface activity to facilitate their choice in a given formulation, although has not been possible in respect of all their areas of applications, has to a large degree proved fertile in establishing such a criteria in the field of emulsions. The HLB of an emulsifier is an expression of its Hydrophile-Liphophile Balance (HLB) i.e., balance of the size and strength of the hydrophobic polar groups and the lipophilic nonpolar groups of the emulsifier. The HLB system developed by Griffin employed certain empirical formula to calculate the HLB number in a range of 0–20 on an arbitrary scale. Hydrophilic surfactants which possess high water solubility and generally act as good solubilising agents, detergents and stabilisers for oil in water (o/w) emulsions lie at the high end of this scale. Surfactants with low water solubility which act as solubilisers of water in oils and good water in oil (w/o) emulsion stabilisers lie at the low end of the scale. The effectiveness of a given surfactant in stabilising a particular emulsion system would depend on the balance between the HLB values of the surfactant and the oil phase involved. For nonionic surfactants with polyoxyethylene (POE) solubilising groups, the HLB is calculated from the formula :

$$HLB = \text{(mole percent hydrophilic group)}/S.$$

In this scheme an unsubstituted polyethylene glycol would have an HLB of 20.

Surfactants based on polyhydric alcohol fatty acid esters such as glycerol monostearate can be handled by the formula :

$$HLB = 20 \, (1\text{-}S/A)$$

where S is the saponification number of the ester and A is the acid number of the acid. A typical surfactant of this type—Tween 20 (POE 20 sorbitan monolaurate) with $S = 45.5$ and $A = 276$—would have an HLB of 16.7. The HLB scale of typical ICI Tween and Span brand emulsifiers is given in Fig. 2.8. The HLB values for typical nonionic surfactant structures are given in Table 2.9. For materials which are not saponified an empirical formula of the type :

$$HLB = (E + P)/S$$

is employed. In the above formula, E is the weight percent of oxyethylene chains, and P is the weight percent of polyhydric alcohol.

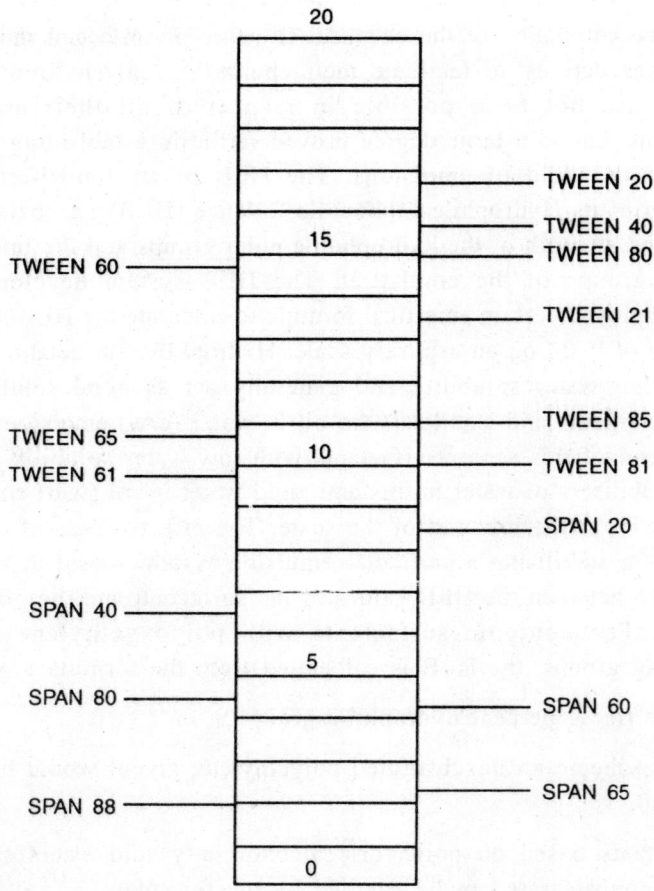

Fig. 2.8. HLB scale of typical atlas emulsifiers.

Table 2.9. HLB value for typical nonionic surfactant structures.

Surfactant	HLB
PEG-4 dilaurate	7.4
PEG-4 dioleate	6.0
PEG-4 distearate	5.0
Sorbitan monolaurate	8.6
POE sorbitan tristearate	10.5
Glycerol monostearate	3.2

(Contd...)

Surfactant	HLB
POE 4 – lauryl alcohol (BRIJ 30)	9.7
POE 2 – oleyl alcohol (BRIJ 92)	4.9
POE 100 – stearyl alcohol (BRIJ 700)	18.8
Sorbitan monooleate	4.6
POE 16 sorbitan tristearate	10.0
POE 8 stearate (MYRJ 45)	11.1
POE 40 stearate (MYRJ 52)	16.9
POE 12 tridecyl alcohol	14.5
POE 10 nonylphenol	13.3
POE 9 octylphenol	13.2
Sorbitan trioleate (SPAN 85)	1.8
Sorbitan monopalmitate (SPAN 40)	6.7
POE 5 stearic acid	9.2
POE 10 oleic acid	12.2
POE 20 sorbitan monolaurate (TWEEN 20)	16.7
POE 20 sorbitan monooleate (TWEEN 80)	15.0
Sorbitan tristearate	2.2
POE 20 sorbitan monooleate	15.0

The Griffin system while proving useful for surfactant formulations had its limitations. In surfactant systems of nonionic mixtures, emulsion stability tests, interfacial tension, gas liquid chromatography, cloud point determinations and phase inversion temperature (PIT) studies have shown that the HLB of the system does not bear a linear relationship as generally assumed by the equation :

$$HLB_{mix} = f_a \, HLB_a + (1\text{-}fa) \, HLB_b$$

where f_a is the weight fraction of surfactant A in the mixture. Several other approaches based on group contributions using group numbers given in Table 2.10 and the formula :

$$HLB = 7 + \sum (\text{Hydrophilic group numbers}) - \sum (\text{Hydrophobic group numbers})$$

and correlations involving Hildebrand solubility parameters and heat of hydration of the surfactant have been evolved.

However, in most general applications, HLB ranges and HLB numbers for several emulsion ingredients as given in Table 2.11 are commonly used.

Table 2.10. Group numbers for the calculation of HLB.

Group	HLB
Hydrophilic	
— SO$_4$Na	38.7
— COOK	21.1
— COONa	19.1
— N (tertiary amine)	9.4
ester (sorbitan ring)	6.8
ester (free)	2.4
— COOH	2.1
— OH (free)	1.9
— O	1.3
— OH (sorbitan)	0.5
Hydrophobic	
— CH	−0.475
— CH$_2$ —	−0.475
— CH$_3$—	−0.475
— CH—	−0.475
— CF$_2$—	−0.870
— CF$_3$ —	−0.870
Miscellaneous	
— (CH$_2$CH$_2$O) —	0.33
— (CH$_2$CH$_2$CH$_2$O) —	0.15

Table 2.11. HLB ranges and numbers for several emulsion ingredients.

Application	HLB range
W/O emulsions	3–6
Wetting	7–9
O/W emulsions	9–18
Detergency	3–15
Solubilisation	15–18
Oil phase	*Normal HLB for O/W*
	(Contd.)

Application	HLB range
Lauric acid	16
Oleic acid	17
Cetyl alcohol	14
Decyl alcohol	14
Benzene	15
Castor oil	14
Kerosene	14
Lanolin	12
Beeswax	9
Carnauba wax	12
Paraffin wax	10
Chlorinated paraffin	8
Pine oil	16
Dimer acid	14
Mineral oil aromatic	12
Mineral oil paraffinic	10

The HLB numbers are most significant in the case of nonionic surfactants. In practice, as oils, waxes and solvents from various sources vary in properties and emulsifying characteristics, a formulator should make a series of trial emulsions with its ingredients using emulsifier combinations of known HLB value and establish the required HLB for this set of ingredients. Apart from knowing the right HLB it is also essential to identify the particular chemical type for maximum effectiveness. This is evident from the fact that an oleate type emulsifier is more suited for unsaturated oils and a stearate type emulsifier for saturated oils. Instead of using a single emulsifier, blends of emulsifiers give most stable emulsions. HLB values of some blended ICI surfactants are given in Table 2.12 and it can be noted that different blends can give the desired HLB. The emulsifier concentration and chemical type are important in attaining the emulsion stability. The importance of HLB can be judged from Table 2.12 which indicates the use of single or blended surfactants in specific industrial applications. While the formulae mentioned earlier for HLB calculation are satisfactory for many nonionic surfactants, those containing propylene oxide, butylene oxide, nitrogen and sulphur show anomalous behaviour. The HLB values of these, as well as ionics which do not follow the 'weight percentage' HLB basis, have to be determined experimentally which is often difficult and time consuming. A rough estimate of HLB in these cases can be made from the water solubility of the emulsifier or its dispersability characteristics on the following basis :

No dispersibility in water	HLB 1–4
Poor dispersion	HLB 3–6
Milky dispersion after vigorous agitation	HLB 6–8
Stable milky dispersion	HLB 8–10
Translucent to clear dispersion	HLB 10–13
Clear solution	HLB 13 and above

Surfactant solubility and the kraft temperature

The nature of surfactant molecules, having both hydrophilic and hydrophobic groups, is responsible for their tendency to concentrate at interfaces and thereby reduce the energy of the system in which they interact. The primary mechanism for energy reduction in most cases will be adsorption at various interfaces. However, when all available interfaces are saturated, the overall energy reduction may continue through other mechanisms. The physical manifestation of one such mechanism is the crystallisation or precipitation of the solute from solution. An alternative is the formation of molecular aggregates or 'micelles' that remain in "solution" as thermodynamically stable, dispersed species with properties distinct from those of the monomeric solution.

Table 2.12. HLB values of some blended surfactants.

Blends	HLB of blend composition						
	100/0	80/20	60/40	50/50	40/60	20/80	0/100
1. Span 20/Tween 20	8.6	10.2	11.8	12.6	13.4	15.1	16.7
2. Span 40/Tween 40	6.7	8.5	10.2	11.1	12.0	13.8	16.6
3. Span 60/Tween 60	4.7	6.7	8.8	9.8	10.8	12.9	14.8
4. Span 80/Tween 80	4.3	6.4	8.5	9.6	10.7	12.8	16.0
5. Brij 30/Brij 35	9.7	11.2	12.5	13.2	13.8	15.3	16.9
6. Brij 52/Brij 58	5.3	7.3	9.4	10.5	11.6	13.7	16.7
7. Brij 72/Brij 78	4.9	6.9	9.0	10.1	11.2	13.3	15.3
8. Brij 92/Brij 98	4.9	6.9	9.0	10.1	11.2	13.3	15.3

1—Laurate. 2—Palmitate, 3—Stearate. 4—Oleate. 5—POE lauryl alcohol, 6—POE cetyl alcohol, 7—POE stearyl alcohol, 8—POE oleyl alcohol Span—Sorbitan fatty acid esters; Tween—POE sorbitan fatty acid esters; Brij—Polyoxyethylene alcohols.

For many ionic materials it is found that the overall solubility of the material in water increases as the temperature increases. That effect is the result of the physical characteristics of the solid phase, i.e., the crystal lattice energy and heat of hydration of the material being dissolved. In the case of ionic surfactants, it is often observed that the solubility of the material will undergo a sharp, discontinuous increase at some characteristic temperature, Tk, commonly referred as the Kraft temperature. Below Tk, solubility of the surfactant is determined by the crystal lattice energy and heat of hydration of the system. The concentration of the monomeric species in solution will be limited to some equilibrium value determined by those properties. Above Tk, solubility of the surfactant monomer increases to the point at which micelle formation begins and the aggregated species becomes the thermodynamically favoured form. The formation of micelles favours an increase in solubility. The concentration of surfactant monomer may increase or decrease slightly at higher concentrations (at a fixed temperature), but micelles will be the predominant form of surfactant present above a critical surfactant concentration. The total solubility of the surfactant, then, will depend not only on the solubility of the monomeric material, but also on the "solubility" of the micelles. A schematic representation of the temperature/solubility relationship for ionic surfactants is shown in Fig. 2.9.

The Kraft temperature of a number of common surfactant types show that Tk can vary as a function of both the nature of the hydrophobic group and the character of the ionic interactions between the surfactant and its counterion. Nonionic surfactants, however, do not exhibit a Kraft temperature, because of their different mechanism of solubilisation. They do, however, have a characteristic temperature/solubility relationship with water in which they become less soluble as the temperature increases. In some cases phase separation and formation of a cloudy suspension results. This temperature is the 'cloud point'.

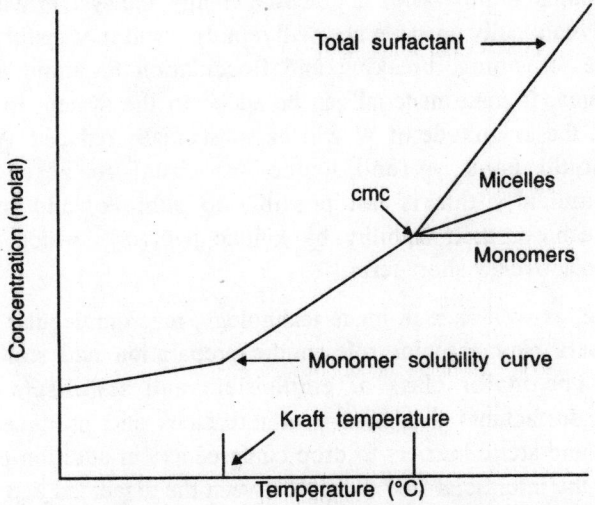

Fig. 2.9. Temperature/solubility relationship for typical ionic surfactants. Data are listed for nonionic surfactants.

Emulsion stabilisation

Emulsions are used in a large number of applications namely, food, cosmetics, medicinals, agriculture, paints and numerous industrial processes. In most of these cases, the emulsion form is used because it represents a convenient means of achieving delivery of some active material to this site, at some convenient concentration and frequently at some convenient area of coverage or controlled rate of release.

An emulsion can be defined as a heterogeneous system consisting of at least one immiscible liquid dispersed in another in the form of droplets whose diameters in general exceed 0.1 mm. Such systems possess minimal stability, which may be accentuated by additives such as surfactants. An emulsion may be either oil-in-water (o/w) or water-in-oil (w/o), where the first phase mentioned is the dispersed phase and the second, the continuous phase.

The preparation of an emulsion requires the formation of a very large amount of interfacial area between two immiscible phases. The work (W) required to generate 1 cm^3 of a new interface is given by :

$$W = \gamma_i \Delta A$$

where γ_i; is the interfacial tension between the two liquid phases and ΔA is the change in interfacial area between the two immiscible phases. If the interfacial tension between oil and water is assumed to be 52 mv/m (as for a hydrocarbon liquid), the reversible work required to carry out the dispersion process will be of the order of 2×10^8 ergs. Since that amount of work remains in the system as potential energy, the system will obviously be thermodynamically unstable and will rapidly result in transformations like coalescence, creaming, breaking and flocculation to attain a minimum interfacial area. If some material can be added to the system to reduce the value of γ_i, the magnitude of W will be substantially reduced. A true stable system should have γ_i (and hence W) equal to zero. However, thermodynamically this is not possible to achieve, and most useful emulsions achieve their stability by kinetic pathways which can play a dominant role over a short term.

In nature, as well as man-made technology, macromolecular emulsifiers and stabilisers play a major role in the preparation and stabilisation of materials. The major class or emulsifiers and stabilisers is that of monomeric surfactants that adsorb at interfaces and produce electrical, mechanical and steric barriers to drop coalescence, in addition to their role in lowering the interfacial free energy between the dispersed and continuous phases. Because of the role of the interfacial layer in emulsion stabilisation, it is often found that a mixture of surfactants with widely differing solubility parameters will produce emulsions with enhanced stability. In terms of HLB (0–20), at the high end of the scale lie hydrophilic surfactants which possess high water solubility and generally act as good solubilising agents, detergents and stabilisers for o/w emulsions. At the low end of the HLB scale are surfactants with low water solubility which act as solubilisers. The effectiveness of a good surfactant in stabilising a particular emulsion system will depend on the balance between the HLB values of the surfactant and the oil phase involved.

Recently, multiple emulsions of the type w/o or o/w have become of practical interest in controlled drug delivery, emergency drug overdose treatment, waste water treatment and separation technology. Such systems are more unstable than conventional emulsion systems and the choice of surfactants to stabilise these systems is complex.

The sensitivity of emulsions to temperature led to the suggestion that 'Phase Inversion Temperature' (PIT) be used as a possible method for emulsion preparation. It is generally found that the same circumstances that affect the solution characteristics of nonionic surfactants (their micelle size, critical micelle concentration, cloud point etc.) will also affect the PIT of emulsions prepared with the same materials. For typical ethylene oxide based nonionics, increasing the length of the polyoxyethylene chain will result in a higher PIT, as will a broadening of the polyoxyethylene chain length distribution. The use of PIT, therefore, represents a very useful tool for the comparative evaluation of emulsion stability. The effects of variables on the relationship between PIT, surfactant structures and emulsion stability have not been clearly defined in a quantitative way. However, it has been found that there is an almost linear correlation between the HLB of a surfactant under a given set of conditions and its PIT under the same circumstances. The process of selecting the best surfactant or surfactants for the preparation of an emulsion has been greatly simplified by the development of the more-or-less empirical approaches exemplified by the HLB and PIT methods. However, each method has its limitations and cannot eliminate the need for some amount of trial-and-error experimentation.

Foam stabilisation

Foams and emulsions have a great deal in common with regard to the basic physical principles controlling their stability. Their major differences lie in the natures of the dispersed phases, liquid or gas, and in the fact that foams will generally involve a much higher volume fraction of the dispersed phase than that encountered in emulsions.

Foams have wide technical importance in the fields of fire fighting, polymeric foamed insulation, foamed rubbers, foamed concrete, mineral separation and electroplating operations. They also have certain aesthetic utility in many detergent and personal care products. Unwanted foams may be a nuisance in sewage treatment, coating applications and crude oil processing.

A foam is produced by the introduction of air or other gas into a liquid phase, during which time the bubbles become encapsulated in a film of the liquid. The thin liquid film separating two or more gas bubbles is referred to as a lamellar film indicating that the film structure has two essentially identical interfaces in close proximity. The ability of a surfactant to perform as a foaming agent is dependent primarily on its effectiveness at reducing the surface tension of the solution, its diffusion characteristics, its properties with regard to disjoining pressures in thin films and the elastic properties it imparts to interfaces. The amount of foam that can be produced in a solution under given conditions will be related to the product of the surface tension and the new surface area generated during the foaming process.

Surfactants play a key role in foam formation, suppression as well as stabilisation. While all foams containing fluid phases are thermodynamically unstable, their degree of stability, commonly referred to as 'Resistance', can vary from seconds to weeks. More or less permanent foams can be produced in presence of extremely small amounts of surfactants. Maintenance of the foam is as important as its original formation. The static aspects of foam stability are related to the action of surfactant molecules and additives at the various interfaces in the system, while the dynamic characteristics are controlled primarily by the dilatational viscosity of the interfacial layers and the bulk rheological properties of the system. The choice of surfactant or surfactant formulations, therefore, must take these factors into account in the context of the system being prepared.

Wetting and related phenomena

The adsorption of surfactants onto a solid surface from solution is an important process in many situations such as detergency, waterproofing and dispersion stabilisation. In these and many other related applications of surfactants, the ability of the surface active molecule to situate itself at the solid-liquid interface and produce the desired effect is controlled by the chemical natures of the components of the system—the solid, the surfactant and the solvent. A similar role can be played by surfactants on essentially infinite surfaces related to wetting, spreading, adhesion and lubrication. Classically, there are three classes of wetting phenomena—adhesional wetting, spreading wetting and immersional wetting. The distinctions amongst these, although may seem subtle, are significant from a thermodynamic and phenomenological point of view.

It has been found that surfactants with symmetrically located internal head group substitution and ortho-substituted alkylbenzene sulphonates are better wetting agents than their straight-chain and para-substituted analogs. The enhanced wetting activity of the nonlinear materials is generally associated with their more compact structure, allowing for further diffusion to the solid-liquid interface and greater relative adsorption efficiency. The presence of additional polar groups in the molecule like ester and amide linkages or POE chains usually result in loss of wetting power. Nonionic surfactants such as polyoxyethylenated alcohols and fatty acids show lesser and lesser wetting power as the POE chain is increased. For a given hydrophobic chain length, the maximum in wetting power will occur for the POE chain length for which the cloud point lies just above the test temperature. As a rule POE nonionics with an effective hydrophobic chain length of 11 methylene units and POE chain length of 6 to 8 will exhibit optimal wetting power. In addition, POE alcohols and thioethers are generally found to be superior to equivalent POE fatty acids.

A number of external factors that can affect the wetting power of surfactants include temperature, electrolyte content, pH and the addition of polar organics and cosurfactants. An increase in solution temperature will generally reduce the wetting power of most ionic surfactants. Electrolytes that cause a reduction in surface tension of a surfactant in solution will produce improvements in wetting power. The addition of long-chain alcohols and nonionic cosurfactants increases the wetting properties of many nonionic surfactants. Solution pH becomes important to wetting characteristics when weakly acidic or basic groups are present in the surfactant molecule.

Wetting plays an important role in many technologically important processes, and an understanding of the part played by surfactants can go a long way towards solving the problem of choosing the proper surfactant for the job. Some of these areas are detergency and soil removal, enhanced oil recovery (EOR) and suspension of solids in liquids.

Dispersion

Dispersion is everywhere. Be it foods, pharmaceuticals, cosmetics, paints or printing inks and many such industries, no new products would have been developed without dispersion, i.e., a process of reducing particles into a uniformly distributed submicron size and distributing them also uniformly in a suitable liquid medium. The paint industry represents a vital area where dispersion processes play a crucial role. Irrespective of quality and quantity of surface treatment received by pigment/extender powders, and no matter how favourable the resin molecular architecture is for dispersion, the wetting agents or surfactants are indispensable to achieve complete dispersion.

The range of surfactants available are so large that it is almost impossible for a paint technologist to choose the right surfactant or surfactant mix and its dosage for achieving dispersion for the given formulation at optimum cost with optimum energy and efficiency. While the role of surfactant in dispersion process is well conceived, its requirements are rigid. The interesting classes of surfactants giving high performance in dispersion process are :

(a) Chemical surfactants based on fluorine and silicon

(b) EO/PO polycondensates

(c) Surpynols (Acetylenic diol based surfactants)

(d) Polymeric hyperdispersants

(e) Innovative class of hyperbranched and dendrite structured surfactants

The evolution of various classes of surfactants became necessary as the dispersion of pigments in hi-tech resins posed problems. Epoxides, polyurethanes, polyesters and powder coatings do not offer ease of dispersion. Polymeric dispersants have been developed for sophisticated low molecular weight, low viscosity, high solid coatings. Unlike alkyds, for achieving dispersions in such difficult resins, dispersants must have anchoring groups with high functionality like carboxylic acids, sulfonic acids, amino acids, amine, pyridyl, acetyl etc. which will have strong bonding to the pigment surface and also soluble polymeric chains more or less identical to the main binder of the paint system. Dispersion has been a challenge to the paint industry from the beginning and this challenge persists with the emergence of new technologies.

Technologies

Technologies of major important to surfactant manufacture are :

(a) Sulphonation and sulphation

(b) Ethoxylation and propoxylation

(c) Esterification

(d) Condensation

Sulphonation and sulphation

The most important processes for the production of anionic detergents are sulphonation and sulphation. The sulphonation and suophation processes can be carried out by the use of a variety of starting materials, namely :

(a) Sulphuric acid and oleum (fuming H_2SO_4)

(b) Liquid sulphur trioxide (SO_3)

(c) SO_3 from oleum stripping

(d) Sulphur

(e) Chlorosulphonic acid ($ClSO_3OH$)

(f) Sulphamic acid ($HSO_3.NH_2$)

Sulphonation with sulphuric acid and/or oleum of various free SO_3 strengths is carried out as follows :

Acid strength	Alkylate acid ratio	Reaction temp. °C	Digestion time
20–22% SO_3 oleum	1:1.1	30 until half addition; 45 at end	2 hours at 40–45°C
12–14% SO_3 oleum	1:1.2	45 half; 55 end	2 hours at 50–55°C
100% H_2SO_4	1:1.5	50 half; 55 end	2 hours at 55–60°C
98% H_2SO_4	1:1.65	45 half; 55 end	2 hours at 50–55°C

In acid sulphonation, process variables decisive for the quality of the finished product are acid strength and purity, alkylate/acid ratio, sulphonation temperature/time of addition of acid, digestion time/temperature. The reaction can be carried out in a glass-lined or stainless jacketed vessel with an anchor type agitator of about 120 rpm. Optimum temperature is maintained during the reaction by cooling with water through the jacket. At the end of the reaction 10 per cent of water calculated or the total of alkylate plus acid, are added to obtain a good spent-acid separation. The water is added as it is or in the form of crushed ice. Under no circumstances the temperature after the addition of all the water should rise above 60°C. Separation takes about 6–8 hours and the resulting spent acid has a strength of 70–80%. In practice it is preferable to aim at a degree of sulphonation which will leave 1.2–1.5 per cent unsulphonated oil calculated on 100 per cent active matter.

The greatest advance in sulphonation technique is the use of SO_3 as sulphonating agent. Whereas sulphuric acid and/or oleum is mainly used for the sulphonation of alkylbenzene, and chlorosulphonic acid is only suitable for detergent raw materials where an OH group is present. SO_3 is suitable for practically all types of detergent raw materials. The most widely accepted method of using liquid SO_3 as sulphonating agent is to employ it in its vapourised form diluted with inert carrier gas, mostly dry air, to give a dilute gas stream of 7–12% by volume SO_3. The air dilution of the vapourised SO_3 can be done by a 'once through' system or a 'closed' system. SO_3 vapourised in partial vacuum has also been employed for the sulphonation, thereby eliminating the need for the carrier gas. In the case of detergent alkylates the amount of SO_3 used must be accurately measured to within ±1% as any greater variation will lead to either over- or under-sulphonated material.

Sulphur is an important raw material for the most modern sulphonation method, which applies converter gas for direct sulphonation of detergent raw materials. The sulphur fed into the sulphur-burner in a converter-gas plant must also be measured exactly. The sulphur is burnt in a stream of dried air to give an optimum concentration of 7–8% SO_3 in the gas reaching the converter. The Sulphurex process of the Ballestra Company of Milan is an important version of this technology. The temperatures normally used for various raw materials to be sulphonated by this process are as follows :

Alkyl benzene (linear or branched)

Sulphonators	50°C
Digestion	55°C
Neutralisation	45°C

Lauryl alcohol

Sulphonators	35°C
Neutralisation	45°C

Tallow alcohol

Sulphonators	55°C
Neutralisation	50°C

C14-C18 Ziegler alcohols

Sulphonators	50°C
Neutralisation	45°C

After sulphonation in the sulphonator, the sulphonic acids pass on to the digestion vessel where they are held to allow the reaction to go to completion. After the digestion, the sulphonic acid stream is passed into another reactor where it is neutralised with caustic. Multitude film reactors of several types are used by different manufacturers in adopting this technology.

Chlorosulphonic acid is widely used for the sulphonation of fatty alcohols, fatty alcohol ethoxylates, alkylphenol ethoxylates and related detergent raw materials. Sulphonation with chlorosulphonic acid requires special corrosion-proof equipment, either glass-lined steel or all-glass. The reaction of lauryl alcohol with chlorosulphonic acid illustrates the chemistry involved.

$$C_{12}H_{25}OH + Cl\ SO_3H \longrightarrow C_{12}H_{25}O\ SO_3H + HCl$$

The hydrogen chloride is absorbed to give a 3% HCl solution as a byproduct. In a typical batch sulphonation process of lauryl alcohol, the reaction is carried out at an optimum temperature of 25°C (not exceeding 3°C).

Sulphamic acid has also been used to sulphate alcohols and ethoxylates. The acid is quite expensive and adds to the cost of the finished product in comparison with those by conventional methods. However, no large investment in plant is needed in this case.

Ethoxylation and propoxylation

Ethoxylation and propoxylation collectively termed alkoxylation are the central processes, which make the production of a large number of custom tailored nonionic surfactants possible. The type of nonionic surfactants having the largest volume is the alcohol ethoxylates, which find use in household detergents. The second largest volume category is the alkylphenol ethoxylates. The basic technology for alkoxylation is the high productivity batch process, which has been widely licensed around the world by Pressindustria S.p.A., Milan, Italy. A newer type of alkoxylation technology available from Buss AG, Basel, Switzerland, involves the continuous absorption of EO vapour into feedstock as it is continuously recirculated. It's enhanced and better product quality, productivity, safety and environmental impact is well recognised and has been acquired by several market leaders.

The various EO and/or PO polycondensation reaction products are obtained by the reaction

$$RXH + (EO\ or\ PO) \rightarrow RX\ (EO\ or\ PO)_mH$$

where R and X represent the hydrophobic group and the connecting link to the active hydrogen, respectively. In the above reaction, chain starter, polyadduct chain length, catalyst, polycondensing PO (in block copolymers or random copolymers) are varied depending on the desired end product. Basic catalysts used in the ethoxylation process are KOH or NaOH, 0.1–0.3% by weight of the final product. Higher concentrations of catalyst do not proportionately increase the reaction rate. They only add to the cost of the investment and operating costs in their removal from the products.

Press industrial alkoxylation technology

The Pressindustria alkoxylation technology has evolved based on the following significant parameters, namely :

(a) Productivity.

(b) Production yield.

(c) Flexibility.

(d) Final product quality.

(e) Performance repeatability.

(f) Operability and automation.

(g) Environmental impact.

(h) Grade switch cleaning and safety.

Figs. 2.10 and 2.11 represent Pressindustria reaction units of the type ETO/2 and ETO/3.

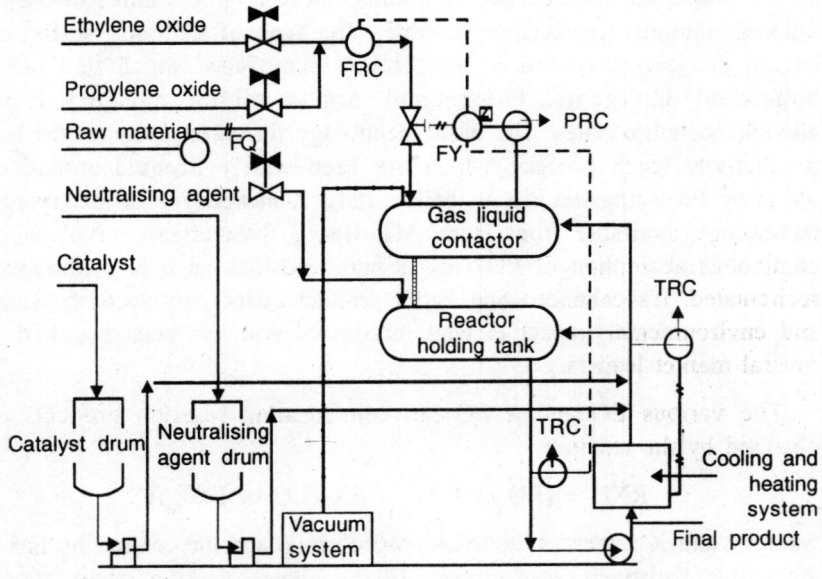

Fig. 2.10. Pressindustria reaction unit type ETO/2.

The reaction system is activated by charging raw materials and catalyst, expulsion of moisture, inert gas pressurisation and heating to initiation temperature. The chain starter and then the product during the reaction are pumped around at a high flow rate in a recirculation loop composed of a holding tank, a pump, a reaction heat removal system and the Pressindustria Gas-Liquid Contactor. Here the liquid suitably sprayed and continuously recirculated passes through a controlled atmosphere of EO or PO and inert gas. Due to the high gas liquid interface there is rapid attainment of equilibrium. During the reaction the polyadduct molecular weight rises reaching the preestablished growth ratio. The heat developed from the exothermic reaction is transferred by an heat exchanger located on the external circuit with automatic control of recirculation temperature, while the EO and/or PO is fed under automatic control through a signal selection of oxide mass flow, reactor pressure and bottom temperature. The flexibility of the Pressindustria system as far as production mix is concerned is very high and the polyadduct growth volumetric ratio which was 7:1 in the second generation units is 50:1 in third generation units in a single batch. This facilitates avoidance of intermediate storages, improvement in final product quality, reduction in production time and cost of high molecular weight polyadducts. It is claimed that the third generation of Pressindustria ethoxylation/propoxylation units also allow production of small batches 500–2500 litres product with a growth ratio upto 7–8:1 v/v, polycondensation of slurries and production of narrow range ethoxylated fatty alcohols with heterogeneous phase catalysts.

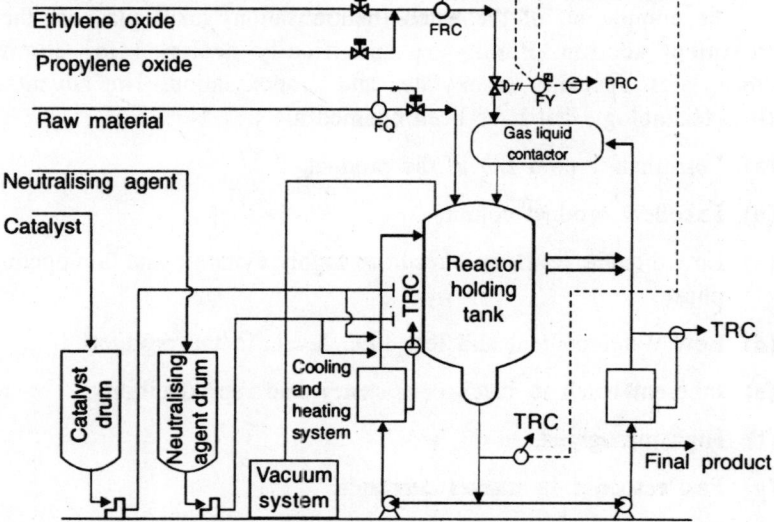

Fig. 2.11. Pressindustria reaction unit type ETO/3.

The Pressindustria ethoxylation technology has been made use of in the production of EO derivatives of fatty acids, fatty alcohols, fatty amines, phenols, alkylphenols, sorbitan esters, glycerol esters, pentaerythritol esters, block polymers, fatty amide derivatives, formaldehyde condensation and several other miscellaneous ethoxylates.

Buss alkoxylation technology

A typical ethoxylation plant consists of three sections : Pre-treatment, Reaction and Post-treatment. The sections are operated in a semi-continuous manner.

In the pretreatment stage, the amount of initiator is accurately charged corresponding to the required product. After catalyst addition, the mixture is dried and preconditioned.

At the reaction stage in the loop reactor, the reaction starts almost instantaneously, i.e., the induction period is greatly reduced or eliminated. The excellent mixing behaviour of the reactor and its accurate temperature control system provide very high reaction rates. The feeding rate of EO is controlled by the operating conditions in the reactor.

In the post-reaction step, the residual EO content up the batch is quickly reduced to less than 1 ppm.

All reaction stages are carried out under inert conditions and hence no ignitable gas mixtures can be formed at any time. The reaction section is operated as a closed system and generates no off-gas.

The completion of the batch (neutralisation) takes place in the post-treatment section. Plants are specifically designed for commodity ethoxylates, specialty ethoxylates and propoxylation. The advantages of Buss technology that have been claimed are :

(a) Less than 1 ppm EO in the product.

(b) Excellent product colour.

(c) Low dioxane levels as a result of highly dynamic and homogenous gas phase.

(d) Low water content and low PEG levels in the products.

(e) Inherent batch to batch consistency and reproducibility.

(f) High throughput.

(g) Fast response to market demands.

(h) Minimum post-treatment operations.

(i) Simple operation and complete remote control.

(j) Easy product changeover procedure.

(k) High flexibility.

Other technologies

Apart from alkoxylation there is a wide range of additional processes that are generally employed as an initial or subsequent stage to the alkoxylation in the synthesis of nonionic surfactants. Important amongst these are esterification (e.g., production of sorbitol esters), amidation (production of alkanolamides), methylation (end group blocking of ethoxylates) and amination/oxidation (fatty amine oxides).

Esterification is a well-established standard technology for the production of a variety of products for several end-use segments and as such is not elaborated.

The production of alkanolamides which have been the workhorse of the surfactants industry is of interest, as they constitute an important class of nonionics. There are two types of alkanolamide products. The first is the Knitchevsky-type liquid products, made by reacting an alkanolamine with a fatty acid or a fatty acid derivative at elevated temperatures in a 2:1 ratio. Such a product contains 60–70% alkanolamide plus some amine esters and diesters, amide esters and diesters and piperazine derivatives that are formed by side reactions. In addition there is significant unreacted alkanolamine.

This excess alkanolamine renders the Knitchevsky-type alkanolamides water soluble. The second type alkanolamide is the so-called superamide prepared by reacting an alkanolamine and a fatty acid ester in a 1:1 ratio. These are generally solid products which have an alkanolamide content above 90%. Some of the same byproducts formed in the case of the Knitchevsky amides also are formed in the case of the superamides but in much smaller quantities. For this reason and the absence of significant quantities for free alkanolamine they are poorly water soluble.

For the Knitchevsky type of fatty acid diethanolamide no catalyst is needed. The high surplus of diethanolamine (DEA) makes this unnecessary. The superamides are generally prepared by reacting methyl esters of fatty acids with diethanolamine using sodium methylate as catalyst. Optimum parameters for the production of superamides are :

(a) Mole ratio of DEA to methyl ester is 1.1:1.

(b) Optimum temperature is 105°C if the reaction is carried out at atmospheric pressure.

(c) Catalyst concentration is 0.15–0.25%, calculated on the batch as a whole and under atmospheric pressure.

The undesirable amine and amide esters are not stable in sodium methylate catalysed reactions and if formed will revert to the amide. With the above parameters the amide content of the product is 90%. Higher amide content products have made by operating under reduced pressure at 60°C with a 0.75% catalyst concentration. Fig. 2.12 gives the plant layout for the production of normal and super alkanolamides.

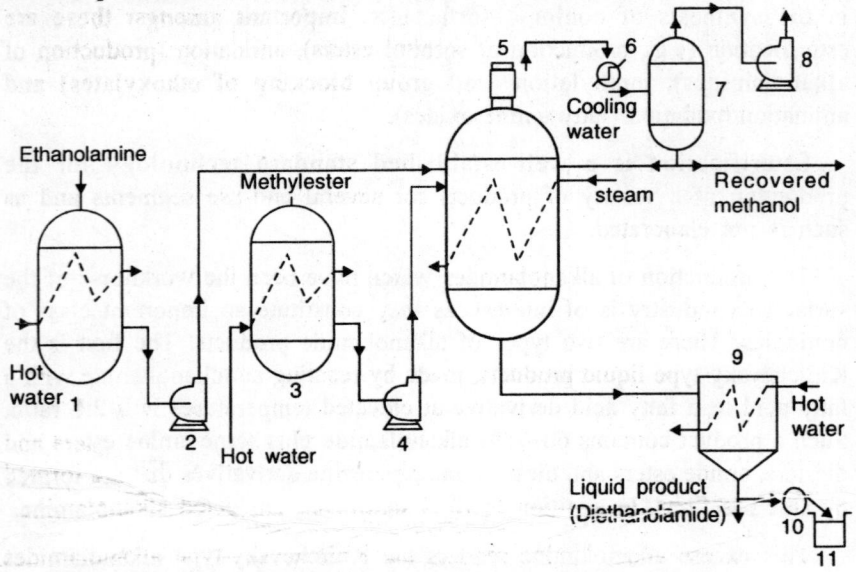

1. Ethanolamine tank; 2. Ethanolamine charging pump; 3. Methylester tank;
4. Methylester charging pump, 5. Reactor; 6. Condenser; 7. Vacuum drum; 8. Vacuum pump; 9. Surge tank; 10. Cooled flaking drum; 11. Flaked product (monoethanolamide).

Fig. 2.12. Plant for the production of normal and super alkanolamides.

MARKETING SURFACTANTS

The general relationships between the properties of surfactants and their applications, principally in the industrial sector are given in Table 2.13. In these applications the various classes of surfactants are used either as such or as blends. From the market analysis standpoint the major surfactant application areas are identified as :

(a) Household sector

(b) Personal care products sector

(c) Industrial and Institutional sector

(d) Agrosector

Table 2.13. Surfactant property/application relationship in some industries.

Industry	Surfactant properties of significance
Petroleum sector	
(a) Drilling fluids	Oil emulsification, solids dispersion, modification of rheology
(b) Workover of producing wells	Emulsification, sludge dispersion
(c) Producing wells	Demulsification, corrosion inhibition
(d) Secondary recovery	Preferential wetting
(e) Tertiary recovery	Wetting, penetration, emulsification, demulsification, viscosity buildup or reduction
(f) Hydrocarbon recovery from tar sands	Solids wetting, demulsification
(g) Oil spills	Wetting, emulsification, dispersion, spreading
(h) Fuel-firing	Emulsification
(i) Cleaning of oil tankers, storage vessels	Wetting, emulsification, demulsification
(j) Refined products (lubricants)	Detergency and sludge dispersion, corrosion inhibition
Food and Beverages sector	
(a) Food products (cooking, baking and frying fats)	Emulsifiers, solubilisers, wetting and suspension agents, viscosity control, antispatter agents
(b) Bakery products	Emulsifiers, crystallisers, solubilisers, antistaling agents
(c) Food processing plants	Detergency, sanitisation, foaming or defoaming action
(d) Beverages	Emulsifiers, solubilisers, foamers, foam stabilisers
Metals sector	
(a) Ore concentration	Wetting and foaming agents, collectors and frothers
(b) Cutting and forming	Wetting, emulsification, lubrication, corrosion inhibition
(c) Plating	Wetting, defoaming

(Contd.)

Industry	Surfactant properties of significance
Textiles sector	
(a) Fiber and yarn making	Detergency, scouring, emulsification, dispersion, lubrication, antistats
(b) Dyeing and printing	Wetting, penetration, solubilisation, emulsification, dye leveling, detergency, dispersion aid
(c) Finishing	Wetting, lubrication, antistatic action
Paper sector	
(a) Pulp treatment	Deresinification, detergency, penetration
(b) Paper making	Defoaming, dye-leveling, dispersion
(c) Calendering	Wetting and leveling
Paints, coatings and inks sectors	
(a) Pigment preparation	Wetting and dispersion of pigment.
(b) Latex paints	Emulsification, dispersion of pigment, suspension stabilisation, latex rheology control
Agricultural sector	
(a) Fertilisers	Anticaking
(b) Pesticides	Wetting, dispersion, suspension stabilisation, emulsification, foaming or defoaming, spreading
Industrial cleaners sector	Detergency, sanitation, wetting, corrosion inhibition, descaling, wax emulsification, foamer or defoamer.
Personal care products sector	
(a) Hair-care	Cleansing, creation or foam stabilisation and viscosity control in shampoos, disinfectant properties
(b) Skin-care	Cleansing, softening, disinfectant properties, emulsifiers, spreading agents, solubilisers
(c) Oral-care	Creating foams, suspension of solids.

In the household sector, the major markets are in laundry detergents, dishwashing detergents and household cleaners.

In the personal care products sector, the important segments are hair care and oral hygiene. The industrial and institutional sector covers a number of industries such as :

(a) Asphalt and road building materials.

(b) Cement and concrete.

(c) Emulsion polymerisation.

(d) Food processing.

(e) Industrial and institutional cleaners.

(f) Petroleum industry.

(g) Paints and coatings.

(h) Pulp and paper manufacture.

(i) Textiles and fibres.

(j) Leather.

(k) Froth flotation and mineral beneficiation.

(l) Coal water slurries.

(m) Antimicrobials.

Household Sector

Surfactants figure largely in the detergent industry in the household sector. In this application they are used in the form of formulations. All surfactant types figure in the detergent formulations for the household sector as well as in the personal care and industrial/institutional sectors. Table 2.14 lists the various classes of surfactants used in detergents and the relevant application areas.

In the household sector, a major value-based market in the 'Industrial Triad', the important segments are laundry detergents, dishwashing detergents, household cleaners and fabric softeners. There are large differences in the chemistry of these products from region to region and the ranges of product formulations fall under several categories. Some of these are 'Premium', 'Price Quality', 'Economy', 'Green' and 'Performance'.

Laundry detergents

The household laundry products segment comprises of three principle types of end products, namely :

(a) Conventional or Universal powders.

(b) Liquids.

(c) Superconcentrates.

Table 2.14. Surfactants in detergent applications.

Surfactant type	Applications
Alpha sulfo methyl esters	
Methyl laurate, sodium salt	Biodegradable surfactants with good detergency, foam properties and mildness, especially in hard water
Methyl cocoate, sodium salt	Used in laundry and dishwashing detergents, bathroom and all-purpose cleaners, bubble baths
Blend of methyl laurate and	In dishwashing liquids partially replacing ether
lauric/myristic monoethanolamide	sulphate for foam boosting and stabilising action
Alkyl benzene sulphonates	
Linear, sodium salt	Basic for household and Industrial & Institutional cleaners. Offers good detergency, high foam, biodegradability, economy
Linear, TEA salt	Base, salt-free detergent for liquid formulations
Linear, calcium salt	Emulsifier, used in combination with nonionics
Branched, amine salt	Drycleaning detergent
Amine oxides	
Lauramine oxide	Foam enhancers, stabilisers and viscosity builders in hard surface
Cetamine oxide	Cleaners and dishwashing liquids
Alkanolamides	
Lauric/myristic, coconut mono-ethanolamides	Foam booster, viscosity enhanced for liquid detergents. Also used in detergent blocks or bars
Lauric, coconut diethanolamides	Foam boosters and viscosity enhancers for liquid detergents
Phosphate esters	
Alkyl ether phosphate	Excellent hydrotropes for nonionics in
Alkylaryl ether phosphate	alkali cleaners; emulsifiers in agriculture; in heavy duty industrial/household alkaline cleaner/soak tank formulations
Alkyl sulphates	
Sodium, magnesium and	For use in rug shampoos, light duty liquid

(Contd.)

Surfactant type	Applications
ammonium lauryl sulphates	detergents
Alkyl ether sulphates	
Sodium and ammonium lauryl ethoxysulphates	In liquid detergents, car wash, dishwash, textile detergent and industrial foaming applications
Alkyl phenol ethoxylates	Detergents and emulsifiers with different ethylene oxide contents. Product with 4 moles EO is most oil soluble and product with 30 moles EO is least oil soluble.
Ethoxylated alkanolamides	Emulsifiers, detergents, wetting agents
Fatty alcohol ethoxylates	Emulsifiers and detergents with varying ethylene oxide contents

Demographic factors and social attitudes have a large bearing in the type of detergent used in a particular region and the formulation chemistry also varies significantly. In the overall analysis liquids and superconcentrates have reduced the predominant position of the flaky low-density powders.

Developments over the years have made it readily feasible to incorporate builders, bleaches, enzymes, brighteners, surfactants and ancillary constituents such as foaming, solubilisation and alkalinity agents in powders. The products thus made have been suitable for use at various wash temperatures and/or differing fibre-mixes and soils. The most effective builder/surfactant ratio has been considered 3:1 and a wash-liquor pH in the 9.0–11.0 range. This type of product permits the use of a relatively low level of surfactant, about 12%, in the warmer washes and 15–17% in the cooler washes, in detergent products. The predominant method of production of detergent powders has been spray-drying requiring high levels of investment, leading to few multi-nationals ultimately controlling the markets. The partial or total replacement of phosphate builders by zeolites in recent years has necessitated the use of higher levels of surfactants in the detergent powders.

Liquids have the advantage of storage convenience, easy dosing, ready solubility, easy manufacture and the facility to innovate on product packaging. However, all builders are not easily soluble in aqueous solutions. It has also been difficult to hold the 3:1 builder/surfactant ratio and the lower proportion of builder, namely 2:1, has called for substantial increases of surfactants in the product. Moreover the pH available for conventional liquids (7.5–8.5) has been below optimum. In case of liquids, which are unbuilt, more common in Europe, very large levels of surfactants have had to be employed to achieve effective detergency. The liquid detergents also pose problems in the incorporation of perborate bleaches and enzymes, which are essential components of a majority of detergents in Western Europe, USA and Japan. The introduction of liquid detergents while posing no problems in the US and Japanese machines, are problematic in the drum type European machines designed for the injection of powders from built-in dispensers at a pre-defined point in the washing cycle. This disadvantage has been overcome by innovative developments of dispensers.

Powder 'Superconcentrates', also referred to as 'Compacts' and 'Ultras' permit the chemistry of powders and also provide ample scope for multi-functionality facilitating the incorporation of sodium perborate bleaches, enzymes and brighteners. The superconcentrates can be characterised by the following features :

(a) Powder of high density.

(b) Has no inactive fillers like sodium sulphate.

(c) High surface-actives content.

(d) Contain performance additives.

(e) Achieve rapid solubility.

(f) Are innovatively packaged in small boxes.

The conventional powder detergent itself has gradually changed in density from 500 gms/litre to 700 gms/litre or more. In terms of consumer use it has thus changed from 1 and 1/3 cup to the washload to 1 cup and smaller. The superconcentrates are in the range 700–1100 gms/litre. This translates to 1/3 cup -1/2 cup dosages. From, the commercial angle, a 2.0 kg box of Proctor and Gamble's *Ariel Ultra* is the equivalent of a regular 3.3 kg box and gives 15 washes per package. Unilever's *Skipmicro* is a still more concentrated product and a 2.2 kg box is the equivalent of a 5.2 kg box of conventional powder with 18–20 washes per box.

The superconcentrates have become popular on various accounts. The advantages of the compact powder have been :

(a) Less storage space.

(b) Availability in smaller packages.

(c) Ability to meet green pressure by minimising on unneeded chemicals.

(d) Alleviation of solids waste problem which is critical to the environment.

(e) Savings accrued due to reduction in shelf space and transportation costs.

(f) Lower energy consumption in the process of manufacture as well as in packaging.

(g) Recycle possibility of the containers.

Dishwashing detergents

Dishwashing detergents are a class of hard-surface cleaners with high volume of usage. They are of two kinds :

(a) The manual dishwashing liquids.

(b) Auto-dishwashing detergents, in powder, tablet and increasingly, liquid forms.

The main components of manual dishwashing products are surfactants, but solubilisers, opacifying agents, foam stabilisers and mildness agents are normally present. Surfactants are predominantly anionic (LAS, AS, AES, SAS and AOS). Henkel has introduced dishwashing liquids with nonionics like APG. Concentrations depend on the quality of the product-premium, economy etc. and is influenced by the eating habits of the population in the region, which account for the proportion of fats, proteins and carbohydrates. The auto-dish products contain very little surfactants.

Household cleaners segment

Specialty cleaners are of several types from oven cleaners, drain cleaners, glass cleaners, bathroom and toilet bowl cleaners. The nonionics reportedly used in this cleaner segment are alcohol ethoxylates, alkyl phenol ethoxylates, fatty alkanolamides and fatty amine oxides. Of these, the alcohol ethoxylates have the major share of the market.

Personal Care Products Sector

Personal care products using surfactants are the following :

(a) Bath and shaving products (bath oils and bath salts; bubble baths; shaving products; toilet soaps).

(b) Skin-care products (foot-care products, skin cleansers, skin tanners, skin-care products, sunscreen preparations).

(c) Hair-care products (hair conditioners, shampoos).

(d) Oral-care products (denture cleaners, oral hygiene products).

(e) Colour cosmetics (lipsticks, blushers, eye-care products).

The personal care sector exploits to a far greater degree than does the household detergent industry, the vast range of properties arising out of surfactant combinations. The number of surfactants used are also larger, with concentrations ranging from the high, as in toilet soaps and shampoos, to small and miniscule but essential concentrations in products such as suntan lotions. At the same time, due to regulatory supervisions, far more stringent controls are exercised in the choice of the surfactants than is necessary in the detergent industry. Table 2.15 gives the various types of surfactants used in personal care products.

Hair-care segment

The hair-care sector is a very large consumer of surfactants, particularly in shampoos. Like toilet soaps, shampoos are cleansers removing particulate dirt, skin debris, pomades and exuded body oils from hair. Unless properly balanced, this cleansing process also removes natural agents like lipids which lubricate hair, make it manageable and give it lustre. Shampoos contain cleansing surfactants together with a whole host of para-synthetic chemical and natural additives which attempt to refat hair and recondition it. Shampoos contain also mildness agents, which take away the harshness inherent in the use of some formulations. A wide variety of anionic surfactants are used in shampoos, with regional preferences. Alcohol ether sulphates and alcohol sulphates are the most significant amongst these. Other surfactants used to control viscosity, create or stabilise foam and make shampoos milder are sulphosuccinates, alkanolamides, amine oxides and amphoteric compounds. Hair conditioners are another class of important products in the hair-care sector. They are used to reverse the swelling of hair and the removal of oils arising out of shampooing. They also aid in improving hair manageability by providing static control agents and improve gloss and lustre through the adsorption of polymers and other agents onto hair. Synthetic quats, as well as quasi-surfactant materials obtained from proteins and amino acids, are the most important surfactants in this application. The quats also provide disinfectant properties, important for the health of the scalp. Nonionic surfactants like alcohol ethoxylates and sorbitan ethoxylates follow the cationic quats in the level of usage in this application wherein they act as emulsifying agents.

Table 2.15. Surfactants personal care products.

Surfactants	Applications
Anionics	
Alkyl sulphates (sodium, ammonium TEA and DEA)	Foaming properties, high viscosity response, low pH and low temperature, clarity are exploited in shampoos, handsoaps, bath products, shaving creams, medicated ointments, dentrifices and tooth pastes.
Alkyl ether sulphates (ammonium and sodium lauryl sulphates)	For low pH, clear, gel or liquid shampoos, bath and cleansing preparations, baby products
Alpha olefin sulphonate (Sodium C-14 to C-16 olefin sulphonate)	Shampoos, hand soaps and bath products. More stable over a broad pH range than alcohol sulphates.
Linear alkylbenzene sulphonate (TEA dodecyl benzene sulphate)	For oily hair shampoos used to remove oil without stripping hair.
Alpha sulpho methyl esters (Sodium methyl and ethyl 2-sulpholaurates)	Mild surfactant used in bubble baths and other products with low formulated viscosities.
Sarcosinates (Sodium lauryl sarcosinate)	For personal care and mild products due to excellent wetting and foaming properties
Sulphoacetates (Sodium lauryl sulphoacetate)	For foaming and excellent viscosity response in powder bubble baths, shampoos, cleansing creams and syndet bars.
Sulphosuccinates	For low irritation shampoos and cleansing products due to good foaming, viscosity building and mildness properties.
Anionic blends	Surfactant concentrates for shampoo and bath products.
Cationics	
(Cetyl trimethyl ammonium chloride Oleyl and stearyl dimethyl benzyl ammonium chlorides)	In hair rinses, skin creams and lotions due to pronounced conditioning, softening and emolliency characteristics. Also as cationic emulsifiers.
Nonionics	
(Lauramide DEA, Lauramide MEA, Cocamide DEA)	As foam boosters, stabilisers, viscosity builders/modifiers for shampoos, hand soaps and bath products
Ethoxylated amides (PEG 3 Cocamide, PEG 6 Cocamide, PEG 6 Lauramide)	Mild effective emulsifiers for fragrance and essential oils. They impart viscosity and foam enhancement in shampoos, hand soaps and bath products.

(*Contd.*)

Surfactants	Applications
Amine oxides	
(Lauramine, Cetamine, Myristamine Olamine, Cocamidopropylamine oxides)	For good foaming and stabilisation with excellent wetting properties in shampoos, bubble baths and liquid handsoaps. Compatible with anionics, cationics and other non-ionics.
Amphoterics	
Betaines (Cocamidopropyl betaine)	-do-
General purpose emulsifiers, opacifiers viscosity builders (Glyceryl stearate, glyceryl distearate, Glyceryl oleate, glyceryl laurate, polyglycerol oleate esters ethoxylated to different degrees)	Creams, lotions, anti-perspirants, hair care products, sunscreens and other cosmetic preparations.

Skin-care segment

Skin-care products employ a very wide range of surfactants, most often in modest quantities. The formulated products themselves have a wide range. Surfactants are used with the objective of cleaning the skin with a low level of irritation, to refat the upper layers of skin and soften it through moisturising action, as body protectors, to soften beard etc. All four types of surfactants figure in this application. Anionics like AES, AS, sulphosuccinates, sarcosinates, taurates are used for their cleaning and foaming actions. Nonionics are used as emulsifiers, spreading agents for oils, and as solubilisers for fragrances. Cationics are used mainly for their antimicrobial activity. The amphoterics tone down the hardness of primary surfactants and are important constituents of products like body shampoos. The skin-care segment comprises a variety of products—skin creams, lotions, tonics, bath salts, bubble and foam baths, shaving creams etc. Only some of these employ surfactants and their presence varies from miniscule amounts to substantial (liquid soaps).

Toilet soaps are categorised in various different ways, namely, regular', superfatted, syndet and combo bars, 'liquid soaps' 'purity bars', and additive bars (complexion soaps). AOS, AES/AS and other surfactants are used in some of the special soaps.

Skin cleansers are the products of affluent societies and have the properties of cleansing the skin, protecting it, beautifying it, providing a sense of well being. They overcome the skin-swelling tendencies of water and the irritation of soaps and provide desirable features associated with skin complexion. Anionic surfactants are the most important of the surfactants used in this area, notable ones being AS, AES, SAS, AOS, alkyl phosphates, sulphosuccinates and ester carboxylates.

Oral-care segment

Oral-care products, namely, tooth pastes, tooth powders, pastes, cream gels, mouthwashes and rinses are used to keep the oral cavity clean. Surfactants are incorporated into formulations for various end needs. In tooth pastes and toothpowders they act to create a foam which aids in suspending solids and thus their removal through rinsing. Most important for this purpose are sodium lauryl sulphate, sodium lauryl sarcosinate, sodium alpha-sulpho laurate and sodium dioctyl sulphosuccinate. The alcohol sulphate is preferred despite its astringent after-taste.

In the personal care products industry the surfactants have a special role as stabilisers of emulsions and foams. The glycerol esters, notably glycerol monostearate, is the most widely used product as an emulsifier and stabiliser in addition to its functions as thickener, opacifier and emollient in cosmetics. The nonionic ethoxylated alcohols function as emulsion stabilisers in creams and lotions. The ethoxylated amines serve as coemulsifiers and stabilisers for emulsion systems. The alkanolamides are excellent foam boosters, stabilisers and viscosity builders for shampoos, hand soaps and bath products. Ethoxylated alkanolamides and sulphosuccinates derived from these also figure in these applications. Amine oxides are also used for a similar purpose. The betaines provide good foaming and stabilisation with excellent wetting properties in shampoos, bubble baths and liquid handsoaps. The lauroyl and oleyl sarcosines are foam stabilisers in bath oils, hand and skin cleaners.

Industrial and Institutional Sector

The Industrial and Institutional cleaners segment is characterised by thousands of products targeted to hotels, restaurants, hospitals, nursing homes, food processing plants, retail establishments, schools and health clubs. The components to this cleaner segment are janitorial services, food processing, commercial dishwashing, metal cleaners, laundry and dry cleaning. Table 2.16 gives the nonionics content of various detergent related formulations.

Table 2.16. Nonionics content of various detergent related formulations.

Formulation type	Percent non-ionics in the formulation
Spray dried general purpose powder	3
Spray dried household low foaming laundry powder	7
Concentrated powders	19–20
Heavy duty fully automatic washing machine powder	10
Heavy duty liquid detergent	2–4
Heavy duty detergent with controlled foam	4
Light duty liquid detergent	1–2
Machine dishwashing powder for	
(a) soft water	2
(b) hard water	3
(c) moderately hard water	1
Dry cleaning detergent	40
Detergent sanitizers	5
Detergent solvent combinations	20–30
Waterless hand cleaner	6
Alkaline metal cleaning compound	1
Food and dairy alkaline detergent cleaners	1
Defoamer for use in detergents	10
Cream shampoo	2
Liquid shampoo	3
Low foaming heavy duty granular detergent based on Pluronic polyols	20
Extra high concentrate of shampoo	10
Aluminium soak tank cleaner/brightener	2
Liquid steam cleaner	5
Powdered steam cleaner	8
Transportation vehicle cleaners	2–5
Household and janitorial cleaners	1–3

Petroleum industry

Table 2.17 gives the various types of surfactants used in the drilling, completion/stimulation and production sectors of the petroleum industry.

Table 2.17. Types of surfactants used in various segments of crude oil production.

Segment			Surfactant type	Surfactant function
1.	Drilling			
a)	Additives for water-based drilling fluids	(i)	N + A	Drilling detergent
		(ii)	Cocodiethanolamide	Drilling detergent, wetting agent
		(iii)	Polyglycol ester	High temperature drilling detergent
		(iv)	Mixture of emulsifiers	Defoamer
		(v)	Fatty alcohol polyglycol ether	-do-
		(vi)	Synergistic mixture of fatty acid esters	Liquid lubricant
		(vii)	N + A	Struck pipe freeing agent
		(viii)	Organic solvents plus surfactants	Rig wash
(b)	Additives for oil-based drilling fluids	(i)	Alknaloamide	Secondary emulsifier
		(ii)	Salts of organic acids + amino alcohols + aninonics	Rig wash
		(iii)	Nonionic + paraffinic hydrocarbon	Cleaner for cutting metal surfaces
		(iv)	N + A	Cleaners for oil contaminated cuttings
(c)	Foam drilling		Sulphated alcohol ethoxylate	Foamer
2.	Completion/stimulation			
(a)	Cementing	(i)	Anionics in hydrocarbons	Cement spacer
		(ii)	Sulphonated naphthalene condensate	Dispersant for cement slurries
(b)	Acidisisng	(i)	N + corrosion inhibitor	Acid inhibitor for H_3PO_4, H_2SO_4, etc.
		(ii)	A + C	Clay swelling preventor
		(iii)	A Emulsifier	Emulsifier for diesel/HCl systems

(Contd.)

Segment	Surfactant type	Surfactant function
(v) Alkylaryl polyalkylene glycol Wetting agent	(iv) Fatty amide with betaine	Suspending agent
(c) Miscellaneous	(i) Fatty alcohol polyglycol ether	-do-
	(ii) Alkyl polyglycol ether	-do-
3. Production		
(a) Corrosion inhibition	(i) Quaternaries + surfactants + solvents	Water soluble corrosion inhibitor
	(ii) Corrosion inhibitor + surfactants	Oil soluble corrosion inhibitor
	(iii) Surfactants mixture	Pipeline inhibitor
	(iv) Mixture of nitrogen containing fatty derivatives and surfactants in aromatic solvent corrosion inhibitor	Oil soluble, water dispersible
(b) Demulsification	(i) Polyoxyalkylene glycols	Demulsifier base
	(ii) Ethoxylated phenolic resins	-do-
	(iii) Cationic polymer formulation	Deflocculant/demulsifier
	(iv) N	Demulsifier base
(c) Paraffin inhibitory in light dispersion	(i) Fatty acid alkanolamide	To reduce Lewis acid carry over end polymerisation
	(ii) Nonionic nitrogen containing EO/PO product	Breaking of waste wash water/oil emulsions

A– Anionic; C – Cationic; N – Nonionic

Surfactants not only depend on petrochemicals from petroleum as feedstock for their manufacture but also have an important role in the oil and gasfield exploration, drilling and production of petroleum. The specialty chemicals used in crude oil production are demulsifiers, corrosion inhibitors, surfactants, biocides, scale inhibitors, wax inhibitors and others. Apart from the surfactants used as such for several functions, the demulsifiers and corrosion inhibitors used in the petroleum industry are also largely based on surfactants.

The major area in the petroleum industry where nonionics dominate is the area of demulsification. Polyoxy-alkylene glycols, ethoxylated phenolic resins and modified quats figure largely in the demulsifier formulations.

Typical commercial surfactants in oilfield applications are given in Table 2.18.

Table 2.18. Typical commercial surfactants in oil field applications

Description	End use/function drilling detergent
Anionic + nonionic	
Modified cocodiethanolamide	Drilling detergent/wetting agent
Polyglycol ether	Drilling detergent for deep drilling
Fatty alcohol polyglycol ether	Wetting agent
Oleic acid sulphonate	-do-
Nonionic	Demulsifier base
Nonionic	Demulsifiers
Phosphate esters	Emulsifiers in drilling/scale inhibitors
Alkylphenol ehtoxylates	Emulsifiers in drilling
Imidazoline	-do-
Quaternary	-do-
Alkanolamide	Foaming agent
Ether sulphates	-do-
Sodium dodecryl benzene sulphonate	Drilling wetting agent
Oleyl and stearyl imidazolines	Corrosion inhibitors
Block polymers	Demulsifiers
Resin ethoxylates	-do-
Surfactant resin blends	-do-
Oxyalkylated phenolic resins	-do-

Apart from their use in crude oil production, the surfactants also figure in cleaning of oil tankers and storage tanks, in containing oil spills and enhances oil recovery (EOR).

Food Industry

The food industry constitutes another large volume industrial sector for surfactants. The food sector is a sensitive sector and the types of surfactants that can be used are limited from the toxicity and possible contamination standpoints. Nonionics acceptable to the food industry are mono and diglycerides, sorbitan fatty acid esters, PEG/PPG esters and polyglycerol esters. These find application either singly or as blends in several food products indicated below :

Cakes and cake mixes	Emulsified shortening
Candy	Fat and oil emulsions
Coffee white	Ice cream
Colour dispersants	Frozen desserts
Dehydrated fruits and vegetables	Peanut butter
Dog food	Toppings

Food emulsifiers are widely used in convenience, snack and microwavable food products. Specifically visible fats and oils routinely need emulsifiers for food product processing, appearance, shelf life, texture and taste uniformity. The monoglycerides and diglycerides are used in the largest amounts in several food preparations. In these products the surfactants exhibit several functional characteristics. Table 2.19 gives the functional effects of various surfactants in food products.

With food habits changing worldover and with increase in demand for fast foods and low fat foods, the surfactant needs of the food industry are continuously on the rise.

Textiles industry

The textiles processing industry is a large user of surfactant materials with an estimated current global consumption of around 600,000 metric tons per annum. The nonionics constitute nearly two-thirds of this volume, with anionics and cationics/amphoterics being around 23% and 11% respectively. The manufacture of textile fabric derived from natural fibres (cotton, wool, silk), regenerated cellulose, petrochemical intermediates and inorganic materials (asbestos, glass, metal, carbon, ceramics) involves the processing of the fibre, spinning it to yarn, processing the yearn for weaving, dyeing, printing and finishing. Surfactants are used in each of these basic manufacturing stages. Salient features of the surfactant related textile operations are :

(a) Spin finishes (yarn and fibre lubrication and antistatic treatment).

(b) Sizing.

(c) Scouring.

(d) Textile assistance.

(e) Finishing.

Table 2.19. Functional effects of surfactants in food products.

Surfactant	Functions*				
Distilled monoglycerides (saturated types)	G (w/o)	VG	S	VG	S
Distilled monoglycerides (unsaturated types)	VG (w/o)	S	S	S	S
Acetoglycerides	None	None	None	VG	VG
Citric acid esters	VG (o/w; w/o)	S	S	S	None
Ethoxylated mono-glycerides	(o/w)				
Diacetyl tartaric	VG (o/w)	G	VG	None	G
Lactic acid esters	G (w/o)	S	None	VG	VG
Polyglycerol esters	VG (o/w)	S	S	G	None
Propylene glycol esters	None	None	None	None	None
Sorbitan esters	S (o/w)	None	None	S	VG

* 1—Emulsion stability; 2—Starch complexing; 3—Protein interaction; 4—Aeration and foam stability; 5—crystal modifications.

w/o—Water in oil emulsions; o/w—Oil in water emulsions; G—Good; VG—Very good; S—Slight (moderate).

Spin finishes: These finishes of a proprietary nature generally contain a lubricant, an antistat and an emulsifier. The function of the lubricant is to facilitate the processing of yarn particularly at high processing speeds and preserve its structure. The antistats are essential in the case of all synthetic fibres, which being nonconductors build up high levels of static during processing. Emulsifiers are needed to ensure the removal of the lubricant and the processing chemicals at a later stage of the spinning and weaving cycle. The surfactant content of modern 'producer finishes' may be as high as 50%, the balance being mineral oils and waxy materials. Nonionic surfactants like polyethylene glycol esters, ethoxylated alcohols, alkylphenols, amides and alkanolamides are used in this operation in the form of blends.

Sizing : The basic objective of sizing is to give the yarn enough strength and smoothness to withstand the conditions of weaving operations, particularly high yarn processing speeds (200–400 metres minute). If the fibres of the yarn do not adhere together or the surface of the yarn is not smooth, then the warp breaks, resulting in the shut down of the loom hampering production. Sizing also stiffens the yarn which is desirable in its mechanical processing. Most filament sizes are surface-active to some extent by virtue of their composition. This, together with the often very low application viscosities generally used, helps to wet the yarn surface and penetrate between the filaments. The size applied to the fibre reduces the water absorption characteristics of the fibre and will interfere in the dyeing and printing operations. Desizing is therefore an essential step.

Scouring agents: Scouring is an essential feature of textile processing at more than one stage to remove impurities in the yarn material (waxes, natural oils etc.) and extraneous contaminants (spinning, coning and other processing/machine oils). Like sizes, scouring agents are formulations made effective by surfactants, which act as wetting agents, penetrants and dirt removers. Typical surfactants used for this purpose are alkylphenol ethoxylates.

Finishing agents: Fabrics are treated in several ways to impart special characteristics. Important amongst these are anti-creasing and durable press treatments, water and oil repellent treatments, flame proofing, fluorescence whitening, softening and providing antistatic properties. Surfactant-based softeners of all ionic types are used by the industry. Although cationic softeners are more effective on synthetics, the most prevalent products are nonionic surfactants due to their low cost. Fatty amine oxides are increasingly being used in 'foam finishing', typically used for permanent press finishes.

In the textile industry mutual repulsion of fibres having similar charges causes difficulties in many textile-processing operations. Repulsion causes filaments in a charged warp to bend away from one another, ballooning of a bundle of slivers, excessive fly during processing and increases in hairiness of the yarn produced. Difficulties arising in folding fabrics into layers is especially disturbing in dyeing and finishing plants wherein long lengths of dry fabrics must be handled. Mutual attraction of oppositely charged fibres also leads to difficulties in carding, drawing, weaving, knitting, roving, spinning etc. due to tangled fibres and yarns. Charged fibres are attracted to adjacent parts of the spinning machinery and thus interfere with processing. Clinging of garments, draperies and other products by attracting oppositely charged particles of dust and dirt from the atmosphere to the charged fabric cause annoyance to the textile consumer.

As such, to overcome the ill effects of static charges on textiles, particularly synthetic fibres, antistatic finishing of textiles is essential.

The antistatic agents used in the textile industry function either by reducing the generation of static charge or by increasing the rate of charge dissipation or by both mechanisms. The antistatic finishes for textiles can be durable or non-durable. Effective finishes must be durable and be capable of withstanding repeated laundering and dry cleaning cycles. The antistatic additives can be incorporated internally in the bulk of the polymer before spinning or applied externally to textiles by padding, exhausting, spraying and coating followed by drying. In the textile industry, antistatic agents are often combined with lubricating oils to allow emulsification of the oils and to facilitate their removal during scouring and dyeing of textile materials. Quaternary ammonium salts, esters of fatty acids and their derivatives, sulphonic acids and alkyl aryl sulphonates, polyoxyethylene derivatives, polyglycols and their derivatives, polyhydric alcohols and their derivatives and phosphoric acid derivatives are the different types of surfactants figuring in various antistatic agents formulations. For durable finishes long chain amines, amides and high molecular weight polyethylene glycol are used. Table 2.20 gives some commercial antistatic agents in the textile industry.

Metal working

In the metalworking industry a wide range of surfactants serve as emulsifiers, coupling agents, lubricity additives and mineral oil replacements. Methyl and butyl esters lubricate beverage can drawing operations. EO/PO esters are water-soluble lubricants useful in high water-based and fire resistant hydraulic fluids. Polyol esters with their unique flash and smoke points are recommended as lubricity additives.

Several PEG esters cover a wide range of HLB values to assist in formulating both macro and micro emulsions. Aqueous solutions of these fluids provide excellent lubricants and heat-transfer agents for use in metal cutting and spinning processes. These polyalkylene glycol fluids can be used alone or combined with other water-soluble products such as amine soaps, amides or phosphate esters. Many polyalkylene glycol fluids are less soluble in hot water than cold, and this property can be exploited to provide a most effective combined coolant and lubricant. Heat generated at the cutting tip or other lathe tools will cause the polyalkylene glycol to come out of solution providing a neat lubricant. As the coolant is circulated and the temperature falls, the lubricant element once again becomes water-soluble. The polyalkylene glycols are also used as high performance quenchants for both ferrous and non-ferrous metals offering considerable advantages over the use of water. They minimise residual stress, distortion and 'soft-spotting', producing clear components ready for further processing.

Table 2.20. Some commercial antistatic agents in the textile industry.

Trade name	Source	Chemical type
Aerotex	American Cyanamid Corpn.	—
Antistatic D		
Aston 123	Lepinex Onyx	Polyamine resin
Cirrasol Z	ICI	—
Nonax 1166	Henkel Corpn.	Polyamine resin
Pernalose T	ICI	Nonionic
Stanax 1166	Standard Chemical Products	Polyamine resin
Ampitol VAC	Dexter Chemical Co.	Quat
Avitex DN	Du Pont	Cationic
Kemamine Q 9702 C	Humko Chemical	Quat
Antistatin D	BASF	—
Arkostat AC	Hoechst	—
Aston MS	Refinex Onyx	Quat
Aston AP	—	Amine condensate
Cassastat	Cassella	Cationic
Cetanac SN	American Cyanamid	-do-
Elfugin UW	Sandoz	—
Igepal CO 430	GAF Corpn.	Nonionic
Neutrostat	Simco Co.	—
Siligen APE	BASF	Quat
Tinorex TC	Ciba-Geigy	—
Zelec DP	Du Pont	Cationic
Zero C	Lutex Chem. Corpn.	Nitrogenous polymers

The alkanolamides are used in chain lubes, cutting and drawing compounds for emulsification, lubricity and corrosion protection. Other nonionics also provide superior wetting, lubricity, coupling and defoaming properties. Table 2.21 gives various surfactants used in metalworking.

Table 2.21. Various surfactants in metalworking applications.

Surfactant class	Applications
Alkanolamides	Emulsifiers and corrosion inhibitors in cutting fluids, drawing compounds and cleaners
POE aliphatic base	Wetting agent, lubricant, coupling agent, defoamer in cutting fluids, drawing compounds and corrosion inhibitors.
POE (36) castor oil	-do-
Methyl esters	Replacements for mineral oils, aluminium beverage can drawing lubricants
EO/PO esters	Lubricity additives and coupling agents, mineral oil replacements in cutting fluids, drawing compounds, fine resistant hydraulic fluids, quanchants.
Fatty esters	Water in oil emulsifiers and lubricity additives for cutting oils and drawing compounds; sperm oil replacement.
Polyol esters	Compounding of 100% synthetic lubricants for low and high temperature applications.
PEG esters	Primary emulsifiers and solubilisers in clear micro-emulsions. Esters of lauric and stearic acids used in aluminium wire drawing.
Amine alkylbenzene sulphonate	Effective wetting, surface tension reducer and defoamer
Ethoxylated nonyl-phenol phosphate	Wetting, lubricity, coupling, defoaming in cutting fluids, drawing compounds
Sodium alpha sulphomethyl laurate	-do-

Emulsion polymerisation

Emulsion polymerisation is another large volume sector where the surfactants have a very important role. This has been the key factor in the development of the emulsion paint industry and some adhesives and sealants. The essence of all emulsion paints is the formation of a fine dispersion or latex of water insoluble organic polymers, which after evaporation of the water will coalesce to form a film-forming binder. The advantage of water-based systems over solvent-based systems are freedom from obnoxious vapour, very much reduced fire hazard in storage and greater ease in clean-up of the equipment and spillage areas. The water-based emulsion paints containing a variety of synthetic copolymers, pigments, thickeners and additives have revolutionised the paint industry in the area of interior decoration as well as industrial finishes. In aqueous media amphipathic adsorption is the most frequent type of adsorption process encountered and as this is by far the most predictable behaviour, the selection of surface active agents for particular uses is not complicated.

When an amphipath is used in aqueous solution, it exists as single molecules or ions and above the critical concentration range, as aggregates or micelles. A monomer if emulsified in such an aqueous solution will distribute itself in these loci, namely :

(a) In the emulsion droplets which comprise by far the largest fraction of the total monomers.

(b) In true solution in the aqueous medium.

(c) Solubilised in the micelles.

In emulsion polymerisation process the polymerisation takes place via a free radical mechanism and water soluble initiators are mainly used.

The stability of lattices when polymerisation is completed also depends on the amount and type of surfactants present. Combination of nonionics and anionics stabilise lattices for emulsion paint systems. EO condensates of fatty alcohols and alkylphenols with alkyl and aryl sulphonates are recommended for this purpose. The sulphonates, alkyl sulphates, alkyl ethoxysulphates, alkylphenol ethoxylate sulphates and sulphosuccinates are the anionics used and alkylphenol ethoxylates the major nonionics used in emulsion polymerisation. The nonionics contribute mechanical stability to latexes and also assist in particle size control. These properties are achieved not only by blending the nonionics with anionics but also by increasing the degree of ethoxylation of the alkyl ethoxylate sulphates and the alkylphenol ethoxylate sulphates and thereby their nonionic character. Sorbitan fatty esters are also used as emulsifiers for emulsion polymerisation of styrene and other water-insoluble monomers. In general the fatty acid derivatives constitute nearly 55% of all surfactants used in emulsion polymerisation.

Choosing the right surfactant for latex manufacture is a critical step. The surfactants that have been used to bring quality latex products to the market place are listed in Table 2.22.

Table 2.22. Surfactants used in emulsion polymerisation for quality latex products.

Surfactant type	Applications
Sulphonates	
Linear sodium dodecylbenzene Sulphonate Isopropylamine branched dodecylbenzene Sulphonate Branched sodium dodecyl benzene sulphonate Linear sodium alphaolefin sulphonate	Used in making styrene-butadiene latexes and in vinyl chloride and vinylidene chloride latexes. They exhibit excellent hydrolytic and thermal stability.
Alkyl Sulphates	
Sodium lauryl sulphate Sodium octyl sulphate Ammonium lauryl sulphate	While useful in most polymer systems, they are specially useful with vinyl chloride. Particle size control can be achieved by varying the alkyl chain length.
Alkyl ethoxylate sulphates	
Ammonium lauryl ether sulphate with 4, 12, 30 moles EO, sodium lauryl ether sulphate with 4, 12, 30 moles EO	Useful in acrylics, styrene-acrylics and vinyl acrylics. Increasing EO content improves mechanical and freeze-thaw stability, usually obtained from nonionic-ionic blends.
Alkylphenol ethoxylate sulphates	Products containing 3 or 4 moles EO with uses same as above.
Nonylphenol ethoxylates 4, 6, 10, 14, 30, 34 or 40 moles EO	They contribute mechanical stability, freeze-thaw stability in latexes. They are useful as components of blends with anionics for particles size control.

Antimicrobials (Biocides)

Institutional establishments such as hotels, restaurants, schools and hospitals are areas where hygiene is a very important factor. Changing eating habits, freezer foods, convenience foods, increased poultry and fish consumption, communal eating, cross contamination in the home due to hurried food preparation and careless cleaning down of surfaces pose great risk to life. Keeping these establishments clean and germ-free requires the use of biocides in formulations having compatible surfactants which are essential for the detergent action and other functions.

Biocides themselves come in a range of chemical types including nitrogen based quaternary ammonium compounds (Quats), ampholytes, biguanides such as chlorhexidine, phenols, chlorinated phenols, iodophors and aldehydes. The quats are both biocide and surfactant at the same time. Their use is increasing in general disinfectant-cleaning formulations, because of their relatively low toxicity. At a normal use dilution of 1000 ppm or less, toxicity problems are unlikely to occur. Long term feeding studies have established that these materials can be used in the food industry without running any risk from toxic residues. Quats have reasonable hard water tolerance, acting very rapidly against both gram-positive organisms such as 'Staphylococcus Spp' and also against gram-negative organisms such as 'Escherichia' and 'Salmonella Spp'.

Surfactant-biocide interactions are of extreme importance in practical applications. Many biocides although possess some detergent action, detergency being a very vital matter in cleaning of surfaces necessitates the use of biocides and surfactants together. Biocide-detergent combinations have shown synergistic as well as antisynergistic effects in respect of the biocidal activity and as such the choice of the surfactant in antimicrobial formulations becomes critical. These formulations are to cater to diverse areas like food processing, brewing, leather tanning, paint preservation, cosmetics, oil drilling and animal health as well, and as such biocide-surfactant combinations should be complementary. The surfactants used as such for antimicrobial activity or in formulations of these products are indicated in Table 2.23.

Table 2.23. Surfactants in antimicrobial applications.

Surfactant type	Applications
Dialkyl dimethyl ammonium chlorides (octyl decyl, dioctyl, didecyl derivatives)	Water soluble quats used as disinfectants, sanitizers and fungicides for hard surface.
Alkyl dimethyl benzyl ammonium chlorides (Benzalkonium and myristalkonium chlorides)	Used for hard surface disinfection, satitisation and deodorisation; algicide and slimicide agents for swimming pool and industrial water treatment; in tablet manufacturing of disinfectants, sanitizers and deodoriser.
Alkyl dimethyl benzyl/ethyl benzyl Ammonium chlorides	Used as hard surface disinfectants, sanitisers, fungicides for hospitals, nursing homes and public institutions; effective algicides in swimming pool and industrial water treatment.
Dialkyl dimethyl ammonium chloride/ Benzalkonium chloride blends	-do-
Nonionic surfactants Blend of alcohol ethoxylates Amine oxide EO/PO copolymer	In ampholyte formulations. In chlorhexidine formulations. -do- and iodine formulations.
Nonyl phenol with & moles EO PEG 4000	In aldehyde formulations. Polyvinylpyrrolidone/iodine formulations.
Alcohol EO/PO condensate	Iodine formulations.
Anionic surfactants	
Alkylaryl sulphonate	As co-surfactant in ampholyte actives.

Other miscellaneous industries

In addition to the major sectors covered so far to identify surfactant markets, there are several other miscellaneous areas wherein surfactants do play key roles. These are :

(a) Paints, printing inks and plastics.

(b) Paper industry.

(c) Pharmaceuticals.

(d) Rubber industry.

(e) Adhesives.

(f) Photographic processes.

(g) Water treatment.

(h) Leather industry.

(i) Cement and concrete.

Certain significant aspects of surfactants in some of these markets are outlined below.

Dyes/pigments are used in paints, printing inks and plastics. Surfactants play a leading role in the dispersion process which involves :

(a) Wetting of the powder.

(b) Breaking up the clusters to form colloidal particles.

(c) Flocculation of the dispersion.

The dispersion efficiency is related to the degree of lowering of the interfacial tension and is independent of the polarity of the liquid. The surfactants provide stability to the system by getting reversibly adsorbed on the particles and modifying the surface electric charge or introducing steric barrier or altering the surface adsorption characteristics to increase the adsorption of polymeric materials in paint and other pigment systems. In printing inks the surfactant in addition to its role as a dispersant has to serve to maintain the rheological behaviour of the product. Mass pigmentation of plastics is carried out by blending the dye/pigment with the monomer using a surfactant and then subjecting it to polymerisation. As in the case of textiles, glycerol ester surfactants serve as antistats in plastics, as internal lubricants, thermal stabilisers and as additive agents for PVC processing.

In the paper industry there are a number of areas where the surfactants have multifunctional uses. They are used as emulsifiers, solubilisers and detergents. Recycling of paper is no longer an environmental issue but a fact of life. Deinking of newspapers by the wash system in USA or flotation system in Western Europe involve surfactants like PEG esters, alcohol ethoxylates, alkylphenol ethoxylates, modified fatty acids and blends of specialty nonionics. Newsprint is moving towards what is called 'Super news'. This paper requires filler to help smooth the sheet, improve the brightness and control the way the ink reacts on the paper. Polyethylene oxide polymers serve as retention aids in conjunction with bentonites. In the paper industry the sizing technology has undergone radical transformation resulting in alkaline sizing. The gradual replacement of the alkyl ketene dimer (AKD) by alkyl succinic anhydride (ASA) in the sizing market has resulted in a boost for nonionic surfactants as ASA needs prior emulsification for use.

There are a large number of foaming problems in textile, fermentation, food, surface coating, sewage treatment, petroleum processing, building, chemical and pharmaceutical industries and the solution to these problems lies in the surface activity of the defomers or antifoams. Some of the foamer specialties are required to generate and stabilise foams in concrete admixtures, fire fighting foamers, wetting and foaming agents for dust control, gypsum board foamers, latex frothing agents, gas and oilfield drilling foams. Alkylbenzene sulphonates, alkyl ether sulphates and proprietary anionic/nonionic blends figure in these applications.

Polymeric surfactants have emerged as vital ingredients in surface coating systems. They are also used in oil based drilling muds, as lubricating oil additives, in ceramics, magnetic tapes and discs and inverse emulsion polymerisation processes.

The surfactant markets in the industrial and institutional sector are thus diverse. However, the detergent industry is the largest outlet with the household products sector of the surfactant market having the largest volume consumption. Table 2.24 gives the choice of surfactants in detergent applications in the household and Industrial and Institutional sectors.

Agrosector

The two main applications of surfactants in this sector are as formulation aids and as adjuvants. In the former they are used as emulsifiers and dispersing agents in emulsifiable concentrates, suspension concentrates, water dispersible granules and wettable powders enabling ready formulation and application of these materials by the end user. Invariably these formulations use surfactant mixtures containing a good percentage of nonionics. As adjuvants, their role is as activity enhancers so that a higher activity is achievable from lower quantities of the active ingredients such as herbicides, insecticides and fungicides. This will also facilitate the reduction in the quantity of toxic ingredients getting released to the environment.

Amongst the general methods of applying agrochemical actives, the most important method is spraying. Here the surfactants are employed to produce a variety of concentrates in water, oil or powder/granular form. Careful selection of a 'balanced pair' emulsifier is necessary to ensure emulsion dilution stability over widely differing climatic conditions and degrees of water hardness. The principal emulsifier used in emulsifier concentrate formulations is 70% active solution of dodecylbenzene sulphonate in butanol. Nonionics are used here as co-emulsifiers to produce the 'balanced pair' and these consist of castor oil ethoxylates (40 moles EO), nonylphenol ethoxylates and EO/PO block copolymers. The emulsifiable concentrate formulations are easy to prepare and apply. However, they have the disadvantage of flammability and problems of phytotoxicity. Nevertheless, they remain the largest single formulation class used in crop-protection today. The suspension concentrates or aqueous flowables consist of an insoluble toxicant in a concentrated aqueous suspension. Essential requirements for these are storage, hydrolytic and viscosity stabilities in the range $-20°C$ to $+50°C$. They are manufactured by dissolving the surfactants in a cosolvent like ethylene glycol followed by addition of the agrochemical active component and water. Flowables are less prone to exhibit phytotoxic side effects and for stable formulations a balanced combination of wetting agents, dispersants, viscosity modifiers and antifreeze compounds is essential. Water dispersible granules (WDG), also referred to as 'dry flowables', contain the toxicant with one or more wetting agents and dispersants, the formulations having the advantage of a high concentration of the toxicant. These compositions exhibit rapid break up in the mix tank giving a stable dispersion or emulsion.

The trends in agricultural formulations are towards the following :

(a) Development of more cost effective methods for formulations.

(b) Elimination of certain solvents due to environmental pressures.

(c) Burning of environmentally unacceptable toxicants.

(d) Development of blends to replace costly 'actives'.

(e) Reduction of dosage per hectare.

(f) Increased interest in additives for synergistic effects.

Table 2.24. Choice of surfactants in detergent applications in the household and I & I sectors.

Application areas	Surfactant types used*			
	A	N	C/Am	B
Household products :				
(a) Laundry detergents				
Powder detergents	1, 4	8	—	—
Liquid laundry detergents	1	8, 12, 14	—	—
(b) Dishwashing products				
Manual dishwash liquids	1, 5	8, 12, 14	21	—
Machine dishwashing products	4	9, 10	—	26
Rinse aids	—	9, 11	—	23
(c) General purpose cleaners				
Hard surface cleaners	1, 3, 4, 5	8, 11, 12	21, 22	—
Cream cleansers	1	8, 12	—	—
Oven cleaners	4, 6	8, 12	22	26
Carpet and upholstery cleaners	1, 2, 3, 5, 6	8, 11, 12, 14	21	—
Window and glass cleaners	2, 3	8, 9, 11	—	—
Disinfectants	1, 5	8	21, 25	—
(d) Sanitary cleaners				
Toilet blocks	—	8, 19, 20	—	—
Acid toilet cleaners	1, 4	8, 9, 16	22	—
Drain cleaners	4	—	—	—
Industrial & Institutional (I & I) products				
(a) Laundry detergents				
Powder detergents	1, 4	8	—	—

(Contd.)

Application areas	Surfactant types used*			
	A	N	C/Am	B
Liquid laundry detergents				
(b) Food, dairy and catering industries	1	8, 12, 14	–	–
Machine dishwashing products	4	9, 10, 15	–	26
Rinse aids	–	9, 10, 11	23	–
Bottle washing	4	9, 10, 15	–	26
Steam cleaners	4	9, 10	–	26
Iodophors	4	8, 9, 11, 13	–	–
(c) Janitorial products				
Hard surface cleaners	1, 3, 5	8, 9, 11, 12, 13	21, 22	–
Cream cleaners	1	8, 12, 13	–	–
Hard gels and cleansers	1, 4, 5, 6	8, 12, 13, 17	21	26
Floor cleaners and strippers	1, 3, 4, 5	8, 11, 13	21, 22	–
Bactericidal cleaners	1, 2, 5, 6	8, 12, 14	21, 22, 25	–
Disinfectants	1, 5	8, 13	25	–
Manual dishwash liquids	1, 5	8, 12, 13, 14	21, 22	–

A – Anionics; N – Nonionics; C/Am – Cationics/Amphoterics; B – Blends

1. - Alkylbenzene sulphonates and their salts; 2.– Sulphosuccinates; 3. – Potassium oleic acid sulphonate; 4. – Phosphate esters; 5. – Fatty alcohol ether sulphates; 6. – Fatty alcohol sulphates; 7. – Nonylphenol ether sulphate; 8. – Fatty alcohol ethoxylates; 9. – Low foam biodegradable fatty alcohol EO/PO copolymers; 10. – Other low foam, biodegradable fatty alcohol based nonionics; 11. – Ethoxylated aromatic co-solvents; 12. – Diethanolamides; 13. – Nonylphenol ethoxylates; 14. – Ethoxylated alkanolamides; 15. – Low foam amine based EO/PO copolymer; 16. – Amine ethoxylates; 17. – Ethoxylated oils and fatty acids; 18. – Polysorbates; 19. – EO/PO block copolymers; 20. – PEG 4000; 21. – Betainers; 22. – Coco-imidopropionate; 23. – Quaternised ethoxylated amine; 24. – Formulated quaternary surfactant; 25. – Benzalkonium chloride; 26. – Blended surfactants/electrolyte soluble)

In the light of the above trends agricultural formulation technology is getting oriented towards water dispersible granules, microemulsions, of active blends, activity enhancers and microencapsulation for controlled release. Of these, activity enhancements and controlled release are areas where surfactants will play a dominant role. Apart from these, surfactant wetting agents have been evaluated as soil additives to improve crop yield and nutrient availability in crops such as corn, potato and soybean. Surfactant foams have been in use to protect crops from frost and in this area EO/PO block copolymers are gaining importance. Surfactants also act as growth simulators in conjunction with the variety of plant growth regulators in vogue.

Chapter 3

Principal Groups of Synthetic Detergents

INTRODUCTION

Synthetic detergents, like soap, are materials which dissolve or tend to dissolve in water and in non-aqueous materials under certain conditions. To achieve this double tendency, these materials include two distinct groupings in their molecular structure. One, which is easily soluble in water, is called the *hydrophilic group*, and the other, which on its own would be insoluble in water, is known as the *hydrophobic group*.

The hydrophilic group is usually, although not always, added synthetically to a hydrophobic material in order to produce a compound which is soluble in water. However, this solubilisation does not necessarily always produce a detergent, since detergency depends on the balance (ratio) of the molecular weight of the hydrophobic portion to that of the hydrophilic portion.

As an illustration, consider the material dodecane, $C_{12}H_{26}$. This is completely insoluble in water. If an OH group is substituted for one of the terminal hydrogens, the new material $C_{11}H_{25}CH_2OH$ lauryl alcohol, is still practically insoluble, but a tendency to solubility has arisen. If now this lauryl alcohol is sulphated to

$$C_{11}H_{23}\overset{\displaystyle H}{\underset{\displaystyle H}{C}}-O-\overset{\displaystyle O}{\underset{\displaystyle O}{S}}-OH$$

we have a material which is completely miscible in water in all proportions, ignoring, for the sake of development of this theme, hydrolysis which will occur when dissolving the acid product. If this sulphuric ester is neutralised with a caustic alkali, ammonia, or organic amines, the material becomes completely soluble in water and in this instance is a very good detergent.

If, instead of sulphating the lauryl alcohol, it is treated with, say, ten molecules of ethylene oxide :

$$H_2 \text{---} C \text{---} C \text{---} H_2$$
$$\diagdown \text{O} \diagup$$

we obtain the material :

$$C_{11}H_{23}CH_2 \text{---} O \text{---} CH_2 \text{---} CH_2(OCH_2CH_2)_8 \text{---} OCH_2CH_2OH$$

This material is again completely water-soluble and a good detergent. If only five or as much as twenty molecules of ethylene oxide had been used, the detergency would fall off, although the materials would still be water-soluble. If less than five molecules of ethylene oxide had been used, the product would be insoluble.

To return to the lauryl alcohol, if it were to be oxidised to lauric acid, that is the original dodecane, with a carboxyl group replacing the terminal CH_3 group, the material is still practically insoluble in water, but if this acid is neutralised with caustic alkalis or certain selected organic amines, the product is again water-soluble, being, in fact, a soap. In this instance it is a moderately good detergent, but if the original chain length had been say 18 carbons, the detergency would be greatly improved. This demonstrates what was pointed out above, i.e., that the hydrophile/hydrophobe balance is important for optimum detergency.

It should be noted that in the previous paragraph no mention was made of neutralisation with alkaline earths. This illustrates one of the failings of soaps. With the magnesium and calcium salts which are normally present in hard water, soaps form an insoluble scum which lowers the efficiency of the soap. This is a basic defect of the carboxyl group and detergents are so designed that this carboxyl grouping is replaced by something else which does not behave adversely in the presence of salts of calcium and magnesium.

To continue with our analogy of dodecane as the starting material, this in itself can be sulphonated by sulphoxidation to dodecane sulphonic acid :

$$CH_3 \text{---} (CH_2)_n \text{---} CH \text{---} (CH_2)_{9-n} CH_3$$
$$|$$
$$SO_3H$$

which is miscible with water. When this is neutralised with caustic alkalis, selected alkaline earths or organic amines, solutions in water are obtained which are detergents.

Now if the dodecane were to be chlorinated and then reacted with trimethylamine the compound obtained is lauryl trimethylammonium chloride :

$$C_{12}H_{25}\overset{+}{N}(CH_3)_3Cl^-$$

again soluble in water and a detergent of sorts.

Finally, if by a rather complicated series of condensations, dodecyl beta alanine :

$$C_{12}H_{25}NHC_2H_4COOH$$

is produced, this is soluble in water, even without neutralisation of the carboxyl group, and also has detergent properties.

Thus, in every example we have discussed, a solubilising group has been added to the dodecane. This group or grouping is hydrophilic, while the original dodecane is hydrophobic, and in every case the resultant product is a detergent which is available on the world market. In practice dodecane itself is not often used as a starting material, very few, if any, of the reactions discussed are as simple as shown. The discussion illustrate how an insoluble hydrophobe can be solubilised and converted into a detergent with varying qualities and reactions are shown schematically in Fig. 3.1. In addition, this diagram includes another very important reaction which reaction consists of chlorinating the dodecane, coupling this product with benzene by means of the well-known Friedel-Crafts reaction, and then sulphonating it to dodecyl benzene sulphonic acid. This process not only provides the most important detergent on the market but is also one of the few cases when dodecane can be used as the starting material.

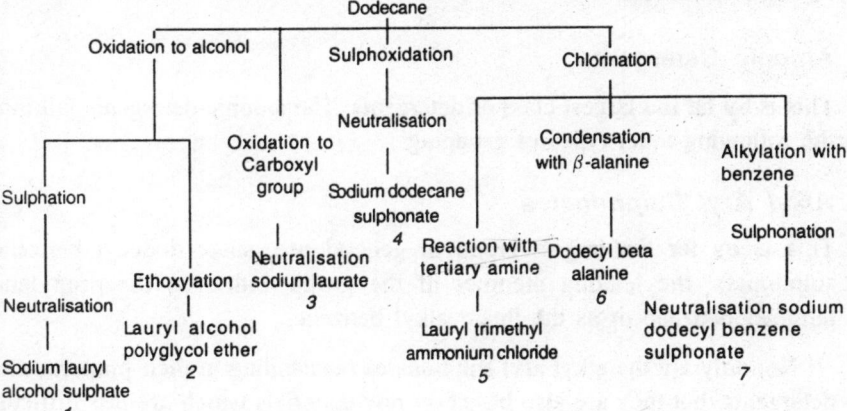

Fig. 3.1. Hypothetical illustration of how dedecane can be solubilised and converted into a detergent with varying qualities.

From the figure it will be noticed that end products 1, 3, 4 and 7 are all acidic and need to be neutralised with a base before being used. No. 5 is a basic material which in the process of manufacture is produced as the chloride (i.e., it is in effect neutralised with hydrochloric acid) and No. 2 needs no neutralisation at all. If the molecular structure of compound No. 6, for which we have taken as an example dodecyl beta alanine, is examined, it will be noted that it contains both a carboxyl (acid) and an amine (basic) group. The carboxyl group can be neutralised with an alkali, or by the amine group from an adjacent molecule. Similarly, the amine group can be neutralised by a strong acid, such as hydrochloric acid, or by the carboxyl group from an adjacent molecule.

CLASSIFICATION

There are four main classes of detergent: anionic, cationic, non-ionic, and amphoteric.

Anionic detergents are compounds in which the detergency is vested in the anion, which has to be neutralised with an alkaline or basic material before the full detergency is developed. In *cationic detergents* the detergency is in the cation, and although in the manufacturing process no neutralisation takes place, the material is in effect neutralised by a strong acid. *Non-ionic detergents* contain no ionic constituents, as their name implies, and *amphoteric detergents* include both acidic and basic groups in the same molecule.

Anionic Detergents

This is by far the largest class of detergents. The anionic detergents fall into the following main types of grouping :

Alkyl Aryl Sulphonates

This is by far the largest group in general use, since dodecyl benzene sulphonate, the leading member of the group, either as the propylene tetramer benzene or as the linear alkyl benzene.

Not only are the alkyl aryl sulphonates outstanding in their properties as detergents, but they are also based on raw materials which are less difficult to obtain and less expensive than the raw materials on which the other types of detergents are based. The main source of alkyl aryl sulphonates is the petroleum industry.

As the name implies, these products are based on aromatic compounds combined with an aliphatic chain bound to the aromatic nucleus. Taking as an example the most important type of alkyl aryl condensate, dodecyl benzene, this has the following structural formula :

whether the side chain is branched or linear.

This compound may be obtained by condensing benzene with a monochlorinated aliphatic chain having about 12 carbon atoms, by the Friedel-Crafts reaction, usually with anhydrous aluminium chloride as catalyst. The source of the aliphatic chain is either a special cut from straight-chain petroleum, for which purpose the petroleum from Pennsylvania is especially suitable, or a synthetic hydrocarbon chain obtained by condensing a lower molecular aliphatic hydrocarbon (generally propylene, which is condensed to its tetramer). This may be coupled with benzene without any previous chlorination.

The aromatic nucleus is usually benzene, but occasionally it is naphthalene, toluene, xylene, or even phenol. When naphthalene is used, the side chain can be as low as propyl. However, the sulphonated product is not suitable as a detergent, because of the low-molecular-weight side chains on the naphthalene nucleus. Nevertheless, by increasing the length of the side chain the detergent properties are increased.

To achieve the full hydrophile/hydrophobe balance when the lower homologues are used as the side chain, two members are necessary. Thus propyl naphthalene sulphonate is unsuitable either as a wetting agent or detergent. Dipropyl naphthalene sulphonate begins to show some wetting properties. Dibutyl naphthalene sulphonate is considerably better than butyl naphthalene sulphonate and, in fact, has fairly good detergent properties. Mono capryl naphthalene sulphonate is a very good detergent. With a side chain over C_{10}, however, the solubility of the alkyl naphthalene sulphonates starts falling off.

Thus, as the molecular weight of the side chain increases, it is necessary to limit the final product to one side chain. Conversely, if the size of the aromatic nucleus is decreased, it is necessary to increase the length of the side chain. As a rough guide, nonyl naphthalene sulphonate is equivalent to dodecyl benzene sulphonate. Thus it can be generalised that, as the side chain on the aromatic nucleus increases, the solubility decreases, and the detergency reaches a maximum and then starts decreasing. In practice, dodecyl benzene sulphonate was found to have the best balance of characteristics.

As stated above, the dodecyl benzene sulphonate seems to give the best all-round performance. However, factors vary in different parts of the world, and also conditions and requirements vary. It should be stressed that when we discuss dodecyl benzene it is meant to be a material whose average molecular structure suggests something close to dodecyl benzene: that is a product with an average molecular weight of the order of 246. This does not, in fact, indicate that the product is exactly that. In fact, if the base material is 'propylene tetramer', the propylene itself could have had small amounts of ethylene and butylene present as adulterants, and when this is polymerised to the tetramer appreciable amounts of the trimer and pentamer of each of these ingredients can also be produced as well as possible co-polymers of two or three of each of the ingredients. Many combinations are now possible and each has several possible stereo-isomers, and when all of these are coupled to benzene, coupling is then possible at many different points of the side chain; so one can appreciate that the total possible amounts of isomers and homologues are enormous. With the switch to linear alkylates, the alkyl portion is less varied, and can be in the range C_{11} to C_{15}.

It is therefore becoming standard practice to refine the raw alkylate to one product with an average molecular weight in the range of 233–45 and to another with the average molecular weight in the range 257–65, the exact figure depending on the manufacturer and his possible sales of the alternate product. The first is called commercially dodecyl benzene and the second sometimes tridecyl benzene. Tridecyl benzene sulphonate, in general, has better detergency and foams better in soft water, whereas dodecyl benzene has a lower cloud point and viscosity in liquid formulations. At approximately 180–200 ppm water hardness, the difference in detergency falls away.

Although the branched and the linear alkylates are both usually dodecyl benzene, it has been common to call the sulphonated branched product ABS and the linear one LAS.

With the changeover to linear alkyl benzene certain changes were observed. In this case as the alkyl portion is straight-chained it has no substantial amount of isomers but again contains homologues. When coupled with benzene, isomers arise in that the coupling can be at any of the positions between 2-phenyl and 6-phenyl, and it has been found that detergency is dependent on the position of the phenyl group. It has been suggested that for optimum light-duty detergency, linear alkyl benzene should have a mean chain distribution of lower than 12, the molecular weight should be in the range 231–41 and it should contain 15–25 per cent 2-phenyl isomers. For heavy-duty formulations they suggest a mean chain distribution of higher than 13, a molecular weight range of 258–66, the 2-phenyl content to be less than 20 per cent and the 5- and 6-phenyl content to be at least 40 per cent.

In the refining of the alkylate a certain amount of polyalkyl benzene is left as 'bottoms' in the still. This is a highly complex mixture of materials with an average molecular weight of 300 or more and on sulphonation it in turn yields a product which is a very good emulsifier, but poor detergent.

Three principal ways to use LAS and ABS are open to the manufacturer of cleaning materials :

(1) He may produce his end product from basic detergent raw materials, such as dodecyl benzene, by sulphonation processes and finishing processes requiring rather high initial investment, but giving him the widest scope for developing a complete range of cleaning materials.

(2) He may use a detergent material that has already been sulphonated, the sulphonic acid. This method still provides a fairly wide scope for development and avoids the necessity of erecting an expensive sulphonation plant which can only be operated by skilled labour and under expert chemical supervision.

(3) He may purchase ready sulphonated and neutralised detergents in a highly concentrated form, either as a paste-slurry or in the form of spray-dried, or drum-dried, powder or flakes. This material has then only to be diluted to obtain liquid products, etc., or be mixed with suitable builders to provide products ready to be marketed. The last-mentioned way of simply mixing detergent concentrates is the least expensive, from capital costs but also the least interesting, and does not, in fact, constitute a genuine manufacturing process. By careful selection of those materials with which he compounds the detergent concentrates, quite interesting and valuable cleaning materials may be obtained. This will become clearer during the description of the composition of cleaning materials.

Long-chain (fatty) alcohol sulphates

More recently, alcohols have been produced by the Ziegler process, a growth reaction using ethylene and trialkyl aluminium. This process yields even-numbered alcohols (those produced from natural fatty acids are invariably even-numbered); any desired length of chain can be produced, and the alcohols can also be blended in proportions not found in nature. Thus myristic alcohol (C_{14}) is not in great supply from natural sources, but can be manufactured in unlimited quantities by this process. Despite the fact that the Ziegler alcohols have to a large extent replaced the lower (C_{12}) natural alcohols, the higher ones (C_{18}) made from tallow are not necessarily more expensive. Several household and industrial detergents have been based either wholly or partially on sulphated tallow alcohols for many years and at the moment it is not clear whether synthetic or tallow alcohols are being used. Also included in this group are the OXO alcohols.

The OXO alcohols, being made by a hydroformylation reaction using an olefine and CO, are mixtures of both even and odd numbered alcohols with some side branching. Catalysts are now being used which limit the branching to 25 per cent 2-methyl to give a structure :

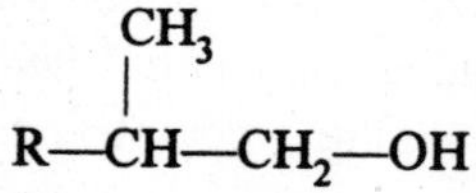

$$\begin{array}{c} CH_3 \\ | \\ R-CH-CH_2-OH \end{array}$$

There was a misapprehension that this side-branched alcohol is a secondary alcohol but as can be seen from the above structure it is an isoalcohol.

OXO-type alcohols are both ethoxylated and sulphated. It must be pointed out that some manufacturers use a type of process which does give some secondary alcohols :

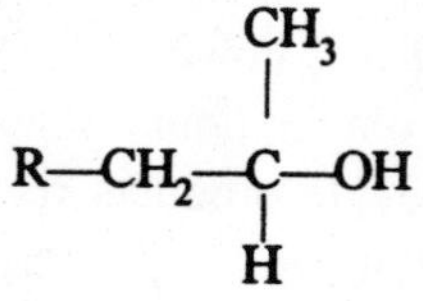

$$\begin{array}{c} CH_3 \\ | \\ R-CH_2-C-OH \\ | \\ H \end{array}$$

and these are not readily sulphated.

Many of the large 'soapers' are now using in their heavy-duty washing powders both LABS and a high molecular (C_{12}–C_{18}) alcohol sulphate. This is achieved relatively easily by a firm which does its own sulphonation, not so easily by a firm that has to buy its detergent raw materials.

Despite the fact that the sulphate bond —C—O—SO_3 is not as stable as the sulphonate bond —C—S—O_3, the stability of the alcohol sulphates is sufficient for all normal purposes, and there is not much to choose between the detergency of, say, cetyl alcohol sulphate and dodecyl benzene sulphonate.

Sulphonated olefins

The first commercial sulphonation of olefins with SO_3 involved the use of an SO_3-organo compound complex, either produced separately or *in situ*. Nowadays best results are obtained by sulphonating with uncomplexed diluted SO_3 in a film reactor.

The reaction between alpha olefins and SO_3 is not straightforward, mixtures of alkene sulphonic acids, sultones, alkene disulphonic acids and sultone sulphonic acids might be formed. The sultones need to be hydrolysed (saponified) to produce hydroxysulphonic acids. The materials sold under the name AOS are all therefore neutralised, usually with soda, and consist of mixtures of :

$$R—CH =CHCH_2SO_3Na \quad \text{and} \quad R—CHOH(CH_2)_nSO_3Na$$

where n can be both 2 and 3.

Despite what was said above about internal olefins, it is claimed that the presence of internal olefins together with alpha olefins gives an AOS with syngergistically improved detergency over pure alpha olefin sulphonates.

AOS has not yet made great strides in the heavy duty laundry field but is being used successfully for light duty detergents, hand dishwashing, shampoos, bubble baths and synthetic soap bars.

There is also controversy about skin sensitising by materials containing AOS. One party to the dispute maintains that after twenty years of commercial use no ill effects have been noted, whereas the other party states that if hypochlorite bleaching is used together with the alkaline hydrolysis, traces of sultones in ppb quantities can remain and cause this skin sensitisation.

The newer film reactors which produce light-coloured olefin sulphonates without bleaching are therefore the answer.

Sulphated monoglycerides

These were a moderate success, but, being based on fats, which are expensive and in short supply, they were quickly superseded by the cheaper alkyl benzene sulphonates. Quite apart from the economic aspect, the sulphated monoglycerides do not lend themselves easily to the manufacture of suitable powder products, although they are still being used occasionally in light-duty liquid formulations. Because of their sensitivity to alkaline hydrolysis, no real advantage is seen for them.

Sulphated ethers

Non-ionic detergents of the ethylene oxide condensate type, are in general excellent detergents. They have two disadvantages, however, in that they produce weak or unstable foams and although completely soluble in water, they exhibit 'invert solubility' in that they are more soluble in cold than in hot water. Thus when solutions of non-ionics are heated, at certain temperatures the solution becomes cloudy and in the limiting instance can separate into two phases. Fatty alcohol sulphates on the other hand are high foamers but their sodium salts do not produce clear solutions unless in very low concentrations. To produce clear solutions of fatty alcohol sulphates, recourse must be made to cations other than sodium with concomitant difficulties.

If a fatty alcohol is ethoxylated the ether produced still has an —OH group at the end able to be sulphated. This class of ether sulphates has become the fastest-growing group of anionic detergents. In practice the alcohol is not ethoxylated to the degree where it becomes in itself a detergent; only two to four molecules of ethylene oxide are added and the unsulphated material is still water-insoluble. The material is then sulphated by chlorosulphonic acid, or SO_3 and neutralised normally by caustic soda.

The sodium salts of these ether sulphates have very low cloud points even in relatively high concentrations. In addition, their wetting properties are lower than other anionic detergents, while their foaming properties are considerably higher, making them excellent raw materials for hair shampoo, and in combination with other anionic or non-ionic detergents, they are being used more and more for household dishwashing. They are used to increase the detergency of liquid detergents based on alkyl benzene sulphonate or n-paraffin sulphonates. Generally 20 per cent of the total active matter is replaced by sulphated ethers.

Lightly ethoxylated alkyl phenols are also being sulphated. In this case, however, the molecule contains two possible points of attack. The terminal —OH group can be sulphated and the benzene ring can be sulphonated. As the hydrophilic group in this class of product is the 'ethoxy sulphate' it is desirable that sulphation and no ring sulphonation takes place. A complete exclusion of ring formation can only be achieved by the use of sulphamic acid, which, however, produces simultaneously with the sulphation the ammonium salt. The use of SO_3 can under the best of conditions give a mixture of 20 per cent ring sulphonate but if great care is not taken di-sulphation, i.e., sulphation of the —OH group and also sulphonation of the ring of the same molecule, can occur. Alkyl phenol ether sulphates have been recommended as bases for toilet preparations but their main use is as components of light duty household detergent liquids, and for emulsion polymerisation.

These lightly ethoxylated alcohols are more complex mixtures than one would consider at first glance. The ethylene oxide content is only a mean and for these low values appreciable portions of the alcohol will remain unethoxylated (25 per cent when 2 moles EO are added, 18 per cent for 3 moles), thus the sulphated product will have alcohol sulphate as well. Figures can vary for different manufacturers so nominal ethylene oxide from two manufacturers will not necessarily have the same properties, not taking into account that the basic alcohols might also vary in their micro-composition. Ethoxylation of alkyl phenols follows a different path, all the phenol is ethoxylated before multiple ethoxylation starts.

Sulphosuccinates

Materials of excellent detergency can be obtained by condensing a hydrophobe containing an —OH group with maleic anhydride :

$$R-OH +$$

to produce the half-ester :

and reacting this with sodium sulphite or bisulphite to produce the sulphosuccinate :

$$
\begin{array}{c}
\quad\quad\quad\ \overset{\displaystyle O}{\underset{\displaystyle \|}{}} \\
\text{C—OR} \\
| \\
\text{H—C—H} \\
| \\
\text{H—C—SO}_3\text{Na} \\
| \\
\text{C—ONa} \\
\underset{\displaystyle O}{\overset{\displaystyle \diagdown}{}}
\end{array}
$$

Sulphosuccinates, depending of course on the choice of the R-group, are excellent high-foaming detergents and ideally suited for toilet preparations as they are non-irritating to the skin.

This is a process which does not require large capital outlays and allows the small or medium sized manufacturer to do his own synthesis. The choice of the hydrophobe is wide, it can be an alcohol, a lightly ethoxylated alcohol or an alkanolamide (which also has a terminal —OH) or a mixture of any of the possible variations of these. The maleic anhydride (the common choice but other unsaturated anhydrides can be used) can be reacted with 1 mol of the (mixture of) hydrophobe(s) to produce the half (or mono-) ester or with two molecules to produce the di-ester. In the case of the di-ester the properties of the finished sulphosuccinate are more wetting than detergent and they are used for industrial purposes. The half-esters are excellent high-foaming detergents, mild to the skin and ideally suited for toilet preparations.

If the di-ester is produced, the sulphonation is performed with sodium bisulphite; for the half-ester sodium sulphite is used.

The sulphosuccinates suffer from one disadvantage, their solubility is not good (depending of course on the choice of hydrophobe). This does not allow of the preparation of clear liquids if the concentration is over 10 per cent active matter.

Solubility can be increased by using monoethanolamine sulphite in place of sodium sulphite. This monoethanolamine sulphite is not an article of commerce but can readily be manufactured by passing SO_2 into a solution of monoethanolamine in water to form either the bisulphite or sulphite as required :

$$HOC_2H_5NH_2 + SO_2 + H_2O \longrightarrow HOC_2H_5NH_3SO_3H$$

$$2HOC_2H_5NH_2 + SO_2 + H_2O \longrightarrow (HOC_2H_5NH_3)_2SO_3$$

Allied to the sulphosuccinates are the sulphosuccinamates, where the maleic anhydride is reacted with a primary or secondary amine :

Maleic acid amide or
maleamic acid

The second carboxyl group can also be amidised to form the di-amide or it can be esterified with an alcohol, ethoxylated alcohol, alkanolamide, etc., to produce the mixed ester-amide.

These are again sulphonated in the usual way, using either sulphite or bisulphite.

The possibilities of all the possible variations of the sulphosuccinamates have not yet been fully explored but they are said to be even kinder to the skin than the normal sulphosuccinates.

Sulphonated methyl esters

In considering the aforementioned and following detergent raw materials, it will be noted that the vast majority are from petrochemical sources.

Fats and oils are self-replenishing and attention is being paid to their use as detergent intermediates. Alcohol sulphates are one such product and now attention is being drawn to sulphonated methyl esters.

They are already being used in powders and a surprising spin-off was noted in that the material has the ability to sequester calcium and magnesium ions, even in the presence of soap, without the use of phosphates.

The methyl esters need to be of high quality with a minimum of double bonds, indicating the use of refined coconut or hydrogenated tallow as starting materials. Production of methyl esters will be discussed later.

The sulphonation reaction is considered to be :

$$R{-}CH_2C{-}OCH_3 + SO_3 \rightarrow RCH_2{-}\underset{\underset{O}{\parallel}}{C}{-}OCH_3 \rightarrow R{-}\underset{\underset{SO_3H}{|}}{CH}{-}\overset{\overset{O}{\parallel}}{C}{-}OCH_3$$

that is, an intermediate complex is formed and this rearranges to the sulphonate.

There are two competing reactions. If double bonds are present, an olefin type of sulphonation will occur, particularly as that rate of reaction is considerably faster than the methyl ester sulphonation. It is, therefore, necessary to produce the methyl esters from material with very low iodine values. The second reaction is a double sulphonation of the ester which on neutralisation produces a disodium sulphocarboxylate, a sodium carboxylate and sodium methyl sulphate.

Fortunately the second reaction can be controlled by sulphonating conditions. However, on neutralising the sulphonated methyl ester neutralisation conditions must be such that no ester hydrolysis (saponification) occurs.

Alkane sulphonates

They were made according to the Reed reaction by chloro-sulphonation of petroleum fractions, but were quickly superseded by sulphoxidation and by alkyl benzenes when these appeared in commercial quantities. When the problem of biodegradation of propylene tetramer dodecyl benzene arose. Thus a new process for sulphonating straight-chain alkane fractions, which uses gamma rays from a cobalt-60 source was developed. The method of synthesis is not straightforward but appears to be the initial formation of a peroxy-sulphonate as an intermediate, and this acts as an initiator for the sulphonation reaction.

Gamma irradiation has been replaced by the less dangerous ultra-violet irradiation system. It is very important that the n-paraffins used are practically free from aromatics or alkyl aromatics, whose presence retards the reaction. The removal of aromatics and alkyl aromatics is best accomplished when the n-paraffins obtained by the molecular sieve process are treated with SO_3 converter gas.

In Europe three large production units for the sulphoxidation process have been set up by Hoechst of Germany. Another sulphoxidation plant by the Société Nationale des Pétroles d'Aquitaine (SNPA) is now at planning stage.

The SNPA process consists of the following basic steps: n-paraffins are reacted with SO_2 and O_2 under ultra-violet radiation with a wavelength essentially between 3300–3600 μ with a power level of several kW. Unfortunately, at the same time as paraffin sulphonate is formed, SO_2 reacts with O_2 and H_2O to form H_2SO_4. The overall reaction can be written as follows :

$$RH + 2SO_2 + O_2 + H_2O \rightarrow RSO_3H + H_2SO_4$$

where RH represents n-paraffin and RSO_3H the sulphonation product on a secondary carbon. Operating temperature and pressure are close to ambient.

The next step is to remove the sulphuric acid and some dissolved SO_2 from the mixture of paraffin sulphonate and unreacted paraffin (the latter to be recycled). SNPA has developed its own separation process which uses a solvent or solvent mixture of a 'slightly polar nature' with alcohol of at least five carbon atoms. No exact details of the SNPA process have yet been revealed. A flowsheet attached to the article, however, shows the basic principle. The paraffin sulphonate is neutralised with NaOH and the product marketed in the form of a 30 or 60 per cent past.

The detergency of alkane sulphonates is very similar to that of LABS. Stability against hardness and biodegradability are even better. Liquid detergents based on alkane sulphonates are being manufactured in combination with ether sulphates, where the ether sulphate constitutes 20 per cent of the active matter. Powders have not been produced so easily by spray-drying detergents solely based on a high percentage of alkane sulphonates because the material is soft and somewhat hygroscopic. This, however, can be overcome by the use of sodium toluene sulphonate and by the addition of colloidal silica.

Another method of producing alkane sulphonates is by the bisulphite addition to alpha olefins. This is a well-known reaction but until recent years complete reaction of the olefin was not attained, necessitating extraction of the unreacted olefin.

Recently in a latest process in which the olefin, the bisulphite solution and an initiator are all reacted together in the presence of a water-miscible organic solvent, which is eventually stripped off. Yields are close to 100 per cent based on the olefin.

Phosphate esters

A series of interesting and highly specialised detergents are the phosphate esters.

Two phosphating reagents can be used, P_2O_5 and polyphosphoric acid. If one takes as an analogy sulphonation with oleum, where oleum is considered to be SO_3 dissolved (or loosely combined) with H_2SO_4, we have the same phenomenon.

P_2O_5 when dissolved in phosphoric acid (or in a dearth of water) gives the complex phosphoric acid :

$$4H_3PO_4 + P_2O_5 \rightarrow 3H_4P_2O_7$$

In this reaction we have shown the production of pyrophosphoric acid, but of course in practice higher condensed acids are also produced and commercial polyphosphoric acid is sold as 115 per cent (calculated as orthophosphoric acid).

The reactions between an alcohol and polyphosphoric acid are :

$$R{-}OH + HO{-}\underset{\underset{OH}{|}}{\overset{\overset{O}{\|}}{P}}{-}O{-}\underset{\underset{OH}{|}}{\overset{\overset{O}{\|}}{P}}{-}OH \rightarrow R{-}O{-}\underset{\underset{OH}{|}}{\overset{\overset{O}{\|}}{P}}{-}OH + H_3PO_4$$

or

$$2R{-}OH + H_5P_3O_{10} \rightarrow R{-}O{-}\underset{\underset{OH}{|}}{\overset{\overset{O}{\|}}{P}}{-}OR + 2H_3PO_4$$

Thus both mono- and di-esters are formed and also free phosphoric acid.

Conditions can be varied but as a general rule when using 115 per cent polyphosphoric acid two-thirds mono- and one-third di-ester are formed. The free phosphoric acid is not normally removed from the ester.

When using P_2O_5 as the phosphating agent the reaction is different :

$$3R{-}OH + P_2O_5 \rightarrow R{-}O{-}\underset{\underset{OH}{|}}{\overset{\overset{O}{|}}{P}}{-}OR + RO{-}\underset{\underset{OH}{|}}{\overset{\overset{O}{|}}{P}}{-}OH$$

and yields equimolecular amounts of di- and mono-esters and no phosphoric acid (in practice a small amount of orthophosphoric acid is always formed).

Depending on the choice of the hydrophobic portion and the phosphating reagent, the possibility of varying the properties of phosphate esters is enormous.

Theoretically any compound containing reactive —OH can be phosphated. More often than not ethoxylates rather than pure alcohols are used. These ethoxylates need not necessarily be the lightly ethoxylated ones as used for ether sulphates, the true detergent grades, both alcohol and alkyl phenol based, are often used. Also if low molecular weight materials, not necessarily water-insoluble, are phosphated, wetting agents and good hydrotropes are obtained.

The phosphate esters possess good detergency, especially on hard surfaces, are low foamers with good acid and alkali stability, and are biodegradable. They find application in metal cleaning and plating, and since they are very soluble in organic solvents they are used in combination with solvents as dry-cleaning detergents, for hard surface cleaning and as agricultural emulsifiers.

Alkyl Isethionates

These are made by reacting ethylene oxide with sodium bisulphite :

$$CH_2\!\!-\!\!CH_2\!\!-\!\!NaHSO_3 \rightarrow HO\!\!-\!\!CH_2\!\!-\!\!CH_2SO_3Na$$
$$\diagdown\,O\,\diagup$$

This sodium isethionate is then dried and the —OH group is esterified with a fatty acid.

Alkyl isethionates, which are gentle to the human skin, have been used successfully in synthetic toilet soap bars and for hair shampoos.

Acyl sarcosides

Early attempts to improve soap were based on the principle of adding hydrophilic groups to the carboxylic acid. One of the first products of this type, and a rather successful one, was acyl sarcoside, in which an amido linkage is interspersed between the carboxyl group and the hydrophilic radical.

A method of proparation is to react a fatty acid chloride with N-methyl glycine (sarcosine), an amino acid :

$$R\!\!-\!\!\overset{\displaystyle O}{\overset{\|}{C}}\!\!-\!\!Cl + HN\!\!-\!\!CH_2COOH \rightarrow R\!\!-\!\!\overset{\displaystyle O}{\overset{\|}{C}}\!\!-\!\!N\!\!-\!\!CH_2COOH + HCl$$
$$\underset{\underset{\text{Sarcosine}}{CH_3}}{|} \qquad\qquad\qquad \underset{CH_3}{|}$$

The reaction is normally carried out in alkaline solution so the sodium salt is naturally produced together with salt as a by-product.

The sarcosides are gentle to the skin and are affected considerably less than carboxylates by hard water ions. They are used for personal care products, as dyeing aids and for textile finishing.

Alkyl taurides

Allied to, but somewhat different from, sarcosides, the taurides are manufactured by a somewhat similar process. Taurine (amino ethane sulphonic acid) or methyl taurine are reacted with a fatty acid chloride :

$$R—\overset{\overset{O}{\|}}{C}—Cl + R'HN—CH_2—CH_2—SO_3H = 2NaOH \rightarrow$$

$$R—\underset{\underset{R'}{|}}{\overset{\overset{O}{\|}}{C}—N}—CH_2—CH_2—SO_3Na + NaCl + H_2O$$

where R' can be hydrogen or either methyl or ethyl groups.

The taurates have all the properties of soaps: foaming power, lathering, kindness to the skin and emulsifiability, without the disadvantages associated with carboxylates. They are used in synthetic toilet 'soaps', in personal care items, in textile finishing and in agricultural wettable powders.

It will be noted that the last two items are manufactured from fatty acid chlorides, with the production of salt, whereas the alkyl isethionates are shown to be made from fatty acids without any salt as a by-product.

Alkyl isethionates can also be made by reacting a fatty acid chloride with sodium isethionate in alkaline solution with the production of salt. Conversely, the taurides and sarcosides can also be made by reacting a fatty acid with the anhydrous, molten sodium salt of taurine or sarcosine.

The difference between the two routes is that using the fatty acid chloride the reaction is done in a water solution at a temperature below the boiling point and goes to completion quite easily. Sodium chloride is formed. Using the fatty acid route, no sodium chloride is formed but temperatures over 200°C are required. It is extremely difficult to reach completion and unless special precautions are taken, the finished product will be of a darkish colour.

Fluoro surfactants

A new development in the surface active field is fluoro chemicals, which are at the moment very expensive but have unique properties :

(a) They lower surface tension of aqueous systems to 2.0 N m^2 (most detergents give a surface tension in the range of 2.8–3.1 N m^{-2}).

(b) They are effective in concentrations of the order of 0.01 per cent.

(c) They show surface activities in organic systems.

(d) They are chemically stable in hostile environments.

Most modern methods of manufacture involve reacting tetrafluorethylene with a fluoride ion (caesium, potassium or tetra alkyl ammonium fluoride) in polar aprotic solvents (for example dimethyl formamide). Polymerisation takes place, and the pentamer (the most abundant) :

$$(C_2F_5)_2 - \overset{\overset{\displaystyle CF_3}{|}}{C} - \overset{\overset{\displaystyle}{}}{\underset{\underset{\displaystyle CF_3}{|}}{C}} = C \overset{\diagup F}{\underset{\diagdown CF_3}{}}$$

is used for further synthesis.

This pentamer is coupled with phenol :

$$(C_2F_5)_2CF_3CC(CF_3){:}C(CF_3)F + C_6H_5OH \rightarrow (C_2F_5)_2CF_3CC(CF_3){:}$$
$$C(CF_3)OC_6H_5 + HF$$

This phenol group can be sulphonated to give an anionic detergent; also it may be chlorosulphonated to give the sulphonyl chloride, then treated with N'N'-dimethylpropanediamine to give a tertiary amine and then quaternarised with methyl iodide to produce a cationic detergent. The tertiary amine can also be reacted with propiolactone to produce an amphoteric surfactant.

Non-ionic fluorosurfactants can also be made from these phenyl compounds or else by reacting the pentamer with ethylene oxide condensates and rearrangement of the molecule (with the removal of one molecule of HF).

Cationic Detergents

The cationic detergents are of relatively limited interest to the cleaning materials manufacturer. Compared with the anionic and non-ionic detergents their detergency is relatively poor.

The term cationic detergents is somewhat of a misnomer. These materials do exhibit surface active properties but are seldom used in cleaning formulations. Their detergency is relatively poor and the cost high. The principal uses for cationic detergents are as germicides, fabric softeners and specialised emulsifiers. Anionic and cationic detergents are mutually incompatible, as the anion and cation precipitate each other.

These cationic detergents are almost invariably amino compounds and the most effective of the group are the quaternary ammonium salts, with one long chain attached to the nitrogen nucleus, or quaternary pyridine based salts. Here the possible permutations are again enormous. For example, cetyl trimethylammonium chloride, which is a typical normal quaternary ammonium salt, can also appear as the bromide and then, instead of the cetyl group, any of the long chains available both from fatty or synthetic sources from lauryl to stearyl. can be used. One methyl group can be replaced by another cetyl or benzyl (or even a naphthyl) radical and then one or all of the methyl groups can be replaced by an ethyl group and so forth.

Non-ionic Detergents

In discussing anionics we mentioned that most of the anionic detergents are sulphonates or sulphates of one sort or another. Similarly, the vast majority of all non-ionic detergents are condensation products of ethylene oxide with a hydrophobe.

This hydrophobe is invariably a high-molecular-weight material with an active hydrogen atom, and the non-ionic material can be one of four main reaction products :

Fatty alcohol and alkyl phenol condensates

The reaction proceeds thus :

$$R—OH + CH_2 \overset{}{\underset{O}{\diagdown\!\!\diagup}} CH_2 \rightarrow R—O—CH_2—CH_2OH$$

Reaction of one molecule of ethylene oxide is not sufficient to produce a water-soluble detergent product and the reaction is continued :

$$R—O—CH_2—CH_2OH + n(CH_2 \overset{}{\underset{O}{\diagdown\!\!\diagup}} CH_2) \rightarrow$$

$$R—O—CH_2—CH_2—(OCH_2CH_2)_{n-1}OCH_2CH_2OH$$

It will be noted that an ether linkage is obtained. Since this linkage is basically a strong one, the material is not subject to hydrolysis and, as there is no possibility of ionisation, it is not affected by metallic ions. In general for an average hydrophobe when n is approximately 6, the product becomes water-soluble and starts assuming detergent qualities.

The above reaction can be continued indefinitely but, in practice, optimum detergency is found in the range of 10–15 molecules of ethylene oxide per molecule of fatty alcohol. The source of the alcohol varies greatly, but natural, ZIEGLER and OXO alcohols of various molecular weights are being used and up to 50 molecules of ethylene oxide are being added.

Although chemically very different from fatty alcohols, the alkyl phenols behave in exactly the same manner as the fatty alcohols. Here again the choice of one particular alkyl phenol (or naphthol) is quite wide, but the preference usually falls on nonyl (sometimes octyl) phenol. With 4 molecules of ethylene oxide, nonyl phenol is still completely insoluble in water; at 6–7 molecules it is completely soluble in water at room temperature; at 8–12 molecules it exhibits excellent detergency.

Two further variations are now also being used, dodecyl phenol ethoxylate for certain agricultural emulsifiers and dinonyl phenol ethoxylate (where problems of non-ionic biodegradation have not yet arisen) as low or non-foaming ingredients of household washing machine powders.

Up to now, the alkyl phenol ethylene oxide condensates have been the most widely produced non-ionic detergent. However, with the recent international preoccupation with biodegradation they have fallen into slight disfavour and fatty alcohol ethylene oxides condensates are gaining. The situation is not completely clear because of conflicting claims as to the biodegradability of the alkyl phenols (which by virtue of their method of manufacture are branched chain), largely because there is not at present an internationally accepted method for chemical determination of non-ionics in micro quantities.

These ethylene oxide adducts are solubilised by the ethylene oxide units forming hydrates with water. Compared with all other materials which dissolve in water, these products show an apparent anomaly: their solutions are completely clear in the cold, but when heated they become turbid and, if the temperature is raised sufficiently, separation into two phases can even take place. This is explained by the fact that at an elevated temperature the hydrates are destroyed; so much so that this temperature is a function of the amount of ethylene oxide molecules present in the molecule.

It should again be stressed that, although the number of molecules of ethylene oxide present is considered a basic property of the material, this number is only an average, and if, say, there are 10 molecules stated to be present, the actual molecular configuration can vary from 4 to 15.

With the advent of polyester fibres for use in clothing and bedding, the use of a non-ionic constituent in laundering increased tremendously as it was found non-ionics will remove soil from these fibres better than the hitherto used anionics, but most laundry powders or liquids use a combination of the two.

With the latest trend to 'cold water' washing a further discovery has been made. As stated above the ethylene oxide distribution follows a Poisson curve around the mean. It has now been found that if 'peaked' ethoxylates are used the performance in cold water is improved.

Normal ethoxylation is done with an alkaline catalyst, usually KOH, sometimes K_2CO_3 or sodium methylate. These give the normal distribution of condensations. Kravetz has indicated that these peaked narrow range ethoxylates are produced commercially in two ways. The first way is to distil off the more volatile elements of the condensate, in which case if the alcohols are the base material the unethoxylated alcohol and the 1 or 2 mol products are 'topped'. The topped portion being returned for further processing. The distribution of a normal and topped ethoxylate is indicated in Fig. 3.2.

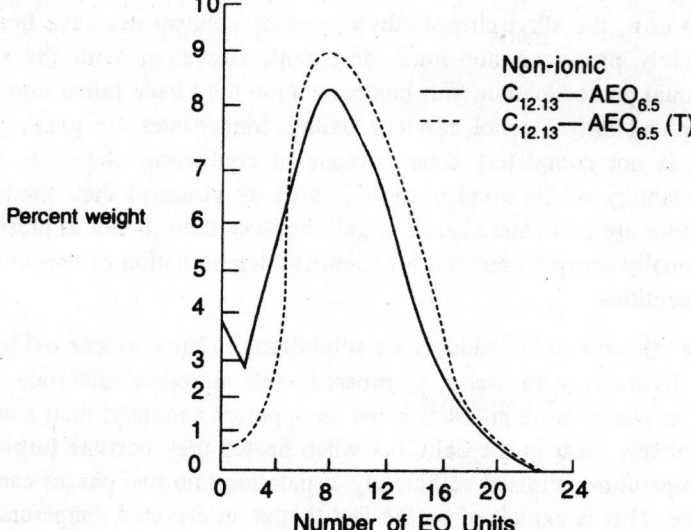

Fig. 3.2. EOD of conventional versus narrow range ethoxylate.

Fatty acid condensates

Another type of non-ionic can be prepared by reacting fatty acids with ethylene oxide. This is the same type of reaction :

$$R—COOH + n(CH_2OCH_2) \rightarrow$$

$$R—\overset{\overset{O}{\displaystyle\|}}{C}—O—CH_2—CH_2—(OCH_2CH_2)_{\overline{n-2}}OCH_2—CH_2OH$$

A second method of manufacture is by the esterification of a fatty acid with a polyethylene glycol, $HOCH_2(CH_2OCH_2)_nCH_2OH$. The difference in these two methods is that the polyethylene glycol has two —OH groups, so that unless very specific catalytic control is taken, non-ionic esters made by esterification will always have a certain proportion of di-esters. This is not necessarily a disadvantage, but makes the esterification ester different from the ethoxylation ester.

Being esters, these materials have one drawback that non-ionics in general do not suffer from. Non-ionic detergents are not in general affected by metallic ions, acids or alkalis and their functions as detergents are only marred by the way these extraneous ions in the water affect the physical properties of the water itself. However, the fatty acid ethylene oxide condensates are readily hydrolysed by acids or alkaline solutions to their component fatty acid and to polyethylene glycol. Thus in strong alkaline solutions these materials are no better than the soap of the fatty acid from which they were made, and in strong acid solution they are ineffectual. However, as components in household detergent powders, they do perform quite well.

Condensation of ethylene oxide with an amine

Two reaction products are possible :

$$R—NH_2 + n(CH_2OCH_2) \rightarrow$$

$$R—NH—CH_2—CH_2(OCH_2CH_2)_{\overline{n-2}}OCH_2—CH_2OH$$

or

$$R—NH_2 + 2n(CH_2OCH_2) \rightarrow$$

$$R—N\Big\langle\begin{array}{l} CH_2CH_2(OCH_2CH_2)_{\overline{n-2}}OCH_2—CH_2OH \\[2mm] CH_2—CH_2(OCH_2CH_2)_{\overline{n-2}}OCH_2—CH_2OH \end{array}$$

In the case of the second reaction, the end product is shown as being symmetrical, but it need not be, and can again be a mixture of many homologues and isomers. Other than specialised items, this class of non-ionic has not been used to a large extent in cleaning compositions. It should be noted, however, that these materials in acid solution can exhibit cationic characteristics, whereas in neutral or alkaline solutions they are non-ionic.

Condensation of ethylene oxide with an amide

We shall later describe the alkylolamide type of products. These are basically amides of an alkylolamine and a fatty acid. Certain members of this class exhibit detergency and others do not. To convert the non-detergent materials into detergents, they can be condensed with ethylene oxide :

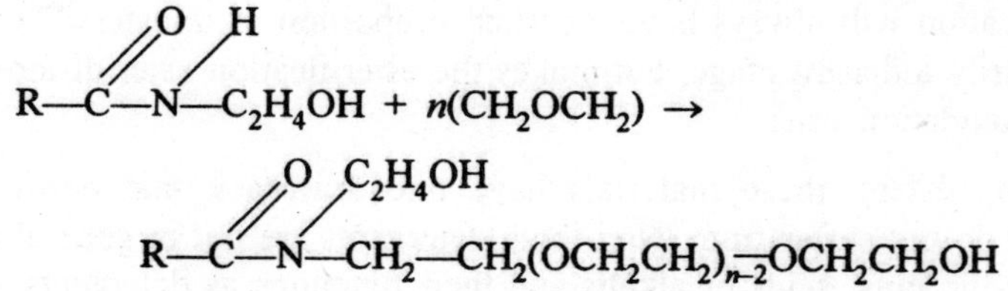

$$R{-}\overset{\displaystyle O}{\overset{\|}{C}}{-}\overset{\displaystyle H}{N}{-}C_2H_4OH + n(CH_2OCH_2) \rightarrow$$

$$R{-}\overset{\displaystyle O}{\overset{\|}{C}}{-}\overset{\displaystyle C_2H_4OH}{N}{-}CH_2{-}CH_2(OCH_2CH_2)_{n-2}OCH_2CH_2OH$$

or

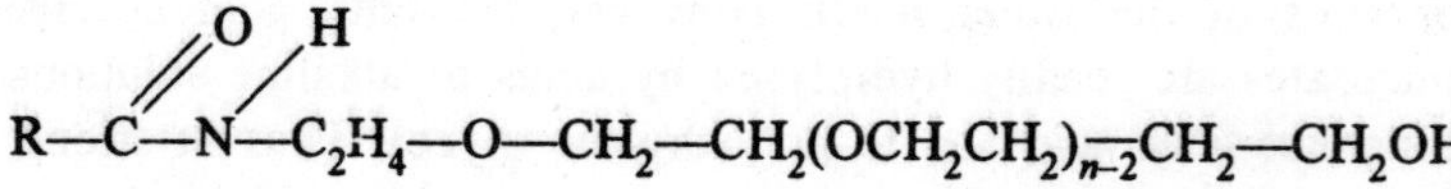

$$R{-}\overset{\displaystyle O}{\overset{\|}{C}}{-}\overset{\displaystyle H}{N}{-}C_2H_4{-}O{-}CH_2{-}CH_2(OCH_2CH_2)_{n-2}CH_2{-}CH_2OH$$

or

$$R{-}\overset{\displaystyle O}{\overset{\|}{C}}{-}\overset{\displaystyle H}{N}{-}C_2H_4OH + 2n(CH_2OCH_2) \rightarrow$$

$$R{-}\overset{\displaystyle O}{\overset{\|}{C}}{-}N\begin{array}{l} C_2H_4{-}O{-}CH_2{-}CH_2{-}(OCH_2CH_2)_{n-2}OCH_2{-}CH_2OH \\ \\ CH_2{-}CH_2{-}(OCH_2CH_2)_{n-2}OCH_2{-}CH_2OH \end{array}$$

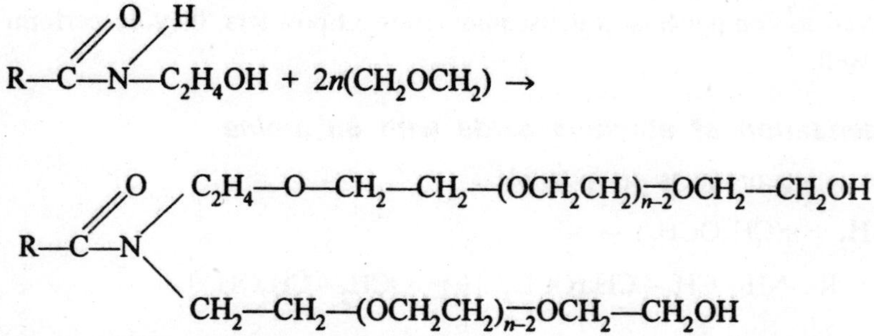

Again, the last reaction product is shown to have the ethylene oxide equally divided between the two possible directions, but there is no absolute reason for this. The ethylene oxide is divided at random in the two directions; also, any amount of either of the first two reaction products may be present. The alkylolamines are themselves complex mixtures and not as simple as shown above, so that an alkylolamide condensed with ethylene oxide consists of many components.

Block polymers

As a general rule glycols are water-soluble. Polyethylene glycols (produced by the ethoxylation of ethylene glycol) are also completely water-soluble, the lower molecular weight members of the class are described as infinitely soluble and at a molecular weight of 10,000 specifications state the solubility to be of the order of 50 per cent.

Polypropylene glycol behaves somewhat differently. The glycol itself and the lower condensation products are miscible in all proportions with water. Thus at molecular weight 400 the solubility is still infinite. Above 400 it starts falling off; polypropylene glycol of molecular weight 900 has a solubility in water of 1 per cent and this becomes even lower as the molecular weight rises.

These high-molecular-weight polypropylene glycols can be classed and serve as hydrophobes (they are soluble in or mix with oils, aromatic and paraffinic solvents) and in addition have two terminal —OH groups :

$$HO—\overset{\displaystyle CH_3}{\underset{\displaystyle |}{CH}}—CH_2—O—CH_2—\overset{\displaystyle |}{\underset{\displaystyle CH_3}{CH}}—O—\;\;...$$

$$...O—CH—CH_2—O—\overset{\displaystyle CH_3}{\underset{\displaystyle |}{CH}}—OH$$
$$\underset{\displaystyle CH_3}{|}$$

Ethylene oxide, added to this molecule, will attach itself at random to either —OH group and then again at random on to the new —OH groups formed. The ethylene oxide chains serve as the hydrophilic portion. Surface activity is attained in the resultant product, called a block polymer, when the ethylene oxide added has reached a definite ratio to the starting hydrophobe.

Block polymers can be and are made with a wide range of hydrophobic/ hydrophilic ratios and molecular weights to give surfactants 'tailor-made' to specific physico-chemical requirements.

Several modifications of the basic block have been developed. One is the 'inverse' type in which the propylene oxide is condensed on to the preformed polyethylene glycol. Here the hydrophilic portion is in the centre of the molecule with hydrophobic tails emerging.

All these materials need an initiator—a core on to which the initial charge of either propylene or ethylene oxide can be attached. In the two examples given above the initiators can be propylene or dipropylene glycol or ethylene or diethylene glycol, all four of which have two reactive —OH groups. Initiators with more reactive hydrogens are also used. On occasion these initiators are condensed with formaldehyde to give a bigger, more compact core with several hydrogen nuclei. One of the more interesting products uses ethylene diamine to which four molecules of propylene (sometimes ethylene) oxide are condensed :

$$H_2N-CH_2-CH_2-NH_2 + 4CH_2-CH-CH_3 \rightarrow$$
$$O$$

$$(HO-CH-CH_2)_2-N-CH_2-CH_2-N-(CH_2-CH-OH)_2$$
$$\quad\quad\; CH_3 \quad\quad\quad\quad\quad\quad\quad\quad\quad\quad\quad\quad\quad\quad CH_3$$

This serves as a base for further growth reactions with propylene oxide to form the hydrophobe and then ethylene oxide to make the surfactant. The molecule has four hydrophilic tails spreading out from the hydrophobe in the shape of an X.

Further variations are possible. In one the propylene and ethylene oxides are pre-mixed before condensation. The reactivity of ethylene oxide is nine times that of propylene oxide, but for all that random distribution of the two is obtained. In another, propylene oxide and ethylene oxide are added alternately to the chains.

It can be seen that the possible permutations and combinations are enormous to give an extremely wide spectrum of properties. Block polymers are manufactured for use in a large variety of processes and products from defoamers and de-emulsifiers through wetting agents, emulsifiers and cosmetics to laundry detergents. One particular use where they have proved very successful is in automatic dishwashing powders where foam is unacceptable.

Sucrose esters

These are products which have had their ups and downs in the detergent world. A future is being predicted for sucrose esters, particularly when production costs can be brought into line with other detergent raw materials.

Production is based on the original Foster D. Snell method which reacts sucrose with a fatty acid ester (methyl stearate) in dimethyl formamide solution. Mono- or di-esters can be formed, depending on the original ratio of reactants. The mono-esters are suited to detergent use.

Recently a new technique has been developed which works with a large excess of sugar, and uses potassium carbonate as a catalyst, operating at 90°C at a pressure between 9.3 and 13.3 kPa. Unreacted sugar is precipitated from the reaction solution by the addition of toluene.

Recently a new method has been developed in which the sucrose is dissolved in propylene glycol and the methyl ester emulsified into this solution. The glycol is distilled off and in the distillation process transesterification takes place. This process eliminates the use of expensive and poisonous solvents used hitherto.

Sorbitan esters

Sorbitol and mannitol (its stereo-isomer), both hexahydric alcohols, can react with fatty acids (usually in the form of fatty acyl chlorides) to form the mono-, di-, tri-, ..., hexa- esters, but these products have not attained great commercial importance.

When sorbitol (or mannitol) is heated with a maximum of four molecules of a fatty acid to a temperature of close to 200°C in the presence of an acid catalyst, esters and an internal ether are formed. In the case of the mono-ester the product could be :

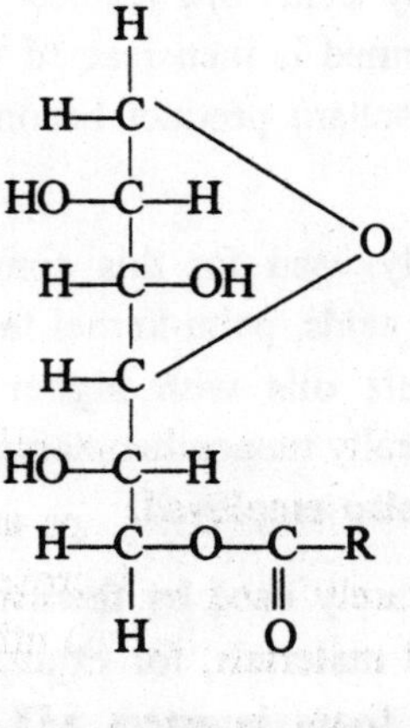

called sorbitan ester. These are water-insoluble, oil-soluble and useful as water-in-oil emulsifiers in the food, pharmaceutical and cosmetic industries. Except for sorbitol hexa-ester or sorbitan tetra-ester they all have at least one reactive hydrogen derived from an —OH group. Ethylene oxide is coupled to these hydrogens to make the material more hydrophilic and in the ultimate water-soluble.

Ethoxylated sorbitan esters are available in varying degrees of esterification, ethoxylation and with different fatty acid tails. They are valuable emulsifiers for cosmetics and polishes, in the food industry, and for specialised treating purposes in the paper, textile and plastics industries. As emulsifiers they behave differently from the conventional non-ionics because of the distribution of the hydrophilic chains. Very rarely is the ethylene oxide attached to one —OH group only. The molecule has thus several (relatively) short ethylene oxide chains rather than one long one, making it more compact.

Alkylolamides

Another group of synthetic detergents which falls into the non-ionic category is the fatty acid alkylolamides. These are made by reacting fatty acids with alkylolamines :

$$R—\overset{\displaystyle O}{\overset{\|}{C}}—OH + H—N\overset{\displaystyle H}{\diagup}C_2H_4OH \rightarrow R—\overset{\displaystyle O}{\overset{\|}{C}}—N\overset{\displaystyle H}{\diagup}C_2H_4OH + H_2O$$

The reaction outlined above is between a fatty acid and a monoethanol-amine. This has no detergent properties nor is it soluble to any great extent in water. The uses of this type of compound will be dealt with below.

However, if certain fatty acids are reacted with a di-alkylolamine as above, and if the amide formed is then reacted with a further molecule of the di-alkylolamine, the resultant product becomes water-soluble and has detergent properties.

The materials commonly used for this condensation are lauric acid, myristic acid, coconut fatty acids, palm-kernel fatty acids, and occasionally the fatty acids of vegetable oils with higher molecular weights. The alkylolamines used are generally monoethanolamine and diethanolamine, but the isopropanolamines are also employed.

The alkylolamides are rarely used by themselves as detergents, but as additives to other detergent materials, for example, with dodecyl benzene sulphonates. They act as foam boosters and increase the detergency synergistically, and in liquid products they increase viscosity without increasing the cloud point. They also have the valuable property of acting as skin-protecting agents: whereas detergents in general tend to de-fat the skin, the emollient effect of these alkylolamides helps to overcome this tendency.

A further use of alkanolamides is as an intermediate for the manufacture of sulphosuccinates. The preference is for the monoethanolamide or isopropanolamide of coconut fatty acid.

The thickening effect of alkanolamides on detergents has been demonstrated to be dependent on inorganic salts present as an effect of the inorganic salt on the micelles. Salts need not necessarily be added, they can be present as the neutralisation products of the excess sulphonating agent.

More recently monoethanolamides, especially those derived by reacting a fatty acid methyl ester with monoethanolamine, are further condensed with ethylene oxide. Depending on the type of fatty acid methyl ester and the number of molecules of ethylene oxide, non-ionic surfactants with excellent detergency power may be obtained.

Table 3.1 shows the change in viscosity of a 12 per cent neutralised dodecyl benzene sulphonate solution with the addition of 2 per cent alkylolamide and the foam height of a 0.1 per cent active solution in hard water of 300 ppm at 45°C.

Table 3.1. Effect of the addition of alkylolamides to dodecyl benzene sulphonate solution.

Alkylolamide	Viscosity at 25°C (Pa s)	Foam height	
		Immediate	After 5 minutes
None	0.075	16	15
Coconut monoethanolamide	0.305	18	17
Coconut monoisopropanolamide	0.305	16.5	16
Coconut diglycolamide	0.960	18.5	17
Coconut diethanolamide	0.650	17.5	16
Oleic diethanolamide	0.840	17	17
Coconut di-isopropanolamide	0.360	17	16

Table 3.2. Effect of the addition of ethanolamide on the foaming power of various detergents.

Type of detergent	Concentration of detergent (%)	Detergent alone		Detergent with ethanolamide*	
		Foam height at 45°C in hard water			
		Imme-diate (cm)	After 5 minutes (cm)	Imme-diate (cm)	After 5 minutes (cm)
Lauryl alcohol triethanolamine salt	0.02	3.8	1.5	11	9
Lauryl alcohol monoethanolamine salt	0.02	3.0	1.2	10.5	10
Lauryl ether sulphate	0.02	16.0	15.0	15.0	14.5
Lauryl alcohol sodium salt	0.02	5.5	5.0	13.5	12.5
C_{10-14} alcohol sulphate sodium salt	0.02	2.0	1.0	4.0	3.5
Octyl phenol, 10 mols ethylene oxide	0.12	14.0	9.0	17.0	15.5
Nonyl phenol, 10 mols ethylene oxide	0.12	13.5	12.5	15.5	14.5
Octyl phenol, 14 mols ethylene oxide	0.12	15.5	14.5	15.5	15.0
Nonyl phenol, 14 mols ethylene oxide	0.12	16.0	14.5	14.5	14.0
Low foam non-ionic	0.12	1.5	1.5	2.0	2.0

* The ethanolamide was added at the rate of 20 per cent of the active matter.

Table 3.2 indicates the increase in foaming power of various detergents at low concentrations when coconut diethanolamide is added. It will be noted that the effect is very marked for alcohol sulphates, the ether sulphates are hardly affected, and the effect on non-ionic detergents varies.

Fatty Amine Oxides

A further type of non-ionic detergent is the group known as fatty amine oxides. These again offer many different possibilities, but basically they are made by treating tertiary amines with hydrogen peroxide. For our purposes, those amines which contain at least one long-chain group, usually from a fat source, are detergents. Two typical reactions are shown :

$$C_{12}H_{25}(CH_3)_2N + H_2O_2 \rightarrow C_{12}H_{25}-\overset{\overset{\textstyle CH_3}{|}}{\underset{\underset{\textstyle CH_3}{|}}{N}}\rightarrow O + H_2O$$

Again these two representative types include the dodecyl grouping.

These amine oxides are in themselves good detergents, but their main use is as foam boosters, viscosity increasers, and skin-protecting agents in liquid detergents. They are sticky materials and are not suggested for incorporation into powdered materials.

Their foam-boosting ability has been compared with lauric diethanolamide in Table 3.3. The method was to wash soiled plates in water of 125 ppm hardness at 46–49°C till the foam disappeared, with a 0.0125 per cent solution of the following formulation :

Sodium linear alkyl benzene sulphonate	18
Sodium lauryl ether sulphate	12
Foam stabiliser	x
Water	to 100

From the table, it appears that the fatty amine oxides are as effective as the alkylolamides as foam boosters, but in smaller concentrations.

Table 3.3. Comparison of foam-boosting ability between amine oxides and lauric diethanolamide.

Foam stabiliser	x	Plates washed to foam end-point
Dimethyl cocoamide oxide	0.48	26
Dimethyl hydrogenated tallowamide oxide	0.48	26
Dimethyl hexadecylamine oxide	0.48	27
Bis (2-hydroxyethyl) stearylamine oxide	0.60	20
Lauric diethanolamide	1.5	25

Amphoterics and Zwitterionics

This class of material contains in the same molecule both an anionic and a cationic group. The amphoterics are usually manufactured so that the anionic (carboxylate or sulphate/sulphonate) group is neutralised by Na and the cationic (quaternary ammonium) has associated with it, depending on the pH, an hydroxyl or chloride anion.

Zwitterionics also contain the same groups but the positive electric charge is neutralised by the negative charge on the same or an adjacent molecule. As will be explained below amphoterics can become zwitterionics under certain conditions of pH.

The chemistry of the syntheses is smothered in trade secrets. One route to the manufacture of amphoterics is the condensation of a fatty acid with a substituted ethylene diamine :

$$R\!-\!COOH + H_2N\!-\!CH_2\!-\!CH_2\!-\!NH\!-\!CH_2CH_2OH \longrightarrow$$

fatty acid   aminoethylethanolamine

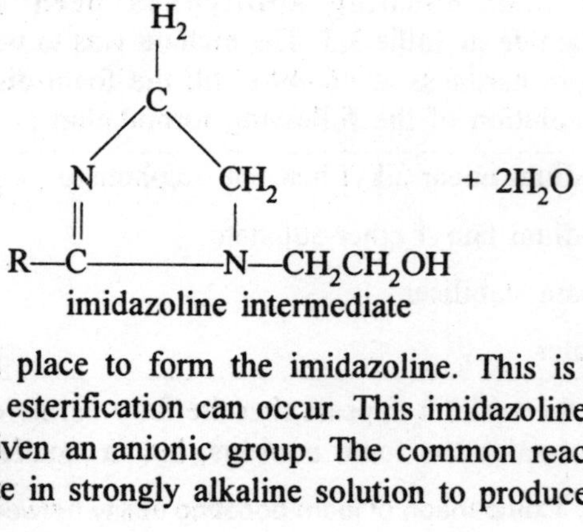

imidazoline intermediate

Cyclosation has taken place to form the imidazoline. This is not the complete picture as some esterification can occur. This imidazoline is now both quaternarised and given an anionic group. The common reactant for this is sodium chloracetate in strongly alkaline solution to produce :

either

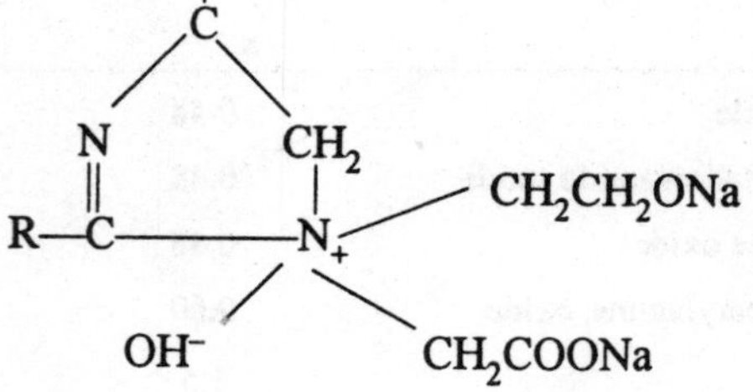

or

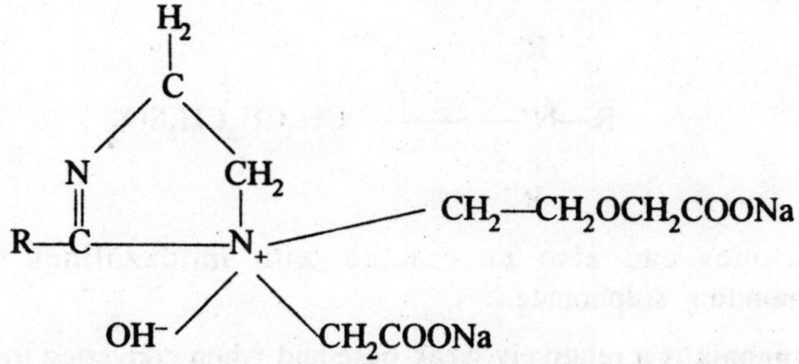

$$
\begin{array}{c}
\overset{\displaystyle H_2}{C} \\
\diagup \quad \diagdown \\
N \qquad CH_2 \\
\| \qquad \quad | \\
R-C-\!\!\!-\!\!\!-N^+-\!\!\!-\!\!\!- CH_2-CH_2OCH_2COONa \\
\diagup \qquad \diagdown \\
OH^- \qquad CH_2COONa
\end{array}
$$

It will be noted that in the first example the caustic soda reacts with the — OH group to produce an alcoholate (in addition to the quaternarisation) and in the second example further chloracetate has reacted with this group to give an ether linkage and a dicarboxylate. In both instances the quaternary N will be associated with an hydroxyl ion unless the pH is lowered.

Variations of the above are possible, mainly in the choice of the fatty acid, but an interesting product on the market has a fatty alcohol sulphate anion to neutralise the positive charge on the quaternary N:

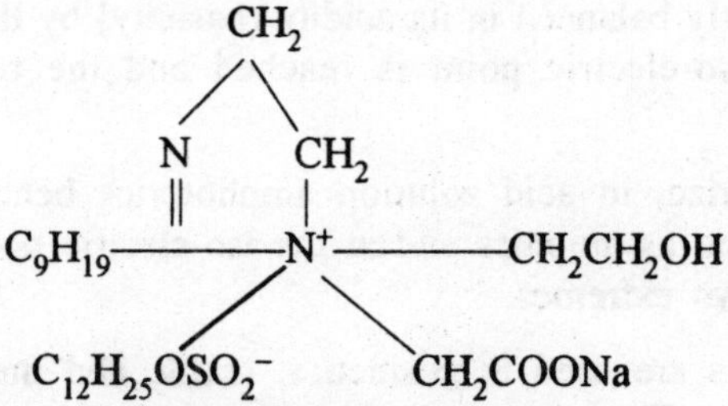

$$
\begin{array}{c}
CH_2 \\
\diagup \quad \diagdown \\
N \qquad CH_2 \\
\| \qquad \quad | \\
C_9H_{19}-\!\!\!-\!\!\!-\!\!\!-N^+-\!\!\!-\!\!\!-\!\!\!- CH_2CH_2OH \\
\diagup \qquad \diagdown \\
C_{12}H_{25}OSO_2^- \qquad CH_2COONa
\end{array}
$$

One method of manufacturing zwitterions is to react a tertiary fatty amine with either chloracetic acid to produce the betaine

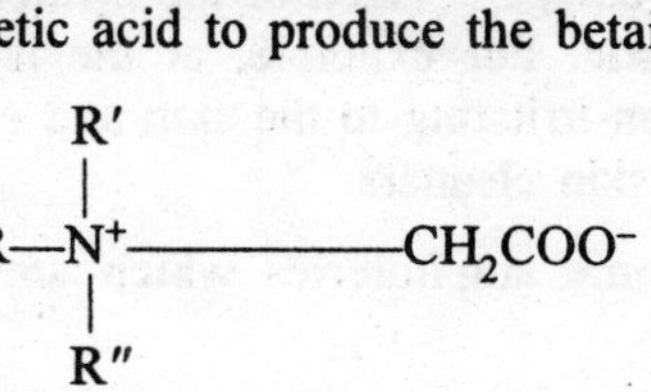

$$
\begin{array}{c}
R' \\
| \\
R-N^+-\!\!\!-\!\!\!-\!\!\!- CH_2COO^- \\
| \\
R''
\end{array}
$$

or with propane sultone

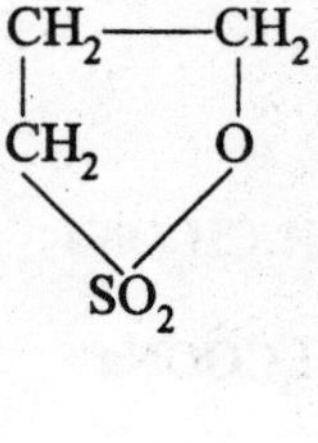

$$
\begin{array}{c}
CH_2-\!\!\!-\!\!\!- CH_2 \\
| \qquad \quad | \\
CH_2 \qquad O \\
\diagdown \quad \diagup \\
SO_2
\end{array}
$$

to form the sulphobetaine :

$$R-\overset{\overset{\displaystyle R'}{|}}{\underset{\underset{\displaystyle R''}{|}}{N^+}}\text{———}CH_2CH_2CH_2SO_3^-$$

Sultones can also be reacted with imidazolines to form the corresponding sulphonate.

Ammonia is a relatively weak base and when converted to an amine the basicity is weakened further, a primary amine being however a stronger base than a tertiary. Also the longer the chain on the N nucleus, the weaker the basic quality of the amine.

Carboxylic acids are weak acids, the higher the molecular weight, the weaker the acid. Sulphonic acids and sulphuric esters are strong acids, comparable with mineral acids.

If the pH of the amphoteric is altered, either the anionic or cationic portion is partially neutralised. When the residual anionic (or cationic) radical is exactly balanced in its acidity (basicity) by the cationic (anionic) portion, the iso-electric point is reached and the resultant becomes a zwitterion.

To summarize, in acid solution amphoterics behave as cationics, in alkaline solution as anionics and at the iso-electric point the behaviour is between the two extremes.

Amphoterics are used in cosmetics, textile and metal cleaners and as fabric softeners. They have excellent compatibility with inorganic salts, acids and bases. Advantage is taken of the zwitterion/pH effect to promote a desired characteristic. For example, at the iso-electric point they are almost completely non-irritating to the skin and eyes, indicating the use in baby shampoos and skin cleaners.

We list below some amphoterics which are available on the world markets :

$$\text{C}_{11}\text{H}_{23}-\text{C}\overset{\displaystyle \text{N}=\overset{\text{CH}_2}{\underset{}{}}\hspace{-1em}}{\underset{\text{OH}^-}{\overset{}{}}}-\overset{\displaystyle \text{CH}_2}{\underset{\text{CH}_2\text{COONa}}{\text{N}^+}}-\text{CH}_2-\text{CH}_2-\text{OCH}_2\text{COONa}$$

$$\text{CH}_3(\text{CH}_2)_{10}\text{CH}_2-\underset{\text{H}}{\text{N}}-\text{CH}_2\text{CH}_2\text{COONa}$$

Derivatives of the basic betaine structure can be:

$$\text{R}-\overset{\displaystyle \text{O}}{\overset{\|}{\text{C}}}-\text{NH}-\text{CH}_2-\text{CH}_2-\text{CH}_2-\overset{\displaystyle \text{CH}_3}{\underset{\text{CH}_3}{\text{N}^+}}-\text{CH}_2-\text{COO}^-$$

Alkylamidopropylbetaine

$$\text{R}-\overset{\displaystyle \text{CH}_2-\text{CH}_2-\text{COOH}}{\underset{\text{CH}_2-\text{CH}_2\text{COO}^-}{\text{N}^+}}$$

Alkylimidodipropionate

$$\text{R}-\overset{\displaystyle \text{O}}{\overset{\|}{\text{C}}}-\text{NH}-\text{CH}_2-\text{CH}_2-\overset{\displaystyle \text{H}}{\underset{\text{CH}_2-\text{CH}_2-\text{OH}}{\text{N}^+}}-\text{CH}_2-\text{COO}^-$$

Hydroxyethyl-alkylamidoethylglycinate

The chain length of the R group being of course dependent on the fatty amine or acid used in the synthesis.

The optimum detergency is obtained with only one long-chain component and at a pH of 7. At a pH of 9–10 there is a considerable drop in detergency. In acid solutions these amphoteric compounds are mainly cationic and in alkaline solutions they behave as anionic detergents. Depending on the type of 'cationic' radical being used they may be incompatible with anionic detergents in acid solution but they are all compatible with cationic and non-ionic detergents.

Some of the amphoterics give very excellent foam, which makes them suitable for foam cleaners which are to work at a neutral pH, such as hair shampoos and rug shampoos.

A very useful property of the amphoterics is that they are stable in highly acid solutions, and have thus found use in acid cleaners based on hydrofluoric acid.

BIODEGRADABILITY

With legislation or agreements having appeared or appearing in many countries, it is important for the chemist to know whether a material he is using or intends using is biodegradable. A test for biodegradability is usually beyond the means of even a well-equipped laboratory as found in most detergent factories, and reliance must be placed on accepted principles or manufacturers' advice. The degree of biodegradability is also to a large extent dependent on the method of test.

As a general rule we can, however, state that it has been found that straight-chain materials are biodegradable. Hence soaps, linear alkyl benzene sulphonates, olefin sulphonates, alkane sulphonates, fatty alcohol and ALFOL sulphates, ether sulphates, sucrose esters, alcohol phosphates and fatty alkanolamides are all sufficiently biodegradable for all standards laid down by legislation or agreements. The position of non-ionic detergents of the ethylene oxide condensate type is obscure. Various claims for the biodegradability of alkyl phenol ethylene oxide condensates have been made but not substantiated. It appears that fatty alcohol and *Ziegler ethoxylates* are biodegradable but this falls somewhat with increasing amounts of ethylene oxide. Some OXO alcohol sulphates and ethoxylates appear to be biodegradable but this again is dependent on the degree of branching.

Chapter 4

Oils, Fats and Their Pretreatment

INTRODUCTION

Oils and fats, belong to the class of compounds called lipids, which are insoluble in water, but soluble in ether and other organic solvents. The lipids have long been considered important as one of the three major classes of food materials. They differ from the other two, proteins and carbohydrates, in that they provide more than twice the number of calories per unit weight when converted to carbon dioxide and water. Oils and fats are usually distinguished from one another by their melting points and, to some extent, by their sources. Oils are generally liquid at ordinary temperatures while fats are solid or semisolid. The fats are usually of animal origin while oils are extracted from plant tissue, fish, or marine animals. Both fats and oils are composed largely of glycerol esters of fatty acids.

THE FATTY ACIDS

The soapmaking oils and fats are essentially glycerides, that is fatty acid esters of the trihydric alcohol glycerol. Recent investigations, have shown that natural oils and fats may contain a wide variety of fatty acids, including some with odd numbers of carbon atoms, branched chains, ketonic groups, or ring structures. Apart from ricinoleic acid, an hydroxy acid which is the major constituent of castor oil, it can be assumed that the important fatty acids all have a straight, unsubstituted, hydrocarbon chain with an even number of carbon atoms, including the one in the carboxyl group. That is, the alkyl groups contain an odd number of carbon atoms. The fatty acids can be divided into several groups. The first series comprises the saturated acids C_nH_{2n+1}–COOH; and members with various chain lengths are listed in Table 4.1. The other series are unsaturated; that is, they contain less hydrogen than the saturated series and what are known as double bonds. The unsaturated acids have lower melting points than the saturated acids with the same numbers of carbon atoms. The main groups of unsaturated fatty acids are :

(i) The mono-unsaturated series, C_nH_{2n-1}–COOH, with one double bond.

(ii) The di-unsaturated series, C_nH_{2n-3}–COOH, with two double bonds.

(iii) The tri-unsaturated series, C_nH_{2n-5}–COOH, with three double bonds.

(iv) Acids with four, or more, double bonds.

Table 4.1. The saturated fatty acids C_nH_{2n+1}–COOH.

Number of carbon atoms	Popular name	Melting point, °C
4 (i.e., $n = 3$)	butyric	–8
6	caproic	–3
8	caprylic	17
10	capric	32
12	lauric	44
14	myristic	54
16	palmitic	63
18	stearic	70
20	arachidic	75
22	behenic	80
24	lignoceric	84

Two types of isomers (that is, compounds with the same empirical formula, such as $C_xH_yO_z$, but with different structures and hence different properties) exist amongst the unsaturated fatty acids :

(i) Positional isomers, in which the difference is in the position of the double bond, or bonds, in the carbon chain.

(ii) Geometric isomers.

The two carbon atoms, which form the double bonds, each have attached to them a hydrogen atom and a carbon chain. The double bond prevents free rotation; and if the like groups are attached on the same side of the molecule the compound is called the *cis*-isomer; if they are on the opposite sides it is the *trans*-isomer. For example, with the very important $C_{17}H_{33}COOH$ acids, possible isomers include :

$$CH_3-(CH_2)_7-CH$$
$$\parallel$$
$$HOOC-(CH_2)_7-CH$$
oleic acid (*cis* form)
mp 15°C

$$CH_3-(CH_2)_7-CH$$
$$\parallel$$
$$HC-(CH_2)_7-COOH$$
iso-oleic, or elaidic acid
(*trans* form), mp 44°C

Many other isomers are possible with the double bond in other positions in the chain.

Natural acids are usually, but not invariably, *cis*. The higher melting *trans* forms are produced when fats are hydrogenated, or otherwise heated, particularly in the presence of certain catalysts.

It will be appreciated that with variations in chain length, the numbers and positions of the double bonds, and *cis/trans*-isomerism at each double bond, there are a very large number of possible unsaturated fatty acids.

A systematic way of identifying a fatty acid is :

(i) the number of carbon atoms is indicated by a Greek prefix, for example dodeca = 12, tetradeca = 14, hexadeca = 16, and octadeca = 18..

(ii) the name of a saturated acid ends in -anoic, so that stearic acid is octadecanoic acid.

(iii) the number of double bonds is indicated by a suffix, so that the name of a fatty acid with one double bond ends in -enoic, one with two double bonds in -dienoic, one with three double bonds in -trienoic, and so on.

(iv) the position of a double bond is given by the number of the first carbon in the double bond, starting from the carbon in the carboxyl group (or, alternatively, by the numbers of the carbon atoms between which the double bond occurs).

Thus, oleic acid is,

$$CH_3(CH_2)_7-CH=CH-CH_2-CH_2-CH_2-CH_2-CH_2-CH_2-CH_2-COOH$$

$$\quad\quad\quad\quad\quad 10\quad 9\quad\; 8\quad\;\; 7\quad\quad\quad 6\quad\;\; 5\quad\;\; 4\quad\quad 3$$
$$\quad 2\quad\;\; 1$$

cis 9-octadecenoic acid, or *cis* -9, 10-octadecenoic acid. The position of a substituent group is numbered similarly.

A few of the more important unsaturated fatty acids are listed in Table 4.2.

The oils from fishes and marine mammals, such as whales and seals, contain fatty acids with a wide range of chain lengths, some of which include three to six double bonds.

Table 4.2. Some important unsaturated fatty acids.

Formula	Scientific name	Popular name	Melting point °C
One double bond			
$C_{15}H_{29}COOH$	cis-9-hexadecenoic	palmitoleic	–
$C_{17}H_{33}COOH$	cis-9-octadecenoic	oleic	15
$C_{17}H_{33}COOH$	trans-9-octadecenoic	elaidic or iso-oleic	44
$C_{19}H_{37}COOH$	cis-eicosenoic	–	–
$C_{21}H_{41}COOH$	cis-13-docosenoic	erucic	34
$CH_3—(CH_2)_5—CH—CH_2—CH=CH—(CH_2)_7COOH$	12-hydroxy-cis-9-octadenoic	ricinoleic	–
Two double bonds			
$C_{17}H_{31}COOH$	cis,cis-9,12-octadecadienoic	linoleic	–5
Three double bonds			
$C_{17}H_{29}COOH$	cis,cis,cis-9,12.15-octadecatrienoic	linolenic	–15

In polyunsaturated fatty acids, a special arrangement of the double bonds, known as a conjugated system, —CH=CH—CH=CH—, is possible. This gives oils such as tung, which contains eleostearic acid, 9, 11, 13-octadecatrienoic acid (an isomer of linolenic acid), specially strong drying properties; that is they can form hard paint, or varnish, films. Such oils are unsuitable for soap.

THE FATTY ACID COMPOSITIONS OF OILS AND FATS

One fat (or oil) can differ from another in several ways, but the most fundamental is in fatty acid composition. Oils and fats are obtained from a wide range of animal and vegetable sources; and it is to be noted that even those from nominally the same kind of source can vary considerably. In the case of animal fats, this can be related to the part of the animal from which the fat came, the food which the beast ate, and the particular breed of sheep or cattle. Similarly, vegetable oils vary with factors such as plant variety and climate. A striking example is the oil known as rape, or colza, obtained from the seeds of certain varieties of *brassica*. Traditionally, this oil contained some 50% of *erucic* (including also eicosenoic) acid, but varieties of seed giving what is sometimes called canbra oil has now been developed which contains only 0–5% erucic acid. Rape oil is widely used for edible purposes, but doubts have arisen regarding the advisability of ingesting much *erucic* acid; and the new types of rape oil meet this requirement. High erucic oils are still valuable for some industrial purposes.

Table 4.3 gives some approximate compositions of important oils and fats, grouped into categories often used by soapmakers. The *hard fats* contain some 45–55% of palmitic plus stearic acids and about 35–50% oleic plus palmitoleic acids. They give firm sodium soaps with excellent detergent properties, particularly at temperatures approaching the boil. Their sodium soaps are not very readily soluble, so that lather is obtained only slowly when a piece of the soap is rubbed in cool water.

The *lauric*, or *nut oils* are characterised by a very high proportion, about 70–80%, of C_{10}–C_{14} saturated acids, hence the name lauric oils. They give hard, freely soluble, sodium soaps.

The *soft oils* (liquid vegetable oils) contain low proportions of saturated acids and high proportions of unsaturated acids, particularly oleic and linoleic. They give sodium soaps which are softer and more freely soluble than those from the hard fats. It will be noted that there are appreciable variations in the compositions of the oils included in this class.

The oils classed here as *soft/drying oils* are more unsaturated than those in the soft class. They include appreciable proportions of acids containing

Table 4.3. Approximate fatty acid compositions of some oils and fats.

	Saturated acids					C_{20} and above	Unsaturated acids			Others
	C_{10} and below	C_{12}	C_{14}	C_{16}	C_{18}		oleic	C_{18} linoleic	linolenic	
Hard fats										
Beef tallow	–	1	5	27	16	1	42	3	–	5 palmitoleic
Mutton tallow	–	1	4	24	27	–	37	4	–	3 palmitoleic
Palm oil	–	–	1	44	5	–	40	10	–	3 palmitoleic
Chinese vegetable tallow	–	1	4	61	6	–	28	–	–	–
Lard	–	–	2	28	14	–	44	–	–	3 palmitoleic
Bone grease	–	–	–	–	–	–	–	–	–	1 C_{20} unsaturated
Goat tallow	–	4	2	26	28	2	38	–	–	–
Lauric or nut oils										
Coconut oil	13	48	19	8	3	–	8	1	–	–
Palm kernel oil	7	52	15	7	2	–	16	1	–	–
Babassu oil	11	45	16	7	5	–	14	2	–	–
Soft oils										
Olive oil	–	–	–	10	2	–	82	6	–	–
Groundnut oil	–	–	–	10	3	3	50	34	–	–
Cotton seed oil	–	–	–	25	2	–	21	51	1	–

(*Contd*)

	Saturated acids					C_{20} and above	Unsaturated acids			Others
	C_{10} and below	C_{12}	C_{14}	C_{16}	C_{18}		oleic	C_{18} linoleic	linolenic	
Maize (corn) oil	–	–	–	13	3	–	36	48	–	–
Soyabean oil	–	–	–	13	4	–	23	52	7	1
Sunflower oil	–	–	–	7	3	1	23	65	1	–
Sesame oil	–	–	–	9	4	–	45	42	–	–
Rapeseed I	–	–	–	3	1	1	15	15	10	55 erucic and similar
Rapeseed II	–	–	–	5	2	1	52	28	12	–
Soft/drying oils										
Linseed oil	–	–	–	6	4	–	22	16	52	–
Tung oil	–	–	–	4	1	–	8	4	3	80 eleostearic
Whale (baleen) oil	–	–	8	17	2	–(73% unsaturated C14-C22, 1–6 double bonds)				
Herring oil	–	–	5	18	2	–(75% various unsaturated)				
Menhaden oil	–	–	12	22	3	–(63% various unsaturated)				
Miscellaneous										
Castor oil	–	–	–	1	2	1	4	5	1	86 ricinoleic

three, or more, double bonds; and are not generally suitable for use in solid sodium soaps, unless the unsaturation is reduced by hydrogenation or, sometimes, polymerisation. Some kinds are used directly in soft soaps.

Rosin, or *colophony*, is not a glyceride, but it was at one time used extensively as a fat in the manufacture of soap. Various grades were available which differed in colour and odour. The better grades were gum rosins, the residue after turpentine has been distilled from the exudate from pine trees.

OTHER DIFFERENCES BETWEEN VARIOUS FATS

Fats can differ from one another in ways other than differences in fatty acid composition :

(i) In glyceride structure, that is the way the various fatty acids are combined with the three hydroxyl groups in the molecules of glycerol. This is of great importance in some uses of fats, but it is irrelevant when the fats are to be saponified, or hydrolysed, to soaps or fatty acids.

(ii) In the nature and proportions of the non-glyceride substances they contain. These ingredients can loosely be classified into three groups :

    (a) Inherent constituents, such as vitamins A, D and E (tocopherol); pigments such as carotene, gossypol and chlorophyll; and various sterols. Some of these which are significant to the soapmaker are pigments which need to be removed and natural antioxidants, including tocopherol.

    (b) Free fatty acids which develop when fats are stored under poor conditions, or extracted from materials which have been allowed to deteriorate. Examples are palm oil and rice bran oil. Palm oil produced under good conditions from plantation fruit should have an FFA (free fatty acid) content of 2–5%. Oil recovered from the fruit of wild palms by villagers may containers 20–30% of FFA, or even more. The oil in rice bran deteriorates seriously in even a few hours if steps are not taken to inactivate enzymes present in the bran.

    (c) Miscellaneous impurities which are found in all soapmaking fats, but particularly in lower grade materials, such as those recovered from wastes. Proteinaceous materials, and coloured and odorous constituents of many kinds, can be present as well as particulate dirt. Deterioration leading to development of a high FFA% commonly also results in the presence in the fat of brownish olouring matter and other undesirable impurities. A relatively odern problem results from the presence in tallows and greases of traces of polythene, which come from plastic film included in f a t t y materials collected from butchers and other sources.

Additional points in relation to the general quality of fats are discussed later.

SOME ANALYTICAL CHARACTERISTICS OF OILS AND FATS

General

Well equipped modern laboratories now usually characterise mixtures of fatty acids by GLC (gas liquid chromatography), or similar techniques, using methyl esters formed from the fat or soap. This gives composition in terms of the percentages of individual fatty acids, from which, if required, deductions can be made as to the blend of fats from which the fatty acids probably came. The older tests outlined in the following subsections are still used for various purposes, including specifications. Before the common availability of GLC, analysts deduced approximate fat compositions from these figures and from certain tests for specific oils or fats, but in a routine laboratory it was often only possible to reach fairly generalised conclusions.

Moisture and Dirt

For routine purposes, moisture can be determined by loss of weight at 105°C. This can conveniently be estimated by heating a sample in a flat porcelain dish over a small flame, using a very short thermometer to stir and to indicate the temperature. This gives total volatile matter; and moisture can be determined more accurately by the Karl Fischer method. Dirt is determined by filtration of a hot dry sample of the oil through a dry filter paper and weighing the residue after fatty matter has been removed with a solvent such as light petroleum (40-60°C boiling range).

Titre

Dry fatty acids are first prepared from the fat or soap. The liquid fatty acids are cooled under standardised conditions; and the titre of the sample is the highest temperature reached when a rise occurs due to the latent heat of solidification. It is, thus, a measure of the melting point of the mixture of fatty acids.

FFA (free fatty acid) Content, Acid Value, Saponification Value, and Ester Value

Estimation of the free fatty acid content of a fat can be expressed in terms of $Na_2O\%$; or it can be converted to percentage of a notional fatty acid. For many fats, FFA is quoted in terms of oleic acid which has an equivalent mass of 282, so that percent. FFA as oleic equals FFA as $Na_2O \times 282/31$, or $Na_2O \times 9.10$. In the case of palm oil it is sometimes expressed in terms of palmitic acid (factor 8.26); and for lauric oils in terms of lauric acid (factor 6.45). FFA content can also be expressed in terms of mg KOH required to neutralise the FFA in one gram of fat; this figure is known as the acid value (AV), or acid number. If n = FFA% as Na_2O, the FFA in lg fat is equivalent to $n \times 1000/100 = 10\ n$ mg Na_2O, or $10 \times 56.1/31 = n \times 18.1$ mg KOH.

Saponification value (SV), or saponification number, is similarly expressed as the number of milligrams of KOH required to saponify completely one gram of fat. The test is carried out by heating under reflux a suitable weight of fat with a known volume of standard alcoholic potash and back titration of the excess potash with standard acid to the phenolphthalein end point. The net quantity of KOH used has reacted with both FFA and glyceride and gives the SV.

Ester value (EV), or ester number, equals SV-AV; that is it a measure of the glyceride present in the sample.

Saponification value (or, more strictly, the acid value of the fatty acids), and the fatty acid ratio are both measures of the average equivalent mass, and hence average chain length, of the mixture of fatty acids in the fat.

Content of Unsaponifiable Matter

The percentage of *unsaponifiable matter* in a fat is determined by extraction with ether after complete saponification with caustic potash. A soap may also contain a proportion of free fat, which should normally be very small except when saponifiable superfatting materials are added deliberately. A sample of soap is extracted and, after the extract has been weighed, it is boiled under reflux with caustic potash to complete saponification, and the unsaponifiable is extracted. The first extract gives total unsaponified and the second unsaponifiable. Unsaponified fat equals total unsaponified minus unsaponifiable.

Iodine Value

The average degree of unsaturation of a fatty material is measured by determination of its *iodine value*, or iodine number. Several methods are available, but the Wijs method, using iodine monochloride in glacial acetic acid as the reagent is usual. The weighed sample of fat is dissolved in carbon tetrachloride and a measured volume of reagent added. After the appropriate time for addition of iodine to the double bonds has elapsed, water and potassium iodide are added and the excess iodine is titrated with standard sodium thiosulphate. The difference between the titration of a blank without fat and that of the sample represents the iodine absorbed, from which the iodine value is calculated as the number of grams of iodine absorbed by 100g of fat. The weight of fat used in the test must give an excess within narrow limits and a repeat test may be needed after an approximate value has been found. Care is needed in interpreting the results with highly unsaturated materials; in particular, oils containing fatty acids with conjugated double bonds do not take up the stoichiometric quantity of iodine, but these do not normally concern the soapmaker. Iodine values can be determined on dry fats, or on dry fatty acids (normally plus the unsaponifiable matter, that is on total fatty matter), but, with a given fatty acid composition, the value will obviously be lower with the fat than with the fatty acids, because of the diluting effect of the combined glycerol.

Reichert-Meissl, Polenske and Kirschner Values

The *Reichert-Meissl, Polenske* and *Kirschner* values of a mixture of fatty acids (derived from a fat or a soap) are determined by distillation under standardised conditions; and are measured in terms of the ml $0.1N$ NaOH needed to neutralise the fatty acids in the distillates. The Reichert-Meissl value is determined on the water soluble fraction, and is a measure of the milk, or butter, fats present. The Kirschner value measures butyric acid more specifically. The Polenske value, of particular interest to soapmakers, determines the water insoluble volatile acids; and is a measure of the lauric oil included, with some indication whether it is coconut or palm kernel.

Specific Gravity or Relative Density

The *specific gravity* of the fatty acids is also useful, particularly when rosin may be present.

Rosin Content and Tests for Specific Oils

Tests are available to show the presence of *rosin* (Leibermann-Storch) and to determine its percentage (McNichol or Wolff). There are also tests to indicate specific fats; for example, the Halphen test for cottonseed, or kapok, oils, although a positive may also be obtained from the fats of animals fed on cotton seed cake.

SPECIFICATIONS QUALITIES OF TALLOWS AND GREASES

The most important for soapmakers are those concerned with tallows and greases, because these fats are the major constituent of most soaps; and because they can vary widely in quality. Now tallows are often of mixed origin and are commonly sold on the basis of specification related to analytical characteristics, including colour and/or colour after a standard bleach. Soapmaker often find that a colour of soap, rather than a colour of fat, test is particularly relevant, although it does not appear in the usual specifications.

A convenient colour of soap test for routine purposes is to take a sample of the tallow, or other fat, bleached if appropriate, in a white enamel cup, add a slight excess of alcoholic caustic soda and boil with constant stirring. The fatty is saponified and the alcohol boiled off; this leaves granules of soap which are assessed by eye by an experienced observer. Other colour of soap tests used instrumental measurements of the colour of a solution of the soap.

Tallows are defined as fats with a titre of 40°C, or above; and an iodine value below 60 calculated on the fatty acids from a determination on the fat (Wijs method). Greases are defined a having a titre between 36°C and 40°C; and an iodine value below 63 calculated on the fatty acids. Tallows and greases produced by solvent extraction must be so described. (Ref. BIS 887:1977 for Animal Tallow and BIS 1780:1961 for Vegetable Tallow).

THE PRE-TREATMENT OF OILS AND FATS; AND THE PRODUCTION OF BY-PRODUCTS USED IN SOAPS

General

Most fats are pre-treated prior to use in edible products, as are many intended to be used in soaps, although large quantities of the better grade fats are made into soap without any pre-treatment. Some of the pre-treatments mentioned below are important to the soapmaker mainly because they produce by-products which can be used in soaps.

Types of Pre-treatment

The main types of pre-treatment are :

(i) *De-gumming* or *de-sliming*. This is usually used with soyabean and other oils with a high content of phosphatides and other unsaponifiables. Warm water, sometimes with about 0.1% of phosphoric acid, is added to hydrate some of the impurities which are then separated in a centrifuge to leave a relatively dry, clean oil. Commercial lecithin is recovered from the other fraction, particularly in the case of soyabean oil. These gums, known generally as 'break' material, can also be separated by heat.

(ii) *Acid refining*, is carried out with concentrated sulphuric acid in the pre-treatment of linseed, and other drying oils, for use in paints. The strong acid chars the phosphatides and related materials. Treatment with sulphuric acid is also used to clean up low grade materials prior to splitting with water (particularly by the Twitchell process).

(iii) *Alkali*, or *soda*, *refining* is normally included in the preparation of fats for edible purposes. It is not commonly employed in the pre-treatment of soapmaking fats, but it can be useful to upgrade certain tallows for use in high quality soap bases. An excess of caustic soda (or sometimes sodium carbonate) solution is used to neutralise the FFA in the fat and to carry away in the lower aqueous soapstock layer other impurities, plus some neutral fat. Careful attention is necessary to avoid an undue loss to the soapstock of good neutral fat.

(iv) *Steam distillation* can be used to reduce the free fatty acid content of a fat; a process which has been employed particularly with fairly high FFA palm oil. The fatty acids separated are used in soap. The relatively high temperatures involved bleach some fats, particularly palm oil; but some colouring matter tends to be 'set', and so more difficult to remove after the treatment.

(v) *Deodourisation* is a steam distillation process carried out under fairly high vacuum to remove odiferous and taste producing impurities from fats. Deodourisation of edible fats is normally carried out on materials which have been soda refined and bleached; otherwise FFA would be removed in this process with other volatile constituents and this is better done by soda refining. The process is sometimes used to remove highly odorous constituents from coconut and palm kernel oils to be included in high proportions in certain toilet soaps. Metal scavengers, such as citric, tartaric and phosphoric acids are often added to fats prior to deodourisation.

(v) *Bleaching*, widely used for soapmaking as well as edible fats, can be by adsorption or by chemical means.

(vi) *Hydrolysis*, or *splitting*, often followed by distillation of the fatty acids, is, with the manufacture of soaps from fatty acids.

(vii) *Hydrogenation*, or *hardening*, in which the unsaturation of a fat is reduced by catalytic hydrogenation. This process is used extensively in the edible fat industry. At one time, particularly when whale oil was plentiful and cheap, it was also used to produce soapmaking hard fats from oils which were too unsaturated for direct use in most soap products. With current prices it is now little used in the soapmaking field, but it may be useful in countries which produce large quantities of unsaturated oils and, for politico-economic reasons, need to use these materials for the production of soap. The hydrogenation process decolourises oils such as palm by converting carotinoid pigments into colourless compounds.

The chemistry of hydrogenation is complex and the product obtained depends upon processing conditions as well as upon the starting material. This is because in the production of fats which retain a substantial degree of unsaturation, as is normally required, the hydrogenation can be more, or less, selective; and isomerisation can occur.

Soapstocks and Acid Oils, or Acidulated Soapstocks

Soapstocks, the by-products from the soda refining of fats, are an important source of fatty matter for soapmaking. They can be used in several ways :

(i)   Directly as the soapstocks.

(ii) After concentration with additional fat.

(iii) After acidification with sulphuric acid to produce acid oils, sometimes known as acidulated soapstocks.

Advantages of using soapstocks directly in soapmaking are :

(i)   The alkali used in the soda refining process is transferred to soapmaking so that caustic soda is saved.

(ii) Sulphuric acid does not have to be used to convert the soapstock to acid oil.

(iii) Glycerol liberated by saponification of glycerides in the soda refining process is transferred to soapmaking and not lost as when acid oil is produced.

Advantages of transfer to soapmaking via an acid oil are :

(i) Transport costs are saved, unless the refinery producing the soapstock is adjacent to the soap factory. Soapstocks should usually contain 30-50% TFM, but may be weaker, particularly when washings are included.

(ii) The water content of the soapstock can have an adverse effect on the efficiency of the soapmaking process which is avoided by the use of acid oil.

(iii) The acidification, and disposal to waste of the aqueous layer, followed by washing of the acid oil, removes various impurities which are not then transferred to soapmaking. If required, the quality of the acid oil can be improved by bleaching, or if necessary, by splitting and distillation. Alternatively, certain chemical bleaching processes can be used with soapstocks. Soapstocks can also be purified by completion of saponification, followed by graining with salt and the discard of the lye. This will result in the loss of some of the free glycerol in the soapstock.

Concentration of soapstock is achieved by mixing the soapstock with a carefully controlled quantity of additional fat at a temperature above 70°C. The quantity of fat needed should be determined by experiment, but it is often found that about 3.8 parts of new fat should be added to soapstock containing one part of fatty matter. Circulation by a pump, with return to the vessel through a fish-tail jet. On settling for 3-4 hours, an aqueous layer should separate, containing perhaps 0.5-1.0% glycerol, which can be introduced in an appropriate manner into the soapmaking system. The upper fatty layer, containing the soap, can be handled as an oil. If conditions, particularly temperatures, are not right an emulsion may be formed and no proper separation achieved. Such emulsions can often be broken by addition of brine, or treated soap lye. Benefits which can accrue from the concentration of soapstock are mainly improvements in soapmaking efficiency.

Acidification of soapstocks is usually carried out with sulphuric acid in wooden, plastic, or lead-lined vats. After sufficient acid has been added to leave free mineral acid, the mixture is boiled with steam until the soap is split and the emulsion broken. After the mass has been settled, the aqueous layer containing various impurities is run to waste; and the acid oil is washed with hot water as many times as necessary.

Crude cottonseed oil is often soda refined twice. The first treatment, plus alkali and water washes, gives washed cottonseed oil and a very dark soapstock, sometimes known as black cottonseed grease. This black grease, which contains substantial proportions of gossypol and phospholipids, is not normally suitable for incorporation in soaps directly, or as an acid oil. Usable, but rather dark, fatty acids can be recovered from it by splitting and distillation. The washed cottonseed oil is soda refined again to yield a soapstock, or acid oil, of reasonable quality for soapmaking.

The fatty matter in soapstocks, and acid oils, has a very similar fatty acid composition to the parent oil, or fat, but the colour and general quality are inferior.

Bleaching of Oils and Fats

Methods used to bleach oils and fats can be divided into adsorption bleaching, used for both edible and soapmaking fats, and chemical bleaching used only for soapmaking fats.

In *adsorption bleaching*, the molten fat, or oil, is brought into contact with a powdered solid adsorbent, which can be natural fullers earth, or activated earth or clay, or, to a limited extent, activated carbon used in conjunction with activated earth. Activated earths are made by the action of mineral acids on suitable mineral substances, such as bentonite or montmorillonite. The adsorbent removes colouring matter from the oil which is then separated by filtration. Sulphuric acid markedly improves the performance of natural fullers earth and may also be useful with activated earths. An addition to the bleaching vessel of about 10% of the weight of natural earth used is often found suitable for soapmaking oils.

The use of 2–3% sulphuric acid, followed by 3–5% earth, in the bleaching of certain dark liver oils is recommended.

Only about half the ratio of earth to oil is required with a good activated earth to give the same result as that obtained with natural earth. Various grades of activated earth are available, some acidic and some more nearly neutral. Acidic earths are usually more effective than neutral types and are commonly used with soapmaking oils. They may cause a very slight increase in the FFA of a soda refined oil and, for this reason, they may not be desirable with some edible oils. The proportion of earth needed to achieve the desired bleach should be determined by experiment. It varies widely from one oil to another, and with the colour of bleached oil required. 2–6% of activated earth, or 4–10% of natural earth, is common for soapmaking oils; the highest proportions often being needed with palm oil.

Batchwise bleaching in single vessels is usual for soapmaking oils, but continuous bleaching processes are sometimes used in edible fat refineries operating on a large scale. Batchwise bleaching can be carried out in open vessels, fitted with closed steam coils and with an agitator to keep the earth in suspension in the oil and to promote heat transfer without local overheating; but closed vessels working under vacuum (50–70 mmHg absolute pressure) are usually used and are better because :

(i)   Moisture is removed more effectively.

(ii) There is less risk of undesirable oxidation of the oil by air, particularly if steps are taken to deaerate the oil as much as possible before it reaches bleaching temperature.

Vacuum bleachers can be of various designs, but vertical, cylindrical, vessels with dished base and top are common. Various types of agitator, including paddles or tubines are possible. Heating is usually by internal steam coils, but in the case of very large bleaching vessels, external circulation through a tubular or similar heater is sometimes used, and is preferable. Coolers may be fitted to cool the oil before it is filtered. In the case of edible oils, this is to minimise risk of oxidation whilst the oil is exposed to air; with soapmaking tallows and greases it is primarily to remove polythene, for which purpose the oil should be filtered at below about 70°C. Bleaching temperature is not very critical; typically, good quality tallows and other oils and fats are bleached at 90–95°C, poorer qualities at about 110°C, and palm oil at 120–130°C. About 15 minutes is said to be sufficient for equilibrium to be reached, but the complete bleaching cycle takes several hours.

The role of water is important. Complete drying of the bleaching earth will destroy its efficiency; and water in the oil charged to the bleacher does no harm as regards final quality. It has been reported that an addition of water can improve the performance of activated earths when bleaching tallow; other experience shows that sufficient water is important when palm oil is being bleached with activated earth, or with natural earth plus 10% acid; and if necessary, water is added to bring the level to about 2%. On the other hand, it is usually found necessary to adjust temperature, pressure and time, so that the oil is substantially dry before the bleaching stage is completed in order to achieve proper adsorption of colouring matter.

Mild steel, or cast iron, is normally used for the vessels and coils of bleachers, even when sulphuric acid is added, but a splash plate near the acid inlet is desirable to direct the acid to the oil rather than the vessel wall. The earth is commonly sucked in from a hopper by the vacuum.

It is necessary to separate the earth from the oil completely and in such a way that the quantity of oil left in the filter cake is minimised. This requires a substantial pressure, so that vacuum filters are not suitable, and facilities are needed to permit the cake to be air blow and/or steamed. Filter presses, manual or, more recently, mechanised, are traditional; but some types of leaf-filter can be convenient and require little labour. The oil discharge arrangements should minimise the risk of fume being dispersed in the work area, particularly during steaming or blowing. Filters are normally cast iron. Various filter cloths, of different weights, weaves and materials are now available; and the following table, provides useful information concerning the main fabric materials :

Material	pH range	Maximum temperature °C
Cotton	6–10	100
Polyamide	7–12	110
Polypropylene	2–12	100
Polyester	2–7	160
Polyvinylidene chloride	2–11	80

Cotton cloths were traditional for bleaching earths and are still widely used; but synthetics, particularly polyester, withstand acidic conditions better and may have several times the life of cotton, which can outweigh the higher initial cost.

Cotton cloths seal the joints between the plates well, but synthetic cloths do not; and special treatment of the areas of the cloths which form the seal, or special mechanical devices, may be needed. Fine weave metal gauzes are used in some mechanised filters. As in other filtrations, the cloth, or other primary medium, must retain most of the solids, but the initial layer of earth residue on the cloths forms the true filter medium, and it is usually necessary to return the first runnings for re-filtration.

A second filtration through a polishing filter is sometimes used to remove traces of fines which pass through the first filter. Paper may be used over the cloths, particularly in polishing filters, to improve the clarity of the filtered oil. When filter-presses are to be used, there is a choice between the plate and frame type and the recessed plate type. Either can be satisfactory for the main filtration; but, contrary to what might be expected, a plate and frame press should be chosen for polishing. This is because in a polishing filter a thin layer of very fine earth builds up, and becomes impervious, before the chamber is nearly full. It is then necessary to drain the oil and to remove the paper and residue. This can be done with a plate and frame press, but not satisfactorily with a recessed plate press having a central feed inlet.

The proportion of oil left in the filter cake after it has been steamed and/ or blown with air as effectively as possible, varies considerably with the type of earth used. With natural earth it is often possible to achieve about 30% on the weight of residue, but with activated earths it is usually higher and 40–50% is common; up to 70% has been mentioned with edible oils. The loss of oil per tonne of crude fat treated is controlled by the percentage of bleaching earth used, as well as by the percentage of oil left in the residue, so that it is frequently less with an activated than with a natural earth, despite the higher retention of activated types. This loss of oil is a major item in the cost of bleaching, often exceeding the actual cost of the earth. Even a small percentage of activated carbon can greatly increase the oil retention of an earth.

One reason why plant bleaches may be more effective than laboratory bleaches lies in the bleaching which occurs whilst the mix is being passed through the cake in the filter. Multi-stage counter-current operation has theoretical advantages, but is not often practised. It is, however, sometimes useful to clean up, and bleach slightly, an oil for a relatively low grade soap by pumping it through full presses. The oily residue discharged from the filters can sometimes be sold for use in animal feedstuffs (in which substantial percentages of some kind of fat are often included). Processes for recovery of oil from these residues include :

(i)   Wet extraction with a solution of alkali, which is not very efficient.

(ii) Extraction with a hydrocarbon solvent. This gives a rather low yield of oil of moderate quality, but requires all the precautions for a process with a flammable solvent.

(iii) Extraction using a mix and settle process with a chlorinated solvent, such as trichloroethylene. This gives a higher yield of a poorer quality fat, because the chlorinated solvent extracts more coloured and oxidised material than the hydrocarbon solvent. Better settling, and an even higher yield, but of a very dark brown oil, can be achieved by steaming the solvent residue mixture. It is often found that it is not worthwhile to recover oil from bleaching earth residues. Disposal of earth residues, extracted or not extracted, can be difficult; particularly if residues from the bleaching of highly unsaturated oils, which can be subject to spontaneous combustion, are involved.

Chemical bleaching normally involves the oxidation of the colouring matter in the oil or fat. The result varies greatly with the kind of fat; and palm oil, and by-products containing palm oil residues, are particularly responsive to oxidation bleaching. In some cases, an undesirable odour develops; and chemical bleaching is not used with edible fats, or, usually, with fats to be made into the better quality toilet soaps.

The oxidation can be carried out with air (or oxygen), with alkali dichromate (plus acid), with sources of available chlorine, or of chlorine dioxide, or with other reagents. Air can be blown through hot palm oil to produce air bleached palm oil. Soap made from this oil has a slight brownish shade and a characteristic odour. These qualities limit its use, but it has been employed in orange/red carbolic toilet soaps. Bleaching of palm oil with chromic acid is very effective. The oil is agitated with air for several hours after addition of sodium dichromate, salt, and hydrochloric acid. After the mix has settled, the aqueous layer is run to waste and the oil is washed several times with water. More recently, most chemical bleaching processes have used chlorine compounds.

Oils and fats, particularly acid oils, can be bleached under acid conditions. Sodium chlorite is usually employed, activated with acid; and the active species is thought to be chlorine dioxide (ClO_2). The bleaching of tallows and greases by using sodium chlorite plus sufficient sulphuric acid reduces the pH to 4, or below. It has been found that no additional activator is necessary if the fatty matter contains sufficient FFA (usually above about 20%); and when the fat is not sufficiently acidic, the preferred activator is phosphoric acid. About 0.5–1.0% of sodium chlorite is used. The process can be carried out in a simple vessel in which agitation is provided by compressed air from a perforated pipe. The vessel can be constructed of mild steel, but stainless steel is preferred. Alternatively, continuous bleaching plants can be designed for use if the scale of operation is sufficiently large. When phosphoric, or other mineral, acid is used, particular care must be taken to avoid addition of reagents when no fat is present; otherwise an explosion is liable to occur. The bleached fat should be saponified as soon as possible. Soaps made from chemically bleached fats tend to deteriorate more than those made from adsorption bleached materials.

Chlorine compounds can also be used under alkaline conditions to bleach soaps, including soapstocks. Soapstocks should be completely saponified, and soaps should preferably be washed to remove the glycerol before the bleaching operation. Sodium hypochlorite has long been used to bleach soap in the pan, but, particularly for soaps containing some palm material, there are advantages in using :

(i)   Sodium hypochlorite activated with hydrogen peroxide.

(ii) Sodium chlorite activated with sodium hypochlorite, although various other activators are possible.

The sodium hypochlorite/hydrogen peroxide process can conveniently be carried out in a continuous reactor in the form of a vertical cylinder with two turbo impellers; one near the base and the other part way up. The soap flows upwards, is mixed with the sodium hypochlorite in the lower impeller, and the hydrogen peroxide is fed to the eye of the upper impeller. Sufficient hot water must be introduced to avoid the grained condition in the reactor, it being remembered that sodium hypochlorite with its associated salt and caustic soda is a strong electrolyte. Hypochlorite to give about 2% available chlorine plus 0.1% hydrogen peroxide, both expressed on TFM, may be found suitable.

As already mentioned, sodium hydrosulphite has been used in the pan as a reduction bleach to improve the colour of good coloured soaps; it is not very effective and is now seldom employed. The hydrosulphite powder is sprinkled onto the surface of the soap when in a 'closed' condition, and the pan contents are boiled for about ten minutes.

SELECTION OF FAT BLENDS FOR VARIOUS SOAPS

Soaps can be made from a wide range of fats; and unusual fats, or blends, are used in some countries for politico-economic reasons.

For many purposes, including conventional toilet soaps and household soaps made by the modern vacuum chilling/extrusion processes, a blend of about 15–25% lauric oil (coconut or palm kernel) and 75–85% hard fat (usually tallow or palm oil) gives a soap base with an excellent combination of properties. The hard fat gives a firm texture and good detergency at high washing temperatures; the lauric oil also gives a firm soap, but improves the rate of solution and freedom of lather. Particularly for household soaps, but also for cheaper toilet soaps, less than 15% lauric oil is sometimes used because of cost. Too high a content of lauric oil tends to give a soap which is rather harsh (irritant) to the skin, unless superfatting is employed. High lauric oil soaps are sometimes used for certain textile operations and when the soap is to be used in sea, or brackish, water; but tablets containing non-soapy detergents may now provide a better answer to the salt water problem.

Assuming that tyre and iodine value for blends are additive, values for a mixture of 80% tallow, of titre 41°C and IV of fatty acids 45, with 20% coconut oil, titre 22°C and IV 10, can easily be calculated :

Item	Titre °C	Iodine value
80% Tallow	0.8 × 41 = 32.8	0.8 × 45 = 36.0
20% Coconut oil	0.2 × 22 = 4.4	0.2 × 10 = 2.0
Blend	37.2	38.0

A titre of 36–38°C and an iodine value of about 37 are often considered ideal for a conventional toilet soap. If a rather hard, high titre, tallow is to be used, a proportion of good quality soft material, such as lard (choice white grease) or groundnut oil, will adjust the titre and iodine value of the blend; and produce a soap with an improved balance of plastic properties. Oils with significant proportions of highly unsaturated acids are liable to give soaps with tendencies to deteriorate through oxidation on long storage, particularly at relatively elevated temperatures. Care must, therefore, be exercised in the choice of soft material for high quality soaps and minor ingredients in the fats can be important. Some years ago an example occurred in which inclusion of a proportion of distilled bone grease fatty acids to soften the blend resulted in abnormal deterioration of the colour of a white toilet soap after some months storage. Olive oil, which has a very high content of oleic acid and little polyunsaturated acids, is a particularly good soft oil. It was a traditional soapmaking material in Mediterranean countries, but, apart possibly from some exceptionally poor, inedible, grades it is now much too expensive for use in soap. The use of preservatives, primarily but not exclusively, to minimise the adverse affects of heavy metal catalysts in producing deterioration on storage.

Tallow and palm oil are the main hard fats. Apart from some difficulty in the production of very white soap, palm oil is an excellent soapmaking material, but it is now normally too expensive compared with tallow. Few years ago, whale oil was relatively cheap and much was hardened (hydrogenated) for use in soap. Special attention needed to be paid to the hardening conditions to produce an acceptable tallow substitute, but hardened whale oil was generally less good than tallow. Little whale oil is now available and the soft vegetable oils are normally more expensive than tallow. Hardening may be useful in countries where there are pressures to use soft oils in soapmaking; and, although not very good technically, hardened fish oils may be economically attractive for low grade soaps under some circumstances.

For household soaps made by the old frame, or water-cooling, processes, a substantial proportion of soft oil is advisable to produce a good lathering soap. Although soft oils can be, and are, used with the modern household soap processes, they are not generally desirable.

Rosin was used extensively in soaps at one time, primarily because it was cheap; but it does also have useful properties. To some extent, it is an alternative to lauric oils as a means by which the solubility and lather of a frame-cooled household soap made from hard fats can be improved; and tallow/rosin soaps were popular. A good quality rosin also gives the soap a characteristic odour which was generally thought pleasant and preferable to that of a tallow soap. At one time, 2–3% of good quality rosin was included in toilet, and other soap blends as a preservative. Soaps with a high rosin content .tend to be sticky, which makes them unsuitable for some purposes, including soap powders.

Particularly for household soaps, the fat is the predominant item in the cost of the product, so that it is very important commercially to use as effectively as possible any relatively cheap materials, such as by-products, which can be obtained. These by-products are often of somewhat uncertain composition, although described as 'soft acid oil', etc. A preliminary assessment of their suitability can be made from an analysis giving fatty acid composition by GLC; or, less precisely, from fatty acid constants, such as titre and iodine value. Colour, odour and stability on storage must also be considered.

FATTY ACID AND GLYCEROL CONTENTS OF FATS, AND RELATED MATTERS, GLYCERINE CREDIT

Initially, it will be assumed that the fat contains only triglycerides and fatty acids; complications due to the possible presence of di- and mono-glycerides will be considered subsequently. Also, the calculations will be based on 100% fatty matter; but the small corrections for the 1-2% of moisture and dirt commonly found in crude fats can readily be made.

As already explained in the saponification value, acid value and ester value of a fat are all expressed in terms of mg KOH per g of fat.

The quantity of caustic soda needed theoretically to saponify the fat is thus equivalent to SV mg of KOH per g fat. This is equal to

$$\frac{SV}{1000} \times \frac{31}{56.1} = SV \times 0.000553 \text{ g Na}_2\text{O / g fat (or t / t fat)}$$

The glycerol content of the fat is proportional to the ester value; and EV = SV – AV.

Glyceride + 3 KOH → $C_3H_5(OH)_3$ + 3 potassium soap.

$$3 \times 56.1$$
$$= 168.3 \quad 92$$

EV = mg KOH equivalent to the glyceride in lg fat

Hence, percentage glycerol in fat

$$= \frac{EV}{1000} \times \frac{92}{168.3} \times 100$$

$$= EV \times 0.0547$$

Estimation of the fatty acid content of a fat is more complicated, but the formula in terms of SV, EV and AV can be derived as follows :

Consider first a pure triglyceride with no FFA and let M = the average equivalent (= molecular) mass of the *fatty acids* it contains.

$$C_3H_5(OOC—R)_3 + 3H_2O \rightarrow C_3H_5(OH)_3 + 3R.COOH$$
$$3 \times 18$$
$$= 54 \qquad\qquad 92 \qquad 3M$$

To make the equation balance, the molecular mass of the glyceride = $3M$ + 92–54 = $3M$ + 38, that is the equivalent mass of the glyceride is

$$\frac{3M + 38}{3} = M + 12.67$$

lg of fat is equivalent to $(56.1 \times 1000)/(M + 12.67)$ mg KOH, and, by definition to SV mg KOH. Hence,

$$\frac{56.1 \times 1000}{(M + 12.67)} = SV, \text{ or } M = \frac{56100 - 12.67SV}{SV}$$

Considering now pure fatty acids, the equivalent mass is M, instead of $(M + 12.67)$ and SV = AV; so that M = 56 100/SV, or 56100/AV. In actual fat containing both triglycerides and free fatty acids, the ratio of combined to total fatty acid is EV/SV (assuming, as is usually approximately the case, that the composition of the free is similar to that of the combined fatty acids). The addition of 12.67 to the equivalent mass of the fatty acids applies only to the combined portion so that, from the equation for pure glycerides,

$$M = \frac{56100 - 12.67\, SV \times \dfrac{EV}{SV}}{SV}$$

or

$$M = \frac{56100 - 12.67\, EV}{SV}$$

The total mass of fatty acids in 1000 mg of fat equals the quantity of KO required to saponify it $\times M/56.1 = SV \times M/56.1$. Hence, from the expression just derived for M, the total fatty acids in 1000 mg fat.

$$= \frac{SV}{56.1} \times \frac{56100 - 12.67\ EV}{SV}$$

$$= 1000 - \frac{12.67\ EV}{56.1} = (1000 - 0.226\ EV)\ mg$$

Hence, 100 parts of fat contain $(100 - 0.0226\ EV)$ parts total FA.

$$FFA\% = \frac{AV}{SV} \times total\ FAs\ in\ 100\ parts\ fat$$

$$\frac{AV}{SV} \times (100 - 0.0226\ EV)$$

If it is required to use a determination of FFA% rather than of AV, the figure in terms of $Na_2O\%$ should be converted to FFA% using an estimated molecular mass for the actual fatty acid blend, rather than a notional one, such as that for oleic acid. In many cases the difference is not great.

It may also be helpful to note that

M = fatty acid ratio $\times 31$

As an example, consider a tallow with SV = 195 and AV = 15, EV = 195 − 15 = 180

$$\%FFA = \frac{AV}{SV}(100 - 0.0226\ EV) = \frac{15}{195}\{100 - (0.0226 \times 180)\}$$

Theoretical requirement of caustic soda for saponification

$$= SV \times 0.000553$$

$$= 195 \times 0.000553 = 0.108g\ Na_2O/g\ fat$$

or $\qquad$ 0.139 NaOH/g fat.

glycerol content = EV $\times$ 0.0547

$$= 180 \times 0.0547 = 9.85\%$$

fatty acid content = (100–0.0226 EV)

$$= 100 - (0.0226 \times 180) = 95.9\%.$$

As mentioned previously, all these figures are calculated for pure fatty matter in the fat, and strictly they should be corrected for its moisture and dirt content. This correction is usually only significant for the fatty acid content.

Table 4.4 shows figures calculated in this way for a number of typical soapmaking fats. It will be seen that :

(i) The theoretical caustic soda requirements fall into two groups, one for coconut (and other lauric) oils, and one for other fats.

(ii) The glycerol content varies widely with both the equivalent, or molecular, mass of the fatty acids and the percentage of free fatty acids.

(iii) The total fatty acid content does not vary much, although it is a little lower for the lauric oils, and a little higher when the FFA is high. After allowance for moisture and dirt, fatty acid content is about 94–96% for most fats.

As fats contain about 95% TFM, 1 tonne of fat yields about 95/63 = 1.51 tonnes of 63% TFM soap. This supports the common approximation that 1 tonne of fat should yield around 1½ tonnes of soap base at 62–64% TFM.

Saponification almost certainly occurs in stages; but mono- and di-glycerides are not normally found in partly saponified fat, probably because they react with caustic soda more rapidly than triglycerides. In the case of hydrolysis of fats with water, substantial proportions of mono- and di-glycerides are found during the process; and some are likely to be present in fats which have deteriorated with development of FFA, although presumably not in acid oils, unless they have deteriorated after production. The following calculations show that this can be important in the estimation of the glycerol content of a fat from its SV and AV.

Consider a low grade tallow, or grease, having an equivalent mass of its fatty acids of 279 (fatty acid ratio 9.0), so that the group (O.CO.R) is equal to 279–1 = 278. The molecular mass of the triglyceride, $C_3H_5(O.CO.R)_3$ is $(3 \times 12) + 5 + 3 (278) = 875$; and it yields $(92 \times 100)/875 = 10.5\%$ glycerol and $(3 \times 279 \times 100)/875 = 95.4\%$ FA. Similarly, the diglyceride, $C_3H_5(OH) (O.CO.R)_2$, has a molecular mass of 614; and yields 15.0% glycerol and 90.9% FA.

Again, the monoglyceride, $C_3H_5(OH)_2 (O.CO.R)$, has a molecular mass of 353; and yields 26.1% glycerol and 79.0% FA.

Table 4.4. Contents of glycerol and fatty acids, and caustic soda requirements for saponification, of some typical fats (no allowance for possible presence of mono- and di-glycerides)

Fat	SV	AV	EV	FFA % (calc)	Theoretical caustic soda requirement Na₂O% on fat	Glycerol content %	FA content %
Tallow	195	5	190	2.5	10.8	10.4	95.7
Tallow	195	15	180	7.4	10.8	9.9	95.9
Grease	192	90	102	45.8	10.6	5.6	97.7
Coconut oil	257	5	252	1.8	14.2	13.8	94.3
Coconut oil	257	12	245	4.4	14.2	13.4	94.5
Coconut acid oil	257	150	107	57.0	14.2	5.9	97.6
Cottonseed oil	194	4	190	2.0	10.7	10.4	95.7

Table 4.5 shows that glycerol and FA contents of various hypothetical fats build up from these components. It will be seen that glycerol content varies considerably between three compositions, all with 10% FFA, but that is before, the fatty acid contents do not vary very much.

Table 4.5 Compositions of hypothetical fats with varying proportions of the different glycerides and free fatty acid.

Composition of fat	Glycerol %	Fatty acid %
100% triglyceride	10.5	95.4
90% triglyceride + 10% FA	9.5	95.9
80% triglyceride + 5% diglyceride + 5% monoglyceride + 10% FA	10.5	94.8
70% triglyceride + 10% diglyceride + 10% monoglyceride + 10% FA	11.5	93.8
100% FA		100

Glycerol and fatty acid contents of the fats used in a soap factory are required for process control and for costing. For many fats the methods described in this section give satisfactory figures, but if substantial quantities of fats likely to contain mono- or di-glycerides are used, it is better to have actual analytical estimations of glycerol and fatty acid contents. A useful compromise is to have a sufficient number of analyses of low grade materials made to permit the compilation of tables, or graphs, relating glycerol and fatty acid contents to AV, or FFA%, for the various materials. Values can then be read off when similar materials are used again.

When a quantity of fat is used to make soap it contains both fatty acids and glycerol; and there is no absolute way in which to decide how much has been paid for each. The lyes from lauric oils and from low grade fats are more difficult to purify than those from a good grade tallow; and, if the lyes were worked-up separately, they would tend to produce a poorer quality crude glycerine. In practice, lyes from the various types of fat are usually treated and evaporated together. On this basis the glycerol in one fat has the same value as that in another; and crude glycerine normally has a reasonably well-defined price.

HANDLING AND STORAGE OF OILS AND FATS, DETERIORATION OF FATTY MATERIALS

As mentioned earlier, the quality of oils such as palm and rice bran can be poor unless the source materials are treated carefully and promptly to minimise the action of air, water, enzymes and micro-organisms.

Oils and fats, particularly those containing the more highly unsaturated fatty acids, are subject to oxidation and development of rancidity. This, and the associated topic of flavour deterioration, is of major importance in relation to fatty foods. Care is also important with soapmaking fats, but the problem is somewhat less critical. The chemistry of rancidity development is complex, and only a few broad generalisations are discussed below :

(i)   Initially oxidation occurs at a double bond with the formation of peroxides. As the process proceeds, the peroxides break down to aldehydes, ketones, hydroxy-acids and other compounds; in consequence the content of peroxides (often determined as a measure of rancidity) passes through a maximum and then tends to fall. The Kreis test for rancidity depends upon the red colour which develops when the fat is reacted with phloroglucinol in ether solution; and is some measure of the aldehyde and ketone content.

(ii) The oxidation starts slowly but, after an induction period of varying length, it accelerates markedly. Pro-oxidants and antioxidants can be present. Pro-oxidants are mainly heavy metal catalysts, such as copper and iron; and one type of antioxidant sequesters, and so inactivates, these metallic catalysts. Other types of antioxidant interfere with the oxidation mechanisms. They can be natural, such as the tocopheraols (vitamin E), or artificial. Sequestrant antioxidants are widely used in soaps, as discussed later, but also with fats, as in the addition of citric acid before deodorisation. Natural antioxidants are more general in vegetable than in animal fats, so that the former tend to keep better than the latter with similar unsaturation. Antioxidants are usually partially removed when an oil is refined; and crude oils often keep better than refined. Beef tallow also tends to deteriorate less than some other animal fats. Various organic compounds can be added to a fat to improve its keeping properties, but some, such as hydroquinone, can lead to the development of undesirable colour if the fat is made into soap. Edible fats at not more than certain specified concentrations, are :

   (i) BHA (butylated hydroxyanisole).

   (ii) BHT (butylated hydroxytoluene).

   (iii) THBP (2,4,5-trihydroxybutyrophenone).

These compounds can also have some adverse affects on the colour of the soap after storage.

(iii) Oxidation bleaching naturally affects unsaturated fatty acids, but such chemically bleached fats are, or at least have been, widely used in soaps, although generally not in the higher grades. Care should be taken not to continue air bleaching of palm oil beyond the optimum point, after which colour deteriorates and undesirable reactions may occur. The high peroxide value of an air bleached palm oil can be reduced by soda refining, adsorption bleaching, or deodorisation, but, apart perhaps from passage through a filter press loaded with spent bleaching earth, these additional processes are unlikely to be commercially attractive.

(iv) Complete saponification is generally considered important to ensure good stability of a soap. It is also commonly considered desirable to leave a small percentage of free alkali in a soap; but superfatting with fatty acids, which eliminates the free alkali in the base, is not unusual.

(v) When solid fats are received at a factory in drums, or barrels, a melting out stage is required; this normally comprises a tank, or tanks, covered with a staging onto which the drums can be put. Steam jets discharge into the bung holes of the drums, which face downwards, and the molten fat and condensed water drip into the tank below. The mixture in the tank is settled, the water run off, and the molten fat pumped away.

Chapter 5

Manufacture of Soap Base by Direct Saponification

INTRODUCTION

The chemical reaction in which an ester is heated with aqueous alkali such as sodium hydroxide, to form an alcohol (usually glycerol), and the sodium salt of the acid corresponding to the ester. The process is usually carried out on fats, (glyceryl esters of fatty acids). The sodium salt formed is called a soap. A typical saponification reaction is :

$$(C_{17}H_{35}COO)_3C_3H_5 + 3NaOH \longrightarrow 3C_{17}H_{35}COONa + C_3H_5(OH)_3.$$

This chapter is concerned with the production of soap base from fats using the fully boiled process, and more modern processes derived from it.

SOME PRINCIPLES

Phase Diagram

Fig. 5.1 shows diagrammatically a simplified phase diagram for a typical soap (say 80% tallow and 20% coconut oil) at about 90–100°C, indicating only phases which are relevant to the soapmaking process. The shapes of the various areas are influenced by the fat blend, particularly its content of lauric oils. The main electrolytes are salt (sodium chloride) and caustic soda (sodium hydroxide); but some sodium carbonate and sodium sulphate (derived from salt returned from the glycerine recovery process) may also be present. Figures quoted by various authorities for the relative salting out powers of these electrolytes vary, but roughly one part of salt is equivalent to 0.85 parts of caustic soda or 2.0 parts of sodium carbonate. The effects of the various electrolytes are additive, and are commonly expressed in terms of salt.

Area A on the diagram represents neat soap, or soap base, the product of the soapmaking operation, from which most products are made. It contains about 67–70% anhydrous soap, or 61–64% TFM.

Area J represents curd, or grained-out soap, sometimes called kettle wax. As discussed later, the high anhydrous soap (or TFM) content indicated on the diagram for this phase is somewhat misleading in that the curd soap produced by settling (or even centrifugal separation) contains a variable amount of enmeshed lye.

Area B represents 'middle soap'. This phase appears as 'water lumps' (possibly augmented by acid soap) if the electrolyte content is allowed to fall too low in any part of the vessel. The area is avoided by experienced soapmakers, as, once formed, water lumps are difficult to eliminate.

Area D is nigre, an impure solution of soap in water.

Area L, an extension of D, represents lye. As the electrolyte content increases, this phase changes from a thin, low soap, nigre, or nigre lye, to an aqueous solution which contains hardly any soap, known as lye.

Area M is a two-phase area which, theoretically at least, can separate into curd and lye. A mixture represented by point x, comprises curd x_1 and lye x_2. The proportions of the two phases are given by the relationship

$$\frac{\text{quantity of curd}}{\text{quantity of lye}} = \frac{xx_2}{xx_1}$$

that is the quantities of the phases are inversely proportional to the lengths of the tie lines from the mixture point x to the phase points x_1 and x_2.

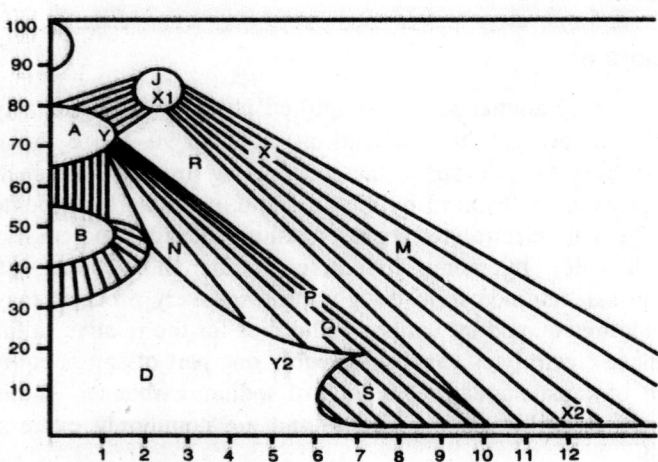

Fig. 5.1. Simplified diagram for phases significant in the production of soap base by the fully boiled process.

Similarly, area N is another two phase area representing mixtures of neat soap and nigre. Mixture of composition y can separate into neat soap y_1 and nigre y_2, in the ratio

$$\frac{\text{quantity of neat soap}}{\text{quantity of nigre}} = \frac{yy_2}{yy_1}$$

Area P represents the neat soap-lye zone. The so-called 'bay area', S, shows the equilibrium between nigre and lye.

The triangular area Q represents mixtures which separate into three phases of fixed compositions represented by the points of the triangle, neat soap, nigre and lye. Similarly, area R.

Saponification

Important practical points concerning saponification are :

(i) Fats and aqueous caustic soda are not miscible at temperatures which can be achieved without pressures well above atmospheric; and the reaction with neutral fats, particularly those other than the lauric oils, does not start readily. It is autocatalytic, that is catalysed by the product of the reaction, soap. Hence, the reaction rate accelerates greatly until most of the fat has reacted, when it slows down again. The soap formed/time curve is, therefore, somewhat S shaped as shown diagrammatically in Fig. 5.2. The slow initial phase is a heterogeneous, surface, reaction helped by vigorous agitation and the formation of an emulsion stabilised by soap. The main part of the reaction is homogeneous, with the fat probably 'dissolved' in the soap micelles. The final stage is slow because the concentration of residual fat becomes low. In factory practice, the slow initial stage rarely causes much difficulty because fatty acids present in most soapmaking fats react readily to provide some soap; and because saponification is commonly carried out in a way which adds the reactants to a mass of soap. As with other chemical reactions, an excess of caustic soda help to minimise the residual unreacted fat, but steps have to be taken to avoid an unnecessary excess of alkali in the lyes which are transferred from soapmaking to glycerine recovery. One way is to carry out the saponification in two stages; in the first an excess of fat is allowed so as to reduce the free alkali in the lye, and in the second the saponification is completed, using an excess of alkali, and the lye containing this excess is passed to the first stage. At one time, soapmakers used drastic treatments to minimise free fat in which the soap mass was boiled for long periods with a substantial excess of caustic soda, using either open, or closed, steam coils. This process certainly produced soaps with excellent keeping properties. Prolonged heating with caustic soda possibly breaks down some components in what is normally regarded as the unsaponifiable portion of the fat.

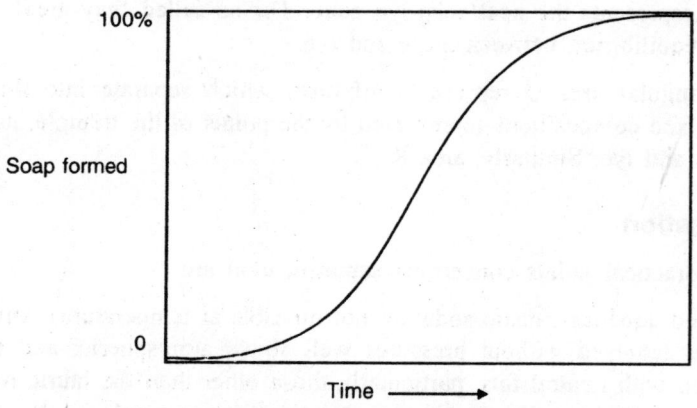

Fig. 5.2. Rate of saponification curve for neutral fat.

(ii) Like other reactions, saponification is accelerated by increases in temperature; and some processes use temperatures which can only be achieved in an aqueous system by the use of elevated pressures.

(iii) When some soap is present, too much electrolyte will salt, or grain, out the soap and greatly slow down the reaction. Too little will cause 'bunching', or thickening, so that it is important to ensure that the electrolyte content of the saponifying mass is within the correct range at all times. The actual requirement depends upon the composition of the fatty acids in the vessel, that is upon the fat blend. Caustic soda is a powerful electrolyte, but allowance must be made for the fact that it is continuously being removed by the reaction.

The '66%' Rule of Wigner; and Related Matters

This work is based upon examination of the results of large numbers of full scale boils of soap, as well as upon laboratory and pilot scale experiments, so-called '66% rule', which says 'a soap curd may be regarded as consisting of soap hydrate containing 66% of fatty acids, and of lye identical in composition with that from which it has separated'. This does not necessarily imply the physical existence of a soap hydrate not containing any salt; and the nearest approach achieved by Wigner in removing additional lye from a curd by settling and centrifuging in the laboratory contained about 64% TFM and 0.3% NaCl. Wigner deduced his rule as an empirical observation from a study of large numbers of boils, which showed that when a pan contents has been grained and settled,

$$\frac{\text{Percentage salt in soap curd}}{\text{Percentage salt in lye}} = 1 - K_s \frac{\text{Percentage TFM in curd}}{100}$$

K_s is nearly constant for almost all soaps, with a range of about 1.50–1.53, average 1.515. If electrolytes other than salt, particularly caustic soda, are present in significant quantities, equivalent salting out figures must be added to the salt concentrations. By substituting salt in soap curd = 0, it follows that for a salt-free curd, the percentage TFM in curd equals $100/K_s = 100/1.515 = 66\%$. Table 5.1 gives some analytical data for full scale boils of different types of soap presented by Wigner in support of his rule.

Wigner also found that the 66% rule could be applied to fitted pans using 'free solution' in place of lye. Free solution is defined as water which is not combined in the 66% TFM hydrate. For the neat soap which separates when a fitted pan is allowed to settle, Wigner says that 'neat soap is a soap hydrate of a composition which varies according to the concentration of salt in the free solution, always containing more water than the hypothetical 66% hydrate, and containing a small amount of salt which increases with the increase in the degree of hydration'. Figures in support are given in Table 5.2. It will be seen that as the concentration of electrolyte in free solution falls and the fit becomes 'finer', or 'closer'.

Table 5.1. Data in support of the 66% rule for soap curds.

TFM % in soap	NaCl % in lye	NaCl % in soap		K_s†	TFM % in soap hydrate
		Found	Calculated		
53.7	9.5	1.76	1.76	1.518	66.0
52.3	9.1	1.87	1.89	1.520	65.9
54.6	9.0	1.44	1.56	1.538	65.0
57.8	8.8	1.10	1.10	1.514	66.0
56.0	8.7	1.33	1.32	1.513	66.1
53.5	9.1	1.69	1.71	1.522	65.7
57.1	8.2	1.17	1.10	1.501	66.6
57.8	9.4	1.17	1.17	1.516	66.0
58.0	9.1	1.12	1.10	1.512	66.2
56.7	9.2	1.34	1.30	1.506	66.3
54.7	9.6	1.68	1.65	1.508	66.3

† The column for K_s has been added by the present author.

Table 5.2. Data in support of the 66% rule in fitted soaps.

Sample	NaCl in free solution %	TFM%	Neat soap NaCl% on sample	NaCl% at 63% TFM	Nigre TFM%
1	11.0	63.2	0.47	0.47	< 1
2	10.9	63.35	0.435	0.43	1.2
3	10.7	63.3	0.425	0.42	3.5
4	10.3	63.2	0.435	0.43	9.6
5	10.25	63.2	0.44	0.44	8.6
6	9.3	62.8	0.445	0.45	21.0
7	8.25	62.35	0.45	0.45	32.6
8	7.85	61.85	0.48	0.49	36.2
9	7.55	61.3	0.53	0.54	37.2
10	7.25	60.85	0.55	0.57	39.3

(i)   The TFM content of the nigre increases very substantially.

(ii) The TFM content of the neat soap falls a little.

(iii) The salt content of the neat soap adjusted to the figures at 63% TFM, or any other fixed level such as 79% for a dried toilet soap, is nearly constant for samples 2–7, but tends to rise as the fit becomes finer. It is also a little higher for sample 1 separated from a nigre lye rather than a nigre.

It is also found that soap given a coarse, or open, fit settles much more rapidly than when the fit is close, or fine; this is due to the greater difference in density between the phases, as well as to changes in viscosity.

These results conflict slightly with what might be expected from the phase diagram as usually drawn, for example in Fig. 5.1, in that the electrolyte content of the neat soap, as well as its TFM content, should fall as the fit becomes finer. As shown in Table 5.3, this latter trend is found in results reported by Palmqvist for centrifugal separation of neat soap and nigre in a De Laval continuous soapmaking plant. Phase equilibrium is not necessarily reached with either gravity or centrifugal separation; and degree of separation can also affect the compositions obtained. Some practical implications are :

(i)   Rapid separation and the low TFM content of the nigre/nigre lye, make open fits attractive whenever they can be used.

(ii) Household soaps made by the traditional frame process require fine fits with long settling times. On the other hand, household soaps made by the modern spray chill/vacuum plodding method usually benefit from an open fit and relatively incomplete separation, which leaves a high salt content in the soap. This type of soap is also generally acceptable for soap powder manufacture.

(iii) Traditionally, bases for toilet soaps and soap flakes were given rather more open fits than those for framed household soaps. This was understood to give lower electrolyte contents.

Table 5.3. Centrifugal separation of fitted soap in a continuous soapmaking plant.

Recorder		Fitting column (that is soap before separation)				From separator Neat soap		Nigre	
Index	Distance from minimum	Composition by volume			NaOH + NaCl %	NaOH + NaCl %	NaOH + NaCl %	TFM %
		Neat soap %	Nigre %	Lye %				
37	0	84	16	–	1.01	–	–	–
41	4	85	15	–	1.12	0.36	4.55	31.6
43	6	85	15	–	1.15	0.43	5.11	29.9
45	8	86	14	–	1.17	0.47	5.54	28.0
50	13	86	14	–	1.23	0.49	6.06	–
53	16	87	13	–	1.25	0.53	6.29	21.2
57	20	87.5	12	0.5	1.29	–	6.70	18.1
62	25	88.5	10	1.5	1.33	0.57	7.23	14.5

(iv) Modern continuous soapmaking plants using centrifuges for the final separation tend to use quite open fits, often with the separation of a very thin nigre, or nigre lye.

Some of these points are elaborated in subsequent sections. Wigner himself mentions that there are certain exceptions to the 66% rule; for example, when a groundnut oil soap is boiled for long periods with a lye having a high salt content in order to produce a curd soap which is as hard as is possible from GNO. The rule is, however, commonly accepted as applicable in all normal soapmaking operations; and some applications are mentioned later.

PRODUCTION OF SOAP BASE BY THE TRADITIONAL METHOD IN SINGLE VESSELS

Plant

Pans, sometimes called kettles (particularly if they are covered), for the manufacture of soap base can have a range of shapes and sizes. Cylindrical pans are common, but one major company standardised on 14 ft (4.27 m) cube vessels which fit conveniently together in groups of four, with alleyways between each group. Whatever the shape, they should have a slightly dished base so that lyes (or nigres) can be run-off completely from the bottom discharge valve, or cock. The heat loss should be as low as possible and the settling distance not too great; and the cube is probably a good compromise. Neat soap and nigre do not separate properly at temperatures below about 70–75°C, so that pans should not be too small and should be lagged for this reason, as well as to minimise steam consumption. Very small pans used for experimental boils should usually be steam jacketed, so that a little steam can be used to prevent unduly rapid cooling while neat soap and nigre are separating. Pan covers are also desirable to reduce heat loss from the soap surface, and to reduce risk of contamination with dirt from the atmosphere; but they are by no means always fitted. Pans are usually made of mild steel, but it has been found useful to lien the upper parts with stainless steel, or other corrosion-resistant material, to reduce contamination of the soap with iron.

The usual fittings, in addition to a bottom discharge facility are :

(i)   An open steam coil to provide both heat and agitation. Closed steam coils are no longer necessary, or usual.

(ii) A skimmer pipe, which can be raised or lowered, and through which the neat soap layer can be withdrawn when the settling is complete and the pan is 'cleansed'; that is the neat soap removed.

(iii) Pipes arranged to discharge into the pan above its contents for fat, caustic soda solution, water, and other materials. Salt can be shovelled over the side of the pan, but it is now commonly added in the form of saturated brine. Particularly in large panrooms making a variety of soap bases, a fairly comprehensive pipe system to permit transfer from pan to pan, and to other departments, is desirable.

Saponification

Traditionally, saponification is carried out by pumping the fat into the pan whilst the soapboiler adjusts the additions of caustic soda, salt, water and steam, so as to maintain the conditions necessary for steady reaction. This is a skilled operation; too much electrolyte can cause the soap mass to grain, and too little to bunch or thicken. If the pan contents are inadvertently grained with too much caustic soda, the reaction slows and great care is necessary when conditions are adjusted to re-establish it; otherwise the heat of reaction may cause the pan contents to boil over.

Nigre and other soap from the previous boil is often left in the pan to provide the soap needed to aid saponification; although it may be found desirable to remove some, or all, the nigre from time to time, for transfer to a lower quality base, or for it to be cleaned by graining, or fitting.

Rosin, which contains no glycerol, is preferably added to the rest of the soap after as much as possible of the glycerol has been removed. Solid rosin can be tipped into the pan, but, particularly when substantial proportions are to be included in the fat blend, it is better to partially saponify it separately. Rosin contains complex acids related to the terpenes, such as abietic and pimaric, which react readily with sodium carbonate solution. When rosin was widely used in soaps, it was common practice to saponify it partially with sodium carbonate in a separate vessel, grain, and discard the dirty lye (which rises to the top instead of sinking to the bottom as with ordinary soap lyes). The partially saponified rosin soap was added to the rest of the soap in the pan at a convenient stage, preferably after most of the glycerol had been removed; and some caustic soda was then needed to complete the saponification.

Saponification in a pan can be made easier by the use of a saponification jet, which can be made quite simply. One design which has been used widely is shown in Fig. 5.3. The partially saponified emulsion of fat, soap and caustic soda solution is discharged through a flexible pipe, which can often be arranged to serve more than one pan; the metal hose being securely clamped to the pan in use. When the jet is used as an aid to saponification in a pan it is only necessary to regulate the flow of caustic soda solution roughly in proportion to the flow of fat. Steam is used in the jet as required, depending upon the temperature of the fat and the reaction rate. The mixture should flow into the pan as an emulsion in which saponification is taking place vigorously. The reaction is completed by boiling in the pan, with additions of caustic soda and salt as necessary.

Graining or Washing

After saponification has been completed, the pan content is grained with salt. When in the grained state, clear lye can be seen between the curds as the pan content boils gently, or in a sample lifted on a paddle. The pan is allowed to settle and the saponification lye is run off. The soap is 'closed' by addition of water and grained again with salt. During this operation sufficient water must be added to produce a wash lye of the desired size, say 0.3 tonnes of lye per tonne of soap curd; care being taken to avoid the formation of water lumps. The process is repeated as many times as is worthwhile, it being appreciated that the glycerol content of each wash is lower than that of the previous wash, so that, using this process, it is uneconomical to wash to reduce the glycerol in soap to a very low level. The capacity of the glycerine recovery plant may also be a limiting factor.

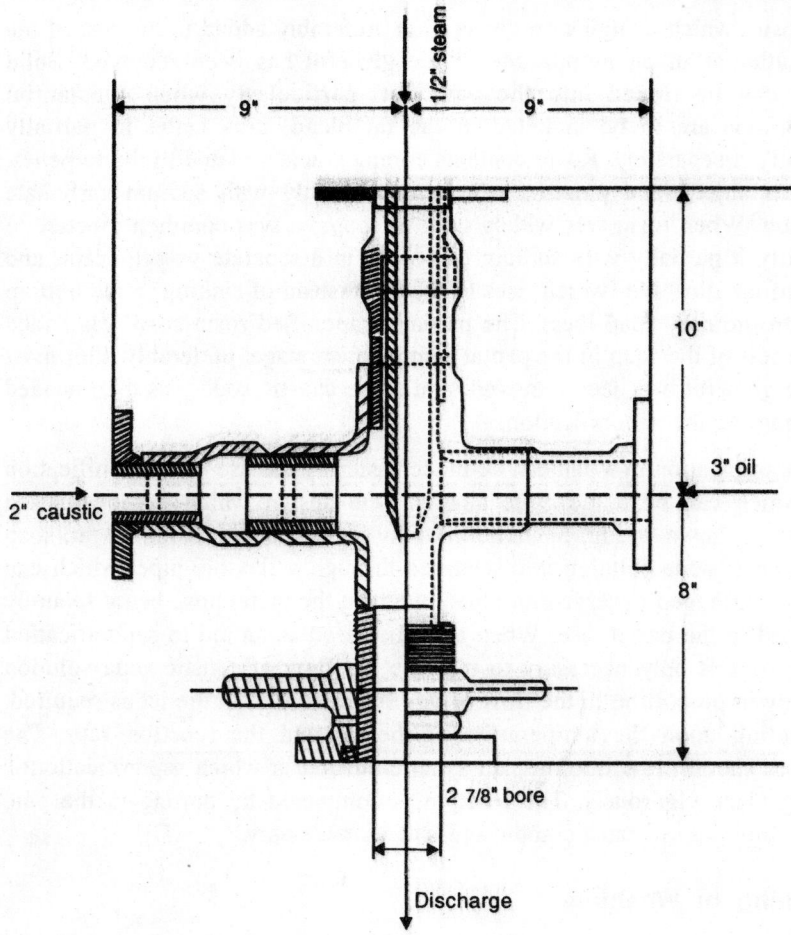

Fig. 5.3. Saponification jet.

Dirt is removed from the soap mass in the washes, but the primary objective is to separate the glycerol. As indicated in the previous paragraph, there is an economic balance between the yield (percentage of the glycerol in the fats which is recovered) and the average concentration of glycerol in the lye produced, because the cost of lye treatment and evaporation is related to the lye concentration; that is to the quantity of lye which must be purified and the quantity of water which must be evaporated to produce a tonne of crude glycerine.

It is found that the glycerol is distributed between curd and lye approximately in proportion to the total water in each layer; and K_G is defined as concentration of glycerol in curd/concentration of glycerol in lye. The lower K_G the more effective the wash.

Alternatively, it can be said that the drier the curd, that is the higher its TFM%, and the less enmeshed lye it retains, the better. This is analogous to the improved removal of dirty detergent solution from domestic washing which results if the articles are wrung, or spun, more effectively between rinses. This important point can be illustrated by some simple calculations. Assume that the lye contains 11.5% salt and 88.5% water plus glycerol. Consider first a very well settled curd at 60% TFM. Its anhydrous soap content is about $60.0 \times 1.085 = 65.1\%$ and, if it contains 1% salt (substantially in accord with the 66% rule), its content of water plus glycerol is $100.0 - 65.1 - 1.0 = 33.9\%$. As water and glycerol occur in constant proportion in the layers,

$$K_G = \frac{\text{concentration of water in curd}}{\text{concentration of water in lye}}$$

$$= \frac{\text{concentration of glycerol plus water in curd}}{\text{concentration of glycerol plus water in lye}}$$

$$= 33.9/88.5 = 0.383.$$

Less well settled curds are equivalent to mixtures of the 60% TFM curd and additional lye. For a mixture of 1 part 60% TFM curd and 0.05 parts lye :

	Soap	Salt	Water	K_G
from 1 part 60% curd	0.651	0.010	0.339	0.383
from 0.05 parts lye	–	0.006	0.044	–
in 1.05 parts new curd	0.651	0.016	0.383	
in 1 part new curd	0.620	0.015	0.365	0.365
				0.885 = 0.412

$$= 57.1 \text{ TFM}$$

Similar calculations and graphical interpolation provide the data for Table 5.4.

A carefully grained and well settled pan should result in a curd at about 58% TFM, giving a K_G of about 0.4, but it is easy to produce less dry curds. It is particularly important to use the minimum concentration of electrolyte which is sufficient to produce lye substantially free from soap, as coarse, overgrained, curds occlude more lye than those which are only just grained. The graining details are given in table 5.5 of a 'washer' type soap.

Table 5.4. Variation in K_G with dryness of soap curd.

TMF% in curd	Soap %	Salt %	Water + glycerol %	Partition coefficient K_G
60	65.1	1.0	33.9	0.383
58	62.9	1.3	35.8	0.405
56	60.7	1.7	37.6	0.425
54	58.5	2.0	39.5	0.447
52	56.3	2.3	41.4	0.468
50	54.2	2.6	43.2	0.488

Table 5.5. Figures for graining of a washer type soap.

NaCl in free solution %	TFM %
10.5	4.15 (a thin nigre)
10.84	0.18
11.4	0.14
11.7	0.17
12.0	0.17
12.5	0.12

It will be seen that nothing is gained by using more salt than about 1.0% over that needed for a very rough fit giving a thin nigre. The actual concentration of salt required depends upon the fat blend. That used in Wigner's 'washer' soap is not quoted, but it may well have been something like 30% hard fat, 30% lauric oil, 30% soft oil and 10% rosin. Other factors equal, the quantity of lye needed to remove a given percentage of glycerol from the soap varies with the number of washes used. Wigner calculates the figures given in Table 5.6.

Table 5.6. Quantity of lye needed to obtain different percentage recoveries of glycerol with various numbers of washes (Wigner).

Number of washes	Parts of lye per part of curd for yield of	
	80%	85%
1	2.47	3.92
2	1.53	2.12
3	1.31	1.83
4	1.24	1.69
5	1.18	1.60
6	1.14	1.54
7	1.12	1.48

As might be expected :

(i) The quantity of lye which has to be produced with the traditional system of washing for a given percentage recovery is reduced very substantially by increasing the number of washes from one to three or four; but the benefit which accrues from more washes becomes increasingly small (also very large individual washes might well be impractical because of pan size).

(ii) With this system of washing yields of glycerol above about 80% become increasingly difficult to achieve.

Settling time is also very important, so that for a given total quantity of lye produced, it may be counter-productive to increase the number of washes further if this leads to an inadequate time to settle each wash.

A useful, but very simple arithmetical calculation shows that if each wash removes, say, 0.4 parts of the glycerol present in a curd when it is washed, so that 0.6 parts are left in the washed curd, n washes will leave 0.6^n in the final washed curd.

If the weakest lye which it is economical to process contains, say, 2% of glycerol, and K_G equal to 0.4 is achieved, the curd from which the last lye came contains $2 \times 0.4 = 0.8\%$ glycerol. It follows that soaps made from fat blends with a high glycerol content and those from fats with a low content, should both be washed down to 0.8% glycerol; so that more lye should be produced from the glycerol rich soaps.

When, as is usual, salt recovered in the process of glycerine lye evaporation is re-used in soapmaking its glycerol content has a considerable adverse effect upon the efficiency of glycerol recovery, or on the quantity of water which has to be evaporated to achieve a given recovery, particularly when the level of recovery is high. With regard to glycerol recovery, it is best to use in the final washes whatever new salt is needed to make up for losses. From the angle of soap quality the new salt is better used in the manufacture of the highest quality bases; but, for convenience, new and recovered salt are often used indiscriminately.

Fitting

The washed curd should be well settled, preferably overnight, before it is fitted, because the quantity of water in the fitted soap prior to settling affects the yield of neat soap, as can be shown by a simple materials balance. Let the total mass of material at the completion of the fit equal 1 and the mass of nigre produced equal x, so that the mass of neat soap equals $(1-x)$. Then if :

a = TFM% of the total pan contents

b = TFM% of the nigre, and

c = TFM% of the neat soap,

a TFM balance can be written

$$1 \times a/100 = bx/100 + c\,(1-x)/100, \text{ or } x = \frac{c-a}{c-b}$$

In words this can be written, nigre produced (as a fraction of the whole mass) :

$$= \frac{\text{TFM\% of neat soap} - \text{TFM\% of pan contents}}{\text{TFM\% of neat soap} - \text{TFM\% of the nigre}}$$

Even more significant than the dryness of the curd, which affects a, is the nature of the fit used, which governs the TFM content of the nigre, b, as discussed earlier; but this has to be decided in relation to quality requirements, including removal of dirt. Table 5.7, shows examples of the influence of different pan drynesses with two types of fit. The significance of type of fit is illustrated further in Table 5.8, with some additional columns from. Recalculation suggests minor errors in some of the figures, but these do not affect the conclusions. The curd is assumed to contain 58% TFM and 1.45% salt, equivalent to 12% in lye. Columns 1–3 are taken from smoothed curves drawn from the data in Table 5.2. Column 4, the TFM% in the total mass after fitting, assumes that only water is added to the curd; and is calculated for the first line by saying that free solution (occluded lye) in 100 parts of curd is $(1.45 \times 100)/12 = 12.08$. This is diluted to $(12.08 \times 12)/9.65 = 15.02$ to reduce its salt concentration from 12.0% to 9.65%. Hence water added = $15.02 - 12.08 = 2.94$; and the TFM% in fitted soap is $(58.0 \times 100)/102.94 = 56.35\%$. Column 5 is obtained by application of the materials balance equation. Percentage nigre in total mass equals $(63.0 - 56.35)/ (63.0 - 18.5) \times 100 = 14.9\%$ (compare 15.2% reported, as shown in the table). Column 6 equals (column 2 × column 5)/column 4 = $(18.5 \times 15.2)/56.35 = 5.0\%$ (4.8% reported). Column 7 equals salt in curd recalculated as a percentage of the new total mass = $1.45 \times 100/102.94 = 1.41\%$. Column 8 equals new free solution as a percentage of new total mass = $(15.02 \times 100)/102.94 = 14.6\%$ (15.0% reported).

Table 5.7. Influence of dryness of pan contents and type of fit on loss of usable soap to nigre.

Average TFM% in pan	Nigre, percentage of pan contents		Percentage of TFM in nigre	
	Coarse fit yielding neat soap at 63% TFM and nigre at 18.5% TFM	Fine fit yielding neat soap at 61% TFM and nigre at 38.5% TFM	Coarse fit	Fine fit
56	15.7	22.2	5.18	15.3
54	20.3	31.2	6.98	22.2
52	24.6	35.6	8.80	26.4
50	29.3	49.0	10.84	37.7
48	33.8	57.8	13.02	46.2

Table 5.8. Additional information concerning the fitting operation (the curd is assumed to contain 58% TFM and 1.45 salt, equivalent to 12% in lye).

Line	Neat soap TFM % (1)	Nigre TFM % (2)	NaCl% in free solution (3)	TFM % on total mass (4)	% Nigre on total mass (5)	% of total TFM in nigre (6)	NaCl% in total mass (7)	% Free solution in total mass (8)
a	63.0	18.5	9.65	56.35	15.2	4.8	1.41	15.0
b	62.5	28.5	8.75	55.52	20.6	9.9	1.39	16.6
c	62.0	33.5	8.15	54.88	25.2	14.6	1.37	17.8
d	61.5	36.5	7.70	54.34	28.8	18.1	1.35	18.85
e	61.0	38.5	7.35	53.89	32.0	21.3	1.34	19.75
f	60.5	40.3	7.10	53.65	34.4	24.0	1.33	20.4
g	60.0	41.5	6.85	53.18	36.8	26.4	1.33	21.2

The composition of the pan contents needed for any selected type of fit can be obtained either by adding water when the soap mass has too high an electrolyte content, or by adding electrolyte solution when it is too low. In addition, it is necessary to adjust the caustic soda content so that, after the electrolyte has been partitioned as indicated in Table 5.3, the neat soap will have the required alkalinity. Because of the large 'loss' (to reprocessing) of soap when a rather wet curd is given a fine fit, it is often advisable to use a rough fit prior to the finishing fine fit. This removes water and electrolyte, together with a relatively small mass of soap because of the low TFM content of the rough fit nigre. Hence, the quantity of the nigre from the final fit is small and its removal does not involve the rejection of very much soap despite its high TFM.

Fitting, as carried out in the traditional way, requires skill and experience; and is best learned by personal tuition by an experience soapboiler. The soap mass is boiled extensively; and the end point is judged by the look of the boiling soap and by the way it leaves a heated trowel. The soapboiler's trowel test is carried out using a bricklayer's trowel. The trowel is heated by immersion in the boiling soap, after which the adhering soap is scraped off. The trowel is then again immersed, drawn out edgewise and tilted to about 60° to see how the soap runs off. A soap which is too close for a typical medium/fine fit travels slowly, is apt to stick to the trowel, and the trailing edge tends to leave streaks of soap adhering to the blade. A soap which is too open slides off the trowel fairly rapidly and breaks into a number of small pieces. In the ideal condition the soap film should be transparent; and should slide slowly and steadily from the trowel, leaving the surface of the trowel clean and bright. It should break up into two, or three, large broad flakes. The trailing edge of the moving film should thin out gradually to an almost imperceptible transparent wafer of soap which may extend to $\frac{1}{4}$ inch behind the main body of soap.

The actual yield of usable neat soap is less than that which might be expected from the loss of nigre, because crusts of solid soap form on the soap surface and on the sides of a pan which has cooled during settling; and so cannot be cleansed with the liquid neat soap.

As mentioned earlier, soap given an open fit settles relatively rapidly, and neat soap suitable for soap powders, and for household soaps made using the modern chilling and extrusion processes, can commonly be cleansed after 1–2 nights settling, instead of the 3–4 nights needed after a medium/fine fit.

Wigner System of Quantitative Soapboiling

The data can be collected from experimental boils for the phase equilibria for the various fat blends used in a factory. If the composition of the pan contents is known, it is then possible to calculate what additions should be made to achieve a desired wash, or fit. A problem in the application of this idea is that steam condenses as the soap is boiled with open steam, so that it is difficult to know how much water is present. Also, the density of the pan contents is influenced by the bubbles of steam and total mass cannot be determined by staffing to measure volume. These difficulties can be over come by the use of a pneumercator type of gauge shown diagrammatically in Fig. 5.4. When air has bubbled gently through the soap and the compressed air supply is shut off, the column in the manometer balances the soap above the air orifice; and the manometer can be calibrated directly in terms of the mass of material in the pan, regardless of its density. Allowance must be made for the material in the pan dish. Water content is determined by difference between the total mass and the known additions. The theory explains in detail how the quantitative method of control of the traditional process can be achieved, together with the use of nomographs to minimise calculations. Possible extended use of this system has been overtaken by other developments to be described, which have made the traditional process largely obsolete, although it is still used by many small, and/or, less technically developed, companies.

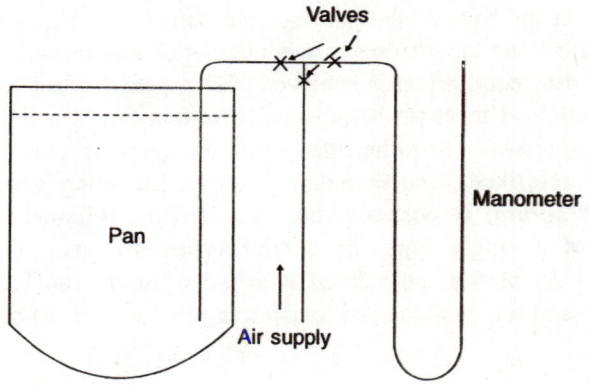

Fig. 5.4. Pneumercator gauge for measurement of pan contents.

DEVELOPMENTS OF THE TRADITIONAL PROCESS

One important weakness of the traditional process is the low glycerol content of the later washes. This results in the production of large quantities of weak lyes; and seriously limits the glycerol recovery which is economical, or even practical. Some weak liquors can be used to dilute the caustic soda, or to dissolve salt, but it has long been appreciated that counter current operation of the washing system is needed, as used in other extraction processes. Counter-current washing is complicated by the addition of water to close the soap and then of salt, or brine, to grain it in each washing stage; and a major advance, few years ago, was the discovery that soap can be washed quite satisfactorily by just mixing curd and lye, and allowing the mix to settle. All modern washing processes wash the soap in the grained condition in this way.

This principle was first applied in pans. Steam jets can be used to mix and transfer the pan contents; or a mechanical system can be used. Fig. 5.5 shows diagrammatically a system which was used satisfactorily for a number of years. Fig. 5.5(a) shows a fish-tail shaped nozzle, sometimes called a lye diffuser, through which lye is pumped into the upper part of the layer of soap curd in a pan. The nozzle is installed at the end of a jointed pipe system arranged so that the discharge is horizontal, which is also able to be adjusted to the level required just below the soap surface. Fig. 5.5(b) shows a group of four pans (diagrammatically in line, but, in practice, conveniently arranged in a square), with the essential pipe connections; and Fig. 5.5(c) shows the operation of one pan. Lye is transferred from one pan to the next by the pump through the lye diffuser; and is then circulated from the lye layer at the base of the pan into the curd layer. The action of the jet of liquid from the lye diffuser causes circulation and mixing of lye with curd, so that near equilibrium is achieved after a period of circulation. The pan is then allowed to settle before its lye is transferred and the curd is ready for another wash, or to be fitted. With an eight-pan unit (two sets of four as just described), the complete process, including several nights settling after fitting, is possible, but this is only suitable for a large production of a single type, or similar types, of base. For smaller productions a set of four pans fitted with lye diffusers can be used as a washing unit, and the washed curd transferred to other pans to be fitted and settled.

AN ADDITIONAL PRINCIPLE

The influence of quantity of water added, and the number of washing stages used, on the efficiency of glycerol recovery have been considered. The way in which the water is added is also important. The principle is simple, it being that fresh water should, as far as possible, be added at the last washing stage. This is because water added at the saponification stage is used only once for the removal of glycerol from the soap; whereas water added at the end of a multi-stage process is used several times. It is also to be noted that although processes can be divided into saponification, washing, and fitting stages, glycerol is transferred from the soap in each stage. Various liquors containing some glycerol, such as those from nigres, or from the concentration of soapstock, often need to be introduced into the soapmaking process. The principle for the most effective return of such liquors is that they should, as far as possible, be introduced into the soapmaking system at the point where the concentration of glycerol in the system is the same as that in the liquor. This discussion explains the point made earlier that it is advantageous as regards recovery to concentrate a soapstock, so that a portion containing the fatty matter, but relatively little water, can be added at the saponification stage, and a liquor containing a small percentage of glycerol can be introduced at a later stage.

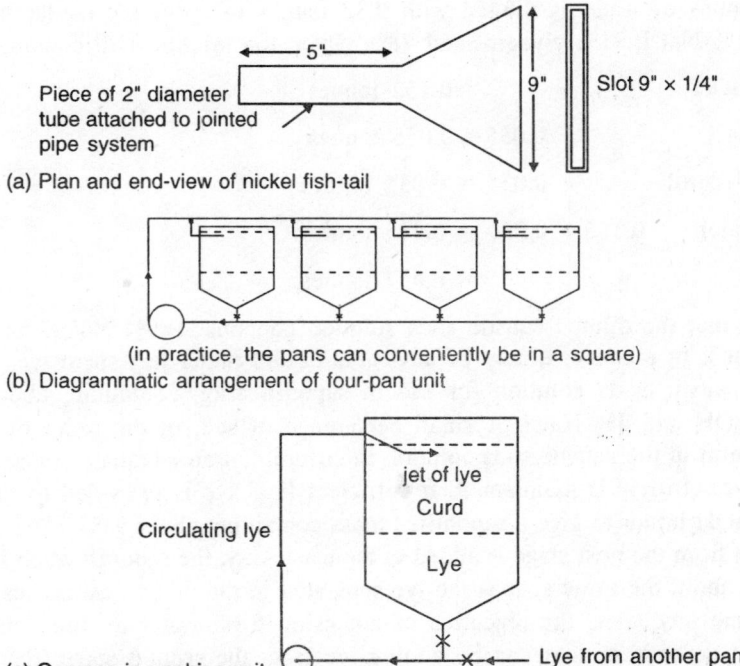

Piece of 2" diameter tube attached to jointed pipe system

9" Slot 9" × 1/4"

5"

(a) Plan and end-view of nickel fish-tail

(in practice, the pans can conveniently be in a square)
(b) Diagrammatic arrangement of four-pan unit

Circulating lye

jet of lye
Curd

Lye

(c) One pan in washing unit   Lye from another pan

Fig. 5.5. Lye diffuser system of washing soap in pans.

The saponification of 1 tonne of a typical 80% tallow 20% coconut oil fat blend require about 0.118 tonnes Na_2O, allowing 2% excess. This is equivalent to 0.317 tonnes caustic soda liquor at 102°Tw, containing 48% NaOH or 37.2% Na_2O. The 1.317 tonnes reaction mixture which is produced from 1 tonne of fat plus 0.317 tonnes 102°Tw caustic soda contains about 0.95 tonnes fatty matter, or 72% TFM. When mixed with liquors and grained, this will produce lye plus a curd containing, say, 58% TFM. To convert the 72% TFM soap to 58% TFM soap the total mass has to be increased to 0.95/0.58 = 1.64 tonnes; that is 1.64–1.32 = 0.32 tonnes of material must be added. This could be, and sometimes is, salt plus water; but water introduced in this way is added at the wrong end of the process and is used inefficiently. It is better to use clean spent lye; a procedure sometimes known as use of the 'lye absorption principle'. It may seem peculiar to return to the fresh soap mass glycerol which has already been separated, but the procedure is analogous to what is done in a stage of a counter-current washing process; and it can be shown to be beneficial both by calculation and in practice. The best way to return lye to the saponification may vary with the soapmaking process in use and various soda liquor containing $0.48 \times 0.317 = 0.152$ tonnes of NaOH and 0.165 tonnes of water is diluted with 0.32 tonnes of spent lye containing, say, 11% NaCl, 11% glycerol and 78% water, the mixture will contain :

NaOH	$0.152 + -$	$= 0.152$ tonnes
NaCl	$- \quad + 0.035$	$= 0.035$ tonnes
Glycerol	$- \quad + 0.035$	$= 0.035$ tonnes
Water	$0.165 + 0.250$	$= 0.415$ tonnes
		0.637 tonnes,

so that the diluted caustic soda solution contains 23.9% NaOH and 5.5% NaCl. In practice, it may be advisable to add rather less spent lye, to give a caustic soda solution for use in saponification containing about 30% NaOH and 4% NaCl. A small percentage of salt, of the order of 4%, is useful in the caustic soda solution to ensure that an adequate concentration of electrolyte is maintained. If sufficient final lye is re-cycled to the 48% NaOH liquor to give a saponified mass containing about 58% TFM and the lye from the next stage is added in the usual way, the saponification lye will be about the same size as the lye separated in the other washing stages. In some processes, the objective of not using fresh water to dilute the 48% caustic soda is achieved by adding lye from the second stage (that is the first wash if the saponification wash is not counted) instead of final lye. In this case, the volume of the saponification lye is less than that from subsequent washes by the volume needed to reduce the soap mass from 72% to 58% TFM.

THE COMPONENTS USED IN MODERN SOAPMAKING PROCESSES

Equipment for Saponification

Modern continuous saponification units use either :

(i) Devices to provide a very intimate mixture of fat and caustic soda solution, followed by completion of the reaction in a mass of soap.

(ii) Reaction in a mass of soap, which is usually recirculated, to which the reagents are added, and from which nearly completely saponified material is blend off.

Reaction in a circulating mass of soap is an application of what is called the dominant bath procedure. The soap may be circulated under atmospheric pressure, which limits the temperature to just over 100°C; or the system may be closed to that higher pressure and temperature can be used.

The saponification unit is shown diagrammatically in Fig. 5.6, uses a colloid mill to bring the fat and caustic soda solution into intimate contact followed by flow through a reaction tube and mixer.

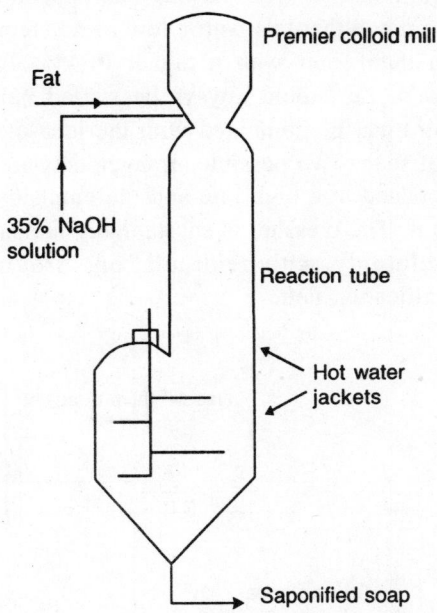

Fig. 5.6. Saponification unit.

Fig. 5.7(a) shows diagrammatically the design of the jet and 5.7(b) that of the saponification tank disclosed in the patent. Fig. 5.7(c) shows a more complex, but superior, design of jet; and it is possible to use several designs of saponification tank with various numbers of baffles.

The solutions containing up to 15% of electrolyte favour the formation of an oil in water emulsion, 15–20% no emulsion, and above 20% a water in oil emulsion; a water in oil emulsion is preferred. Conditions, including design of jet, temperatures and electrolyte content (including caustic soda) must be adjusted to give a good emulsion. If the fat blend is nearly neutral, some free fatty acid, or soap, can be added if thought necessary. The Fig. 5.7 also discloses the following example; a jet with $1\frac{1}{2}$ inch (38 mm) feed pipes and a $\frac{1}{4}$ inch (6.4 mm) steam jet fed with 70 psig steam is capable of emulsifying 7–8 tonnes of mixed fats containing 10–15% FFA per hour using a caustic soda solution containing 27–30% NaOH and 3% NaCl.

of emulsifying 7–8 tonnes of mixed fats containing 10–15% FFA per hour using a caustic soda solution containing 27–30% NaOH and 3% NaCl. Temperatures should be arranged so that the emulsion only heats to just about 100°C when left in a small vessel to saponify. Table 5.8 from the patent gives results obtained with various fats containing a range of FFA contents. It will be noted that fats with a low FFA% required more time for self-saponification than other with a higher FFA%. To avoid risk of jet blockage the flow of fat should always be started before that of caustic soda. The principles can be combined with the idea of saponification in a circulating mass of soap. Two possible arrangements are shown in Fig. 5.9. The soap is recirculated at a high rate and the emulsion from the jet falls, or overflows, into it. The pressure is substantially atmospheric. The jet and tank system, preferably with recirculation, provides a sound, but inexpensive, saponification unit.

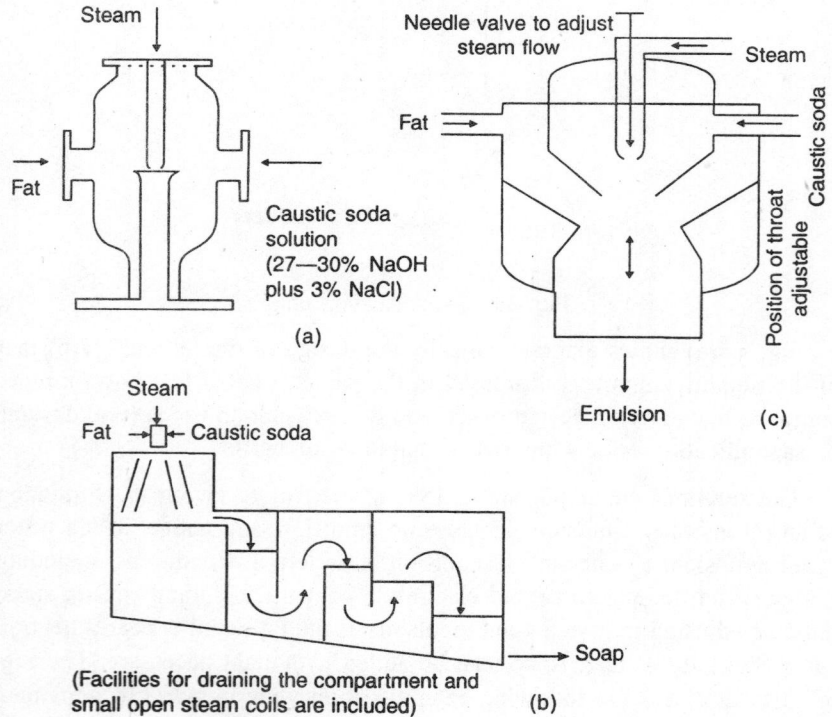

Fig. 5.7. Jet and tank system of saponification.

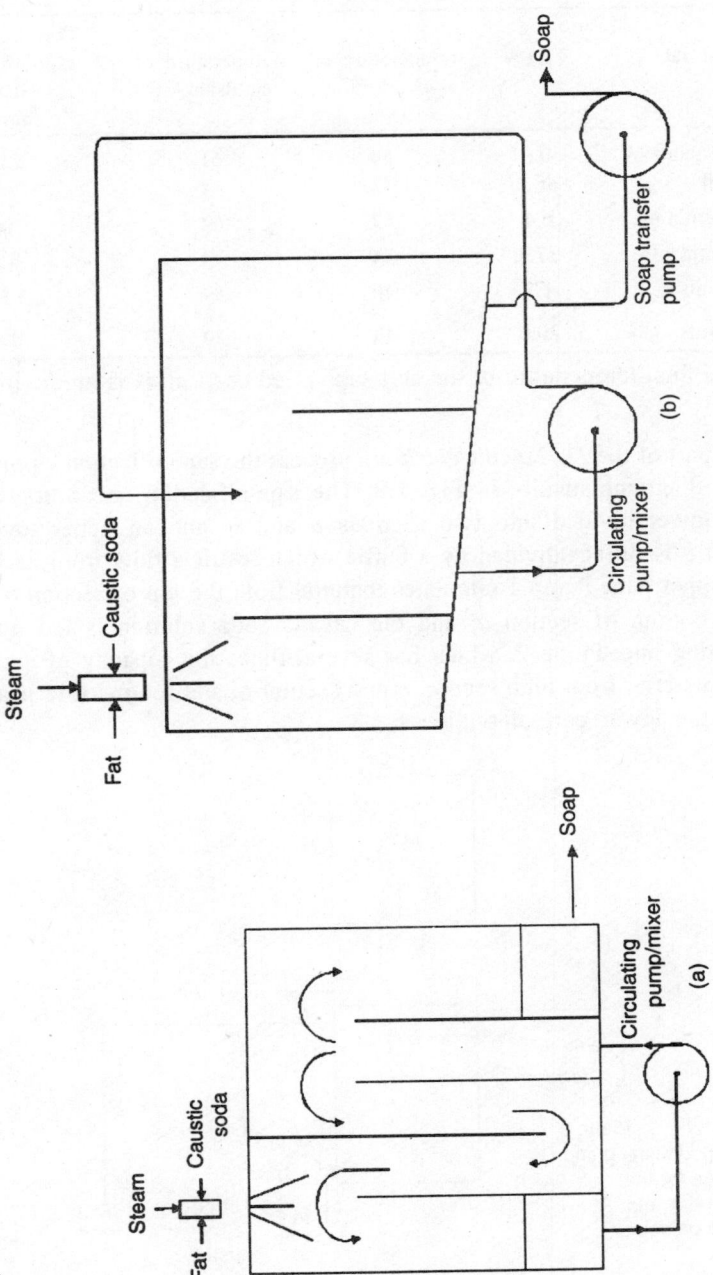

Fig. 5.8. Jet saponification using tank with recirculation.

Table 5.9. Data from continuous saponification unit.

Type of fat	FFA%	Temperature of of oil, °C	Temperature of emulsion, °C	Time for self-saponification 98-100% (mins)
Neutral palm oil	0.6	54	61	23
Palm oil	46.2	52	85	2
Palm kernal oil	6.4	52	62	5
Groundnut oil	17.4	54	74	4
Mixed oils	1.7	46	59	3.5
Mixed oils	20.0	43	70	1

The final temperature of the self saponified soap mass is approximately 100°C.

As part of the *De Laval Centripure* process the saponification equipment shown diagrammatically in Fig. 5.9. The saponification unit comprises a lower tower divided into two sections *a* and *b*, and an upper tower *c*. Section *b* is further divided by a baffle which restricts flow from its lower to its upper part. Pump 1 circulates material from the top of section *b* back to the bottom of section *a*; and the caustic soda solution is fed into its circulating line. Pump 2, which has several times the capacity of pump 1, draws material from both section *a* and section *b*; and returns it to near the top of the lower part of section *b*.

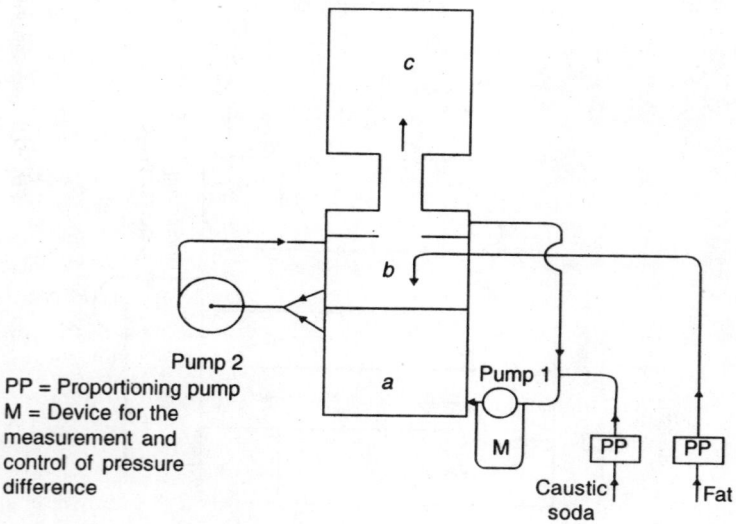

Fig. 5.9. Saponification unit.

The fat is introduced into section b. The whole Centripure plant is hermetically sealed; and the saponifier is understood to operate at up to about 125°C under sufficient pressure to keep most of the water liquid. Contact parts are of stainless steel. After only a few minutes in section b the soap passes to the upper tower c, which is a holding vessel. Saponification is said to be about 99.8% complete on transfer from b to c, and over 99.95% complete when the soap leaves c, with only about 0.1% free alkali. Earlier designs used caustic soda solution at about 28% NaOH, giving a saponified mass at about 62% TFM, and with this design lye absorption can be practised by dilution of the strong caustic soda with spent lye.

A more robust unit which can produce soap at 72–73% TFM by the use of 48% NaOH caustic soda. Spent lye is then added to the mass leaving the saponifier to achieve both lye absorption and cooling. An important feature of saponifier is a control system based upon changes in viscosity which occur when the quantity of caustic soda added varies slightly from the stoichiometric quantity (that is the theoretical requirement for complete reaction). Pump 1 is of the positive rotary, gear, type which transfers soap at a constant rate, so that the pressure rise induced by the pump increases as viscosity increases. This is measured and controlled by a pneumatic controller, indicated diagrammatically in Fig. 5.9 as a U tube manometer. Fig. 5.10 shows roughly how viscosity, measured as a differential pressure, varies with excess, or deficit, of caustic soda at three different TFM levels. The minimum, or maximum, point can easily be determined by simple manipulation of the flows. Operation is normally on the right-hand side of the point of inflexion; and the device is said to be capable of control to ±0.01% NaOH. From Fig. 5.9 it will be appreciated that the caustic soda content at the point of measurement is not the same as that in the soap leaving the saponifier. The De Laval saponifier is excellent, but very costly.

The saponifier and part of the washing section, are shown diagrammatically in Fig. 5.11. The first wash section is shown here because it also forms part of the saponification process. Sharples number the units in their plant 1, 2, 3 and 4, starting at the saponification end. They also call the liquor leaving a separator lye n; but when this liquor is used in the previous section, after addition of caustic soda, or other incoming liquors, as necessary, it becomes reagent $(n-1)$. The fat and the first reagent, comprising second lye plus 48% caustic soda, are introduced into saponification vessel 1. The soap is circulated externally by pump and a portion is bled-off into the holding vessel, saponification vessel II, from which it passes to the centrifuge S_1. Saponification is not complete at this stage and the spent lye separated is low in free alkali. The soap from S_1 is mixed with reagent 2 in mixer 2 where saponification is completed with liquor containing an adequate excess of caustic soda. Additional 48% caustic soda can be added to R_2 as indicated, but sufficient usually remains from additions later in the process and the provision for addition to R_2 is not normally included. It will be noted that the *Sharples process* uses two-stage saponification for the reasons. Actually provision is made for the completion of saponification in the washing stages of most processes, even though free fat may not be left deliberately after the main saponification. The *Sharples process* also includes the version of the lye absorption principle which uses second stage lye.

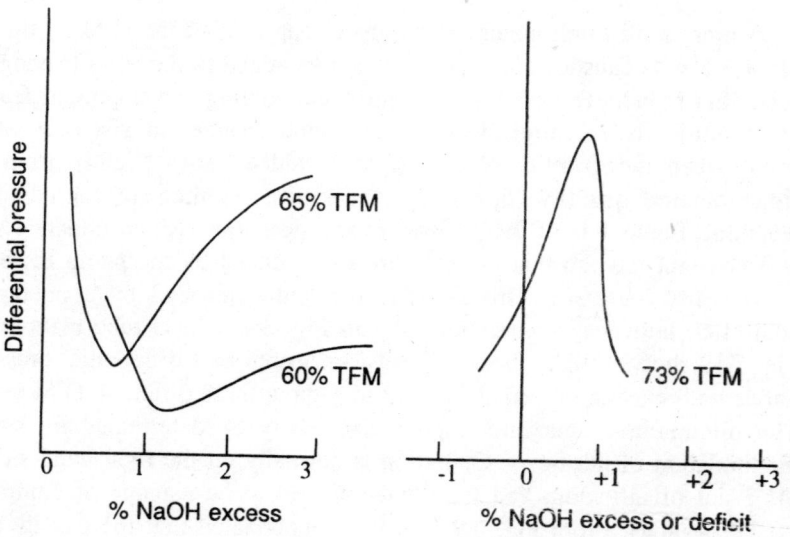

Fig. 5.10. Variation in viscosity, or differential pressure, with excess, or deficit, of caustic soda – De Laval saponifier.

Maccaniche Moderne and *Mazzoni* continuous soapmaking plants use very similar saponification systems shown diagrammatically in Fig. 5.12. Saponification takes place in a four compartment autoclave operating at about 120°C under two atmospheres pressure; and soap is re-cycled. The soap then moves to the mixer/cooler, which has four compartments with axial mixers. Saponification is substantially completed in this vessel. The first two compartments have water cooling coils and the temperature is reduced to about 90°C; this reduces solubility of soap in lye, and the two phases separate so that they can be parted in the static, gravity, separator to which the mixture is passed.

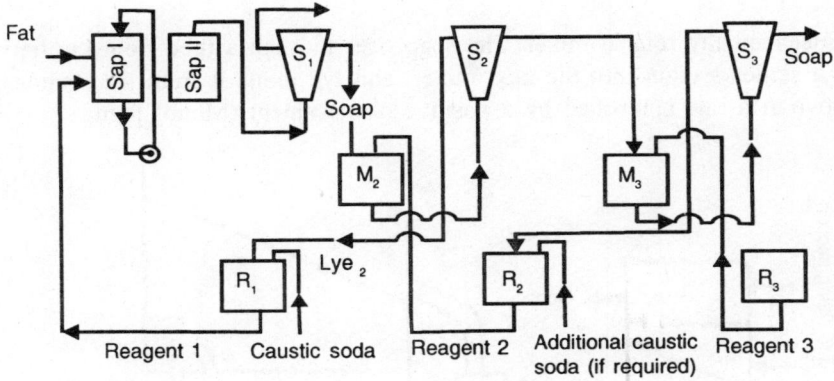

Fig. 5.11. Sharpless soapmaking plant. Saponification including part of washing system.

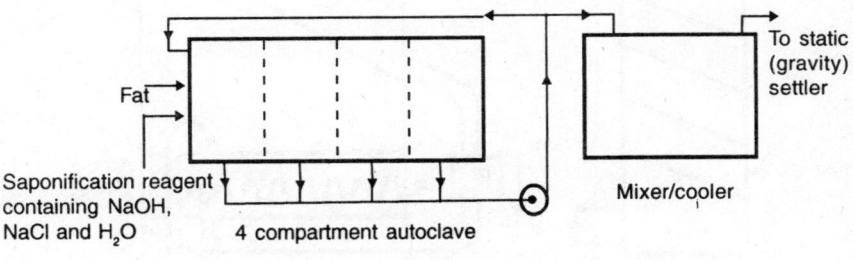

Fig. 5.12. Saponification system for Meccaniche moderne and mazzoni SCN plants.

Equipment for Washing

As mentioned earlier, all modern soapmaking plants wash the glycerol out of the soap by counter-current transfer of glycerol from soap in the grained condition to the lye. Many types of equipment can be used, but all fall into one of the following kinds :

(i)   Mixers and static, or gravity, settlers.

(ii) Mixers and centrifugal separators.

(iii) Truly continuous counter-current contactors.

A tray-type multi-stage washer, somewhat similar to the washing tower is shown diagrammatically in Fig. 5.13. Six stages of mixer/settler washing are included in a single washing tower. Each stage comprises a settling section with inclined top and bottom, through which is inserted a tube with a mechanically rotated mixer. The soap rises through a tube from the top of a settler section into the next mixer; and lye joints it from the section above at a rate controlled by a positive displacement (Mono) pump.

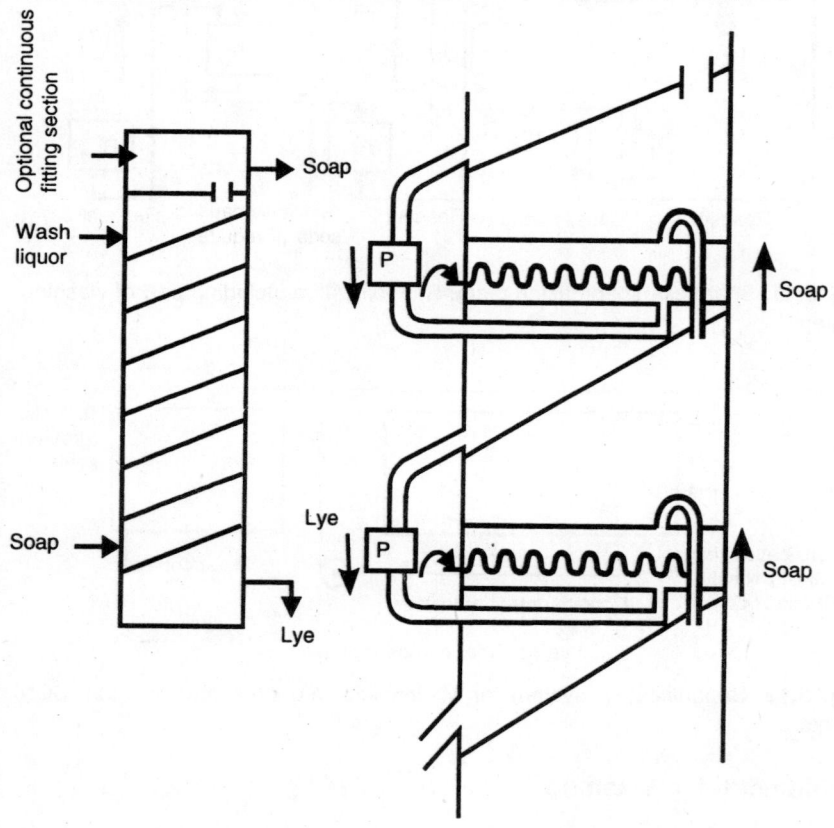

Fig. 5.13. Monsavon washing tower.

A continuous washing unit, known as a *DPU (divided pan unit)*. The main features are shown in Fig. 5.14. In the original unit a 14 foot (4.27 m) cube pan is divided into six sections, but a tank divided into any reasonable number of sections and arranged to suit a required layout can be used. The flows are those standard for a mixer/settler system. Open inpeller centrifugal pumps are used to mix lye and soap curd, and to return the mixture to the middle level in the next stage. Recirculation round the pumps to increase mixing to improve performance. Large floats designed to position themselves at the curd/lye interface control the outflow of lye by actuating gate, or treacle, type valves. To minimise the force needed to

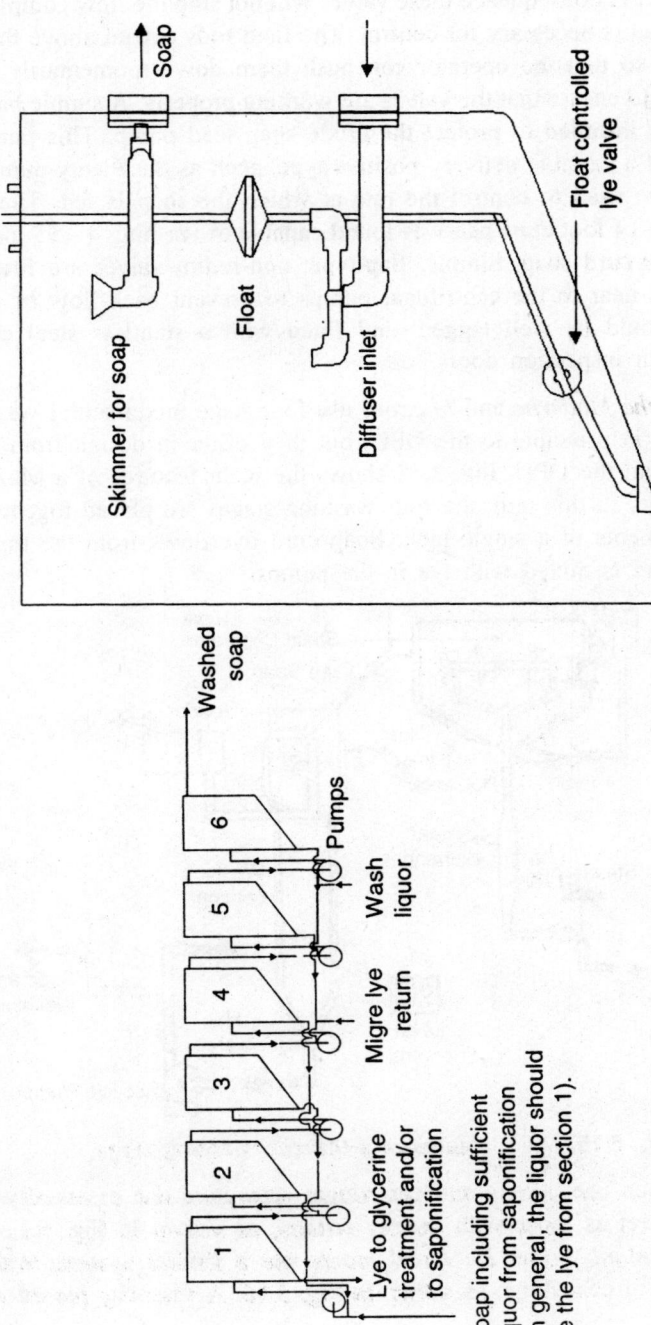

Fig. 5.14. Divided pan type washing unit.

move the valve disc, a small clearance is left between the stationary and the moving plate; in consequence these valves will not stop the flow completely, but do all that is necessary for control. The float rods extend above the top of the unit, so that the operator can push them down momentarily from time to time to ensure that the valves are working properly. A simple basket-type filter is installed to protect the crude soap feed pump. This pump is preferably of a variable delivery positive type, such as the Plenty pump, so that it can be used to control the rate at which the soap is fed. The unit made from a 14 foot cube pan was found capable of washing 4 - 15 tonnes/hour of 58% curd soap. Simple, flap-type, non-return valves are fitted in the lye lines near to the centrifugal pumps to prevent back-flow of soap. The unit should be well lagged; and fitted with a stainless steel cover, provided with inspection doors.

Maccaniche Moderne and *Mazzoni* use four stage mixer/settler washing units similar in principle to the DPU, but they differ in design from each other and from the DPU. Fig. 5.15 shows the main features of a Mazzoni washing stage. In this unit, the four washing stages are placed together in the compartments of a single tank. Soap curd overflows from the tops of the stages and is mixed with lye in the pumps.

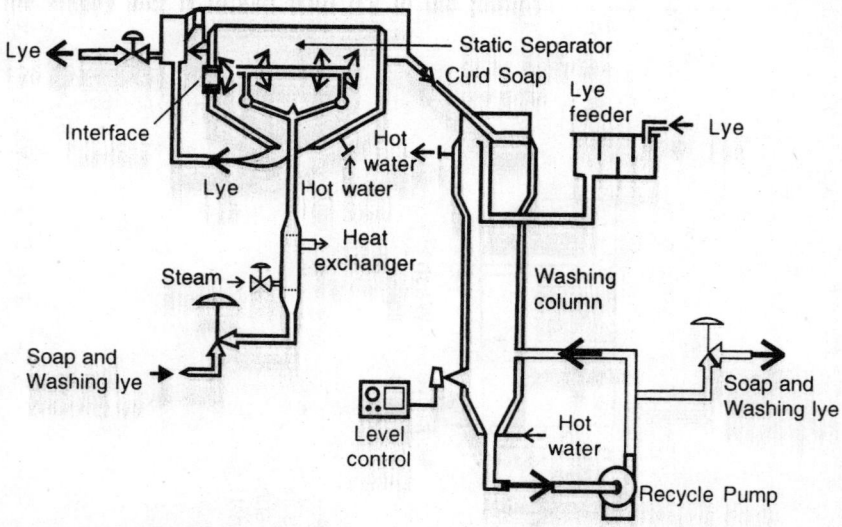

Fig. 5.15. Main features of a Mazzoni washing stage.

Plants which use mixers and centrifugal separators use essentially the same flow sheet as those with gravity settlers, as shown in Fig. 5.11 for the *Sharples plant*. Some *De Laval plants* use a similar system, with a different type of centrifuge, as shown in Fig. 5.16. A viscosity recorder, is used to aid the control of the grain; the proportions of water and concentrated brine being varied manually. Because centrifuges normally separate more efficiently than gravity settlers, less washing stages are required; and often only two are included, apart from the centrifuges used in the saponification and fitting stages. Centrifuges also collect most of the particulate dirt in the material processed; and the bowls need to be cleaned when they become as full as is permissible.

The *rotating disc contactor*, shown diagrammatically in Fig. 5.17, can provide a particularly effective unit for the washing of soap. The shaft rotates at about 80 rpm; and the diameter of the rotating discs is slightly smaller than that of the holes in the fixed discs, so that the shaft can be withdrawn. Lye recirculation to use the lye absorption principle can be arranged in various ways to suit the saponification system employed.

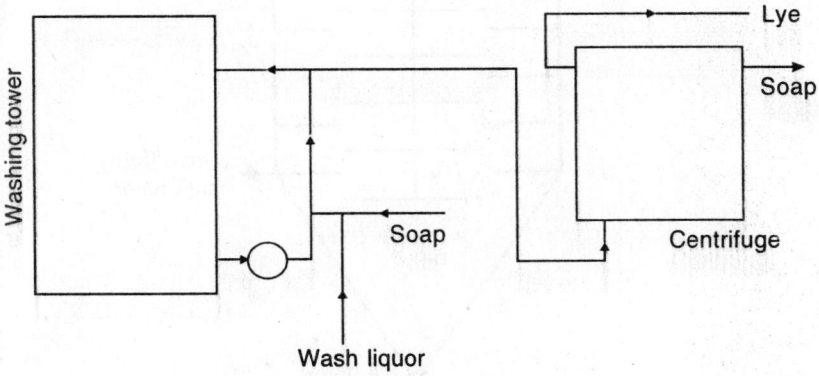

Fig. 5.16. Unit in De Laval washing system.

It must be re-emphasized that, whichever system of washing is used, it is particularly important to ensure that the soap is only just grained sufficiently; overgrained soap occludes excessive amounts of lye in the curd and hence reduces efficiency. Concentrated brine and water in the proportions required for the wash liquor can be blended in various ways; some systems use an automatic density controller, others blend manually with the aid of flow measurers, such as rotameters.

Equipment for Fitting and Final Separation

In the traditional process fitting was a complex operation involving much boiling, but it has been found that it is only necessary to mix the washed curd with the water and, if required, electrolyte needed to give the correct composition for the desired type of fit. An early example of mechanical fitting in a pan used the lye diffuser described, when it was found that soap fitted mechanically settled more quickly than soap which was boiled in the conventional way, probably because it was not filled with bubbles of steam.

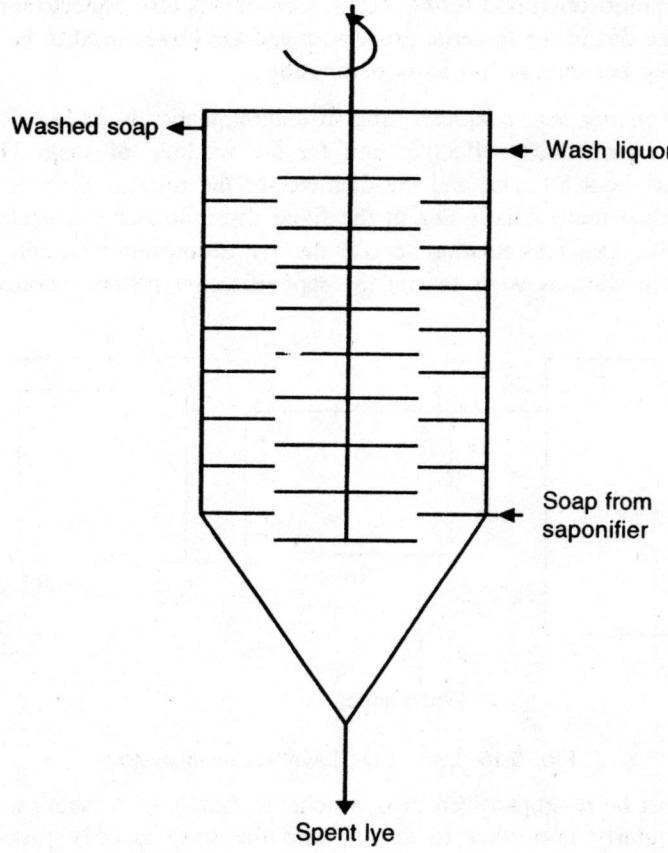

Fig. 5.17. Rotating disc contactor.

Some modern continuous processes use only a simple line mixer, or even just injection of fitting liquor into the pipeline through which the soap flows. Others use a holding and mixing column into which the fitting liquor is injected. With recirculation of the soap. The De Laval process uses such a column with circulation and the addition of fitting liquor is again controlled by differential pressure measurement directly related to viscosity. The mixture passes to a second column intended to ensure full establishment of equilibrium.

Obviously, long settling under gravity is not practicable in a continuous process; and centrifuges are always used for the final separation. Open fits giving a thin nigre, or nigre lye, are also common; and arrangements are usually included for the continuous return of the nigre lye to the washing system.

COMPLETE SOAPMAKING PLANTS

It is, in general, possible to combine any method of saponification with any of washing, and any of fitting, to form a complete soapmaking plant. It is now usual for the whole plant to be continuous, with the elements integrated so that the plant operates substantially automatically. However, it is also possible to use a combination of batch and continuous elements, for example:

(i)   Batch saponification, followed by continuous washing, and batch fitting and final settling.

(ii) Continuous saponification and washing, followed by batch fitting and final settling in pans.

Such combined systems provide some, although not all, of the advantages of a proprietory continuous plant, as compared with the traditional pan process. They might possibly appeal to some small manufacturers who do not feel able to justify the initial cost of a continuous plant; or to any who require the quality of neat soap produced by gravity settling after a fine fit. Additional advantages from the purchase of a proprietary plant are :

(i)   The elements are balanced and connected with the controls needed for simple automatic operation.

(ii) The plant manufacturer is responsible for design and commissioning.

(iii) Some aspects of efficiency are likely to be greater.

(iv) Less space is required than if the soap is settled in pans.

Fig. 5.11 discussed in the section on saponification equipment also shows the arrangement of a Sharples washing stage. In a complete *Sharples* plant, the soap curd from centrifuge S_3 is mixed with fourth stage reagent (water, brine and 48% caustic soda) in a fitting mixer, equipped for recirculation by external pump, and the mixture is centrifuged to give neat soap and thin nigre which is returned to R_3 where it forms part of the third reagent. Two centrifuges in parallel may be used for the final stage, because a lower rate of throughput is needed for proper separation of neat soap than for curd.

PLANT PERFORMANCE

Apart from matters of quality and capacity, the performance of a soapmaking plant is judged by the efficiency with which it separates the glycerol using a given quantity of water. Both these important criteria can be expressed in various ways. The most direct way in which to express recovery is in terms of percentage yield, that is glycerol in spent lye as a percentage of glycerol in fats. It is also often reported in terms of the glycerol content of the soap produced (usually calculated at 63% TFM, or at 100% TFM). The relationship between the two measures depends upon the glycerol content of the fat blend used; and Table 5.10 gives figures calculated by simple arithmetic for (a) a high glycerol fat blend containing 10.5% glycerol, and (b) an acid oil type of blend containing 6% glycerol. The quantity of water used can be expressed as lye/fat ratio, lye/63% TFM soap ratio, or as the concentration of glycerol in the final spent lye. Some figures showing the relationship between these criteria for the two fat blends are given in Table 5.11. The relationship between percentage recovery and lye/fat ratio depends upon the process used and is a measure of its efficiency. The glycerol content of the wash liquor used, which depends upon the glycerol added in recovered salt and in any weak liquors introduced at the final wash stage, affects recovery and becomes increasingly significant as the level of recovery increases and the glycerol in soap decreases. In general, it is possible to improve recovery at the expense of lye/fat ratio (or concentration of glycerol in lye) and vice versa. Just as with simple washing in pans, the efficiency of a mixer/separator washing stage is directly influenced by the dryness of the separated curd, as can be measured by determination of K_G. With gravity settlers, values of K_G around 0.4 have been found which indicates that the mixing of lye and curd is as effective as washing by the traditional method of closing the soap and re-graining at each stage. Centrifuges usually produce curds containing less enmeshed lye than gravity settlers; and so provide more efficient washing stages with lower values of K_G. Operators can check the performance of a plant by estimation of K_G for the various stages, or even by just determination of the TFMs of the various curds from time to time. Inadequate settling time (that is too great a throughput for the particular plant) can result in wet curds; but, as indicated earlier, poor performance is often due to over-graining.

Glycerol cannot be separated until it has been liberated by the chemical reaction, so that it is important to ensure that saponification is as complete as possible at the earliest possible stage. Broadly, it can be said that around 95% recovery is common with a lye/fat ratio of, say 0.7; which, from Table 5.11, is equivalent to a lye containing about 14% glycerol from a fat blend containing 10.5% glycerol.

Table 5.10. Relationship between glycerol percentage in soap and percentage recovery from (a) fat containing 10.5% glycerol and (b) fat containing 6% glycerol.

Percentage glycerol in 63% TFM soap	(a) Fat blend containing 10.5% glycerol		(b) Fat blend containing 6% glycerol	
	Percentage of the glycerol in the fats left in the soap	Percentage transferred to lyes	Percentage of the glycerol in the fats left in the soap	Percentage transferred to lyes
0.1	1.4	98.6	2.5	97.5
0.2	2.9	97.1	5.0	95.0
0.3	4.3	95.7	7.5	92.5
0.4	5.7	94.3	10.0	90.0
0.5	7.1	92.9	12.5	87.5
0.75	10.7	89.3	18.8	81.2
1.0	14.3	85.7	25.0	75.0
1.5	21.4	78.6	37.5	62.5

No allowance for any losses is made in this table.

It is assumed that 1 tonne of fat produces 1.5 tonnes of 63% TFM soap.

Comparison between the modern and the traditional pan processes are :

(i) The old process took about 5 - 8 days, so that much space was required. Material passes through the modern processes in an hour or so; or even less.

(ii) The continuous process normally re-cycle the nigre, or nigre-lye, and are completely self-contained.

(iii) In addition to the large saving in steam, and in capacity needed for lye treatment and evaporation, which results from the higher concentration of glycerol in lyes, the modern processes use much less steam than the pan process itself.

(iv) The modern processes use more power than the old, but the cost per tonne of soap is not high.

(v) Comparison of quality is less clear. Advantages for the modern processes are that :

  (a) There is little opportunity for airborne dirt to contaminate the soap; and centrifuges remove particulate dirt effectively.

  (b) There may be less risk of damage by oxidation and prolonged heating, but it seems doubtful whether this difference is substantial.

Table 5.11. Relationships between criteria for quantity of water used in soapmaking under various circumstances.

Ratio lye/fat	Ratio lye/63% TFM soap	Concentration of glycerol in lye			
		Fat containing 10.5% glycerol		Fat containing 6% glycerol	
		Recovery = 95%	Recovery = 85%	Recovery = 95%	Recovery = 85%
0.4	0.27	25.0	22.3	14.3	12.8
0.5	0.33	20.0	17.9	11.4	10.2
0.6	0.40	16.6	14.9	9.5	8.5
0.7	0.47	14.3	12.8	8.1	7.3
0.8	0.53	12.5	11.2	7.1	6.4
0.9	0.60	11.1	9.9	6.3	5.7
1.0	0.67	10.0	8.9	5.7	5.1
1.1	0.73	9.1	8.1	5.2	4.6
1.2	0.80	8.3	7.4	4.8	4.3

On the other hand :

(i) Very thorough saponification is possible with the pan process.

(ii) Soap from properly operated continuous plants is entirely satisfactory for modern finishing processes. The neat soap produced is also satisfactory for many purposes, but it showed nigre spots when frame cooled unless the centrifuge throughput was very low. Further work might show that any such problem can be avoided by adjustment of conditions.

PRELIMINARY TREATMENT OF SPENT LYES IN THE SOAPMAKING

Spent lyes contain variable quantities of free alkali and soap (particularly low molecular weight soaps), as well as colouring matter, particulate dirt and other impurities. The system of two-stage saponification, in which the final spent lye is separated from soap containing some free fat, produces lye with a low free alkali content; and in modern counter-current processes this is the only lye sent to the glycerine department. Some soapmakers prefer to saponify as completely as possible in the first stage, which, as well as having possible quality advantages, ensures that the glycerol is all liberated as early as possible. With the old process most lyes are transferred for recovery of crude glycerine and the traditional procedure was to leave the lyes in tanks below the pans where they cooled. Soap separated on the surface, was skimmed, and then returned to a suitable boil of probably low grade soap. The cooled and skimmed lyes are transferred to the glycerine department where they are acidified to a pH of about 4.6 with mineral acid, preferably hydrochloric to avoid unnecessary introduction of sulphate into the system. The tank contents was agitated with compressed air and allowed to settle. The lower layer of acidified lye is transferred to the first treatment process. When a sufficient layer of fatty material had collected from several acidifications, it is skimmed off and either discarded or returned to a low grade soap, possibly after filtration. The value of this scum fatty acid is doubtful. It is a poor soapmaking material and contained low molecular weight and hydroxy acids, both of which produce rather soluble soaps liable to be rejected to the next lyes; and so to build up in the system. The acidified lye is well suited to the rest of the lye treatment process. The process had important disadvantages :

(i) It involves a loss of alkali and the need to purchase additional mineral acid to neutralise this alkali.

(ii) It lead to a loss of heat which had to be replaced later, although some of this loss is liable to occur anyway, if a substantial buffer stock of lyes is held in the factory.

(iii) It is a messy procedure which occupies considerable space and labour.

(iv) Unless particular care is taken to keep the alkaline part of the system clean, and mud is generally deposited in the lye tanks, it could provide conditions favourable to the growth of the micro-organisms which destroy glycerol and produce rimethylene glycol (TMG, 1,3-propane diol), a substance which is troublesome in the glycerine refinery, as well as an indication that glycerol has been destroyed.

It has been found that lyes could with advantage be neutralised with fatty acids, or with fats containing a substantial proportion of free fatty acid. The operation, known as lye neutralisation, can be carried out in various ways, but it is necessary to ensure intimate contact between fatty acid and lye.

Chapter 6

Manufacture, Purification of Fatty Acids and Production of Soap Base

INTRODUCTION

Fatty acids have been produced, purified and separated into fractions for very many years. The solid fraction, known as stearin (which actually usually contains a substantial proportion of palmitic as well as stearic acid), was used particularly in candles; and the liquid fraction, known as red oil or olein, in the processing of wool, although there were, and are, other uses for both materials. Relatively minor quantities of various fatty acids were used in soap; and high quality stearin was, and is, used in shaving soaps.

The purchase of fatty acids simplifies the soapmaking process and obviates the need for lye, or sweet water, treatment and evaporation equipment. It also transfers to the fatty acid maker the problems of economical purchase and pre-treatment of crude fats. Many small producers of toilet soaps, the so-called remillers, have for many years gone even further and have bought dried soap base to which they add perfume and other ingredients, and just complete the processing and packaging.

Fatty acids can be produced for special purposes by the acidification of soap with mineral acid, usually sulphuric; and this process can be used to recover fatty acids from soapstocks after saponification has been completed and, if required, the soap has been grained and/or fitted to purify it. This procedure is used in the laboratory, but, on the plant scale, it is expensive and obviously does not provide and alternative method for the manufacture of soap. All important methods for the production of fatty acids from natural fats involve hydrolysis, or splitting, with water, with, or without, a catalyst. Processes can be divided into those carried out at atmospheric pressure and those which use high pressures and temperatures. The processes at atmospheric pressure are :

(i)   Acid splitting.

(ii) Twitchell type splitting.

(iii) Splitting with lipases, the fat splitting enzymes.

The high pressure processes use water alone, or water aided by a basic catalyst, usually zinc oxide. Higher temperatures, and/or longer times, are needed when no catalyst is employed, but processes without catalyst avoid the expense of the catalyst and of removing it from the hydrolysis products, and are now usual in modern installations. When a zinc oxide catalyst is used, most of it reaches the fatty acid still residues. High temperature and pressure processes can be batch or continuous. As explained later, the normal high temperature processes require the presence of a liquid aqueous phase and this makes high pressures also necessary to avoid vapourisation of the water.

FAT SPLITTING

Principles of Fat Splitting

The hydrolysis of fats takes place in stages and the reactions are reversible:

$$C_3H_5(OOCR)_3 + H_2O \rightleftharpoons C_3H_5(OH)(OOCR)_2 + R\text{---}COOH$$

triglyceride   water   diglyceride   fatty acid

$$C_3H_5(OH)(OOCR)_2 + H_2O \rightleftharpoons C_3H_5(OH)_2(OOCR) + R\text{---}COOH$$

diglyceride   monoglyceride   fatty acid

$$C_3H_5(OH)_2(OOCR) + H_2O \rightleftharpoons C_3H_5(OH)_3 + R\text{---}COOH$$

monoglyceride   glycerol   fatty acid

By the usual principles of mass-action, the degree of split can be increased by the removal of a product from the reaction zone; and in practice this is the glycerol. Batch processes use two or more stages, in which the sweet water is removed and replaced by fresh water, or by a sweet water with a concentration of glycerol lower than that in the sweet water removed. This is analogous to the way in which soap is washed in a number of stages; and, for the same reasons as in soapmaking, counter-current contact is desirable in order to reach as a high a degree of split as possible with the minimum quantity of water. In the case of splitting, however, the need to recover the maximum yield of glycerol in as concentrated a form as possible is augmented by the need to liberate the maximum percentage of fatty acid from the fat.

The solubility of water in fatty acids and fats increases with rise in temperature. At temperatures up to 100°C it is slight; between 100°C and 200°C it increases slowly, and then, at temperatures above 200°C the rise is rapid. Ittner and Mills, in patents concerned with continuous splitting, report that at 235°C coconut fatty acids dissolve about 17% of their weight of water when excess water is present, and the pressure is sufficiently high to keep it liquid. At 245°C coconut fatty acids dissolve about 20% of water and tallow fatty acids about 11%. At 287°C (with a pressure in excess of about 1050 psig) coconut fatty acids blend freely with more than their own weight of water to form a single clear liquid phase. Fatty acids of low molecular mass dissolve more water than those of higher molecular mass, and fatty acids more than the corresponding fats, possibly greatly so. In the high pressure processes, the main reactions occur almost entirely in the fatty phase. Free glycerol is partitioned between the fatty and aqueous phases, and , as already explained, two or more changes of water are normally used to achieve a high degree of split in batchwise processes. Each replacement of sweet water by fresh water (or by a sweet water low in glycerol content) causes a transfer of glycerol from the fatty to the aqueous phase and so disturbs the hydrolysis reaction in the fatty phase in the required direction. In a counter-current splitting tower, a similar effect is achieved by a transfer of glycerol from the fatty to the aqueous phase, as discussed in more detail later.

In the Eisenlohr process, fat and water, without catalyst, are pumped through Inconel tubes at 3,500 psig and heated to 330°C. These conditions with concurrent contact give rapid reaction, presumably in a single phase, but the equilibrium is not very favourable, because there is no removal of liberated glycerol from the reaction zone. So far as is known, this process is not used commercially.

The data rate of splitting at various temperatures is shown in Table 6.1. Temperature and catalyst have very substantial effects, but there is little difference between the various fats tested. Fig. 6.1 shows that the reaction between a neutral fat and water at 240°C starts slowly, accelerates and then slows down as equilibrium is approached; the curve being very similar to that for the saponification of a fat with caustic soda. The shape of the curve, and the fact that the hydrolysis starts more rapidly if the fat contains more than about 4% FFA, confirms that the reactions are autocatalytic and suggest that the slow early stage with a neutral fat is because the reaction is then mainly at the fat/water interface. Fig. 6.2 shows how the degree of split at which the reaction rate becomes negligible (that is, equilibrium is nearly reached) varies with the percentage of glycerol in the water phase. These facts show that a fat which contains more FFA than about 4% starts to react more rapidly than one with less. Also, that to maintain an adequate reaction rate whilst producing a reasonably concentrated sweet water, the water must be changed at least before the split is 85 – 90% complete. The actual figures varies with the glycerol content of the fats being split and the ratio of fat to water which is used.

Table 6.1. Influence of temperature on splitting time.

Temp °C	Time to reach equilibrium (mins.)			Half-life time (mins)		
	Beef tallow	Coconut oil	Groundnut oil	Beef tallow	Coconut oil	Groundnut oil
225	156	158	156	62	58	50
240	82	85	85	34	33	25
260	47	46	53	18	23	16
280	34	33	33	8	10	10
260 (with 0.2% ZnO catalyst)	21	–	–	4	–	–

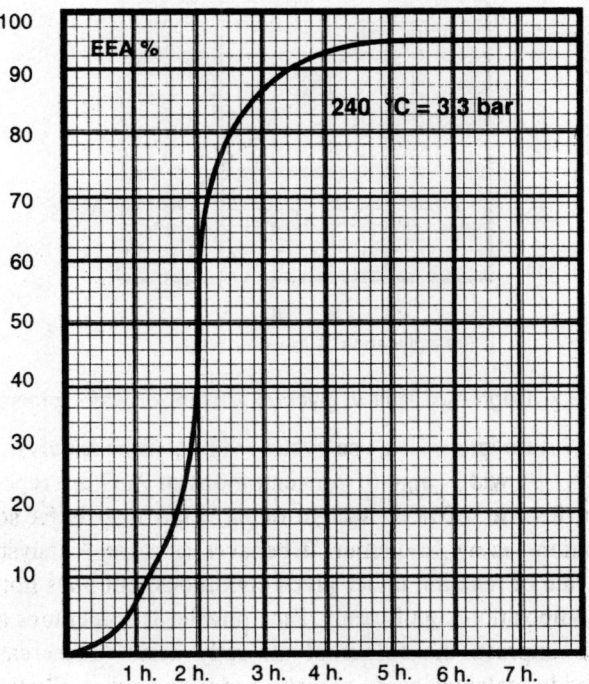

Fig. 6.1. Rate of hydrolysis curve for neutral fat.

Splitting processes carried out at atmospheric pressure use acidic conditioner. With the fat-splitting enzyme from castor beans, a pH of about 5 is maintained with acetic acid; and a small proportion of manganese sulphate is said to be added. The other atmospheric pressure processes are carried out with addition of sulphuric acid to give strongly acidic conditions. The earliest process, used concentrated sulphuric acid (around 80% H_2SO_4) at 110–120°C, which previously had been washed and dried. Severe discolouration occurred; and organic sulphuric or sulphonic acids formed may have helped the acid catalyst by aiding emulsification. A modification in which a sulphonic acid was used in conjunction with water and sulphuric acid; and the mixture was boiled for considerable periods. The original Twitchell reagent, or catalyst, was made by reacting a fatty acid (usually oleic), and naphthalene, and/or benzene, with concentrated sulphuric acid. Most reagents used more recently, such as the 'Kontakt' type, are complex sulphonates, which are by-products from the petroleum industry.

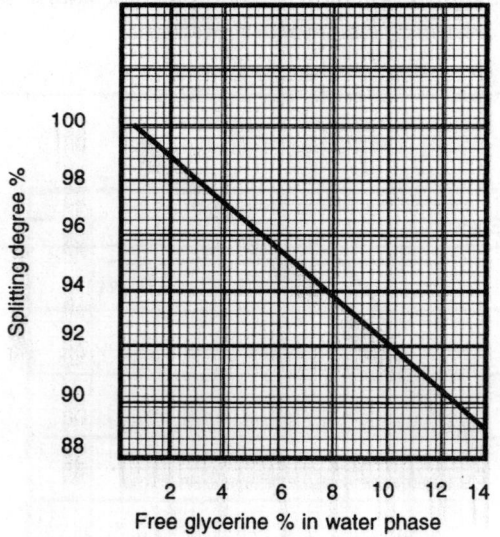

Fig. 6.2. Degree of split v. glycerol content of water phase.

Medium-pressure processes use water with a basic catalyst, normally zinc oxide, ZnO. A wide range of concentrations of ZnO are reported, from 0.2%, or even less, to 4.0% on weight of fat. Zinc soap is fat soluble and probably functions as an emulsifier. It behaves as a true catalyst in that it increases the rate of reaction under given conditions, but does not affect the equilibrium composition significantly. Zinc powder is sometimes also added and is said to improve the colour of the fatty acids. Other catalysts are magnesium oxide, calcium oxide, caustic soda, and other alkalis; but zinc oxide has been found to be far the most effective. Despite the catalytic effect of calcium and magnesium oxides, it is usually considered advisable to use soft, or even demineralised, water in the pressure splitting, as well as the Twitchell, process. Water hardness can affect emulsification in the Twitchell process, and cause problems with insoluble soap scums in the high pressure processes.

Splitting Plant and Processes

Unless the fats to be split are of high quality, pre-treatment is usually necessary, or at least desirable. Break material can cause emulsification in towers which may interfere with the counter-current operation. Polythene in tallow is converted into black plastic masses in towers, or other autoclaves, which can block sparge pipes and other plant items. As already mentioned polythene can be removed by the use of bleaching earth with filtration at below about 70°C. If a dilute sulphuric acid treatment is used prior to high pressure splitting, it is very important to remove the mineral acid by thorough washing to avoid risk of corrosion damage to the high pressure plant.

The enzyme process operates best at about 35°C and is therefore most suited to the splitting of liquid oils. It does little damage to sensitive materials and gives fatty acids of good colour, but it is currently of little, or no, commercial importance, although it may be used to produce highly unsaturated fatty acids in the laboratory. Enzyme developments may possibly change this situation.

The Twitchell process is carried out in vats which are lead lined, or otherwise resistant to aqueous sulphuric acid. Open steam is used to heat and agitate the mix; and two, or three, 'boils', with change of water, are normally given. With water equal to 40% of the weight of the fat, 1–5% of catalyst, and sufficient sulphuric acid (about 0.3–3%), a split of 80–85% can be obtained after the first boil. This is followed by a second boil which should raise the total split to at least 95%. Alternatively, three stages can be used giving, say, 60%, 80–85% and over 98% total splits. The Twitchell process is simple and does not require complex plant, but it gives dark coloured fatty acids. It has now largely been replaced by medium and high-pressure systems.

The high pressure processes received a great impetus from the development of continuous splitting columns. Its core is the splitting column shown diagrammatically in Fig. 6.3. Designs from different plant manufacturers may vary, but the main features are :

(i) The fat is pumped in near the base of the column, through the perforations in a sparge pipe. The droplets rise through the sweet water zone; and the sweet water, substantially free from fat, leaves from the bottom of the settling section.

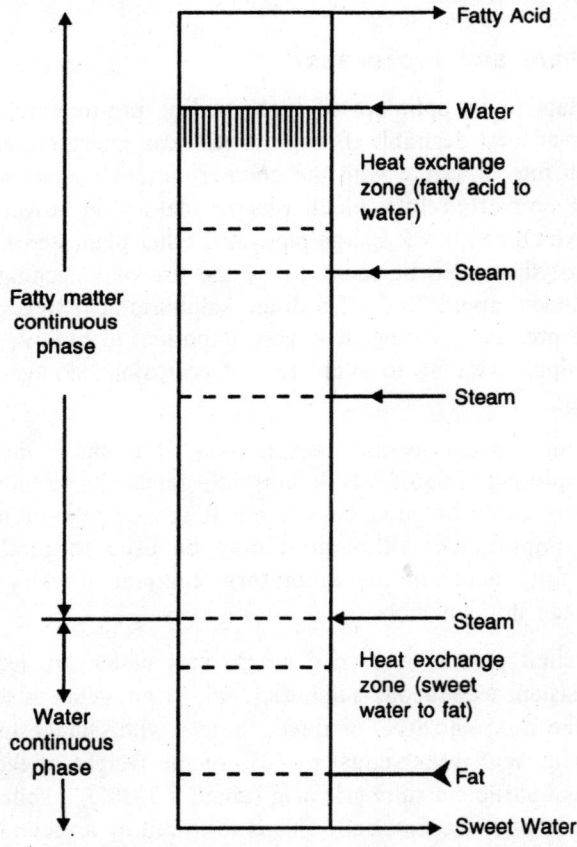

Fig. 6.3. High pressure splitting column.

(ii) Similarly, the water is introduced near the top of the column and is suitably distributed into droplets which fall through the fatty matter phase. The fatty acids leave after passage through a settling zone.

(iii) Fairly low down in the column there is an interface above which the continuous phase is fatty, and below which it is aqueous.

(iv) Internal heat exchange, from fatty acid to incoming water near the top and from sweet water to incoming fat near the bottom, is important, to the discharge of the fatty acids at high temperature to a neutralisation mixer for the production of soap powders. At least in some designs, special heat transfer devices are included in appropriate parts of the column.

(v) High pressure steam is introduced at suitable levels to establish and maintain the required reaction temperature. As an alternative to a sufficiently high pressure boiler it is possible to use a steam compressor to increase the pressure of the available steam.

(vi) The fatty acids are further cooled and dried by a reduction in pressure which allows dissolved (and suspended, if any) water to 'flash' off. Similarly, the sweet water is partially concentrated by allowing it to flash. In some installations, the flashed vapour is used in multiple effect to continue the concentration of the sweet water; and other heat exchange may be possible to improve thermal efficiency. Condensate is collected for re-use as process water, made up with demineralised water.

A sufficiently high operating temperature is selected to give adequate solubility of water in the fatty phase, but not so high as to eliminate the aqeuous phase when the desired ratio of water to fat is supplied. The pressure is substantially above that just necessary to avoid vapourisation.

An adequate height of tower, and proper dispersion of the phases, is necessary to enable the water to extract the glycerol effectively from the fatty phase. This is needed to enable the reaction to proceed substantially to completion, but also to prevent removal of some free glycerol in the fatty phase instead of in the sweet water.

The quantity of water required for the chemical reaction is directly proportional to the ester value, EV, of the fat. One molecule of water (HOH) is equivalent to one molecule of KOH in the test, so that the water required for the reaction equals $EV/1000 \times 18/56.1 \times 100 = 0.032$ EV% of the mass of fat.

The total quantity of water, including condensed steam used in the process, must be sufficient to provide the amount needed for the reaction, plus enough to dissolve in the fatty phase and to leave sufficient to form an aqueous phase to wash the glycerol out of the fatty phase. Table 6.2, obtained by simple arithmetic, illustrates these points.

The first figures for the glycerol content of the sweet water assume that the fatty acids leave the column saturated with water at the reaction temperature, but that the glycerol is all transferred to the aqueous phase. The second set assume that the fatty acids are cooled in the heat exchange zone to a temperature at which they contain a negligible percentage of dissolved water and that the separated water joins the new water descending the column.

In practice, the glycerol content of the sweet water leaving a tower seems commonly to be about 12–18%, depending upon the type of fat and the conditions used. Provided the sweet water leaves the tower at a temperature sufficiently high for it to flash when the pressure is reduced, further concentration then occurs; and the flashed vapours may be used to produce additional evaporation.

The main manufacturers of fat-splitting plants, Lurgi, Gianazza, Mazzoni, and Wurster and Sanger, each make several types of plant. Plants based upon counter-current splitting columns are now usual for large-scale operations, but batch, or multistage, types may be more suitable when the production required is small. High temperature processes lead to a reduction in iodine value when highly unsaturated oils are handled, and milder methods may need to be selected if this is important. The choice of diagram is somewhat arbitrary; and it should be noted that plants similar in flow sheet, although doubtless different in engineering detail, are often offered by the other makers.

Table 6.2. Water requirements of fat-splitting columns and glycerol concentrations in sweet waters produced.

(Basis 1 tonne fat)

	80% tallow plus 20% coconut oil. 50% water on fat	Grease 50% water on fat	Grease 40% water on fat
Fatty acid content	0.954t	0.977t	0.977t
Glycerol content	0.111t	0.056t	0.056t
Water required for reaction	0.064t	0.033t	0.033t
Water dissolved in the fatty phase at 245°C (from solubilities given earlier and mass of FA)	0.122t	0.107t	0.107t
Total water	0.500t	0.500t	0.400t
Water in aqueous phase in reaction zone (by difference)	0.314t	0.360t	0.260t
Percentage glycerol in sweet water, if water dissolved in fatty acids leaves with FA at reaction temperature	35	15.5	21.5
Percentage glycerol in sweet water if FA is cooled before leaving the tower to a temperature at which solubility of water in it is negligible	25.5	12	15

Fig. 6.4 shows the *Lurgi* batch-splitting plant. A feature is the continuous circulation of sweet water by external pump from the base to the top of the autoclave. Standard operating conditions are 230°C and 28 bar (406 psi); and it is stated that 95–96% split should be achieved with two changes of water (the split will also be influenced by the quantities of water and fat used, and the glycerol content of the fat). In the past, Lurgi has made continuous plants comprising three autoclaves of this general type linked in series for counter-current flows of fatty matter and water, but this equipment is no longer offered.

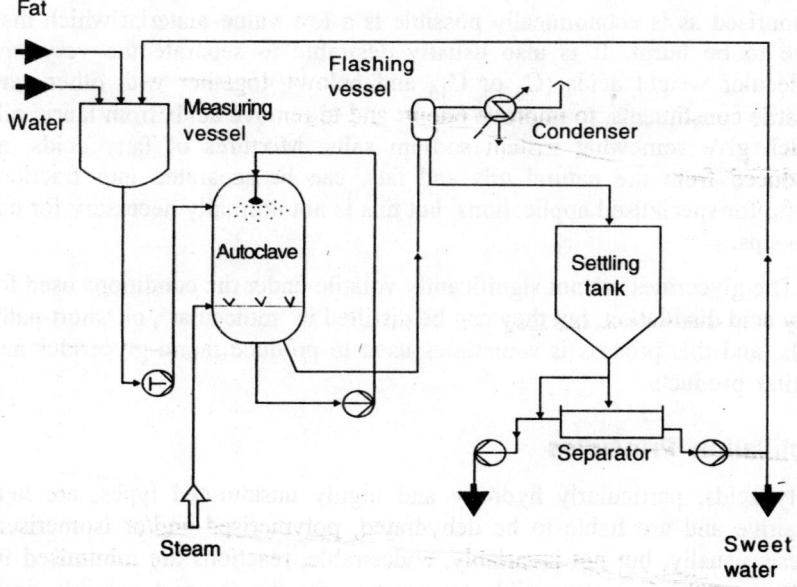

Fig. 6.4. Lurgi batch-splitting plant.

DISTILLATION OF FATTY ACIDS

General

Fatty acids directly from the splitting plant can be, and sometimes are, made into soap, but it is often necessary to improve their quality by distillation. Apart from the requirements of high grade soap bases, this is because one of the merits of the fatty acid route to soap is the possibility of using lower quality, and hence cheaper, raw materials than would otherwise be possible, by taking advantage of the upgrading which can be achieved by distillation of the fatty acids. Simple distillation of fatty acids using direct fired stills initially operated at atmospheric pressure, but later under vacuum, with considerable open steam, has been used for over a hundred years. These stills caused considerable damage to the fatty acids and are largely obsolete.

Distillation separates the volatile fatty acids from non-volatile glycerides and other compounds. If justified, the residue can be re-split with the liberation of more fatty acids, from which a low grade distillate may be obtained (or the re-split material may be returned to the main stream of crude fatty acids, although a residue of non-volatile, non-hydrolysable material must eventually be discarded). A high degree of split in the main hydrolysis process minimises the amount of residue left on distillation and is important. The ultimate pitch left after as much fatty acid has been vapourised as is economically possible is a low value material which may have to be burnt. It is also usually desirable to separate the very low molecular weight acids (C_8 or C_{10} and below), together with other very volatile constituents, to improve odour; and to remove acids from lauric oils which give somewhat irritant sodium salts. Mixtures of fatty acids, as produced from the natural oils and fats, can be separated into fractions useful for specialised applications, but this is not normally necessary for use in soaps.

The glycerides are not significantly volatile under the conditions used for fatty acid distillation, but they can be distilled in 'molecular', or 'short-path' stills; and this process is sometimes used to produce mono-glycerides and similar products.

Distillation Principles

Fatty acids, particularly hydroxy and highly unsaturated types, are heat sensitive and are liable to be dehydrated, polymerised and/or isomerised. These, usually, but not invariably, undesirable, reactions are minimised by the use of the lowest possible temperature for the shortest possible time. Time and temperature are interdependent, so that a very short time at a relatively high temperature may be better than a longer time at a lower temperature; and vice versa.

Vapour pressure/boiling temperature data for some important fatty acids are given in Table 6.3 and for the region of greatest interest in Fig. 6.5. Little information appears to have been published for unsaturated acids other than oleic.

The compositions of vapour and liquid in equilibrium at various temperatures for a binary (two component) mixture are shown diagrammatically in Fig. 6.6, drawn for a fixed total pressure. If a liquid of composition x is heated, boiling commences at temperature t_1 and the composition of the first vapour given off is y, that is it is richer in the lower boiling component A than the liquid from which it comes. This means that the liquid residue left is slightly richer in B, the higher boiling component. As vapourisation proceeds, the boiling temperature rises and compositions change along the liquid and vapour curves as indicated by the arrows marked b (boiling). At temperature t_2 the whole of the liquid is vapourised and the mixed vapours have the original composition x. At intermediate temperatures, such as t_3, a vapour of composition v is in equilibrium with liquid of composition l; and the quantities are inversely proportional to the distances from composition x.

Table 6.3. Vapour pressure/boiling temperatures for some fatty acids.

Fatty acid	Vapour Pressure mm Hg						
	0.01	0.1	1.0	2.0	5.0	10.0	100.0
	Boiling temperatures °C						
Saturated C_4	–	–	26.0	35.5	50.0	61.5	107.5
Saturated C_6	11.5	34.0	62.0	72.5	86.5	98.5	146.5
Saturated C_8	33.5	57.5	87.5	98.5	113.5	126.5	177.0
Saturated C_{10}	55.0	79.0	110.0	121.5	137.0	150.5	202.5
Saturated C_{12}	73.0	98.5	131.0	142.0	158.0	172.5	226.5
Saturated C_{14}	89.0	116.0	149.5	161.0	177.5	192.5	249.0
Saturated C_{16}	105.0	132.5	167.0	179.0	197.0	211.5	–
Saturated C_{18}	119.5	147.5	183.5	196.0	213.0	–	–
Oleic	–	–	198	204	215	225	–

This is the basis upon which fractional distillation is founded. The separation in one stage cannot be more than that for equilibrium at the relevant temperature; and a number of stages are required to effect more than a limited degree of separation. This is usually carried out in a continuous fractionating still shown diagrammatically in Fig. 6.7. The column comprises a number of plates on each of which rising vapour and descending liquid are brought into contact, so that some of component A is transferred from the liquid to the vapour in an approach to the new equilibrium relevant to the temperature of the plate in question. Instead of plates, which give a stepwise, stage by stage, separation, various types of packed columns are often used in which counter-current contact is continuous. Reverting to Fig. 6.6, if vapour of composition x is cooled, condensation will start at temperature t_2 and the first drop of liquid will have composition z. As cooling is continued, the composition of the vapour moves along the vapour curve and that of the liquid along the liquid curve, as indicated by the arrows marked c(condensation). At temperature t_3 the last drop of liquid condensed under equilibrium conditions has composition *l* and leaves a residual vapour of composition *v*, to be condensed separately; this is the basis of fractionation by partial condensation.

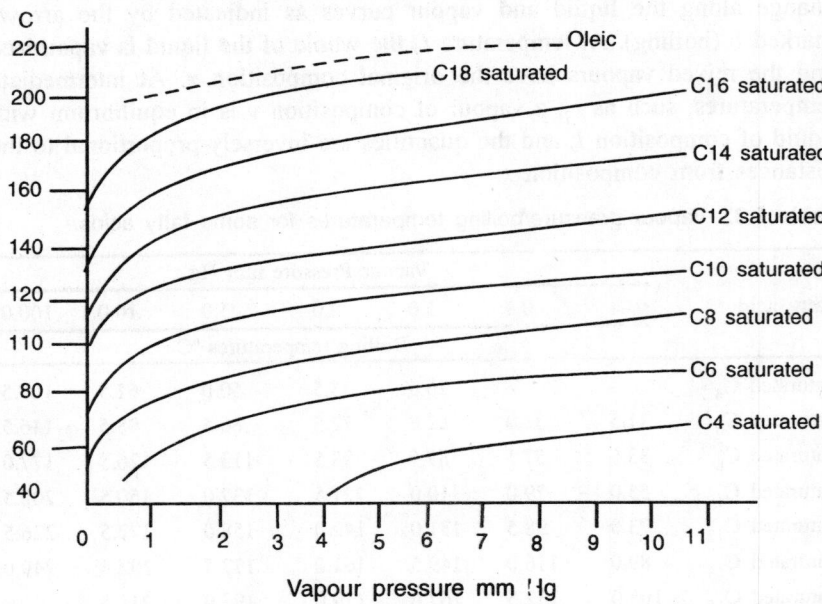

Fig. 6.5. Vapour pressure-boiling temperature curves for some fatty acids.

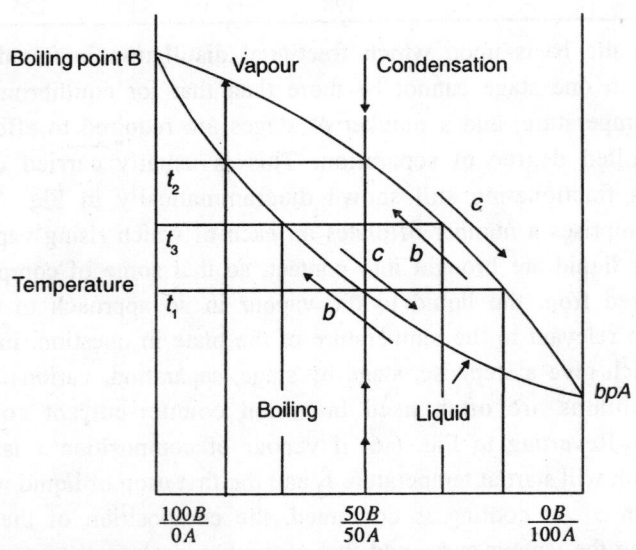

Fig. 6.6. Vapour-liquid equilibria for a binary mixture at a given pressure.

Theoretical analysis of the fractionation of multi-component mixtures is complex, or even impossible, but some useful generalisations can be deduced from information concerning binary systems:

(i) The function of the part of the column above the feed point, called the rectifying section, is to remove less volatile components from the rising vapours which become the distillate.

(ii) The function of the part of the column below the feed point, called the stripping section, is to strip more volatile components from the descending liquid which becomes the residue.

(iii) A fractionating column can only work if liquid reflux is returned to the top, and a reboiler provides heat and vapour to the bottom.

(iv) In general, only one substantially pure component can be separated from a single column. Hence an n-component mixture requires $(n-1)$ columns to separate it into its n individual compounds. More than one complete unit, each with column, reflux and reboiler, can sometimes be included in one shell. In the case of fatty acids, it is seldom required to separate the mixture into individual compounds.

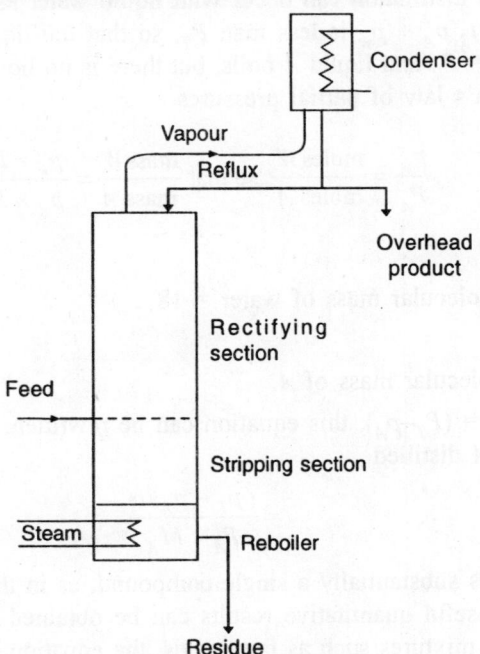

Fig. 6.7. Fractionating still for a binary mixture.

Fractionating columns can be fed with liquid or vapour, or a mixture of liquid and vapour. Liquid feed increases the flow down the stripping section of the column, whereas vapour feed increases the flow of vapour up the rectifying section.

From Table 6.3 it will be appreciated that very high temperatures are required to vapourise the fatty acids from a fat under atmospheric pressure; and this would result in major damage. It is, therefore, necessary to reduce the boiling temperature by the use of reduced pressure (that is vacuum) and/ or a carrier fluid, normally steam. The open steam used as a carrier fluid commonly also provides the agitation needed for good heat transfer and the avoidance of local overheating. Consider first the distillation of a pure liquid A which is immiscible with liquid water (W). The vapour pressure of A is plotted as shown in Fig. 6.8 and its boiling point under total pressure P_T is t_A. The lower P_T the lower t_A in the usual way; and as illustrated in Table 6.3. The vapour pressure of water is plotted downwards from the total pressure point P_T, so that the boiling point of water under pressure P_T is t_W

P_A represents the partial pressure of A in the vapour at any temperature and p_W that of water. At t_1, $p_A + p_W = P_T$; and this is the only condition under which steam distillation can occur with liquid water as well as liquid A in the still. At t_3 $p_A + p_w$ is less than P_T, so that the liquids do not boil. At t_2, $P_A + P_W = P_T$ and liquid A boils, but there is no liquid water in the still. From Dalton's law of partial pressures

$$\frac{p_w}{P_A} = \frac{\text{moles } W}{\text{moles } A}, \text{ so that } \frac{\text{mass } W}{\text{mass } A} = \frac{p_w \times M_w}{p_A \times M_A}$$

where

M_w = molecular mass of water = 18

and

M_A = molecular mass of A.

Since $p_w = (P_T - p_A)$, this equation can be rewritten, mass in lb of open steam/lb of A distilled

$$= \frac{(p_T - p_A)18}{P_A \times M_A}$$

When A is substantially a single compound, as in the case of glycerine distillation, useful quantitative results can be obtained from this equation. For complex mixtures such as fatty acids, the equation provides qualitative indications of the significance of changes in total pressure, use of open steam and supply of indirect heat. The indirect heat provides the latent heat of vapourisation plus other heat requirements, such as sensible heat to raise the temperature of the feed to its boiling point and to replace heat lost by radiation and convection from the still. The actual distillation conditions establish themselves so that both the vapourisation conditions just discussed and the condition of heat balance between heat input and requirements are met. Consider a still in which heat transfer conditions and a certain flow of open steam are set. A boiling temperature will be established so that (a) $p_A + p_w = P_T$, and (b) the heat input balances the requirements. This also determines the rate of vapourisation and the ratio of open steam to fatty acids vapourised.

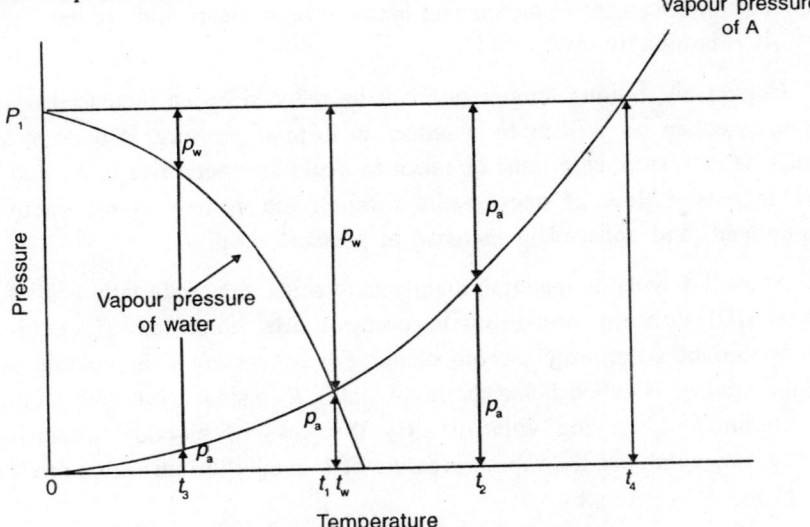

Fig. 6.8. Steam distillation of a pure liquid.

In the discussion so far, it is assumed that the partial pressure of A in the vapours equals the vapour pressure at the distillation temperature; that is the vapourisation efficiency, defined as P_A actual/p_A theoretical, is 1. If contact between open steam and boiling liquid is good, so that the steam is nearly saturated with organic vapour, this is often approximately true. It has been found that vapourisation efficiency, E, varies with the substance being distilled as well as with physical conditions in the still.

With the steam dispensed through small holes at an adequate depth below the surface of the liquid, he quotes 0.9 –0.95 for organic compounds of molecular mass below 100, but only about 0.5 for the steam distillation of lubricating oils of molecular mass greater than 250. Data for fatty acid distillation are not available. If a figure for E is known, or assumed, Ep_A should be substituted for p_A in the expressions in this section.

It will be noted that the usage of open steam per lb fatty acids distilled is reduced by :

(i) Reduction in the operating pressure at the point of vapourisation. This depends upon the pressure drop through the system as well as the absolute pressure induced by the vacuum maintenance system. In the limit, no open steam is used and the mix boils at t_A.

(ii) An increase in the boiling temperature which increases p_A. This results from higher indirect heat input relative to the flow of open steam.

(iii) An improvement in the contact between open steam and boiling liquid if E is initially low.

Conversely, boiling temperature can be reduced by an increased usage of open steam, as well as by a reduction in total pressure. If open steam usage is increased, care must be taken to avoid an undue rise in P_T due to the increased flow of open steam through the system to the vacuum equipment, and consequent increase in pressure drop.

As well as volatile material, mainly fatty acids, the crude fatty acids fed to a still contain non-volatile components, such as glycerides, unsaponifiables and proteinaceous matter. For convenience, the volatile part of the mixture is called V and the non-volatile R (residue), but both include compounds of varying volatility. By the laws of physical chemistry concerning solutions, the vapour pressure of V in equilibrium with a mixture of V and R is given by

$$pv \text{ mix} = pv \times \frac{Mv}{M_V + M_R}, \text{where}$$

$$M_V = \text{moles of } V \text{ present in the mixture}$$

and

$$M_R = \text{moles of } R \text{ present in the mixture}$$

Hence, as V is removed from the liquid by vapourisation, pv mix at a given temperature falls. In a batch, or semi-batch, still the more volatile components are vapourised first; and the residue in the pot becomes increasingly rich in the higher boiling and non-volatile components. The boiling temperature and the ratio of open steam to material vapourised rise until it is not economical to proceed further. In a continuous still, as now normally used, it is necessary to design the vapouriser so that the residue discharged is as free as possible from volatile matter.

The direct heat can be supplied in two ways :

(i) By transfer through a surface to the liquid at its boiling point under still conditions; or approximately so.

(ii) In an external heater in which the material is heated under a relatively high pressure so that the temperature rises, after which the hot liquid is discharged into the still where the lower pressure results in flash evaporation by conversion of sensible into latent heat. Under still conditions, the specific heat of fatty acids appears to be about 0.7 (0.5 at 25 – 100°C) and latent heat of vapourisation about 100 CHu/lb. Hence, roughly, $1 \times \Delta t \times 0.7 = x \times 100$ where Δt°C is the fall in temperature when the mixture is flashed and x is the fraction vapourised.

From this :

Δt°C	x
50	0.35
100	0.70
143	1.0

Distillation Equipment

Vapourisers

In addition to points just elaborated, the requirements for a good vapouriser can be summarised as follows :

(i) The best combination of time and temperature, with both as low as possible, should be achieved. This means that any reservoir of boiling liquid should be as small as possible.

(ii) The pressure under which the liquid boils should be as low as possible. This is primarily controlled by the vacuum equipment, and by the pressure drop through the condensing system and connecting ducts, but it can also be affected by the vapouriser. In particular, the hydrostatic head under which the liquid is heated is important. One foot of liquid of specific gravity 0.8 produces an additional pressure of about 19 mmHg which is very significant when the pressure in the vapour space is only, say, 4 mmHg.

(iii) If open steam is used, the contact between steam and liquid should be intimate, so that E is as near 1 as possible.

(iv) Velocity over the heating surface should be sufficient to avoid local over heating and to promote good heat transfer.

Semi-batch stills commonly use *simple pot vapourisers,* with open steam to agitate the boiling liquid and to act as carrier fluid; and direct heat supplied from a furnace, or via a carrier such as Dowtherm. A simple still of this type is shown in Fig. 6.9. As the initial charge of crude fatty acids is distilled more feed is introduced. The concentration of non-volatile matter in the still liquid increases until it is necessary to stop the feed and to work down the residue as completely as is economically possible. This type of still fails to meet the criterion regarding time and temperature. It is possible to collect the distillate in separate fractions, and an impure first fraction (and possible also last fraction) may be separated. Otherwise, the only fractionation is by slight partial condensation; and distillate quality and yield tend to be poor because of thermal damage. The design is obsolete, but may still be used in some small capacity plants.

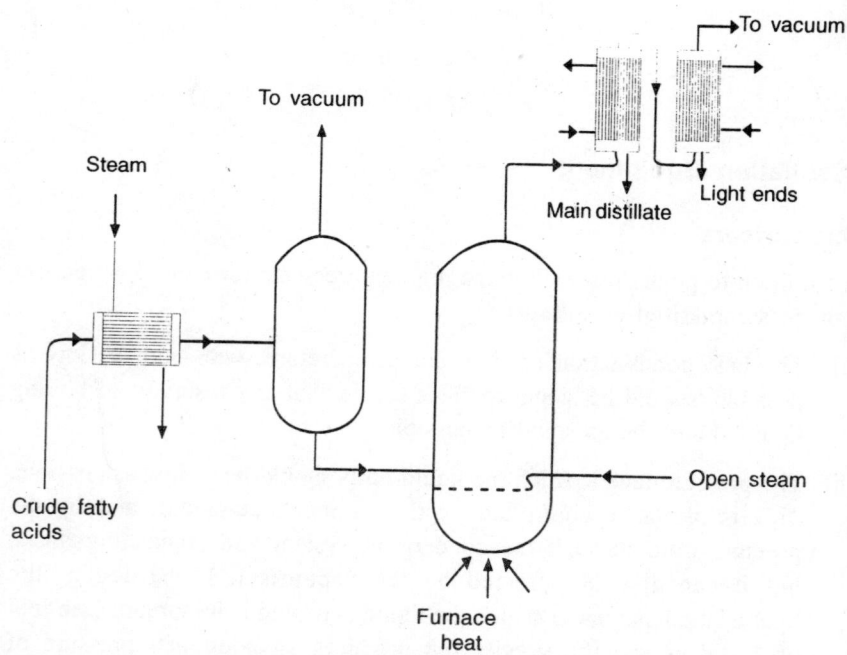

Fig. 6.9. Simple semi-batch still.

The Lurgi Vapouriser, shown diagrammatically in Fig. 6.10, uses a number of compartments to give a continuous vapourisation system. The crude fatty acid feed is partially vapourised in each compartment; and the residual liquid flows from compartment to compartment. Volatile components are removed progressively, so that the residue leaving the last compartment is substantially free of volatiles. In addition, the boiling temperature rises gradually from compartment to compartment as the concentration of non-volatiles increases, so that the bulk of the vapourisation occurs at the lowest possible temperature. The annular heating elements ensure a positive flow of high pressure steam, or other heating medium. The open steam produces a 'vapour-lift' effect, similar to that in an air-lift pump, which causes a good flow of the fatty matter over the heating elements and minimizes the adverse influence of hydrostatic head. As shown, the mixture of steam and fatty matter impinges on a baffle, the vapours separate, and the liquid drops back to be recirculated, or to pass on to the next compartment. 5–10 compartments are used, depending upon the capacity of the plant. Some other plant manufacturers use a broadly similar type of vapouriser.

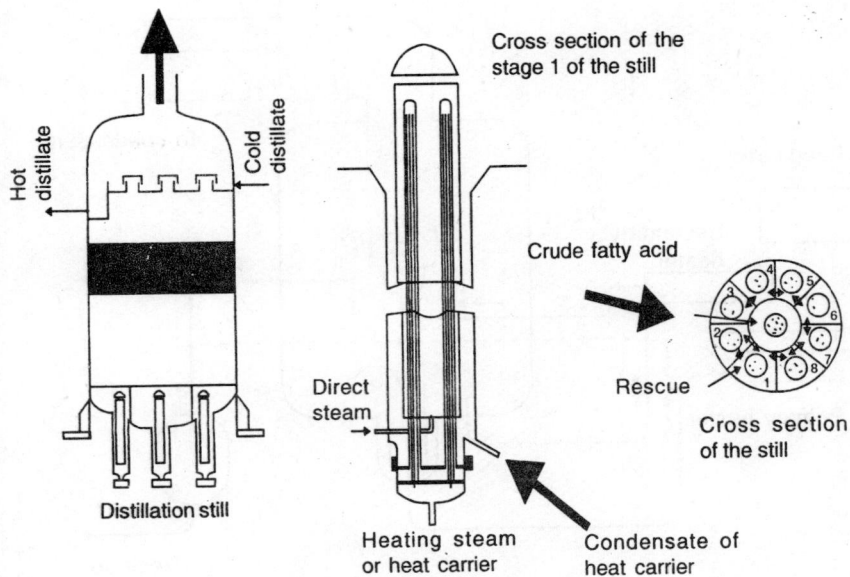

Fig. 6.10. Lurgi vapouriser.

The still shown diagrammatically in Fig. 6.11, was patented by *Mills*. Here external heaters are used to raise the temperature of the fatty matter so that it flashes when discharged into the still body which is maintained at about 3 mmHg pressure. As shown, the small volume of liquid in the still is circulated by a vapour-lift effect. The liquid which remains after the first stage is reheated in the secondary vapourisation unit. The boiling point of a mixture of stearic, oleic, and palmitic acids, as from a tallow or grease, is about 200°C at 3 mmHg pressure, disregarding the influence of the non-volatile components which, in practice, seriously impede the later stages of vapourisation. The temperature to which the mixture is heated in the primary heater is said to be about 450°F (232°C), so that if no vapourisation occurs in the heater, the equation in section 5.4.2 indicates that the fraction vapourised when the material is flashed equals $(32 \times 0.7)/100 = 0.22$. Similar treatment of the liquid fraction (0.78) in the secondary stage would vapourise another 0.17, making a total of 0.39 (39% of the feed). This may have been increased by substantial vapourisation in the heater, but facilities are also provided for recirculation round each stage, as shown dotted. As discussed earlier, the vapour pressure falls to a very low level as the residue is stripped nearly free from volatiles; and this makes flash distillation alone difficult.

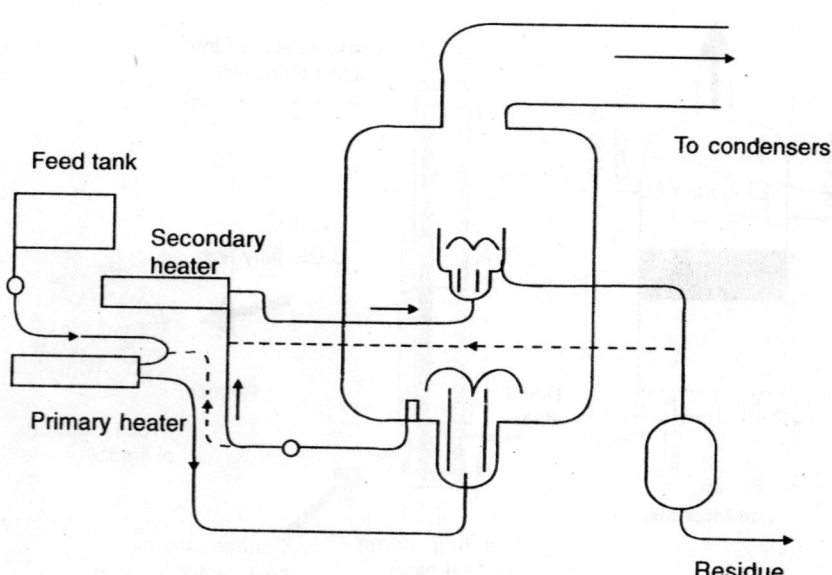

Fig. 6.11. Victor Mills still.

A vapourising column comprising bubble cap trays heated by Dowtherm vapour as shown in Fig. 6.12. The liquid flows across the trays and the unvapourised portion passes through the downcomers to the next tray below. Just as in a fractionating column, the vapour through the slots in the bubble caps produces violent agitation and good contact between liquid and vapour. A separate vapour outlet is provided above each pair of trays to minimise pressure drop. Open steam is injected into the liquid in the reservoir at the base of the column, and below the lower tray in each higher pair.

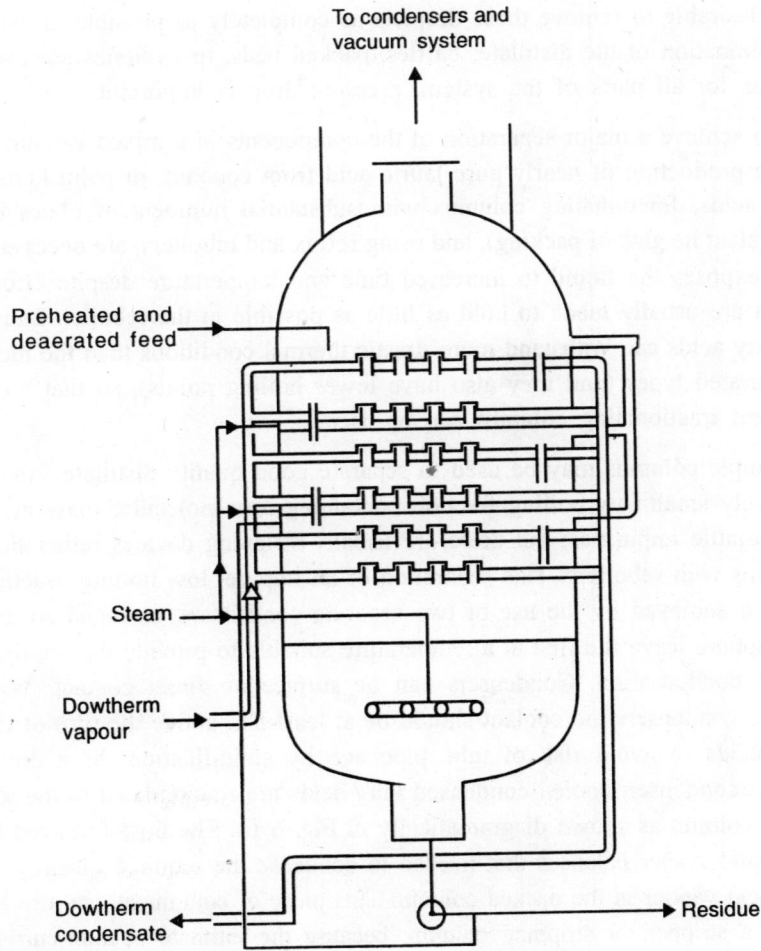

Fig. 6.12. Wurster and Sanger vapouriser.

Falling film vapourisers are used by some plant manufacturers when the fatty acids to be distilled are highly unsaturated and very heat sensitive. In some designs, the vapourisers are tubular heat exchangers with devices to cause the liquid to flow down as annular films inside the tubes. It might also be possible to use scraped wall vapourisers for glycerine distillation, provided the residue is sufficiently liquid to emerge satisfactorily from the base of the unit.

Fractionators, condensers and strippers

If the vapours leaving the vapouriser are liable to carry droplets of liquid, it is desirable to remove these droplets as completely as possible to avoid contamination of the distillate; baffles, packed beds, or cyclones are used, but, as for all parts of the system, pressure drop is important.

To achieve a major separation of the components in a mixed vapour, as in the production of nearly pure lauric acid from coconut, or palm kernel, fatty acids, fractionating columns with substantial numbers of plates (or equivalent heights of packing), and using reflux and reboilers, are necessary. This exposes the liquid to increased time and temperature despite efforts which are usually made to hold as little as possible in the reboiler. Lauric oil fatty acids can withstand more drastic thermal conditions than the more unsaturated types (and they also have lower boiling points), so that more efficient fractionating columns can be used.

Simple columns may be used to separate good quality distillate from a relatively small low boiling fraction containing low molecular mass acids and volatile impurities; but these are usually stripping devices rather than columns with reboilers. This separation of an impure, low boiling, fraction is often achieved by the use of two separate condensers, arranged so that the vapours leave the first at a temperature suitable to provide the required partial condensation. Condensers can be surface or direct contact. With surface condensers the coolant should be at least 5°C above the titre of the fatty acids to avoid risk of tube blockage by solidification. In a direct contact condenser, cooled condensed fatty acids are re-circulated to the top of the column as shown diagramatically in Fig. 6.13. The heat extracted in the liquid cooler balances that needed to condense the required quantity of fatty acid vapour in the packed column. This piece of equipment can also be called a stripper, or stripping column, because the intimate counter-current contact between rising vapour and descending liquid permits a transfer of components. Some of the more volatile components of the liquid fed to the top of the column are vapourised and replaced by less volatile components from the vapour. A similar effect can be achieved in other devices.

Because of the ease with which some fatty acids can be oxidised, distillates should be adequately cooled before they are exposed to atmospheric air.

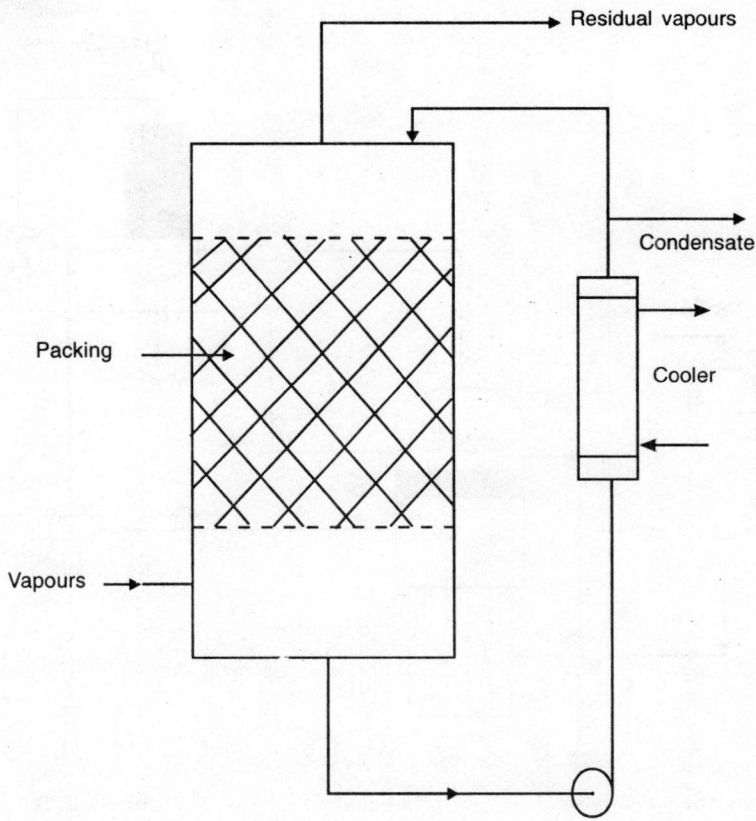

Fig. 6.13. Direct contact condenser/stripping column.

Complete stills

Many different types of fatty acid still have been, or still are, made and used.

Fig. 6.14 shows the *Lurgi simple distillation plant.* This uses the multi-compartment vapouriser (Fig. 6.10), a surface condenser for the main distillate, and a direct contact tower to strip the most volatile fraction from the steam. A special feature is the bubble cap tray in the top of the vapouriser on which the more volatile components are stripped from the main distillate by the rising vapours.

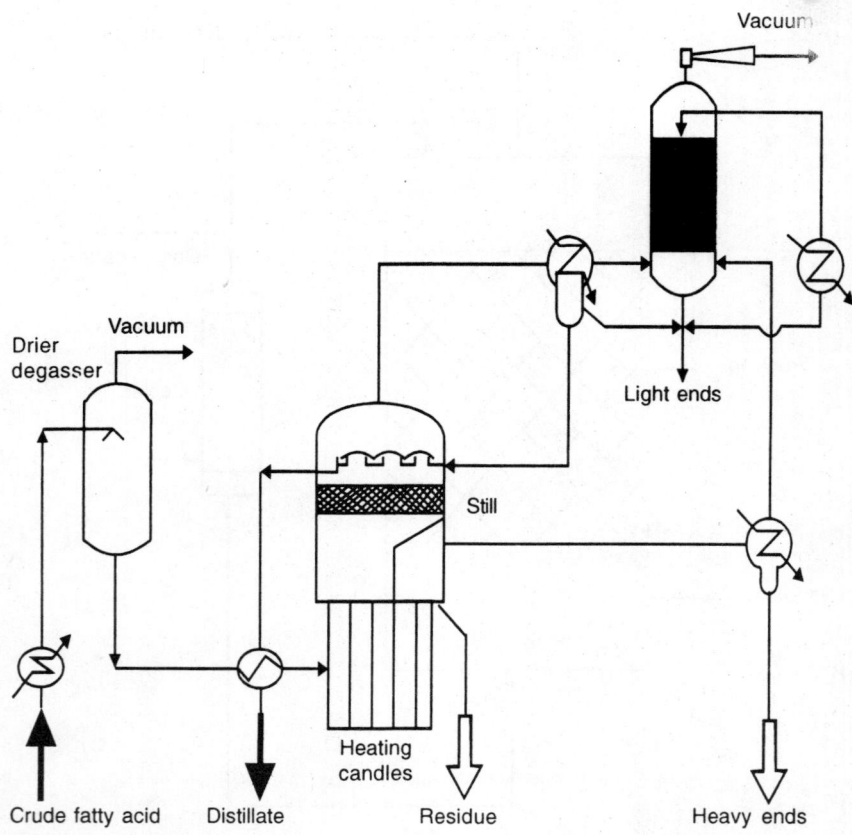

Fig. 6.14. Lurgi simple distillation unit.

The Lurgi fractionating fatty *acid still* is represented in Fig. 6.15. This includes two fractionating columns each with reflux and reboil; and the reboilers are said to have low liquid volumes to minimize thermal deterioration of the fatty acids being processed. The main condensation is in surface condensers, but direct contact columns are again used to condense the most volatile components. Open steam is used as required to aid vapourisation, depending upon the material being distilled. The plant shown produces three fractions, plus the light ends, in fully continuous operation. For smaller plants, other arrangements are possible in which fractions are produced semi-batchwise from a single column.

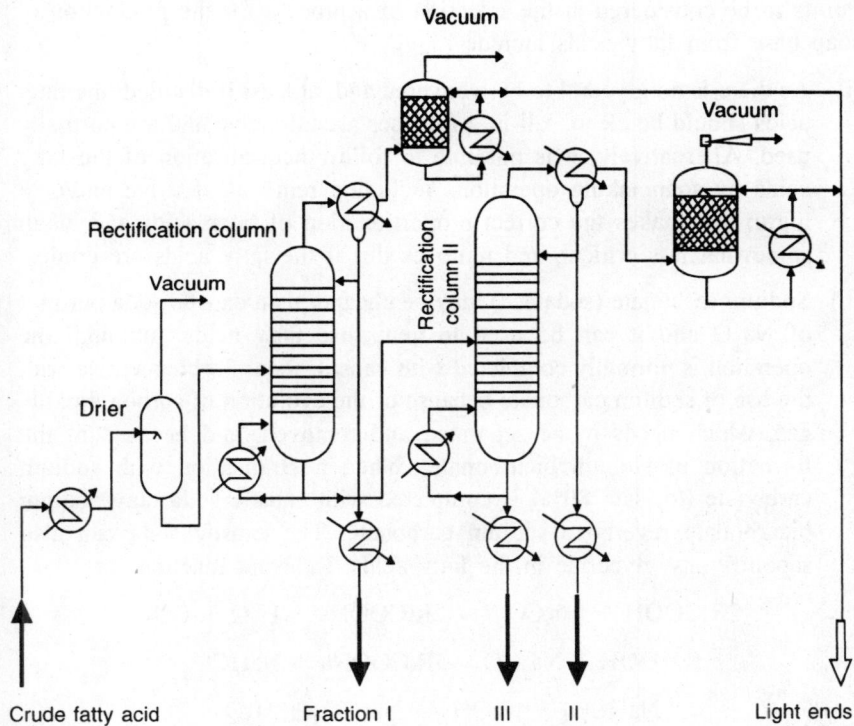

Fig. 6.15. Lurgi fractionating still.

SEPARATION OF SOLID AND LIQUID FATTY ACIDS

The separation of fatty acids from fats such as tallow into a solid fraction, stearin, and a liquid fraction, red oil or olein, is the concern of fatty acid manufacturers rather than soapmakers (although stearin is used in shaving soaps); and the processes will only be mentioned briefly. The traditional method is to crystallise the fatty acids in trays, wrap the blocks in cloth and press. This gives single pressed stearin, and reprocessing with heated presses gives double and triple pressed stearin, which contain less unsaturated material than the single pressed grade. A more recent method involves crystallisation from a solvent; which is methanol in the Emersol process. Lurgi has a third type of process in which stearin crystals are wetted with a solution of wetting agent to give an aqueous dispersion which can be separated centrifugally from the non-dispersed olein phase.

PRODUCTION OF SOAP BASE FROM FATTY ACIDS

Points to be considered in the selection of a process for the production of soap base from fatty acids include :

(i) As there is no glycerol to be recovered and, at least if distilled, the fatty acids should be clean, 'all-in' processes are attractive and are normally used. Alternatively, it is possible to follow neutralisation of the fatty acids by soapmaking operations including removal of a lye and/or a nigre; this makes the correct proportionation of fatty acids and alkali somewhat less critical, and removes dirt if the fatty acids are crude.

(ii) Sodium carbonate (soda ash) may be cheaper than caustic soda per unit of Na_2O and it can be used to neutralise fatty acids, although the operation is normally completed with caustic soda. Problems arise with the use of sodium carbonate because of the evolution of carbon dioxide gas, which needs to be separated and removed, and because of the formation of sodium bicarbonate. When neutralisation with sodium carbonate (to, say, 80%) is completed with caustic soda, any sodium bicarbonate reverts to sodium carbonate. The caustic soda can also saponify any glyceride in the fatty acids. Relevant reactions are :

$$2R.COOH + Na_2CO_3 \rightarrow 2R.COONa + H_2O + CO_2$$

$$R.COOH + Na_2CO_3 \rightarrow R.COONa + NaHCO_3$$

$$NaHCO_3 + NaOH \rightarrow Na_2CO_3 + H_2O$$

(iii) Neutralisation of fatty acids with concentrated (48% NaOH) caustic soda gives a soap at well above 63% TFM unless additional water is used and, if the soap is to be dried for the production of toilet soaps, it appears attractive to save heat in the drying process by neutralisation at a high TFM. Consider a blend of 80% tallow and 20% coconut oil (fatty acid ratio of blend about 8.55) neutralised with caustic soda solution containing 48% NaOH, or 37.2% Na_2O. Assume also an addition of 20% brine sufficient to give a salt content of 0.5% NaCl and of refined glycerine to give 0.4%, both calculated on TFM. Salt is normally added, and glycerine sometimes, to give a composition similar to that of neat soap.

1 tonne fatty acid requires 1/8.55 tonnes Na_2O, or $1/8.55 \times 100/37.2 =$ 0.314 tonnes of 48% caustic soda solution. Add 0.4/100 tonnes glycerine and $0.5/100 \times 100/20 = 0.025$ tonnes of brine, to give a total mass of 1.000 + 0.314 + 0.004 + 0.025 = 1.343 tonnes of soap with a TFM of 100/1.343 = 74.5%. If it is desired to produce soap at 63% TFM, the total mass must be increased to $1.343 \times 74.5/63 = 1.588$ tonnes which requires additional

water equal to 0.245 tonnes, or a dilution of the caustic soda solution to about $48 \times 0.314/0.559 = 27\%$.

The simplest way to neutralise fatty acids with sodium carbonate is in a pan. An approximately saturated sodium carbonate solution is brought nearly to the boil and the fatty acids are added gradually. Sufficient time must be allowed between additions to permit the carbon dioxide to escape without the pan contents boiling or effervescing over. The saponification is completed with caustic soda.

COMPARISION BETWEEN SOAP MANUFACTURE BY DIRECT SAPONIFICATION AND BY THE FATTY ACID ROUTE

The comparison depends upon a number of circumstances which vary from factory to factory; and some important points are:

(i) The cost of the fatty raw materials. Splitting and distillation often permits the use of lower grade materials than are needed to produce soap of the required quality by the direct saponification process, which may lead to substantial savings. The availability of suitable low cost by-products from edible oil plants may be important.

(ii) Possible use of sodium carbonate with the fatty acid route.

(iii) The yields of fatty acid converted into soap and of glycerol, taking into account material downgraded into lower quality soaps and to pitch discharged from fatty acid stills. Losses in pre-treatments, such as earth bleaching.

(iv) Initial capital costs; and capital and maintenance charges. High pressure splitting plants and fatty acid stills are expensive. Some continuous direct soapmaking plants are costly, but efficient direct soapmaking plants can probably cost substantially less than modern splitting and distillation equipment.

(v) Cost of glycerine recovery. One of the factors which led to the development of the fatty acid route is the low concentration of glycerol in soap lyes then common; but, this difference has now been eliminated or greatly reduced. The fact that sweet waters do not contain salt is significant.

(vi) Other operating costs, including that for the high pressure steam, or other heating medium, used in the fatty acid route.

In addition to the cost comparison between the two routes as the major method of soap manufacture :

(i) Fatty acids may be needed for some products, such as certain modern soap powders

(ii) Companies which produce low grade by products, such as black cotton grease, may find it worthwhile to split and distil such materials, even if their main soapmaking is by direct saponification.

Chapter 7

Manufacture of Solid, Household and Toilet Soaps

INTRODUCTION

Genuine household type soaps can be made from soap base in various ways; that is by allowing the soap to cool in frames or water-coolers, or by extrusion, which is now almost universal, apart from some small traditional manufacturers who still use the old processes. Similar processes are used for filled household soaps.

Toilet soaps can also be made in several ways, but for very many years extrusion processes have been dominant. Extrusion processes for household soaps and for toilet soaps have features in common; and it is convenient to discuss some general aspects of the two processes after the older household soap processes have been considered, but before extrusion processes for household and toilet soaps are discussed more specifically.

THE OLDER PROCESSES FOR HOUSEHOLD SOAPS

Genuine soaps are produced by solidifying soap base without major additions; but minor additions of colour, perfume, preservative, etc., and of cresols in the case of carbolic soaps, are frequently made in simple mixing pots called crutchers. The oldest and simplest process is then to run the soap into moulds in which it is left to cool. Almost any containers can be used, but rectangular boxes with detachable sides, known as frames, are usual. Frames can be static, arranged in rows and fed with liquid soap by an arrangement of troughs and gates, or mobile, that is fitted with wheels, so that they can be put under the outlet from the crutcher and then moved elsewhere to cool. Solidification in production sized frames often takes about six days. Two operations should usually be carried out during the solidification of soap in a frame :

(i)   hand crutching whilst the soap is still substantially liquid. This is mixing by a man who normally stands on the sides of the frame using a hand crutcher, which is a perforated steel plate with a shaft and handle at right angles to the plate, pushed up and down in the soap in the frame.

(ii) knocking out, in which the bolts which hold the sides of the frame are loosened and the still plastic, or partly liquid, soap is pressed down with a pummel. In this way the soap which has contracted on cooling is pressed out to fill the frame. Otherwise some bars from zones which have contracted will be short.

The frame joints are liable to leak; and gaps are caulked with paper pressed into the joints as necessary.

When the frame contents are completely solid, the sides of the frames are removed and the blocks are cut into slabs with a mobile cutter using a frame of wires drawn through the soap. The slabs are then cut into bars with another wire cutter. The bars are usually kept for some time in an 'open pile', that is, each layer of bars is placed with spaces between the bars, and each layer is laid at right angles to the one below. This allows air to penetrate the stack; and a slightly dried-skin is formed on most of each bar. The bars are trimmed and, usually, given a simple imprint with a stamper struck with a mallet (or a mechanical equivalent) before being packed in cases as bar soap. Alternatively, the bars are cut into billets and stamped into tablets using dies and shells as described later for toilet soaps.

Mechanical, or hand, stampers can be used; and the tablets may be wrapped and/or cartonned. A considerable quantity of unusable soap is produced in the frame process. In particular, the top and bottom slices and other off-cuts have to be re-boiled, or re-melted, before being returned to a crutching pot with new soap.

To reduce the time (and hence space) needed to convert liquid soap into the solid form, machines were devised in which cooling by water was used. Fig. 7.1 shows a slab cooler, sometimes known as a Jacobi cooler, although types were made by various firms. The liquid soap is run from the pan, or from a crutcher if additions are made, into a pressure vessel from which it is forced by compressed air into the cooler. The cooler is something like a filter-press, in which water-cooled plates alternate with frames in which the soap is solidified. The plates are faced with stainless steel, or other corrosion-resistant metal; and baffles are provided as shown inside the

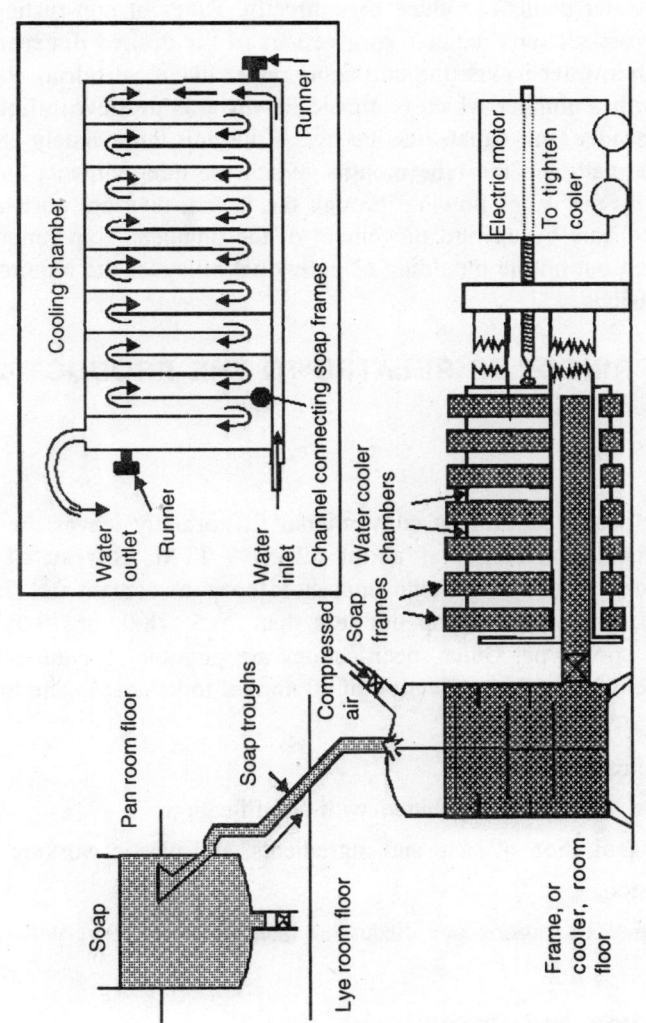

Fig. 7.1. Water cooler – Slab type.

corrosion-resistant metal; and baffles are provided as shown inside the plates to cause the water to flow at a good velocity over the whole of the cooling area. Solidification occurs in a few hours, and the slabs of solid soap are handled like slabs from the frame cooling process.

Other water coolers produce bars directly. Tubes of non-rusting metal, with the cross-sections required to give bars of the desired dimensions, are arranged in a water-jacket (the unit being rather like a calandria). Each tube is fitted with a plunger which is moved downwards to draw in liquid soap from the header tank situated at the top of the unit immediately above the upper tube plate and the tube mouths. When the tube contents have been solidified by the water flowing through the jacket, they are pushed out in the form of bars by upward movement of the plungers. Experiments have been carried out on the moulding of individual billets of the required shape and dimensions.

SOME PRINCIPLES RELATED TO THE PRODUCTION OF EXTRUDED SOAP

General

Soap base produced using a conventional fit normally leaves the settling pan, or final centrifuge, at about 62–64% TFM. Bureau of Indian specifications require a genuine household soap to contain not less than 62% TFM, and a toilet soap not less than 76.5, 78.0, or 79.0% TFM, depending upon type. Other specifications are possible, but the soap base needs to be dried in the production of all normal toilet soaps. The following operations involved :

(i)   Solidification.

(ii) Drying, now often combined with solidification.

(iii) Homogenisation of soap and ingredients, and plastic working.

(iv) Extrusion.

Additional operations are discussed later in connection with specific processes.

Solidification and Drying

In the traditional toilet soap process, the 63% TFM soap base is solidified and dried in separate operations. Frame cooled, or water cooled, soap is cut into chips, which are then dried on trays with mesh bases in a warm air cupboard. This method has long been obsolete except for experimental batches and very small scale production.

The next method, employed for many years, uses a chilling drum and a band drier in which ribbons of soap are conveyed slowly through a chamber in which warm air is circulated to give a continuous drying process. Several bands, through which the air can flow, are normally used, arranged so that the soap falls from one band to the next below. Some moist air must be vented and replaced by fresh air, but internal and/or external recirculation of air is necessary for a good rate of drying and proper thermal efficiency. The soap ribbons are dried from their surface with diffusion of water from the interior to the surface; and they become drier on the outside than in the interior. Some driers include an intermediate mill to homogenise the soap part way through the operation, and so improve the later stages of drying. The fully dried soap is usually milled to crimped chips, or converted into noodles, to make it more compact for the next process and/or for transport and storage, as well as more homogeneous. Particles of overdried soap can impair the smooth texture of a soap tablet and make it feel sandy, or gritty. Special care is needed to avoid soap ribbons with irregular protruberances, or undue thinness, which can become overdried; and to remove dust whenever necessary. The removal of dust applies also to conveyors. The chilling drums used to convert liquid soap base to solid ribbons are large diameter water cooled drums on which the soap is spread in a thin layer. This is often done by providing a small roll to work close to the upper part of the drum. A small reservoir of liquid soap is held in the 'nip' between the two rolls and a layer of soap is formed on the cooled drum which is subsequently removed as ribbons by a serrated knife followed by a full-bladed knife. The large drum into which the cooling water is sprayed can operate half-full of water, or a discharge pipe introduced through the trunnion and bent downwards can keep the drum nearly empty.

Band driers have now largely been replaced by driers in which the liquid soap is heated in a heat exchanger, usually tubular or plate type, and then flashed by reduction in pressure. Three types are used which in principle differ only in the pressure to which the soap is flashed :

(i) Tubular, or plate, driers in which the liquid soap is heated and the pressure then reduced to atmospheric by passage of the fluid through a throttle. Vapour is discharged and the concentrated liquid soap is solidified on a chilling drum, or other device.

(ii) Vacuum flash driers in which the hot liquid soap is flashed into an evacuated chamber. Flashing to a vacuum also chills the soap and converts it to a solid form.

(iii) Driers in which partial drying by flash to atmospheric pressure is followed by a second stage in which the residual liquid soap is reheated and flashed to vacuum.

The simplified sketch for countercurrent flow in a spray-drying tower is shown in Fig. 7.2(a) and Fig. 7.2(b) shows vacuum drier chiller unit, is a key element in modern plants for the production of both household and toilet soaps.

When used to produce *household soap*, the heater is not needed, except as a pipe. The liquid soap enters the spray chamber at about 80–90°C and sufficient evaporation occurs to cool and solidify the soap by conversion of sensible to latent heat of vapourisation.

The liquid soap is sprayed onto the walls of the evacuated chamber and is scraped off by blades on the same shaft set at 180° to the nozzles. Water vapour is drawn through dust collectors by the vacuum system. The solidified soap falls to the base of the chamber and is removed continuously by the first worm extruder (refiner). The noodles of soap fall into the vacuum chest from which they are extruded in the form of a continuous bar by the second extruder (plodder). The two extruders are shown in Fig. 7.2(b), but only one in Fig. 7.2(a).

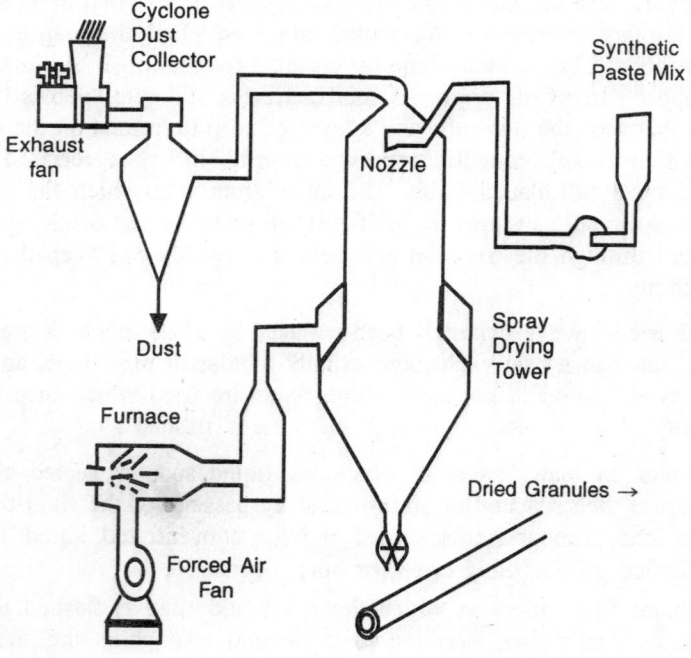

Fig. 7.2 (a). Simplified sketch for countercurrent flow in a spray drying tower.

The worm extruders, which are considered in detail later, also serve to seal the chambers so that vacuum can be maintained.

When the unit is used as a drier to produce *toilet soap noodles*, heat is supplied in the heat exchanger to increase the evaporation, and the second extruder is changed from a plodder to a refiner by removal of the cone and substitution of an end with a perforated plate and rotating knives to give noodles of dried soap.

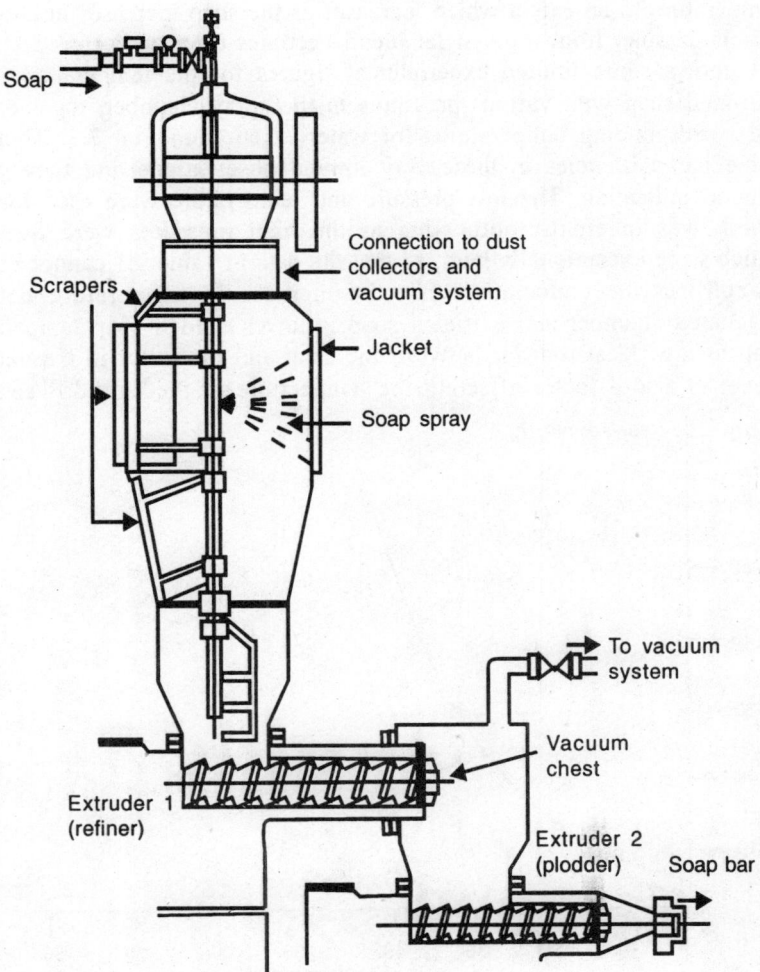

Fig. 7.2 (b). Vacuum drier/chiller unit.

Fig. 7.3 shows the vapour pressure/temperature curve for water, which also indicates the temperature to which water is chilled when warm water is sprayed into an evacuated chamber in which equilibrium between liquid and vapour is reached. Because of the elevation of the boiling point (depression of the vapour pressure) due to dissolved substances, the curve for soap of any given water content indicates a somewhat higher temperature than that for water at the same pressure; but the curve for water shows roughly how the temperature to which the soap is cooled varies with the absolute pressure in the evacuated chamber. Mechanical work done on the solid soap in the extruder, or extruders, raises the temperature to an extent which increases as the soap increases in hardness, that is as soap from a given fat blend becomes drier and/or colder. Table 7.1 shows some limited experimental figures for the temperature of the extruded soap with various pressures in the spray chamber, together with the corresponding temperatures for water taken from Fig. 7.3. There are some inconsistencies in these very limited observations, but they give a general indication. The low pressure and temperature were used for soap which was inherently soft, whereas the high pressures were for soaps which were exceptionally hard, so that the data in Table 7.1 cannot be used to confirm the expectation that the increase in temperature between evacuated chamber and extruded soap is lower at high soap temperatures than at low. Heat transfer between the soap and the water in the extruder jacket is also a factor affecting the temperature of the extruded soap.

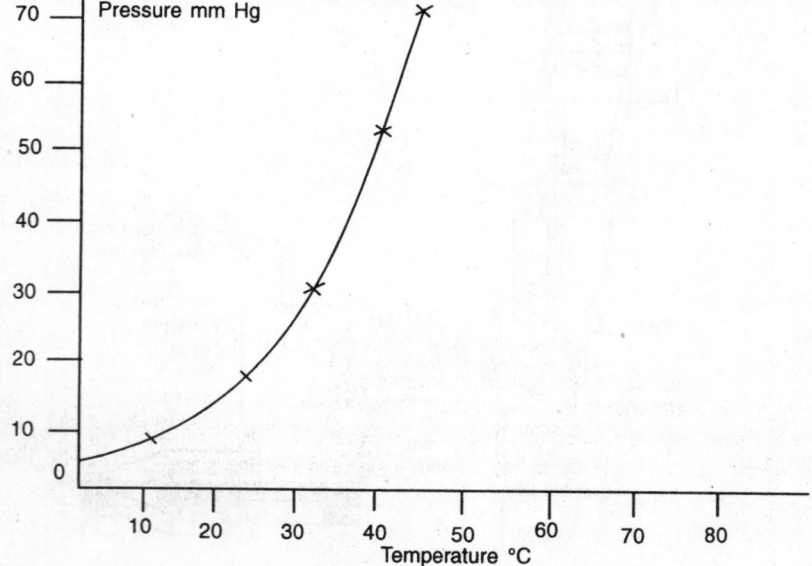

Fig. 7.3. Vapour pressure/temperature curve for water.

Table 7.1. Temperatures and pressures for soap leaving a vacuum spray drier/chiller.

Absolute pressure in spray chamber (mmHg)	Experimentally determined temperatures of extruded soap (°C)	Water temperatures corresponding to pressures (°C)
10	18–24	10
20–35	–	22–26
35	35	31
39	39	33
42	35	35

A vacuum system as shown in Fig. 7.2(a) with no booster jets, can only reduce the pressure to the vapour pressure of water at the temperature at which the cooling water leaves the main condenser. If cooling water at 15°C is used and the quantity is such that it leaves at 25°C, the pressure in the soap spray chamber will be 24 mmHg plus a few millimetres for the non-condensible gas present and the pressure drop through the vapour system; say 28–30 mmHg. This is sufficient for some reasonably firm soaps at about 63% TFM, but lower pressures of 10–20 mmHg are often better for household soaps. To obtain lower pressures than those which can be produced by a condenser and air pump (mechanical or jet) alone, it is necessary to include in the vacuum system a booster jet, or thermocompressor, to compress the vapours before they reach the main condenser. A booster jet is essential when the cooling water is warm, or only available in limited quantity; also when filled, or other soft, soaps are being processed. No steam need be used in the booster jet, if fitted, for the production of toilet soap noodles. If very hard soaps, such as those from 80–100% coconut oil, are to be processed it is usually necessary to use a higher pressure in the spray chamber but, often preferably, not in the vacuum chest. This can be achieved by a partially closed valve, or similar throttle, in the relevant vapour line. As well as necessary for the production of extrudable soap, this higher pressure and temperature avoids the inclusion of specks of extra hard soap. These specks can be removed by fine gauzes in the extruders; but frequent stoppages to clean the gauzes are then needed. The output with 80–100% coconut oil soaps is well below that with soaps of more normal composition. Chilled water for use in the jackets round the extruder worms is also sometimes provided by partial evaporation under vacuum, a process known as spray chilling. The use of chilled water is often desirable with soaps that are soft.

An approximate indication of the amount of drying which results from the use of various liquid and solid soap temperatures can be obtained by the same method as that used for flash evaporation in fatty acid distillation. Assume initially that no vapourisation occurs in the soap pipe prior to the spray nozzles. The specific heat of soap is about 0.75; and the latent heat of vapourisation of water is about 566 CHu/lb (1019 Btu/lb) for the range 90–20°C, 545 CHu/lb (981 Btu/lb) for the range 150–20°C, and 523 CHu/lb (941 Btu/lb) for the range 150-100°C. For 1 lb of liquid soap, let the evaporation be y lb, so that the dried soap produced equals $(1-y)$ lb. The heat balance for the case in which liquid soap at 90°C is chilled to solid soap at the base of the chamber at 25°C is approximately $1 \times 0.75 \times (90-25)$ $= y \times 566$ or $y = (0.75 \times 65)/566 = 0.086$, or 8.6%. If the liquid soap fed contains 63% TFM, the solid produced contains $(63 \times 100)/(100-8.6) =$ 68.9% TFM.

Table 7.2 shows the results of similar calculations for various conditions. In the estimation of the temperature of the extruded bar (or noodles), experience has been used, with allowance for the heating effect of the mechanical work in the extruders and the influence of likely barrel cooling water temperatures. In the production of household soap (with no heat added by the heater) a feed at 63% TFM emerges at about 69% TFM in good agreement with the figures calculated in Table 7.2. As shown in Table 7.2, the feed must be reduced to about 57–58% TFM, if a bar at about 63% is to be produced, provided the liquid soap is sprayed at about 90°C. This dilution of the feed with weak electrolyte (brine or sodium silicate) can be carried out in a crutching pot, or, more conveniently, by direct injection into the soap feed pipeline. The evaporations, and hence final TFMs, estimated in Table 7.2 for toilet soap base are low; this is doubtless primarily because some vapourisation occurs in the heater, so that additional heat is added without a further increase in soap temperature.

In one series of experiments, the dust collected from the vapours when making soap at about 63% TFM was 0.08% of the throughput, whereas it rose to 1.0–1.2% when toilet soap noodles were being produced. Soap lost in the cooling water for the vacuum system was negligible.

Steam and water requirements of flash driers/chillers, can vary considerably. The sensible heat in the liquid soap is utilised in vacuum units, as compared with processes in which the soap is solidified on chilling drums, but steam and/or power is required for vacuum maintenance, particularly when booster jets are needed to achieve the required soap temperature.

Table 7.2. Evaporation and TFM of chilled soap under various conditions of flash drying (no vapourisation prior to nozzle assumed).

TFM	Feed temperature (°C)	Pressure in spray chamber (mmHg)	Temperature of water equivalent to pressure (°C)	Evaporation (%)	Extruded soap (°C) (est.)	Extruded soap TFM) (%)
63.0	90	30	29	8.1	34	68.6
63.0	90	25	26	8.6	32	68.9
63.0	90	20	22	9.0	28	69.2
63.0	90	10	11	10.4	23	70.3
62.0	90	20	22	9.0	28	68.1
57.0	90	25	26	8.6	30	62.4
63.0	125	25	26	13.6	31	72.9
63.0	140	25	26	15.6	31	74.6
63.0	150	25	26	17.0	32	75.9
63.0	160	25	26	18.4	32	77.2
63.0	140	atmospheric	100	5.7	-	66.8
63.0	160	atmospheric	100	8.6	-	68.9

Reasons why flash drier/chiller units have largely replaced other types of plant for both household and toilet soaps are :

(i)   they are versatile, although, as explained later, some different ancillary equipment is needed for toilet and for household soaps.

(ii) they take up much less floor space, and involve much less soap in process, than the frame, or water cooling, processes for household soaps; or band driers plus the necessary finishing equipment for toilet soaps.

(iii) relatively little labour is needed.

(iv) more steam, power and cooling water are needed than for some of the older alternatives; but the overall cost comparison is normally favourable.

(v)   as discussed later, there are commonly quality advantages, although house-hold soap is softer and more readily dissolved than that made from the same fat blend, and at the same TFM%, by the older processes. This may be an advantage with tallow now usually the cheapest good quality fat.

Extrusion, Homogenisation and Plastic Working

Worm type extruders are used for two purposes :

(1) To homogenize and work soap in order to get it into the required plastic condition. Such machines are often called refiners. They push the soap through gauzes backed by perforated plates; and it emerges in the form of rods which are cut into noodles by knives which rotate with the worms. Refiners are also used to convert soap ribbons into noodles for pneumatic, and other, transport.

(2) To form the soap into bars of the required cross-section which can be cut into bars for sale, or into billets to be stamped into tablets. Machines which carry out this task are called plodders.

The quality of an extruded soap is considerably affected by plodder performance which will now be considered in some detail. Fig. 7.4 shows diagrammatically the main features of the most common type of non-vacuum plodder, with a single worm and single orifice. Small pieces of homogenised and worked soap, in the form of noodles, or crimped chips, are fed into the hopper and then pushed along the barrel by the worm. Air is squeezed out; and the cylinder of soap is forced along like a nut on a screw, arranged so that the screw can only rotate and the nut only move longitudinally. The soap is pushed through the holes in the perforated plate in the form of rods which are welded together in the cone; and a bar of the required cross-section emerges continuously from the eye plate. A gauze, say 16 mesh BS (0.385 mm holes), in front of the perforated plate is often advantageous, but is not always used, and it does tend to reduce throughput.

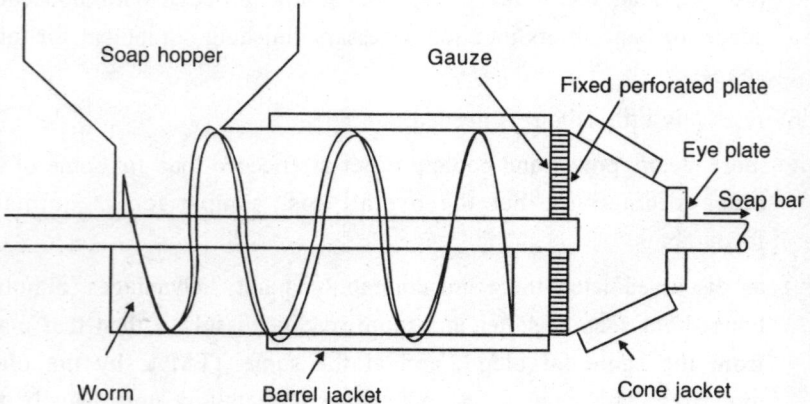

Fig. 7.4. Non-vacuum plodder.

An important advance during the last few years is the vacuum plodder. The soap hopper which feeds the plodder is enclosed and soap is fed into it by a refiner, roughly as shown in the lower part of Fig. 7.2(b). The hopper, or vacuum chest, is connected to a vacuum system, so that the air between the pieces of soap taken into the plodder worm is at a pressure much below atmospheric and the mass of air is much lower than when an open hopper is used.

Under some circumstances it is possible to use an eye-plate with two holes, so that two bars are extruded simultaneously. Plodders, called duplex or twin-worm, are also available in which two worms and cones are fitted in a single casing.

Cold, or cool, water is usually used in the jacket to cool the plodder barrel; and warm water or other heating medium in the jacket on the cone. When collar, or shell, type dies without cutting, or shaping, boxes are used to stamp tablets, the mass (weight) of the tablets is controlled by the cross-section of the billet, the length to which the billets are cut, and the density of the soap. It is often convenient to use an eye plate in which small adjustments can be made to one dimension of the hole with a slide, as shown diagrammatically in Fig. 7.5. In order to obtain smooth surfaced bars and tablets, it is normally necessary to use a heated eye-plate and a small electrical heater, of about 100 watts power, is a convenient way to achieve this. It is important to ensure that the edges of the eye plate are free from defects as lines on extruded bars do not usually stamp out.

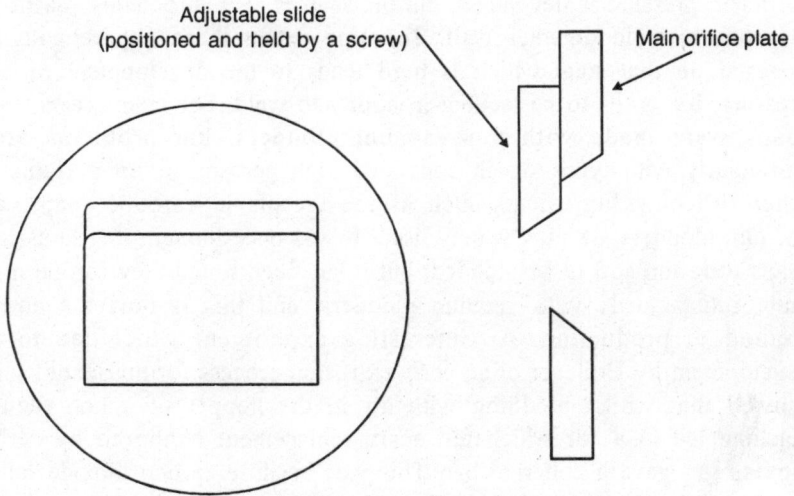

Fig. 7.5. Eye plate with adjustable slide.

Worm extrusion is an apparently simple operation, but results can be influenced by a number of factors in a rather complex way. One variable tends to interact with others, and it is helpful to consider first some general principles. As already described, the noodles, or crimped chips, are fed into the worm with air between the pieces (much less in mass when the hopper is under vacuum than when it is open to atmosphere). This air should be forced back as the soap is compacted by the worm in the barrel; and the effectiveness with which this is done is influenced by the pressure developed in the plodder. The presence of undue air in the bars, or tablets, shows up in the form of blisters when the pieces of soap are allowed to stand for a few hours. Two convenient tests for this defect, sometimes known as blister, or thumbnail, cracking are (1) to cut tablets longitudinally with a sharp knife, or wire, and to leave the halves cut face upwards for a period, after which thumbnail-shaped cracks develop in faulty soap; and (2) to determine the density of the soap, which will be lower than normal if air is present.

The rods of soap formed by the perforated plate must weld properly in the cone to produce as substantially homogeneous bar. All plodded bars are weaker in the direction of extrusion than in the cross-direction, but the difference should be relatively small. A useful test is to take a length of plodded bar and to twist it backwards and forwards in the hands to ascertain how well the rods have welded together and how brittle the soap is. The aim is to produce a bar which is uniform; and free from streakiness and tendencies to blister and crack. This requires conditions under which sufficient pressure is developed, but the soap is also sufficiently plastic for the rods to weld together well. To some extent these requirements are opposed, in that soap which is hard leads to the development of high pressure, but tends to be lacking in ability to weld. For many years, toilet soaps were made with non-vacuum plodders, but problems arose particularly with types which contained high percentages of perfume, or other softening ingredients such as the cresols in carbolic soaps; and vacuum plodders are now widely used. It was once thought that household soaps were too soft to be plodded; but it has been found they can be made quite satisfactorily with vacuum plodders; and this is now the normal method of production. An interesting experiment which led to the development by Unilever of an early extrusion process for household soaps showed that whilst plodding with air in the hopper of a non-vacuum machine led to a bar which fell apart, replacement of the air by carbon dioxide gas gave a coherent bar. This was because carbon dioxide left in the bar was absorbed chemically, whereas air was not. The use of a vacuum plodder was a practical next step. The household soap, but at about

68–69% TFM and made from 100% tallow base, extruded satisfactorily without vacuum, but these conditions are not normal for household soap.

When non-vacuum plodders are used to produce toilet soaps it is particularly important to ensure that the worm is kept well covered with properly compacted soap.

Apart from the additional cost, the only disadvantage of vacuum compared with non-vacuum plodders is that there can be some additional loss of the more volatile components of a perfume.

Plodding is affected by soap factors and by machine factors :

Soap factors include :

(i) Composition of the fat blend.

(ii) Moisture, or TFM, content.

(iii) Salt content.

(iv) Soap temperature.

(v) Nature and quantity of added ingredients. As already mentioned, high contents of perfume, cresols, and some superfatting agents, soften the soap and can cause problems even with toilet soaps at around 79% TFM.

Machine factors include :

(i) Use, or non-use, of vacuum.

(ii) Worm (and barrel) dimensions.

(iii) Clearance between worm and barrel, which increases with time due to wear, until the machine has to be reconditioned.

(iv) Worm speed.

(v) Temperature of barrel cooling water.

(vi) Use of gauzes behind the perforated pressure plate.

It is important to balance various factors so as to achieve the best results. For example, if the soap is of an inherently soft composition, it may be advisable to dry it to a lower moisture content, to cool the noodles to a lower temperature, or, in the case of household soaps, to include a higher content of salt. Conditions in the layer of material adjacent to the plodder barrel wall are important. Plodder performance tends to fall when the clearance between worm and barrel increase, but it may be possible to mitigate this to some extent by the use of very cold water in the barrel jacket which hardens the layer of soap adjacent to the barrel wall.

Before a further discussion of machine factors, it is convenient to present some design and performance data for a number of plodders tested some years ago. Most of these plodders are no longer available; and new machines are now usually obtained from the main producers of soap finishing plants.

Experiments have shown that small diameter plodders are much more sensitive to the temperature of the cooling water in the barrel jackets than those of large diameter. When this temperature is found to be important, it is best to arrange a circulating system to which cold water is added as required to maintain the required temperature and the surplus water is allowed to overflow. The normal pattern is for output to increase as barrel jacket water temperature increases up to near the point at which churning starts with a fall in output. This again emphasises the significance of the soap layer near to the barrel wall. Even when the required output can be obtained from a small diameter plodder, it is usually found desirable for quality reasons to use a large diameter machine when bars of large cross-section are to be made.

Special perforated pressure plates which rotate with the worm, or are fitted in the final flight of the worm, have been described, but it is doubtful whether they are beneficial.

Plodder worms are often made with the pitch greater at the hopper than at the discharge end in order to allow for the increase in bulk density and consequent decrease in volume as the noodles are compacted and the air squeezed out. This seems logical, but limited tests failed to show an advantage.

It is possible to gain some insight into the working of a plodder by stopping the machine, removing soap from the hopper, filling up again with similar soap of a different colour, and running the machine for a short time until the second coloured soap has reached a point in the cone. The cone of soap is then removed as a whole and cut longitudinally, or into a series of slices. The boundaries between the two soaps provide interesting information as to the flow of soap through the machine.

When ingredients, such as perfume, pigments and colour solutions, are added in appropriate proportions to dried toilet soap noodles the ingredients are spread on the surface of the soap. It is then necessary to homogenise the mixture, and to work the material; this can be done with either refiners or mills, or a combination of the two.

As already indicated, refiners are worm extruders, basically similar to plodders, except that the cone is omitted and a multibladed knife is fitted to rotate with the worm and to cut off the rods of soap as they emerge from the perforated plate in the form of short pieces called noodles. In continuous production lines units of three refiners in series are often used. The three barrels are usually arranged in a stopped horizontal fashion, or inclined. In either case, this saves space and the soap which emerges from one barrel drops into the hopper of the next stage. Refiners use gauzes behind the perforated plates, to improve the homogenisation and working of the soap; and to strain out any particles, such as hard soap, dirt, or fibres. The gauze in one stage may be finer than the gauze in the preceding stage because the soap is warmed and made more plastic as the operation proceeds. Too fine a gauze in the first stage would tend to result in excessive pressure and power consumption; and/or reduced throughput. The percentage of hole area in a gauze depends upon the wire diameter, as well as the number of holes per inch (mesh size), but the wire must be sufficiently strong to withstand the pressure on the unsupported areas over the holes in the perforated plate. A 52 mesh gauze (0.3 mm hole) is about the finest likely to be used. In some cases the fine gauze can be supported by a coarse gauze, say 12 or 16 mesh (1.4 or 1.0 mm hole), between it and the pressure plate. The higher temperature induced by a fine gauze may be compensated by increased barrel cooling, that is by the use of more, or colder, water.

It should be noted that vacuum flash driers provide one, or two, stages of refining (before some important ingredients have been added); and a vacuum plodder one additional stage of refining.

Throughput 2.25–2.5 tonnes/h depending upon the build-up of debris on the meshes, which are usually cleaned about every two hours. The total power consumption is about 26 kWh per tonne of soap at 78% TFM. The soap enters the first stage at 30–35°C and leaves the last at 38–42°C.

The barrels are water cooled in series using 4800–6000 lbs/hr. Water temperature rises from 13–14°C to 21–28°C.

Mills can be used as an alternative to refiners; and they were so used many years before refiners were developed. Mills can have 3, 4 or 5 rolls. For continuous toilet soap lines sets of three three-roll mills in series are often used, but large new installations now more often select refiners. At one time, granite rolls were common, but now cast iron rolls (often with a chilled skin), at least some of which are water cooled, are almost universal. Granite rolls are sometimes considered desirable if medicated soaps containing corrosive ingredients, particularly mercuric iodide, are to be processed. Roll arrangements for 3, 4 and 5 roll mills are shown diagrammatically in Fig. 7.6. Each roll rotates at a substantially higher speed than the proceeding one, so that :

(i) The soap transfers from roll to roll, and so moves through the mill as indicated in the diagram.

(ii) There is a shearing action at each nip which homogenizes the mix and tends to break down any particles of hard soap.

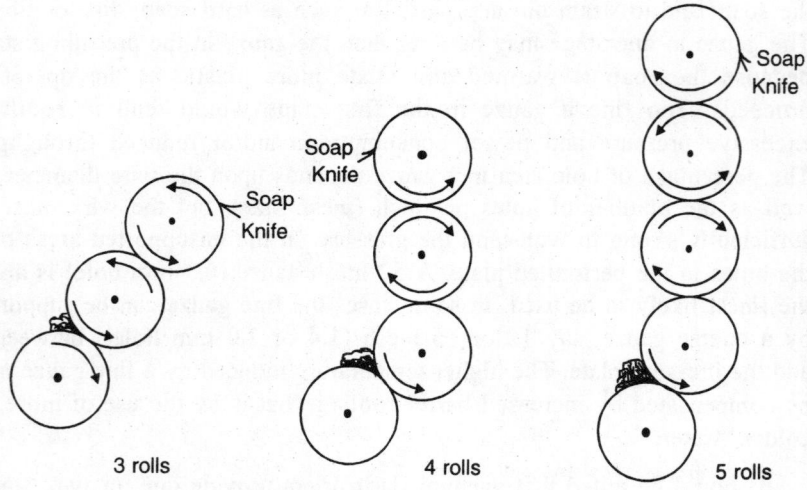

Fig. 7.6. Roll arrangements in mills.

Traditionally, the final roll was fitted with two knives. The first was serrated and removed alternate ribbons of soap; and the second was a complete blade which removed the remaining ribbons. These knives were often followed by crimping bars arranged to impede the flow of the ribbons and so to produce compact, crimped, pieces of soap. The more modern procedure is to use a comb-like knife which produces conveniently crimped pieces of soap directly. With some mills the knife, or knives, is/are positioned mechanically; in others, pressure is maintained hydraulically. The position of the middle roll is fixed and the top and bottom rolls are adjusted to give the required clearances. In the mill this is done manually with hand wheels; in some more modern mills, the gaps are set and held hydraulically. Because of the increase in roll speed upwards, the gap must decrease from nip to nip in order to maintain a full ribbon on the rolls.

The most satisfactory way to measure the true running clearance between a pair of rolls is to insert a piece of lead wire whilst the mill is running normally and to measure it after it emerges. Neither the thickness of the soap flake produced, or that of a piece of lead wire passed through the mill when it is empty, gives the correct answer. One test showed :

lead wire-rolls empty	0.004 inches (0.1 mm)
lead wire-normal operation	0.011 inches (0.3 mm)
soap flake on the roll	0.020 inches (0.5 mm)

Typically, the lower gap is 0.010 inches (0.25 mm) and the upper gap 0.006 inches (0.15 mm).

Whilst sets of three by three roll mills are convenient for continuous large scale production, small productions may require the use of one mill. Methods by which more than one milling operation can be achieved with only one mill are :

(i) The mill operates in two halves, with chutes arranged so that the soap passes through one half and is then diverted back to pass through the second half of the mill.

(ii) The mill is equipped with two sets of knives, and with upper and lower storage hoppers. On the first pass the soap is removed by the back knives and held in the upper storage hopper. When a batch has been collected, the once milled soap is dropped into the lower hopper and is given a second milling. This process is repeated if a third milling is required. When the soap has been milled sufficiently, it is taken off by the front knives and leaves the machine.

Points in a comparison between milling and refining of toilet soap are :

(i)   with proper operation soap quality is similar; but mills have an advantage if overdried soap has to be incorporated.

(ii) studies some years ago showed little difference in cost.

(iii) mills are more easily cleaned for a change from one soap to another; and so have an advantage when the plant is to be used for short runs of different varieties of soap. Mills are also preferable if the plant is only operated intermittently, as the soap in refiner worms cools and hardens (and hence has to be removed) on standing for several hours.

(iv) refiner systems can be substantially enclosed to minimise loss of perfume by vapourisation and pick-up of air-borne dust.

(v) it is sometimes claimed that the noodles from a refiner pack into the plodder worm better than crimped chips from a mill. This is of little significance if a vacuum plodder is used, because the actual plodder worm is fed with noddles from the refiner part of the machine.

The use of three refiners, plus refining in a flash vacuum chiller/drier and in a vacuum plodder, is typical modern practice, but some plants use less working and a few more. Some operators still consider the inclusion of a milling stage desirable. If mills and refiners are used in series, it seems sensible to refine first, in order to strain out unwanted particles, before any such particles are broken by the shearing action in a mill. The converse arrangement, in which the mill comes first, might impede effective straining in the refiner by breaking down the particles.

So as to avoid damage, mills, refiners and plodders are usually provided with a replaceable link which will break if the machine is overloaded.

PROPERTIES OF SOLID SOAPS

Important properties of a piece of solid soap are :

(1) hardness, including sufficient firmness to avoid damage to the tablet whilst it is being handled in the factory, in transit and in use. Soaps dry out and become harder on storage.

(2) lathering properties, which are closely related to the rate at which soap is removed from the tablet surface when it is rubbed, for example, with a damp cloth or the hand. This rate is more important than the true solubility of the soap.

(3) freedom from undue tendency to form mush when the tablet is left damp, particularly when it is left on a damp surface. Much formation seems to be related to tendency to swell in water and to disintegrate; that is to crack in use.

(4) tendency to crack on storage, or, more particularly, in use.

(5) smoothness to the touch in use; that is freedom from grit, or sandiness.

(6) freedom from deterioration of odour and of colour, including the development of spots or patches. Also freedom from undue distortion.

(7) detergent properties. Differences between one soap and another can be significant when bar soaps are used for laundry purposes, but they are largely unnoticed when tablets are used for personal washing, which is now the major application in the more developed countries.

An experienced person can assess the hardness of a piece of soap by pressing it with a thumb, but an instrument is desirable to provide more reliable and comparative data. One type is a penetrometer, in which the force necessary to push a sphere (or hemisphere) of given size a set distance into a soap surface is measured. The method used determines the quantity of water which has to be run into a can on a lever arm, as shown diagrammatically in Fig. 7.7(a). Another type of instrument sometimes called a sectilometer measures the force needed to cut a piece of soap of known width with a wire of given diameter drawn through it at a steady speed. In one version the wire is held on a structure connected to the frame of the instrument by springs, and the extension of the springs, that is the position taken up by the wire, is a measure of the force being exerted. A trace as shown in Fig. 7.7(b) is obtained, and the height is an arbitrary measure of hardness. The initial peak, due to the extra hardness of the outer skin, can normally be disregarded. The cutting wire, attached to a spring balance, was drawn through the soap and the reading of the balance was a measure of hardness.

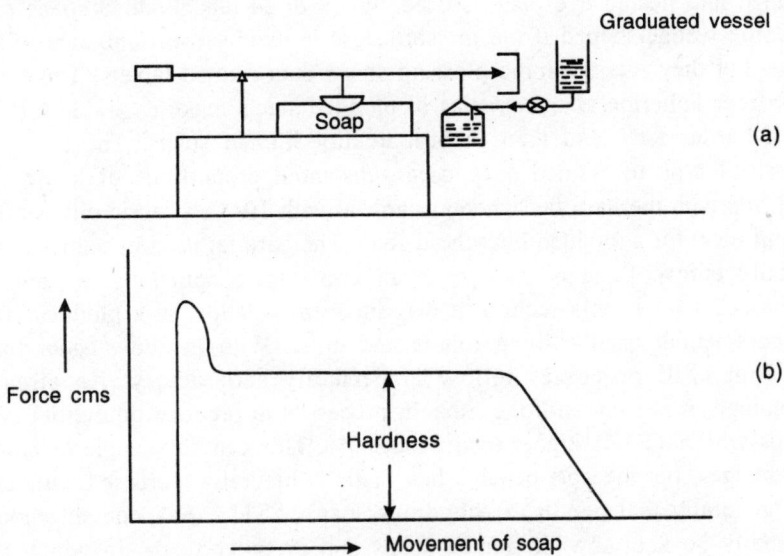

Fig. 7.7. Measurement of hardness. (a) Penetrometer (an electrical contact in a circuit with flashlamp bulb and battery, can be used to indicate the end-point). (b) Sectilometer trace.

Apart from the influence of chemical composition, the properties of a piece of soap can be affected by its physical state. There is an extensive literature relating to phase differences, crystal compositions, crystal size, and crystal orientation. It is, however, conflicting; and published information is not readily applied to practical situations. Many operators find an empirical approach more useful, although observations concerning factors such as temperatures and conditions under which the soap is worked may apply only to specific soaps under a limited range of circumstances.

There are major differences between soaps produced by slow cooling in a frame, and to a lesser extent in water coolers, and those made by the modern chill, work and extrude methods. Fractional crystallisation, with solidification of sodium stearate and sodium palmitate before that of the unsaturated and lower molecular mass soaps, can occur when soap cools slowly in a frame, to give a very different structure from that in an extruded bar. Framed soaps are much harder and slower lathering than plodded household soaps of similar composition. Unless filled, they also have a rough surface, showing what is called 'feather', and they tend to distort as they dry out. Plodded soaps show a smooth surface, provided a properly shaped and heated eye plate is used. When dried out, plodded soaps may become wedge-shaped if the moisture loss is mainly from one side of the case, but they retain a more pleasing shape than framed tablets. The softer and freer lathering characteristics of plodded soaps make desirable the use of a harder fat blend than in best quality framed soaps. The so-called 'washer' type of framed soap used substantial proportions of lauric and soft oils with the hard fat, whereas hard fat with 10–15% lauric oils is often found ideal for a plodded household soap. The hard fat used in soaps is now usually tallow. Palm oil was once an important soapmaking fat and, as discussed later in this section, it was superior to tallow in a plodded soap process which used chilling drums and mills. With the now usual flash vacuum chill processes tallow is perfectly satisfactory. As already explained, it is easy with the modern processes to produce household type soaps with a TFM% higher than about 63%. This can have some technical advantages, but the cost per lb when made is naturally increased, although not generally that per lb of anhydrous soap or TFM. Salt content should normally be kept low in framed soaps and in toilet soaps, in which too much salt is liable to cause brittleness and a tendency to crack in use. In plodded household soaps a somewhat higher level of salt than in framed soaps is often desirable as it hardens and improves the texture, particularly if soft oils have to be used in the fat blend. In one factory using the early chilling drum type of plodded bar process, it was observed that soap which had settled in a pan for a shorter period than normal behaved better than the longer settled equivalent which would have been superior if framed.

This was particularly the case when the hard fat was tallow rather than palm oil. It was later found that the same advantage could be achieved by an addition of an appropriate quantity of brine to well settled soap. Genuine Hard Soaps, allow up to 1.25% salt in extruded soap, but only upto 0.75% in batch (that is framed or water cooled) soaps. It has been found that, within the range of compositions tested, hardness, swelling and cracking after immersion in water did not vary greatly. Mechanical tests on rate of foam generation in hard water with sebum present showed a definite peak at 85/15; but subjective assessments by panels of users carried out under controlled conditions showed that preference increased with increase in coconut oil content up to the maximum tested. The panel members used more soap with the higher coconut oil soaps; 0.5 g per use with the 95/5 soap and 0.8 g with 75/25. Slushing (much formation) also increased linearly with coconut oil content.

When it is required to estimate the composition when made of competitive, or other stored, soaps it is important to allow for the fact that salt and caustic soda migrate into the core of the tablet as drying takes place from the outside. As already explained, an average sample for analysis can be obtained by taking an aliquot portion of a solution of the whole tablet, or of an eighth cut to represent the whole tablet. Sodium carbonate concentration tends first to increase and then to decrease on moving inwards from the outside of a tablet. This may be associated with carbonation of caustic soda by carbon dioxide absorbed from the atmosphere.

INGREDIENTS ADDED TO HOUSEHOLD AND TOILET SOAPS

General

Many types of ingredient are added to soaps, particularly toilet soaps. The main classes of additives are :

(1) preservatives.

(2) perfumes.

(3) colours, opacifiers and optical brighteners.

(4) germicides and other medicinal ingredients.

(5) superfatting agents.

(6) scum dispersants, leading to NSD tablets and bars.

(7) alkaline fillers.

(8) miscellaneous.

Preservatives

Following points should be considered for the stability of a soap :

(1) selection of a fat blend without substantial proportions of highly unsaturated fatty acids. At one time 2–3% of rosin was included in toilet soap and other bases to improve keeping properties, but rosin is little used now.

(2) preservation of fats by natural and artificial antioxidants.

(3) the need for thorough saponification, although much less drastic treatment with caustic soda than was once common is now usual.

(4) the need to minimise the presence in the soap base of heavy metals, particularly iron and copper, and perhaps, nickel.

The most important soap preservatives are substances which negate the adverse catalytic effects of heavy metals which often cause brown spots, or patches, as well as more general deterioration :

(5) organic sequestrants, such as EDTA and, less usually NTA. Too much EDTA can form coloured (yellow) complexes with iron; and it is often better to use EHDP (ethane hydroxydiphosphonic acid) to sequester the iron, plus a small concentration of EDTA to deal with the copper. Concentrations of these preservatives should be determined by experiment to suit local conditions.

$$\begin{array}{ccc} & CH_3 & \\ O & | & O \\ \| & | & \| \\ HO-P-C-P-OH \\ | & | & | \\ OH & OH & OH \end{array}$$

(6) Magnesium silicates which can also be used to inactivate heavy metals in soaps other than those superfatted with fatty acids; presumably they work by adsorption. Magnesium silicate can be made separately and added as such, but it is usual to add the ingredients magnesium sulphate and sodium silicate solutions to the soap so as to form magnesium silicate *in situ*. When modern driers which include tubular or plate heaters are used, it is advisable to inject the silicate after the soap has passed through the heater; otherwise scale is liable to form on the heating surfaces.

Typical additions are :

for toilet soaps :

0.33% of a 50% solution of $MgSO_4.7H_2O$ (equals 0.027% of MgO) plus 0.6% of 57°Tw (1.285 specific gravity) neutral sodium silicate (equals about 0.12% SiO_2).

for household soaps, say, 0.004% magnesium sulphate expressed as MgO plus 0.006% sodium silicate expressed as SiO_2.

Magnesium silicate preservative is not used in soaps superfatted with fatty acids because it causes grittiness. Some technologists consider that it may suffer this defect with conventional toilet soaps; but it has been widely used for many years and has the merit of cheapness and ready availability of the ingredients. Antioxidants of the BHT type are sometimes used in soaps as well as in fats.

Perfumes

Between 0.5% and 2.0% of perfume is usual in toilet soaps although higher concentrations are sometimes used, with about 0.8–1.0% common in the mass produced popular brands. Great care is necessary in the selection of a perfume, both as regards appeal to users and technical factors. It is usually advisable to approach one, or more, companies with experience in the field of soap perfumery and to give them as much information as possible regarding requirements :

(i)   The perfume should accord with the general image possessed by, or desired for, the brand.

(ii) Some perfumes are more substantive to the skin than others; and it is important to decide whether substantivity is desirable and how far it can be achieved (or avoided).

It is necessary to make up soaps containing the perfumes to be considered at the concentration, or usually series of concentrations, which seems appropriate. Storage tests, usually including accelerated tests at temperatures higher than ambient, should be carried out; and these should include exposure to strong light both naked and in the proposed packaging. Consumer tests should be carried out on soaps which have been matured for an adequate time. Perfumes for toilet soaps are normally blends made from a large number of ingredients many of which contain a large number of different chemical compounds. Perfumers with adequate experience in the soap field should only submit blends which are sufficiently stable to the alkalinity of soap, light, etc; and which will not deteriorate excessively in colour and/or odour on storage. Additives such as EDTA can affect the alkalinity of a soap and proposed perfumes should be checked by the soap manufacturer under conditions relevant to his particular product and market. Vacuum plodders are more liable than simple plodders to remove very volatile components from a perfume. In addition to possible deterioration of the soap. It is important to ensure that the soap containing deterioration of the soap. It is important to ensure that the soap containing a perfume under consideration is substantially free from allergic, or irritant, reactions by potential users. This requires extensive tests but, again, experienced perfume houses have built up knowledge and should be able to provide blends for which they can give adequate assurances. The small soap manufacturer is not normally in a position to carry out adequate tests.

Household soaps are often made without perfume. When perfume is added, it is usually a relatively cheap but powerful type, such as citronella, at a low concentration. Many years ago oil of myrbrane, nitro-benzene, was also popular, particularly for tropical markets, despite its toxic nature.

Colours, Opacifiers and Optical Brighteners

Colours used in soaps are of two types, dyes soluble in water or in an organic solvent, commonly the perfume, and pigments which are added as suspensions. As for perfumes, various checks are necessary in addition to an assessment of aesthetic appeal before a colour system is accepted as satisfactory :

(1) *adequate stability.* When the soap is stored and/or exposed to light (naked and packaged). Some colours have been used which are not very stable to alkali and which could only be used with dried soap base exceptionally low in free caustic soda; such dyes should be avoided if possible.

(2) *safety for the user.* Some dyestuffs, or impurities they contain, have carcinogenetic properties; and many countries have strict regulations concerning colouring matters which are used in cosmetics, as well as in foods and drugs.

Opacifying agents are also added to soaps to obtain the desired appearance. Titanium dioxide (of which there are two major types, rutile and anastase) is now the normal material used; and 0.2% is a typical addition to give an opaque toilet soap. Zinc oxide has also been used, but is less effective than titanium dioxide. At one time magnesium carbonate was employed as an opacifier in filled household soaps. Extruded soaps, particularly household types, are liable to show a pattern of relatively translucent soap on the surfaces cut at right angles to the direction of extrusion, known as end marks. If the end marks cannot be eliminated by adjustment of the conditions of plodding, a small addition of titanium dioxide will largely obscure the pattern and make the soap more attractive in appearance. A little opacifier also helps to minimise changes in colour as a soap dries out and becomes more translucent.

As already mentioned special types of optical brightener, such as Tinopal are used to improve the colour and brightness of white soaps. It is also possible to use a low concentration of cotton substantive optical brightener to improve slightly the performance of household soaps in laundry work; but care is necessary to avoid risk of patchy effects when the soap is rubbed on extra dirty parts of articles.

Germicides and Other Medicinal Ingredients

Many brands of soap have been sold on the basis of germicidal, or freshness, claims; and a number of different bacteriostatic, or bacteriocidal, ingredients have been and/or are used to justify such claims. The traditional ingredient was a mixture of cresols and related phenols, often known as cresylic acid, which give 'carbolic' soaps. Carbolic soaps, commonly, but not necessarily, coloured red, usually contain about 2% of phenols when intended for the domestic market. It is necessary to select cresylic acid which is reasonably free from pyridine, and other nitrogenous bodies, which can lead to unpleasant smells when the cresylic acid is included in alkaline soap. Naphthalene is often dissolved in the cresylic acid to improve the odour of the soap. Phenols tend to be held in the soap micelles when the soap is dissolved, so that domestic type carbolic soaps do not show much greater anti-germ properties than soap alone. Special soaps with much higher concentrations of phenols have been made for hospital and similar duties.

More recently various more potent ingredients have been used; and anti-germs properties have been demonstrated on human skin. Freshness claims require the further step that use of the soap can reduce the odour from body areas such as the armpits by reducing, or preventing, the decay of stale perspiration to produce unpleasant odours. Unfortunately, many of these ingredients have been found liable to have undesirable side effects on some users; and great care is necessary to select one which is permitted and fully safe. Apart from straightforward toxicity (mainly where a gross excess has been included in the product in error), and sensitisation, or allergic reaction, problems have arisen through :

(1) the presence of impurities in the germicide as bought.

(2) breakdown to produce damaging compounds when the soap is stored for long periods at relatively high temperatures, and, particularly, when scrap is re-boiled and returned to the product.

(3) photosensitisation, that is skin, and possibly more general, reaction when a person who has washed with the soap is exposed to bright light.

Superfatting Agents

Many superfatting agents, including lanolin, petroleum jelly, coconut oil, and stearin (commercial stearic acid), are used in soaps; they are intended to increase mildness in use. As mentioned earlier, an important type of superfatted soap uses a fat blend with a high proportion of coconut oil to tallow; and is superfatted with 5–10% of fatty acids similar to the main fat blend. The maximum effect upon the lather is obtained with about 8% FFA. An unsuperfatted soap made with 30–60% of coconut, or palm kernel, oil tends to be more irritant (less mild) than a conventional toilet soap, but the inclusion of the free fatty acid, and consequent elimination of the free alkali, brings mildness back to about that of a 20% CNO 80% tallow soap.

Scum Dispersants, Leading to NSD Tablets and Bars

Much technical and marketing effort has been expended on the development of toilet tablets in which soap is replaced partially, or completely, by non-soapy surfactants and other ingredients. Objectives are :

(1) to eliminate scum and 'bath-tub ring'.

(2) to produce an immediate lather on bath, or wash basin, water.

(3) with some formulations, to produce a personal washing tablet which is slightly acidic rather than slightly alkaline.

Problems are :

(1) cost, with many consumers not prepared to pay the higher price relative to soap.

(2) to produce tablets which are not too dense and which do not absorb excessive water and become soft.

Many formulations have been patented and/or tried. Some are more, or less, soap-free, but others contain substantial proportions of soap. Stearin is an important constituent of some tablets.

NSD bars for use instead of household soap bars have been successful in some countries. Typical ingredients are sodium alkylbenzene sulphonate, sodium tripolyphosphate, sodium silicate, and sodium sulphate; plus a plasticiser. Waxes and starch may be included to improve texture and feel. The blends are milled and plodded.

Miscellaneous Ingredients

Many other ingredients are, or have been, added to toilet soaps to improve texture, feel, lather, and benefits to the skin of the user; the real merits of some are dubious. Recent developments which include the use of citronellylsenecioate (also known as the citronellyl ester of 3,3-dimethyl acrylic, or 3-methyl crotonic acid), as a non-germicidal deodourant.

EFFLORESCENCE

After storage some soaps develop a surface deposit known as efflorescence. This can be of several types, in particular :

(1) A heavy white deposit, often fibrous in appearance, which comprises inorganic salts, probably mainly sodium sesquicarbonate, and which develops on soaps run with alkaline fillers. Changes in the fillings used may help to reduce this efflorescence. Glued-up soaps are said to be particularly prone to the defect, with fillings based on neutral silicate worse than those using alkaline silicate.

(2) A white fibrous deposit which can form on extruded genuine household soaps. This deposit comprises filaments of soap which are pushed out through weak spots in the tablet surface; that is the filaments grow from the base like hairs. This type of efflorescence develops when there is a fairly rapid rate of drying from the soap surface; and it can usually be avoided by the use of well sealed wrappers of a material with a low moisture vapour permeability. A useful accelerated test for proneness to develop soap hair efflorescence is to cause a rapid flow of water through a relatively thin piece of the soap. This can be done by closing the end of a large test-tube which contains some water with a small piece of soap and storing the tube in a vertical position in a dry atmosphere. In this way, the upper surface dries out in the low humidity atmosphere, whilst the lower surface picks up water from the nearly saturated air below it. Traces of heavy metals on the soap surface, from stamping lubricant or contact with materials such as carton board, can tend to induce the effloresence.

(3) A thin white film which can arise on toilet and other soaps. One cause is the use of excessive quantities of stamping lubricant containing salt; and some versions of magnesium silicate preservative are also sometimes thought to have an adverse tendency.

MODERN PROCESSES FOR THE PRODUCTION OF HOUSEHOLD SOAPS

The main unit of the now standard plant for the production of household soap is the vacuum chiller/extruder shown in Fig. 7.2(b). Broadly in similar plants suitable lengths can be cut from the continuous bar extruded by the vacuum plodder using either a wire cutter swung backwards and forwards at the plodder nozzle, or a simple device shown in Fig. 7.8, sometimes known as a 'hacker-off'. This comprises a cutting wire on a small frame which is connected by a rod of the required bar length to a small plate; the cutter and the plate being at right angles to the rod. The operator holds the plate against the moving end of the bar which is supported on a short piece of narrow wooden roller conveyor, and when sufficient has been extruded, pushes the cutter through the soap. Bars cut in one of these ways can be handled like bars produced from framed soap. Fully mechanised lines, including billet cutter, conditioning tunnel, stamper, wrapping and/or cartonning machine, case packer and case sealer, are now common; variations to suit individual needs are possible.

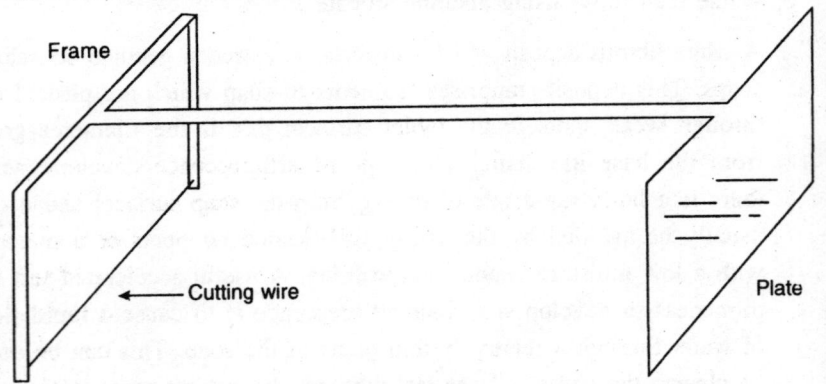

Fig. 7.8. Hacker-off.

Unilever developed independently a process for the production of household soaps which also used a vacuum plodder, but in which the soap was solidified on chilling drums and worked in mills. This process performed well with high palm oil fat blends, but was less satisfactory when the hard fat was tallow. A rather high salt content was important; and it was found that a tubular heater to give about 1% evaporation prior to the chilling drum was markedly beneficial with high tallow blends. It is not clear whether this was due to the small reduction in moisture content of the soap, or whether the temperature from which the liquid soap was chilled was the major factor. The more versatile vacuum chilling units are now preferred; and these do involve rapid chilling over a fairly large temperature range.

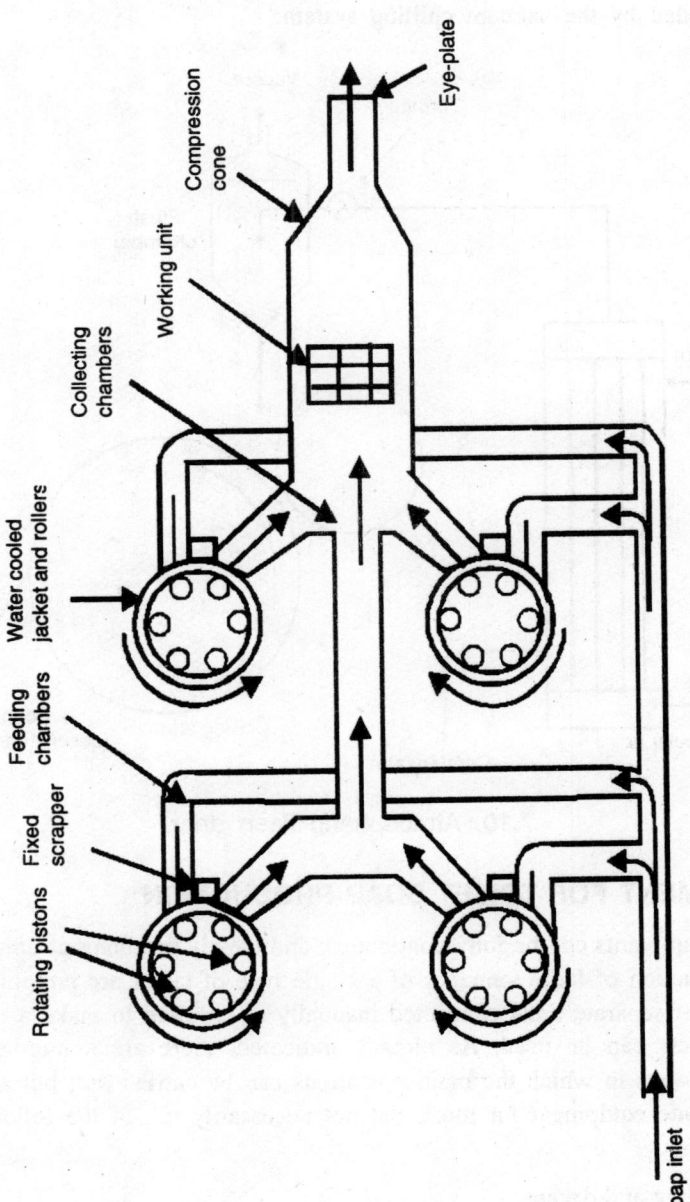

Fig. 7.9. Meccaniche extrusion/chiller for household soap.

Eye-plate

Compression cone

Working unit

Collecting chambers

Water cooled jacket and rollers

Feeding chambers

Fixed scrapper

Rotating pistons

Soap inlet

An extrusion unit shown diagrammatically in Fig. 7.9. In this the liquid soap is cooled and forced out through a barrel, cone and eye plate by rotors which acted as pumps as well as coolers. This plant has also been supersceded by the vacuum chilling system.

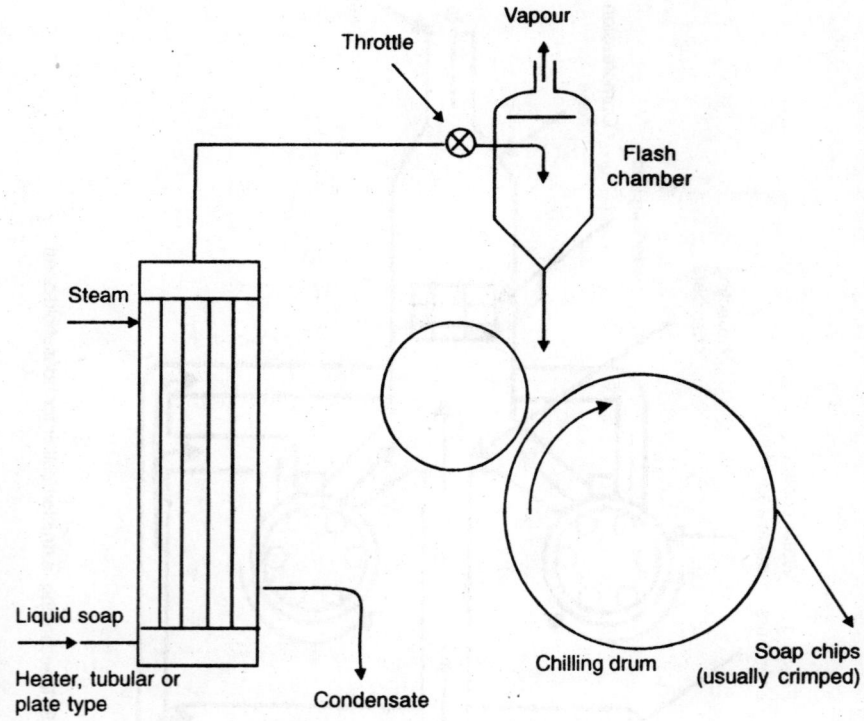

7.10. Atmospheric flash drier.

EQUIPMENT FOR TOILET SOAP PRODUCTION

Toilet soap plants can be fully continuous; and largely mechanised units for the production of large tonnages of a single type of tablet are possible; or a series of separate units connected manually as required to make a range of products can be used. As already indicated, there are a number of possible ways in which the main operations can be carried out; but a line will include equipment for most, but not necessarily all, of the following operations :

(i) Chilling and drying.

(ii) Rough mixing of dried base with perfume and other ingredients.

(iii) homogenisation and plastic working.

(iv) cooling of homogenised chips or noodles.

(v) extrusion of a continuous bar by a vacuum, or non-vacuum plodder.

(vi) cutting into billets.

(vii) cooling and conditioning of billets.

(viii) stamping to produce tablets.

(ix) packaging, which usually includes wrapping and/or cartonning, as well as final assembly in cases.

As explained soap is now usually dried in some form of flash drier. The commonest type is the vacuum chiller/drier shown diagrammatically in Figs. 7.2(a) and (b). Atmospheric flash driers can be built-up as indicated in Fig. 7.10 and Fig. 7.11 shows the Mazzoni two-stage drier, in which the soap is heated and flashed to atmospheric pressure, followed by a second stage in which it is re-heated and flashed to vacuum. Advantages to off-set the additional cost of a two stage drier are :

(i)   a large overall reduction in moisture content is possible, which is particularly useful with some specialised products, for example some which contain NSD ingredients.

(ii) cooling water requirement is about half that for a single stage drier, because of the vapours vented to atmosphere.

(iii) Mazzoni claim less powder entrainment and less need to clean the heaters and condenser.

The various ingredients are added to the dried soap and roughly mixed in a conventional arm, or ribbon, mixer. In small scale operations a known mass of soap is charged to a batch mixer and weighed, or measured, quantities of the ingredients are added by hand. An adequate guard, often in the form of a grid with a trip switch, is important to avoid risk of accidents. A similar system, with the addition of an autoweigher and conveyor which provides a charge of soap at the touch of a switch, is commonly used in larger installations. The modern trend is to mechanised continuous toilet soap lines for which two alternative systems of blending are available :

(i)   the automatic batch system in which increments of soap, liquids and solids are measured by weight or volume, mixed, and passed on to a hopper.

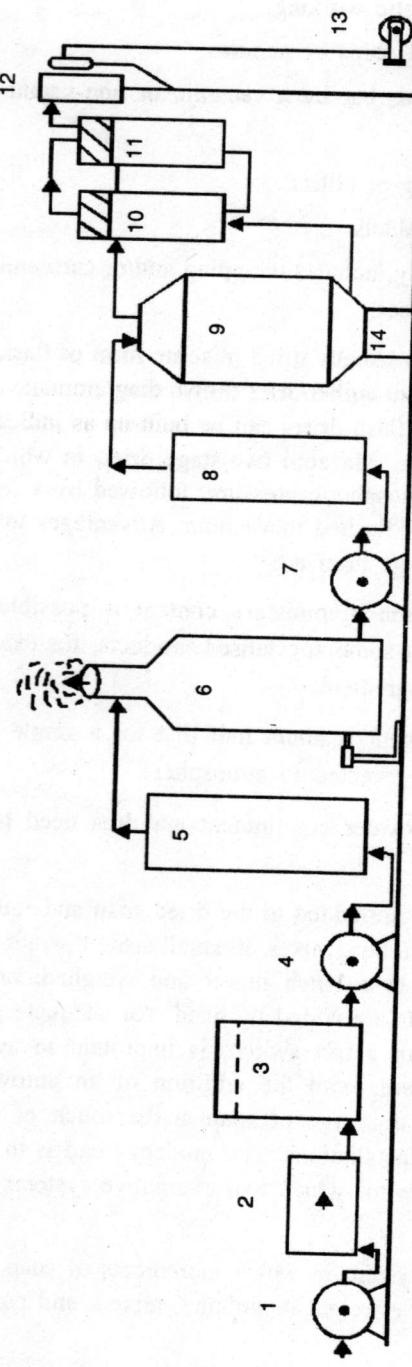

Fig. 7.11. Mazzoni multistage drier; 1. Filtrate a pump; 2. Filter; 3. Feed tank; 4. Stage feed pump; 5. First stage heat exchange; 6. Flash chamber; 7. Second stage feed pump; 8. Second stage heat exchanger; 9. Vacuum spray chamber; 10. First stage separator; 11. Second stage separator; 12. Vacuum pump; 13. Barometric hotwell; 14. Soap base extrusion.

(ii) Relatively long worm type mixers through which the soap is passed continuously at a predetermined rate and to which streams of the various ingredients are added by dosing pumps for the liquids and weigh band feeders for the solids.

The automatic batch system is more accurate and has advantages when critical ingredients, such as germicidal slurries which are difficult to dose continuously at the required low rate, are included. The continuous system is probably rather simpler. To produce good toilet soap, it is particularly important to ensure complete homogeneity; and this includes the incorporation of returned scrap which may be drier and cooler than the rest of the material.

The next operation is to cut the extruded bar into billets of the required length. In small scale operation bar lengths can be cut by hand, and then converted to billets by being pushed through a wire frame. Some early automatic cutters were mechanised versions of this principle, but modern cutters, of the type sometimes known as the Simon cutter, but now made by several plant manufacturers, use the principle outlined in Fig. 7.13. Solid knives on a chain descend gradually and, with a correct type of drive, produce vertical cuts. Some machines are driven electrically with a hydraulic coupling to compensate for variations in the rate of extrusion; others use air motors, with adjustment of the supply of compressed air to serve a similar purpose. As well as single track machines, types are available to cope with two bars extruded in parallel.

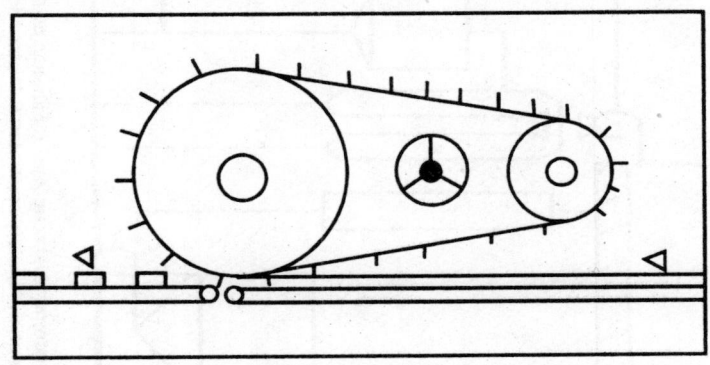

Fig. 7.12. Billet Cutter

Different billet lengths are required for different soaps and there are cutters in which :

(i) Adjustment within a certain range is by a hand wheel which can be used whilst the machine is running.

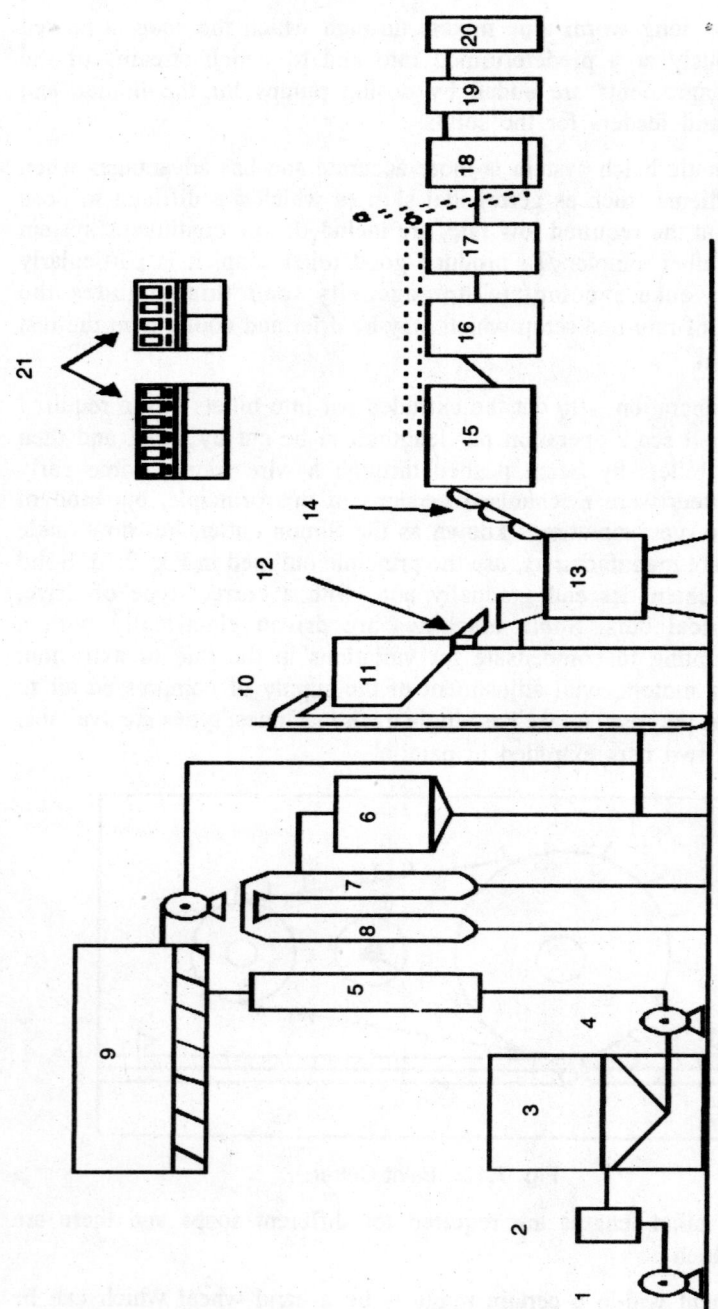

Fig. 7.13. Meccaniche Moderne toilet soap line. 1. Filtration pump; 2. Soap filter; 3. Feeding chamber; 4. Feed pump; 5. Heat exchanger; 6. Vacuum drying chamber; 7. Main separator; 8. Auxiliary separator; 9. Water cooling unit; 10. Bucket elevator; 11. Storage unit; 12. Vibrating dosing unit; 13. Mixer refiner; 14. Conveying belt; 15. Milling machine; 16. Vacuum plodder; 17. Cutting machine; 18. Stamping machine; 19. Refrigeration unit; 20. Wrapping machine; 21. Control panels.

Different billet lengths are required for different soaps and there are cutters in which :

(i) adjustment within a certain range is by a hand wheel which can be used whilst the machine is running.

(ii) The chain must be removed from the machine and adjusted link by link. The extreme form is a cutter in which the chain must be changed for each billet length.

Conditioning tunnels carry billets through a chamber in which they are subjected to currents of air, either atmospheric or chilled. Opinions differ as to the need for conditioning units for normal toilet soaps, but they are particularly useful for billets which are sticky because of non-soapy detergent ingredients or heavy superfatting, and for laundry soaps. The conditioning process :

(i) allows changes in internal structure to occur. This is particularly important for household soaps, in which increases in hardness tend to take place in the first few minutes after extrusion.

(ii) cools the billets with consequent increase in hardness, although the low thermal conductivity of soap means that considerable time is needed for much change in the interior of the piece.

(iii) dries, and cools, the surfaces, which reduces their stickiness.

Bars of laundry soap are often given an imprint on one surface which does not significantly change the dimensions of the bar. This can be done manually with a die of the same dimensions as the bar surface which is hit with a mallet. Similar results can be achieved by machines. A simple device is a wheel with the required design which embosses the lettering, and other imprint, on the soap surface as it rotates with the moving bar in a way similar to the action of a modern cutter. Tablets of household and of toilet soap are produced using dies and stampers which can be divided into two major types :

(i) those which produce tablets with a side-band by the use of a pair of dies operating inside a collar, or shell. With this method the whole of the soap in the billet actually fed to the stamper remains in the tablet, so that the weight of the tablet is the same as that of the billet, apart from any evaporation between the production of the billet and the stamping operation.

(ii) those which produce tablets without a side band, called capacity, pin, or squeeze dies. These dies use billets larger than the final tablets, come almost together, and separate an off-cut which must be removed from the tablet and the machine. Traditionally, pins locate the dies so that they meet correctly and ejectors operated by the stamper usually help to push out the tablets. The volume of the tablet is controlled by the capacity of the cavity enclosed by the two dies when they come together; and its weight is determined by this volume and the density of the soap. The gap between the dies in the closed position is about 0.002 inches (0.05 mm); and this can be varied a little for weight control purposes. Larger changes in weight normally require re-sinking of the dies, although it may be possible to make a small reduction by grinding the die edges. So far as is known capacity dies are not used for household soaps.

A wide range of stamping machines, or stampers, is available, ranging from hand stampers to high speed automatic machines. For many years machines known to those in the industry as the Jones 'K' type were almost universal for mechanised stamping of both household and toilet tablets. This type, and similar ones now made by various manufacturers, uses a shell with four pockets which rotates in the vertical plane in steps of a quarter turn. The billet is pushed into a pocket and is transferred to the next station where it is shaped by a pair of dies which move in horizontally. Modern machines normally give a double blow, that is the dies come together, move a little apart, and then come together again. This helps the escape of air and improves the quality of stamping. The double blow is quite different from the fault of double stamping in which a second billet enters a pocket from which the stamped tablet has not been removed. After it has been moved to the third station, the stamped tablet is pushed out onto the discharge conveyor. The last station may be used for cleaning. The dies are loaded with powerful springs; and should fit reasonably closely into the pockets in the shell. In this way the soap is squeezed out to fill the cavity formed by the two dies and part of the pocket wall. If the fit between dies and shell pocket is not good an unattractive fringe will appear on the tablets. With expensive soaps any fringe, and more particularly, that which is liable to be produced with capacity dies, may be trimmed by hand, but this is not possible with popular brands produced by high speed, mechanised lines. On the other hand, too tight a fit between dies and shell does not permit the escape of trapped air, unless other arrangements are made for its release. An escape route can be made by the provision of a small gap between die and shell, on the central part of the long side. Alternatively, the part of the shell through which the dies move, that is not the central part which actually shapes the tablet, can be provided with escape grooves. Some operators drill fine holes (vents) in parts of the dies where air could be trapped, such as lettering which is raised on the tablet and hence recessed in the die. Small threads of soap are squeezed out of the vents, but the amount should be small and the 'pimples' which are left insignificant. Venting in this way increases the sharpness of the lettering. It is important to keep dies clean.

A lubricant, commonly a weak brine, or a solution of glycerine and salt, may be used to reduce any tendency for the soap to stick to the dies, but it should only be applied sparingly to the billets. Excess salt bearing lubricant is liable to cause surface cracking and a type of efflorescence mentioned earlier. A 10% solution of sodium lactate in water is said to be effective and to avoid risk of surface cracking. Modern stampers now usually have facilities for the circulation of refrigerated fluid through the dies; and this cooling, with billet conditioning if required, often makes the use of lubricant unnecessary. Dilute acetic acid was once used as a stamping lubricant for household soaps.

At one time it was thought necessary to select a nearly brick-shaped tablet for production with 'K' type stampers, and this type of tablet is the easiest to handle, but it has been found possible to widen the range of shapes provided they have a side band.

Until fairly recently, capacity dies could only be used with hand, or certain slow speed machine, stampers. Now fully automatic, high speed, machines are available for this duty. They often work vertically and very important features are effective mechanisms for the discharge of the tablets and the removal of the 'flash' or off-cut. The scrap may amount to some 30% of the billet, and it has to be returned to the plodder, usually via a refiner, for re-processing. The development of these machines has enabled cushion shaped tablets, without any side-band, to be handled on high speed mechanised lines and so facilitated the selection of such shapes for mass produced popular brands. Hand, pedal operated, or pneumatic stampers are still used for small production lines and for very fancy, novelty, shapes.

Bronzes, such as gun metal or silicon bronze, are used for dies and shells, often quite satisfactorily, but with high copper alloys there is some risk of slight contamination of the soap surface with copper which is a powerful catalyst for degradation reactions. Stainless steel dies in monel shells are often considered the best materials for the duty. They are expensive initially, but have a long life. Stainless steel dies in stainless steel shells have been avoided by some companies because of the risk of 'galling', or chafing, when one piece of stainless steel is rubbed on another, but it is understood that the combination has been used successfully. Some authorities favour the use of die surfaces which are slightly matt, and like to 'condition' dies by a period of contact with soap. Others prefer to use highly polished stainless steel surfaces.

The best shape of billet to produce a given design of tablet is controversial and a matter for experiment. One school of thought favours the use of a billet as near as possible in shape to the final tablet, others prefer to use billets which have to be completely re-shaped in the stamping operation. The latter may select a cylindrical billet, or, preferably, an oval with flat areas on the long sides. Some factories feed the cut billets into the stamper directly, but others use cutting, or shaping, boxes through which the billets are pushed on their way into the stamper shell. The use of a shaping box permits the use of final billets which fit closely into the shell pocket. This avoids the development of the flow lines which are often seen when straight ended billets directly from a cutter are stamped to tablets with bowed ends, or rounded corners, for which such billets must be shorter than the final tablet. Shaping boxes are also said to improve substantially the uniformity of weight of tablets from 'K' type stampers, but they produce a trim which has to be removed and reprocessed.

As already mentioned, toilet soap plants range from small units with the minimum amount of equipment to large departments including mechanised lines devised to suit particular requirements. A small unit, for which dried soap is purchased, may comprise a mixer, mill, plodder, stamper, and wrapping machine; or a small number of some, or all, these items. Chips, or noodles, are stored between operations in bins. Trays are used for the transfer of billets and tablets from stage to stage; and to deal with fluctuations in the throughputs of the various units. All the major plant manufacturers supply complete mechanised toilet soap units, tailored to suit individual needs; and many variations are possible, including the duplication of some items in parallel when necessary to balance throughputs. An important requirement is a facility for the diversion of billets, or tablets, which are surplus to the immediate needs of the next machine; and to return this material, plus any offcuts, for reprocessing. Fig. 7.13.

TRANSPARENT AND TRANSLUCENT SOAPS

Transparent soaps have been known in many countries for a long time; indeed, in outline, the traditional Pears process uses the following stages :

(i) all-in saponification of a tallow, rosin and lauric oil blend in industrial alcohol with caustic soda.

(ii) solidification to bars of the required shape (oval for Pears) in water cooled bar coolers.

(iii) cutting into billets, often after the bars have been grooved to provide chamfers on the billet edges, followed by partial removal of alcohol and water in stoves, prior to stamping to tablets with flat faces.

(iv) prolonged maturing in stoves, with continued evaporation of alcohol and water. During this process the tablets develop the traditional concave shape of Pears soap.

Alcohol is removed from the air extracted from the various drying stages in activated carbon adsorbers; and the weak alcohol liquor stripped from the carbon is rectified in a normal type of still for re-use. Some ingredients included are : castor oil in the fat blend; and sugar solution and other ingredients are included in some final mixes. Cooling is in frames in which the soap is left for long periods, but more alcohol appears to be left in the soaps than in Pears. This is also the case with saddle soaps.

Translucent, and even transparent, soaps can also be made from a toilet soap base by extensive working under the right conditions on mills, or in machines somewhat similar to refiners. Transparent soaps made by a heavy working and extrusion processes may have a tendency to produce an opaque mush in use which can be unslightly, particularly with concave tablets such as Pears on which it has been called the 'fried-egg' effect.

FLOATING SOAPS

Several methods have been used to produce floating soaps :

(1) The level of soap in a crutcher is arranged so that air is whipped in; and in the case of a crutcher with a central worm, the direction of rotation of the worm may be reversed.

(2) In a process, partly dried soap is heated and flashed into a chamber from which it is pumped to a Votator after air has been injected into the pipe line. The Votator is a brine cooled, scraped wall, chilling unit, from which the aerated soap is discharged as a continuous bar.

(3) In a process, soap pellets at about 70°C are worked and aerated in a 'converter' with intermeshed screws from which the mass is discharged continuously to a moulding machine.

(4) A process in which hollow capsules containing air are inserted into normal soap bars plodded with a central hollow.

(5) A process in which aluminium powder is mixed with the soap to form hydrogen by reaction with alkali.

Filled, or Run, Soaps

Low-priced bar soaps can be made by addition to neat soap of water containing a suitable concentration of sodium silicate, or other electrolyte. The required concentration varies with the fat blend and with the nature of the fit used; and for frame or water-cooled soaps it is common practice to make up a few small-scale mixes from each boil, although simple graphs can be produced for each type of soap which at least give a good approximation.

The filled soaps can be classified as :

(1) *Liquored soaps containing more than about 50% TFM.* These products are essentially homogeneous neat soaps. For example, to produce a 61% TFM soap from base at 63% TFM of the composition. 100 parts of 63% base yield $100 \times (63/61) = 103.28$ parts of 61% TFM soap. Neat soap produced by a fit which gives 61% TFM contains 0.545% NaCl, so that 103.28 parts contain $0.545 \times (103.28/100) = 0.563$ parts NaCl. Hence salt to be added $= 0.563-0.434 = 0.129$, and liquor to be added $= 103.28 - 100.00 = 3.28$, so that concentration of salt solution to be added $= 0.129/3.28 \times 100 = 3.9\%$. Such soaps can be produced either by a suitable fit, or by addition of the appropriate salt solution to base at, say, 63% TFM. The direct fitting method can be used down to about 60% TFM, but the mixing method is possible down to around 50% TFM. The figures quoted as an example are for a soluble, washer-type, base and much weaker salt solutions are needed for a less soluble, high tallow or palm oil, base. If alkaline sodium silicate is used instead of salt, 1 part of Na_2O in the silicate is roughly equivalent to 1.3 parts of NaCl.

(2) *Liquored soaps containing less than about 50% TFM, usually 28–42%.* Liquored soaps run to less than about 50% TFM are not homogeneous, but on settling for a few hours at crutcher temperature separate into layers equivalent to neat soap and nigre. When the correct concentration of sodium silicate is used there is only a small neat soap layer which can be dispersed satisfactorily in the bulk of the saturated solution (nigre). Adequate hand crutching in the frames is required to minimise separation. Sodium silicate, which has colloidal properties, appears to stabilise this dispersion; and this type of run soap cannot be made with salt. If too high a concentration of silicate is used, unacceptable separation will occur in the frames and/or the soap will be rough in texture; whereas with too low a concentration the soap will

be soft and pasty. In other words, just sufficient electrolyte is needed to ensure that the nigre phase is saturated with soap. Tallow/rosin, or palm oil/rosin soap bases are particularly suitable for the manufacture of this class of soap. More concentrated silicate solutions are required than for the higher TFM class.

Glued-up soaps

These soaps are made from fat blends containing a high proportion of rosin (say 35% rosin, 65% tallow), often at about 56–58% TFM. Strong alkaline, or neutral, silicate, together with soda ash (anhydrous sodium carbonate) and sometimes sugar, form the fillings. The mixture becomes very viscous and special facilities, such as portable frames which can be placed to be filled directly under the crutcher, may be required. The soap is not homogenous and crutching down to a fairly low temperature is necessary. As mentioned earlier these soaps are liable to effloresce.

LAUNDRY SOAP

Though bar laundry soap is rapidly disappearing, it is still manufactured. The procedure is the same as described above for toilet soap, but the raw materials are, for example, 4 parts tallow and grease, and 2-3 parts rosin. The latter is added, after the grease has been saponified, in the form of sodium resinate, which is made in a separate kettle by the action of soda ash on the rosin, which consists chiefly of abietic acid rather than glycerides, the formation of sodium resinate is a neutralising reaction.

The kettle mixture for laundry soap is pumped from the kettle to a crutcher which is a smaller tank fitted with a special agitator and with a steam jacket; one of the following materials is added : silicate of soda, upto 30 per cent; soda ash 2 to 5 per cent, either alone or with borax, 1 per cent; or trisodium phosphate, upto 5 per cent. The mixture is crutched until homogenous, and then run off at the bottom of the crutcher into a frame. The contents of such a frame, of which there may be hundreds, is 1000–1200 Ibs of soap. The contents of frames hardens to a solid block in 3 days, the sides of the box may then be removed. The block is dried some what and then cut into slabs which are, in turn cut into bars and pieces. The water content of the finished soap is 8–10 per cent.

Improvements in Laundry Soap

Many small and medium scale laundry soap producers manufacture the soap from soap stock or acid oil available from vegetable oil refineries. Due to the different properties of vegetable oils and their fatty acid composition attention is drawn to several minor seed oils/fats, which can be used in combination with the by-product soap stock/acid oil to produce good quality laundry soap.

Some principles of soap formulation

For obtaining consistent quality of laundry soap, it is necessary to blend various oils and fats in appropriate proportions. Depending upon their availability to do this successfully, the soap manufacturer must have thorough knowledge of the properties and fatty acid compositions of all the available oils and fats. The properties i.e., saponification value (S.V.), Iodine value (I.V.) and Titre of various oils and fats are given in Table 7.3.

Table 7.3. Properties of oils and fats.

Oil/Fat	S.V.	I.V.	Titre °C
Coconut	251–264	7–10	20–24
Palm kernel	244–254	14–20	20–28
Palm	196–202	48–56	43–47
Groundnut	188–195	84–102	26–32
Cottonseed	189–198	99–113	30–37
Til (Sesame)	188–195	103–116	20–25
Mustard/Rapeseed	170–180	97–108	11–15
Soyabean	189–195	120–141	23–27
Ricebran	181–189	99–108	24–28
Mahua	188–200	53–70	36–45

The saponification value of oil/fat is related to its molecular weight. The iodine value indicates the unsaturation, i.e., double bonds present in the fatty acid molecules. Titre gives the melting characteristics of fatty acid mixture obtained from the respective oils/fats. The unsaturation in fatty acids decide the 'softness' or hardness' of the oils/fats. The higher the iodine value of an oil/fat the lower is the titre and melting point. In other words oils/fats with high iodine values are 'soft' and those with low iodine values are 'hard' oils/fats.

The average fatty acid compositions of major oils/fats are given in Table 7.4.

Table 7.4. Fatty acid composition.

Oil/Fat	Fatty Acids (wt %)							
	Caprylic	Capric	Lauric	Myristic	Palmitic	Stearic	Oleic	Linoleic
Coconut	5–9	4–9	45–52	13–19	7–10		5–8	
Palm kernel			46–52	14–17	7–9		13–19	
Palm					40–60		39–45	7–11
Groundnut					6–11		52–71	13–26
Cottonseed					20–23		23–35	42–54
Til (sesame)					7–12		35–50	35–50
Mustard/Rapeseed			(Erucic	40–50)			10–20	10–20
Soyabean					7–12	2–6	19–30	48–58
Ricebran					12–18		40–50	29–42
Mahua					20–25	16–21	40–50	10–15

Only major fatty acid contents are listed here. Out of the fatty acids mentioned in the Table 7.4, oleic and linoleic are unsaturated fatty acids and they give the iodine value and hence 'softness' to the oil/fat. All other fatty acids are saturated ones with zero iodine value and give 'hardness' to oils/fats. Thus, the oil charge for soap making should contain appropriate quantities of these fatty acids so as to attain desired hardness of soap. The 'hardness' of a given oil/fat may be increased to desired level by hydrogenation. If natural 'hard' oils are not available, the soap manufacturer can hydrogenate available 'soft' oils and use them in required proportion. Ideally, the oil/fat charge for soap manufacture contains a mixture of nut oils, hard oils and soft oils. Generally the percentages of these components do not exceed the following values: Nut oils 30%, Hard oils 50% and Soft oils 30%.

However, for small and medium scale soap producer's, it is not economical to use costly nut oils in laundry soap. Therefore, they have to use formulations based on hard and soft oils/fats only.

Use of minor oils/fats

Minor oils/fats available in the neighbouring region can be used in soap. They should have high melting points and low iodine values. Several such 'hard' oils are suggested which are available from nearby oil extraction units. Their characteristics and fatty acid compositions are also given. The major hard oils/fats which can be used for soap making are: Dhupa, khakan, kokum, pisa, mango kernel, salt dharambe. Out of these, khakan fat contains major quantities of lauric and myristic acids, while, pisa fat contains more than 90% of lauric acid and is even better than Coconut oil and palm kernel oil. The characteristics and fatty acid compositions of these minor oils are given in Table 7.5 and 7.6 respectively.

Table 7.5. Characteristics of minor oils.

Oil/fat	S.V.	I.V.	M.P., °C
Kokum	189	34–36	41–42
Dhupa	187–192	36–43	30–40
Khakan	233	6	41
Pisa	256	11	43–44
Mango kernel	190–195	32–61	34–43
Sal	182–195	31–44	34–37
Dharambe	198–200	48–54	33–37

Table 7.6. Fatty acid composition of minor oils.

Oil/fat	Lauric	Myristic	Palmitic	Stearic	Oleic
Kokum	–	–	–	52–56	39–42
Dhupa	–	–	10	41	42
Khakan	21	53	19	–	–
Pisa	96	–	–	–	4
Mango kernel	–	–	6–18	24–49	33–53
Sal	–	–	2–10	32–55	36–53
Dharambe	–	–	–	35–40	55–60

However, this list is not exhaustive. Many other fats are available from nature which can be used in soap making without much pretreatment. The quantity of these 'hard' fats to be added will depend on their own as well as the properties of the acid oil taken for saponification. Detailed descriptions of these minor oils and fats are presented here.

There are large number of such minor oils and fats available throughout India. They have varying but useful properties and fatty acid compositions. Depending on the availability of these oils and fats near the laundry soap manufacturers, appropriate selection should be done to include these oils and fats in the laundry soap formulation. In the long term, it will be definitely beneficial to develop these oils and fats since they are of free-origin and are locally available every year. If required, a backward integration into production of these minor oils and fats will be profitable for the small or medium scale laundry soap producer in the long run.

Mowrah (Mahua)

Mowrah is also called as Mahua or butter tree. Its botanical names are *Madhuca indica, M. latifolia*, and *Bassia latifolia*. It is a medium sized to large tree usually with short bole and large rounded crown. Its berries are ovoid, upto 5 cm long greenish turning reddish yellow to orange when ripe. There are 1 to 4 seeds that are ovoid and brown coloured. It is a tree of dry tropics and is found throughout greater parts of India. It is common throughout Central India, Maharashtra, Andhra Pradesh and Madhya Pradesh. The seeds yield yellow oil (20–60%) with an unpleasant taste. The oil has following characteristics :

Specific gravity	0.856–0.875
Refractive Index at 40°C	1.458–1.462
Saponification value	187–200
Iodine value	53–70
Acid value	5–50
RM value	0.7–3.6
Unsaponifiable matter	1–3%

Major fatty acids (%) : Oleic 44.6, stearic 24.5, palmitic 17.4, linoleic 9.3, arachidic 3 and myristic 1.2.

The oils is used for edible and cooking purposes in some rural areas. Refining and hydrogenation yield on edible fat. It is sometimes used as an adulterant for ghee. The oil is also used for soap making. Refined oil is used in the manufacture of lubricating greases and fatty alcohols. It is also used for candles, as batching oil in jute industry and as a raw material for the production of stearic acid. It has emollient and emetic properties. Both expelled and solvent extracted oils are available commercially.

Kokum

Kokum is a slender tree having drooping branches. It is also called as Mangosteen oil tree, or Brindonia tallow tree. Its botanical name is *Garcinia Indica*. Kokum fruits are globular or spherical enclosing seeds.

Kokum trees are found in the forests of Western Ghats, in Southern Maharashtra and slopes of Nilgiri hills. Kokum kernels contain about 44% fat. The fat has following characteristics :

Melting point	40–43°C
Titre	60°C
Refractive index	1.4565–1.4575
Saponification value	187–191.7
Iodine value	25–36
RM value	0.1–1.0
Unsaponifiable matter	2.3%

Major fatty acids (%) : Stearic 56.4, oleic 39.4, palmitic 2.5, linoleic 1.7%.

Kokum fat is mainly used for edible purpose, adulterant of ghee. It is considered nutritive, demulcent and emollient. Alongwith Mowrah (Mahua) butter, it is used in chocolate and confectioneries. Kokum fat is also suitable for ointment, suppositories and other pharmaceutical preparations. It is used in Ulcers, dry lips and hands etc. It is also suitable for candle and soap making, sizing of cotton yarn and lighting. Presently the production of kokum fat is about 200 to 300 tonnes and much of it finds a ready export market for use in cocoa butter substitutes.

Dhupa

Dhupa is also called as White dammer or Indian copal tree. Its botanical name is *Vateria indica*. It is a large, evergreen tree. Its capsules (fruits) are ovoid, pale brown, fleshy and are single seeded. The seeds are reddish white or creamy. Dhupa tree is indigenous to forests of Western Ghats from North Kanara to Kerala. It is also planted in Karnataka as avenue tree. The average fruit yield per tree is 500 kg. The seeds of Dhupa yield a greenish yellow to white, semi-solid fat to the extent of 22 to 27% by weight. The fat is fairly soft and has a slightly pleasant odour. It can be easily bleached by exposure to sunlight. Dhupa fat has following characteristics :

Melting point	30–40°C
Specific gravity at 25°C	0.9120
Refractive index at 40°C	1.456–1.459
Saponification value	187–192
Iodine value	36–43
Acid value	1.4–13.0
Unsaponifiable matter	0.6–2.5%

Major fatty acids (%): Oleic 42.2–47.8, stearic 38.9–45.1, palmitic 9.7–13.0, linoleic 0.2–2.3, arachidic 0.4–4.6.

Dhupa fat after refining is used for edible purposes. It is used in confectionery and as an adulterant of ghee. In other applications, it is used in candle and soap making. Medicinally, it is applied in rheumatism and allied problems.

Pisa

The botanical name of pisa is *Actinodaphne augustifolia* or *A. Hookeri*. It is a shrub or medium sized tree. The berries are globular or eliptic red when ripe and are pungently aromatic. Pisa is found in the evergreen forests of Eastern and Western Ghats, Karnataka, Orissa, Sikkim and North East India upto an altitude of 1500 m. It is particularly abundant in Ahmednagar, Nasik and Satara districts of Maharashtra. The seed contains 63% kernel and 31% shell. Both these parts – shell and kernel contain oil, kernel contains 32 to 75% fat called as 'Pisa fat'. It has the following characteristics :

Specific gravity at 40°C	0.9209
Refractive index at 40°C	1.4403
Saponification value	249.5–255.5
Iodine value	7.1–12.3
Acid value	4.0–25.9

Unsaponifiable matter 1.2–7.3%.

Major fatty acids (%) : Lauric (C_{12}) 88.6–96.0; capric 1.8–4.9, myristic 0–0.8 and oleic 0–5.7.

Pisa fat is used for lighting lamps and as an external application in sprains of joints. It is an excellent source of lauric acid, better than coconut or palm kernel oils. The soaps prepared from lauric acid-rich fat have excellent detergency, hardness and solubility. Pisa fat produces soap with excellent wetting and lather formation and detergent qualities. The fat can be used for commercial production of lauryl alcohol which at present is imported to meet the requirements of cosmetic, textile, pharmaceutical, perfumery and syndet industries. The oil can also be used in the manufacture of confectionery, agarbattis and candles. It can be used as lubricating agent for watches and other delicate machines, and as wetting agent for insecticidal preparations. The seed shell gives about 20 to 25% oil. It has following characteristics :

Refractive index at 40°C	1.4590
Saponification value	209.7
Iodine value	62.4
Acid value	33.2
Unsaponifiable matter	7.6%

Major fatty acids (%) : Oleic 54.5, palmitic 25.5–25.7, lauric 12.2–13.3, linoleic 6–7.4.

One survey estimated a potential of 10,000 tonnes of pisa seeds. However, the present collection of fat is estimated to be around 100 tonnes.

Mango kernel

The mango tree i.e., *Mangifera indica* occurs throughout India and the high density areas are Andhra Pradesh, Uttar Pradesh, Maharashtra, Orissa, Bihar and West Bengal. The area under cultivation is around one million hectares. The seed (stone) inside the fruit consists of a hard fibrous outer shell and an inner kernel – the source of mango kernel fat. The dry stone accounts for 10% of the fruit. The dry kernel constitutes 75% of the stone and oil content of the kernel is about 10% to 12%. After removing the outer shell the kernels are recovered and are subjected to solvent extraction for recovery of fat.

Mango kernel fat is pale yellow or cream in colour. The characteristics of Mango kernel fat are as follows :

Acid value	6.4–12
Iodine value	32–60.7
Saponification value	190.1–195.1
Unsaponifiable matter, %	0.75 – 2.4
Specific gravity at 30°C	0.901
Refractive index at 40°C	1.455–1.457
Titre °C	51
Melting point °C	34–43

Major fatty acids (%) : Palmitic 6–18, stearic 24–49, oleic 33–53, linoleic 1–13, arachidic 1–2.6.

The refining of mango kernel fat can be easily carried out by using conventional processes. The main application of mango kernel fat is as a raw material for making cocoa butter substitutes. It can also be used as ointment bases.

The biggest hurdle in the mango kernel fat production is the procurement of stones. Therefore, the current yearly production of mango kernel fat is too low compared to the potential.

Sal

The botanical name of sal tree is *Shorea robusta*. The total area of sal forests in India is estimated at 111,500 sq. km. The forests occur in two principal regions – Central Indian belt and at the foot of the Himalayas. Sal bears a profusion of fruit about the size of a pea with five segments of fruiting calyx enveloping it and forming three wings (2 inches each) and two narrower smaller segments around the fruit. In the mature fruit the enclosed nut remains green but gradually turns brown at storage. The typically dry fruit is composed of 23% wing, 30% pod and 47% kernel. The size of the kernel is small, on an average 4 to 4.5 g. The oil content of sal kernel is about 14%. Therefore, the preferred method of recovering this oil is solvent extraction.

The characteristics of sal fat are :

Refractive index at 40°C	1.4560–1.4619
Titre °C	48–53.4
Melting point °C	34.5–37.0
Acid value and saponification value	182–195
Iodine value	31–44.3
Unsaponifiable matter, %	0.9–2.2

Major fatty acids (%): Palmitic 1.7–10.4, stearic 32.6–55, oleic 36.5–52.5, linoleic 0.3–2, arachidic 1–8.6.

Degumming and neutralisation of sal fat can be carried out by conventional methods, sal fat is dark green and requires slightly higher proportions of bleaching earth (4–6%) for effective bleaching. (Pretreatment with phosphoric acid in conjunction with aqueous sodium chlorite, followed by neutralisation and bleaching (3% earth, 0.3% carbon), yield light coloured oils without a greenish tinge). The bleached fat can be deodorised by usual methods. Sal fat is widely used in preparation of cocoa butter substitutes, large quantities of sal fat are exported to UK, Japan and other countries. It is believed that the sal tree bears a good crop every alternate year.

Dharambe

Dharambe is a small to medium sized tree with horizontal or dropping branches. Its botanical name is *Garcinia Camboqia*. The family is same as that of kokum tree. The fruit of Dharambe is ovoid and seeds are surrounded by a succulent aril. It is common tree in the evergreen forests of Western Ghats, from Konkan to Travancore. It also grows in Shola forests of Nilgiris. Dharambe seeds yield a white granular fat to the extent of about 31 to 50%. The characteristics of the fat are :

Melting point	36.6°C
Titre	51.2°C
Specific gravity at 40°C	0.9003
Refractive index at 40°C	1.4583
Saponification value	203.5
Iodine value	52.5
Acid value	5.0
Hydroxyl value	11.1
Unsaponifiable matter	1.0%

Major fatty acids (%) : Oleic 56.2, stearic 38.4, palmitic 1.6, arachidic 1.5 and linoleic 1.1.

Dharambe oil is used for edible purposes. It may be exploited as a good source of oleic acid.

DRIERS AND METALLIC SOAPS

Metallic soaps are a group of water-insoluble compounds containing alkaline-earth or heavy metals combined with monobasic carboxylic acids of 7 to 22 carbon atoms. They can be represented by the general formula $(RCOO)_xM$ (neutral soap), where R is an aliphatic or alicyclic radical and M is a metal with valence x. They differ from ordinary soap by composition and their insolubility in water. Their solubility or solvation in a variety of organic solvents accounts for their many and varied uses (See Table 7.7). Driers are the most important group of metallic soaps promoting or accelerating the drying, curing, or hardening of oxidisable coating vehicles such as paints.

Composition and Properties

The acid or anion portion of a metal soap can be varied. Typical anions currently used are rosin and tall-oil fatty acids, saturated and unsaturated naturally occurring long-chain monocarboxylic fatty acids with 7 to 22 carbon atoms, and naphthenic, 2-ethylhexanoic, and the newer synthetic tertiary acids.

Acid soaps contain free acid (positive acid number), whereas neutral (normal) soaps contain no free acid (zero acid number). The basic soap is characterised by a higher metal-to-acid equivalent ratio than the normal metal soap. Particular properties are obtained by adjusting the basicity.

Properties are furthermore determined by the nature of the organic acid, the type of metal and its concentration, the presence of solvents and additives, and the method of manufacture. Generally, metals of low atomic weight form soaps with higher melting points (e.g., lithium). Higher melting points are also characteristic of soaps made of high molecular weight, straight-chain, saturated fatty acids. Branched-chain, unsaturated fatty acids form soaps with lower melting points. Light powders are characteristic of precipitated soaps, whereas soaps made by fusion are usually hard solids. Liquids and pastes are mixtures or solutions of soaps in hydrocarbon and/ or nonvolatile solvents. Liquid soaps are manufactured as fluid as possible compatible with maximum metal content. Newer techniques are available in which liquid soaps of very high metal content (18–24%) are obtained which still possess excellent fluidity. Both liquid and paste soaps are manufactured to strict viscosity and flow specifications.

The odour of a soap is determined by its organic constituents. Colour, another important property of metal soaps, is determined by the type and amount of metal used, the colour and quality of the organic raw materials, the method of manufacture, and the care in processing.

Table 7.7. Applications of metal soaps.

Application	Metal
Stabilisers for plastics	Ba, Cd, Sn, Sr
Fungicides	Cu, Hg, Zn
Catalysts	Co, Cu, Mo, Mn, Cr, Ni
Driers	Bi, Ca, Co, Fe, Pb, Mn, Zn, Zr
Fuel additives	Ba, Fe, Mg, Mn, Pb

Manufacture

Metallic soaps are manufactured by three methods: the precipitation or double-decomposition process; the fusion process; and the direct metal reaction (DMR) process. The choice of process and variation depends upon the metal, the desired form of the product, the desired purity, raw material availability and cost, metal content, etc. The desired final composition determines the stoichiometry of the starting materials, i.e., whether an acid, neutral, or basic soap is obtained.

In the case of metallic soap solutions, the metal is assumed to be the active part and the acid portion serves as carrier for the metal, conferring oil solubility, water insolubility, and compatibility with other components of the system in which it is to be used. Therefore, it is economically advantageous to incorporate as much metal per unit of acid as possible, providing the resulting soap is oil soluble and fulfills the requirements of the application.

Health and safety, factors

The hazards encountered in the manufacture, processing, handling, and use of metal soaps are associated for the most part with the inherent toxicity of the metals and solvents present. In general, the acid portion of the metal soap is low in toxicity. However, naphthenic acid is highly irritating to the skin on prolonged contact. High concentrations have a narcotic effect.

Metallic soaps are used in driers, waterproofing agents, fuel additives, rubber, greases, lubricants, and chemical thickeners, plastics, catalysts, fungicides, and cosmetics and pharmaceuticals.

DRYCLEANING AND LAUNDERING

Drycleaning

Drycleaning can be defined as the cleaning of fabrics in a substantially nonaqueous, liquid medium. This process has evolved during the last 60–70 years into a highly effective, low cost, safe method of removing soils from all types of textiles. The industry enjoyed steady growth until the introduction of wash-and-wear garments, and after a period of adjustment, again started to expand and diversify. Recent growth has been aided by the development of new equipment, cleaning techniques, and applications.

The important distinction between dry and wet solvents is that the latter (principally water, glycols, and other hydroxylic compounds) swell the hydrophilic textile fibres, but the drycleaning solvents do not. Dimensional changes that fibres undergo as they swell in water are transmitted throughout the textile structure and can cause serious fabric damage. These changes can be local distortions (wrinkles) or more extensive, causing shrinkage. Drycleaning solvents do not swell the textile fibres and thus do not cause wrinkles or shrinkage. This is one of the principal advantages of drycleaning processes over laundering. Another advantage is that drycleaning solvents remove oily soils at low temperatures, whereas a high temperature, colloidal suspension mechanism is needed in wet laundering processes.

The main difference between drycleaning and laundering processes is the solvent used. Since both methods remove soil by the same mechanisms, similar operations are used. Garments are tumbled in the cleaning liquid in horizontal washers to separate soil and fabric as thoroughly as possible without damage. Detergents are used in both cases to emulsify or solubilise the soils insoluble in the cleaning liquid. Special detergent ingredients peptise and suspend insoluble soils.

The drycleaning operation

The following sequence of operations is used in the normal drycleaning process :

(1) Soiled garments are marked and sorted.

(2) Prespotting is performed when required.

(3) Garments are rotated in a tumble-type washer containing the drycleaning solvent.

(4) Solvent is drained from the tumbler and most of the residual solvent removed by centrifugal extraction.

(5) The small amount of remaining solvent is removed in heated dryers.

(6) Dirty solvent from the wash cycles is continuously passed through diatomaceous earth and activated carbon or disposable cartridge filters to remove as much of the fugitive dyes and insoluble soils as possible.

(7) Soils not removed by filters, diatomaceous earth, or activated carbon can be removed from the solvent by distillation.

(8) Dry, solvent-free garments are inspected and spot cleaned a second time, if necessary.

(9) A garment may be "wet cleaned" at this point, but this is usually omitted when operating a charged-solvent system.

(11) After a final inspection, the garments are assembled according to customer's order.

(12) Garments are bagged and placed on racks ready to be picked up. Mechanical conveyors aid in fast recovery of a customer's garments.

Many drycleaning operations provide special services, such as garment repair, dyeing, and moth- and waterproofing treatments. Many operators add optical brighteners, antistatic agents, and sizing to improve the brightness and feel (hand) of a garment.

Materials

Drycleaning solvents remove oily soils easily by simple solution, and hence detergents are not required. However, detergent addition greatly enhances the removal of the many other soils that are generally present. Detergents probably inhibit redeposition of soil by being adsorbed on the soil particles to cause a lowering of the interfacial energy between them and the solvents, at the same time increasing the interfacial energy between particle and fibre surface.

A modern procedure called the charged system uses a detergent that forms a stable colloidal solution in the solvent. The term colloidal solution, as used here, means a solution containing dispersed aggregates called micelles. The key to the success of the charged system is the reduced effective vapour pressure of water dissolved in detergent micelles. This prevents the solubilised water from being completely removed from solution by the garment.

Sulphonates of mixed petroleum hydrocarbons, system detergents, have now been replaced by purer synthetic detergents, particularly the amine and sodium salts of alkylarenesulphonic acids. Drycleaning detergents are not formulated products similar to household laundry detergents but principally surfactants or a mixture of surfactants, with concentrations of active ingredients ranging from 40–90%.

Water control

Sensitive and reliable control systems are required to maintain the 70% rh in a drycleaning unit. These controls are designed to measure the moisture, inject water into the system, and close the injection valve at optimum rh. Some water also enters the unit with the garments.

Two types of water-control systems are widely used. The direct type uses an electric hygrometer to measure relative humidity in the atmosphere above the solution. An indirect method based on the electrical conductivity of the solution has proved the most successful.

Solvents

An acceptable drycleaning solvent must be an effective solvent for fats and oils, sufficiently volatile to permit easy drying, easily purified, and of low toxicity. Solvents should not weaken, dissolve, or shrink the ordinary textile fibres or cause bleeding of dyes. They must be noncorrosive to metals commonly used in drycleaning machinery and have a flash point above 37.7°C or be nonflammable.

The main advantages of chlorinated hydrocarbon solvents over the petroleum solvents are their nonflammability, permitting their use in many areas where zoning laws prohibit the use of a flammable solvent.

Perchloroethylene (tetrachloroethylene) is the most widely used drycleaning solvent.

Solvent purification, distillation

The principal method of purifying drycleaning solvents is distillation, which has displaced the caustic clarification process. There has been an unfortunate tendency in recent years for cleaners to omit distillation and depend entirely upon activated charcoal and fatty acid-adsorbing sweetener powders for solvent purification. Lower quality cleaning results from this cost-cutting practice.

Carbon recovery

Devices called sniffers should be an essential part of any drycleaning operation to protect workers from exposure to drycleaning solvents and increase operating life.

Industrial drycleaning

In the last decade a new service, known as industrial drycleaning, has become significant in the drycleaning industry. With improvement in fabric appearance, colour retention, hand, odour, and moisture penetration, customer resistance to fabrics of synthetic fibres has disappeared. Rental and maintenance service of career apparel, industrial garments, and linen supply items became economically feasible. In the most successful processes, high concentrations of water were added to the cleaning solvent or a 100% water cycle incorporated.

Laundering

Laundering is the process in which soils and stains are removed from textiles in an aqueous medium. Together with drycleaning, which accomplishes a similar result in nonaqueous solvents, commercial laundering constitutes a multi-billion dollar (10^9) industry. However, the largest portion of consumer textiles is laundered in the home.

Commercial laundering

Commercial laundry practice differs in many ways from home laundering. For example, in the commercial and industrial laundry plant, ion exchange is always used as water softener. Caustic and silicated alkalies are usually employed as detergent builders, and tallow soap is much more widely used than in home laundering. Classification of articles is generally by soil intensity.

Home laundering

The laundry is first sorted, primarily by colour. Further sorting by type of fabric and garment construction is advisable. Pretreatment is often effective, loosening stains that might be firmly set at the higher temperatures employed in the regular washing cycle. The regular washing cycle for white and coloured fabrics, from heavily soiled work clothes to lightly soiled lingerie, normally employs a heavy-duty detergent. Water softening agents are added in areas where water supplies are unusually hard, in addition to bleaches such as liquid sodium hypochlorite, dry chlorine bleach, or oxygen bleach. Fabric softeners impart softness and fluffiness to finished fabrics and reduce static cling, especially in synthetic fabrics. Some household detergents include a fabric softener, combining both laundering and softening in a single application.

Materials

Textiles are either from natural or synthetic fibres. More than 90% of the natural fibres encountered in laundering are cotton; the only other natural fibre laundered of any importance is wool. The chemical constitution of these fibres is quite different, and therefore they must be handled differently in laundering.

Water is the most important material used in laundering, assisted by detergents. Water provides the medium through which mechanical action is transmitted to the fabrics to be cleansed and serves as a wetting agent to penetrate the soil-fibre interface and displace the soil from fibres. It carries chemicals to the soiled area and removes suspended soil. The quality of the water used for washing is crucial to the quality of the finished product.

Chemicals

Chemicals used in laundering are classified as follows : alkalies, detergents, bleaches, sours, and finishing specialties.

Chapter 8

Recovery and Refining of Glycerine

INTRODUCTION

As explained in earlier chapters, the glycerol in fats is usually recovered in either soap lyes or sweet waters. Normally, these materials are purified and most of the water they contain is evaporated under conditions which do not vapourise any significant quantity of glycerol to give respectively soap lye crude glycerine or saponification, or hydrolyser, crude glycerine. Crude glycerine is normally purified by distillation is sometimes followed by bleaching and other treatments. Alternative processes which use ion exclusion and/or ion exchange instead of distillations are discussed in this chapter.

SOME ANALYTICAL TESTS AND SPECIFICATIONS

It is convenient at this point to outline the main analytical tests used for glycerines and to consider some implications of the results which they give.

Glycerol Content

Glycerol content can be estimated by several methods. The standard method is now oxidation with periodic acid (*sodium periodate* in acidic solution) Glycerol is oxidised to formic acid according to the equation :

$$CH_2\text{—}CH\text{—}CH_2 + 2H10_4 \rightarrow 2H10_3 + 2H \cdot CHO + H \cdot COOH + H_2O$$
$$\underset{OH}{|} \quad \underset{OH}{|} \quad \underset{OH}{|}$$

| glycerol | periodic acid | iodic acid | formaldehyde | formic acid | water |

The quantity of glycerol originally present in the sample is determined by titration with standard caustic soda of the formic acid formed under standard conditions. This test estimates as glycerol any compound which contains a groups —CH—CH—CH—with three adjacent carbons each with a

$$\underset{OH}{|}\quad\underset{OH}{|}\quad\underset{OH}{|}$$

hydroxyl group attached. Such compounds include sugars and sugar alcohols (sorbitol, etc.) which are sometimes found in glycerines which have been contaminated with glycerine substitutes used in some food industries, or added in an attempt to defraud. On the other hand, TMG (trimethylene glycol, 1,3 dihydroxypropane, or propane 1,3 diol) and polyglycerols are not estimated as glycerol by this test. As considered later, TMG appears in some crude glycerines made from low-grade fats which have been stored badly, or, more particularly, from lyes which have been stored under conditions which permit destruction of glycerol by micro-organisms. As well as indicating a destruction of glycerol, TMG in a crude glycerine can cause problems in the refining operations. Polyglycerols are formed by heating glycerol under certain conditions, particularly in the presence of alkaline, and other, catalysts. The reactions are of the type

$$\underset{\substack{|\\OH}}{CH_2}-\underset{\substack{|\\OH}}{CH}-\underset{\substack{|\\OH}}{CH_2}+\underset{\substack{|\\OH}}{CH_2}-\underset{\substack{|\\OH}}{CH}-\underset{\substack{|\\OH}}{CH_2}\rightarrow\underset{\substack{|\\OH}}{CH_2}-\underset{\substack{|\\OH}}{CH}-CH_2-O-CH_2-\underset{\substack{|\\OH}}{CH}-\underset{\substack{|\\OH}}{CH_2}+H_2O$$

glycerol   glycerol   di-glycerol

Tri-glycerols, and higher members of the polyglycerol family, are formed by similar condensation reactions in which a molecule of water is eliminated between two hydroxyl groups to form an ether linkage. The reaction can occur between two primary alcohol, —CH$_2$OH, groups as shown, or a secondary alcohol group >CH–OH can be involved, so that many isomeric polyglycerols are possible.

Formation of polyglycerols, or, more strictly, avoidance of their formation, is important in glycerine distillation, and estimation of polyglycerols in still residues (foots) is often useful.

For many years before the development of the periodate method, the *acetin method* was standard for the estimation of glycerol. This involves acetylation of the glycerol with acetic anhydride and determination by hydrolysis with standard caustic soda, of the quantity of the ester tri-acetin formed. The acetin method includes as glycerol other hydroxyl compounds likely to be present, such as sugar, TMG and polyglycerols. Because of the elimination of hydroxyl groups when polyglycerols are formed, the acetin method does not return polyglycerols as the full amount of glycerols from which they were produced. Thus :

(i)   Two molecules of glycerol contain 6 hydroxyl groups, but the diglycerol formed from them has only 4 hydroxyl groups.

(ii) Similarly tri-glycerol produced from three molecules of glycerol with 9 hydroxyl groups, contains only 5 hydroxyl groups.

The third important procedure for the estimation of glycerol is the *dichromate method*. This method, used particularly for soaps and lyes, involves removal of soapy, and other, impurities, followed by estimation of oxidisable matter calculated as glycerol. It includes all oxidisable compounds not removed in the preliminary purification of the sample. Because the condensation reactions in which polyglycerols are formed by elimination of water do not change the oxygen requirement for complete oxidation to carbon dioxide and water, the figure obtained for glycerol percent includes any polyglycerols expressed in terms of the mass of glycerol from which they were formed.

For many purposes, the glycerol content of refined glycerines is estimated by determination of *specific gravity* at 15°C, or *relative density* measured at 20°C, and reference to tables giving the properties of pure glycerol/water mixtures. The presence of other substances can interfere with the accuracy of results obtained.

Non-volatile Inorganic and Organic Residues, Particularly in Crude Glycerines

The content of inorganic matter is determined as *percentage ash*. In the case of refined grades a little sulphuric acid is added during the determination to give *sulphated ash*.

For many years, important organic impurities in crude glycerines were estimated by determination of *total residue* %, the residue left after vapourisation of the glycerol under carefully controlled conditions in an oven at 160°C. The *organic residue* (*or*) was then total residue % – ash %.

More recently, organic residue has largely been replaced by *MONG* % (matter organic non-glycerol) which is defined as 100 – (ash + glycerol + water) %. Glycerol is determined by periodate and water by the Karl Fischer method. Determination of total residue is tedious and rather subject to error, so that MONG is more convenient. The figures found for OR and MONG are very similar for many crude glycerines, but there are differences which are significant under some circumstances :

(i)   TMG is included in MONG, but not in OR, because it is more volatile than glycerol. Hence MONG should equal OR + TMG.

(ii) Any volatile alcohols, occasionally present in samples of glycerine including material from the alcoholysis of fats to form methyl, or ethyl, esters of the fatty acids, are included in MONG, but not OR.

(iii) Sugars, and other polyhydroxy compounds which increase the apparent glycerol content of a sample, artificially reduce MONG.

(iv) Polyglycerols and other relatively non-volatile organic compounds should be included in both MONG and OR, but substantial proportions of diglycerol, which is slowly volatile under total residue conditions, may make difficult the determination of the total residue end point (that is the stage at which the residue is deemed to have reached constant weight).

Nitrogen

Organic nitrogen which comes from proteinaceous materials is sometimes found in quantities up to 0.5% as nitrogen in poor quality saponification crude glycerines made from low grade animal fats, although 0.025 – 0.15% is more usual. Crude glycerines with a high organic nitrogen content tend to give distillates with objectionable odours and colours. Many nitrogenous compounds have been found at low concentrations in such distillates, with 2-piperidone, a toxic substance, the most important. Washing the fats with sulphuric acid before they are split does not help much, but earth bleaching is often useful.

Arsenic

A limit for arsenic content is found in many glycerine specifications. At one time it was troublesome because of use in the factories which produced the crude glycerines of sulphuric acid contaminated with arsenic; but there is now seldom any problem.

Sodium Chlorate

Sodium chlorate is an unusual, and unexpected, constituent of crude glycerines, but it has sometimes been found in significant quantities which have caused explosions during distillation in some types of still. It arises from the use of caustic soda made in electrolytic cells which are incorrectly designed or operated. If crude glycerine containing sodium chlorate has to be processed, the chlorate can be reduced under strongly acidic conditions, for example with spent pickling acid derived from steel works which contains ferrous chloride and hydrochloric acid (see later for the possible use of such material in the purification of soap lyes).

Silver Tests for Reducing Substances in Refined Glycerines

Most specifications for refined glycerines include a test for reducing substances, such as aldehydes, which is based on the darkening which occurs with ammoniacal silver nitrate. In some cases comparison is made with the darkening produced by a standard quantity of dextrose (glucose). Different silver tests vary in sensitivity, so that conditions laid down in a particular specification must be followed precisely. Substances such as soluble phosphates derived from some activated carbons, can interfere with certain tests.

Fatty Acids and Esters in Refined Glycerines

The fatty acid and ester content of a refined glycerine is usually important; and most specifications include a test to measure the quantity of caustic soda reacted when the sample is boiled with an excess under standard conditions. For chemically pure grade this was known as soda absorption expressed in ml of N/10 NaOH absorbed in a standard test. Now it is commonly called saponification equivalent for all grades of refined glycerine and expressed as percentage Na_2O absorbed, or as milli-equivalents per 100 g. Fatty acids and esters in a distillate usually affect its odour and taste adversely.

Chlorides in Refined Glycerines

Most specifications include a test for sodium chloride based on reaction with silver nitrate. This may be expressed as maximum parts per million permitted, or in terms that the turbidity produced in the test is nil, or less than in a control. The sensitivity of this test makes important the almost complete exclusion of entrainment of still liquor when soap lye crude containing a high percentage of salt is being distilled. For specifications of glycerine (ref. BIS specification No. 12590-1988) and Glyceryl monostearate (ref. BIS specification No. 4236-1985).

Other Tests

Apart from tests for colour, odour, sugars, lead, and other heavy metals included in general specifications, some purchasers of glycerine impose tests specially relevant to the purpose for which they use the material.

PURIFICATION OF SOAP LYES AND SWEET WATERS

As already discussed soap lyes are now commonly neutralised with fatty matter before they leave the soapmaking department. They still contain some alkali, as well as soapy matter and other organic impurities, and a substantial percentage of salt (around 10%). If the older system of cooling and skimming, but not neutralisation, is used, the lyes contain a higher percentage of free alkali than in neutralised lyes. Sweet waters from high pressure splitters operated without catalyst only contain organic impurities of a fatty nature, plus some other compounds from the fats used. If the fats are split with the aid of a zinc oxide, or other alkaline, catalyst, the zinc, or other metal, leaves in the fatty acids, as a soap, and the sweet waters are little affected. Twitchell sweet waters contain some sulphuric acid, or sulphates, as well as organic impurities. The purification, or treatment, of the lyes, or sweet waters, is intended to remove organic compounds other than glycerol as completely as possible; and the success with which this is achieved is reflected in the MONG, or organic residue, of the crude glycerine produced.

The steps in the usual treatment of soap lyes are :

(i) *Acidification* to a pH of about 4.6 – 4.8, although this is often omitted if the lyes have been well neutralised with fatty acids. Acidification is preferably with hydrochloric, rather than sulphuric, acid to avoid introduction of sulphates into the system. It liberates fatty acids and those which are insoluble in water, although of relatively low molecular mass, separate on standing. They are removed from the settler, often periodically, and are sometimes returned to a low-grade soap, perhaps after filtration. These skimmed fatty acids are poor soapmaking material; they are low in molecular mass and contain hydroxy acids produced by oxidation, and they tend to give sodium salts which have little value and/or again dissolve in the next lyes.

(ii) *Purification* by addition of a soluble iron or aluminium salt, a stage normally called *first treatment*. The pH is kept at, or adjusted to, about 4.5–5.0 (often 4.6 – 4.8); and the mix is well agitated by compressed air blown in through perforations in a pipe near the base of the tank. This air also ensures that if iron is used it is oxidised to the ferric state and blowing should, if necessary, be continued until the mix is brown rather than greenish black. Oxidation is particularly necessary if sodium hydrosulphite is used in the soapmaking process. The precipitate may be ferric hydroxide, or aluminium hydroxide, on which fatty acids, proteinaceous matter, and other impurities are adsorbed; or basic salts of the low molecular mass fatty acids may be present, the precise chemistry is obscure. The two main precipitants are solutions of ferric chloride, $Fe\,Cl_3 \cdot 6H_2O$, or of aluminium sulphate, sometimes called alum, $Al_2(SO_4)_3 \cdot 18H_2O$. Ferrous chloride with excess acid, which is the residue left after hydrochloric acid has been used to pickle steel, can be used, preferably after the iron has been oxidised to the ferric state by bubbling in chlorine gas carefully through a porous diffuser. Excess acid can be reduced in concentration by allowing the spent pickling acid to stand in contact with scrap iron wire when additional iron goes into solution. It is preferable to use chlorides, rather than sulphates, in all additions to soap lyes, because otherwise sodium sulphate will build up in the recovered salt which is re-used in soapmaking, and because sulphates are liable to increase problems with scale in the evaporators. Aluminium chloride is, however, not a convenient material to handle; and when aluminium treatment is used the aluminium is usually added as a sulphate. Sodium aluminate can also be used as a source of aluminium, sometimes in conjunction with ferric chloride. Opinions differ as to the relative merits of iron and aluminium for the purification of soap lyes, but iron has the advantage that it can usually be added as chloride. The quantity of precipitant needed varies with the composition of the lyes, but about 0.1–0.2% of the hydrated ferric salt (added as a concentrated solution) is common, although more, up to about 0.4% may be required. The completion of the first treatment process is often checked by the filtration of a small sample and addition of a few drops of diluted precipitant to the clear filtrate. If the treatment process is complete, no precipitate should be produced, but a good end-point is not always obtained; the pH at which the test is carried out is important. The adequacy of the treatment is proved ultimately by the MONG of the crude glycerine produced. Lyes from soaps with a high proportion of good tallow are more easily treated than those from high lauric oil blends, or those from low-grade fats. A very high glycerol concentration in the liquor can also make effective treatment more difficult.

(iii) The first treatment mud is removed by filtration, normally with filter presses or leaf filters. It is often advantageous to allow the tank contents to settle, to filter first the lower layer, and to put the nearly clear upper layers through a clean press, or presses.

(iv) The filtrate is made slightly alkaline with caustic soda, in what is known as the *second treatment*. A pH of about 9 is common when the precipitant is iron, but a slightly lower alkalinity may be preferred with aluminium. Mixing and oxidation are again with air. A relatively small quantity of precipitate separates in the second treatment. It contains the excess iron or aluminium, silica, and a little organic matter.

(v) Finally, the treatment process is completed by a second filtration.

Variations in the details of the treatment process are possible, including estimation of the quantity of precipitant required by small scale tests. Single treatment processes have been devised in which, after precipitation of impurities with iron, or aluminium, as in the double treatment process, the mix is made slightly alkaline with sodium carbonate, or sodium bicarbonate, before being filtered once only.

Calcium chloride, followed by a second stage using sodium carbonate, has been used to treat soap lyes, but the purification is less good than with iron, or aluminium.

EVAPORATION OF PURIFIED LIQUORS TO CRUDE GLYCERINE

General

Many variations are possible in the design of plants for evaporation to crude glycerine. Points to be considered in the design of an evaporation plant to produce crude glycerine include :

(i)   The capacity required in terms of water to be evaporated, crude glycerine produced, and, for soap lyes, salt to be separated.

(ii) Thermal efficiency and maximum consumption of steam.

(iii) Type of heating unit.

(iv) Arrangements for separation of salt, when relevant.

(v)   Avoidance of undue losses of glycerol.

(vi) Vacuum equipment and other ancillaries.

(vii) Method of control.

(viii)Materials of construction.

Evaporation of Water, Separation of Salt and Production of Crude Glycerine

A simple procedure for the estimation of the relationships between the relevant quantities is illustrated by an example of a soap lye containing 4% glycerol, 11% salt and 85% water. Minor ingredients and losses are neglected.

	Lye	Crude glycerine	Lye to produce 100 parts of crude glycerine contains	To be separated
Glycerol	4	83	83	–
Salt	11	8	$\dfrac{11 \times 83}{4} \times = 228$	$228-8 = 220$
Water	85	9	$\dfrac{85 \times 83}{4} \times = 1764$	$1764-9 = 1755$
Total	100	100	2075	1975

Hence, to produce 1 part of 100% glycerol in crude :

(i) use $\dfrac{2075}{83}$ = 25.0 parts lye

(ii) make $\dfrac{100}{83}$ = 1.2 parts crude

(iii) evaporate $\dfrac{1755}{83}$ = 21.1 parts water

(iv) separate $\dfrac{220}{83}$ = 2.7 parts salt

$\left.\phantom{\begin{array}{c}a\\a\\a\\a\end{array}}\right\}$ 25.0 parts

The results of similar calculations for lyes containing 4, 6, 10 and 15% glycerol respectively, plus 11% salt, evaporated to give a crude glycerine containing 83% glycerol, 8% salt and 9% water are shown in Table 8.1. The figures show how dramatically even a small increase in lye strength reduces the quantity of water to be evaporated, with consequent reduction in steam usage and evaporating capacity needed. It also reduces the quantity of salt to be recirculated. Similar calculations can be used to estimate quantities for the evaporation of treated lyes to semi-crude strength, or of sweet waters which do not separate salt.

Table 8.1. Quantities required and/or separated to produce 1 part of 100% glycerol in crude glycerine at 83% glycerol.

% glycerol in lye	Parts lye used	Water evaporated	Salt separated	Crude glycerine produced
4	25.0	21.1	2.7	1.2
6	16.65	13.7	1.75	1.2
10	10.0	7.8	1.0	1.2
15	6.65	4.8	0.65	1.2

Thermal Efficiency

In addition to steps to reduce the quantity of water to be evaporated, methods are available to minimise the usage of steam required to carry out the necessary evaporation. One kg of steam will evaporate approximately 1 kg of water fed at its boiling point, but less than 1 kg when the feed is at a lower temperature and sensible heat, which does not immediately produce vapour, has to be supplied before boiling commences. The boiling point of water, or any other liquid, falls as the pressure under which it is held is reduced. These simple facts provide the key to several ways in which the thermal efficiency of an evaporation system can be improved. Thermal efficiency is often and conveniently expressed in terms of kg of water evaporated per kg of steam supplied.

The main ways in which thermal efficiency can be improved are :

(i) *Multiple effect operation* in which the vapour from one evaporating vessel, or 'effect', becomes the heating steam in another effect operating under a lower pressure. In this way, the latent heat is used more than once.

(ii) *Vapour recompression* in which the vapour from a vessel is compressed, by a steam jet or by mechanical compressor, and returned as heating steam to the same evaporating unit. It will be appreciated that for the latent heat to be re-usable the pressure must be increased; and that a simple addition of superheat to the vapour will not do.

(iii) Selection of the optimum *method of feeding* the liquor to the evaporator; and/or the use of the heat exchangers. As discussed later, heat exchangers can be used internally between effects, or to recover heat from outgoing streams of vapour, condensate, or product.

(iv) Integration with other plants in the factory.

Before these procedures are considered, it is necessary to discuss boiling point elevation and temperature differences. Boiling point elevation is the difference between the boiling point of the solution in an effect and the boiling point of water under the pressure in the vapour space of the effect. It is due to two causes :

(i) *Boiling point elevation due to dissolved substances* in relation to fatty acid distillation, which increases as the concentration of dissolved substances increases. For semi-crude soap lye glycerine at 40–45% glycerol it is about 15°C; and for soap lye crude at 83% glycerol about 37°C.

(ii) *Boiling point elevation due to hydrostatic head.* When a liquid is heated at a distance below its surface, the pressure is that in the vapour space plus that equivalent to the head of liquid above the point at which the heat is supplied. 3 ft of glycerine of specific gravity 1.2 is equivalent to a hydrostatic pressure of $3 \times 12 \times (1.2/13.4) = 3.2$ inches of mercury, so that with a pressure of 2 inches Hg in the vapour space, the pressure at a depth of 3 feet is 5.2 inches Hg. The relative effect is clearly smaller when the pressure in the vapour space is higher. Boiling only occurs when the temperature reaches the boiling point under the prevailing pressure, but it is possible for the liquid to be heated to above its boiling point at the pressure in the vapour space and for it to flash when the liquid is circulated to a zone of lower pressure.

The apparent temperature difference, Δt_{ap}, is defined as the difference between the condensing temperature of the steam in the heater and the temperature at which water boils under the pressure in the vapour space. Consider, for example, a single effect evaporator in which the heating steam condenses at 5 psig, 109°C, and the vapour space has a pressure of 2 inches mercury at which water boils at 38°C. The apparent temperature difference is 109–38 = 71°C. If the liquid being boiled is crude glycerine with a boiling point elevation, bp_{elev} including any hydrostatic head effect, of 37°C, it will boil at 38 + 37 = 75°C; and the actual temperature difference Δt_{act}, is 109 – 75 = 34°C (or, alternatively, 71 – 37 = 34°C, the apparent temperature difference minus the boiling point elevation). This situation is depicted as case 1 in Fig. 8.1. If a double effect evaporator is used in which vapour from the first effect, which contains partially concentrated liquor with a boiling point elevation of 10°C boiling at 91°C, is used to boil crude glycerine under a pressure of 2 inches Hg in the second effect, the conditions are as shown as case 2. The various temperatures and pressures adjust themselves automatically to fulfil both heat transfer and heat balance requirements, so that the figures shown in Fig. 8.1 are approximate only and vary with conditions.

A positive Δt_{act} is essential in each effect in order to achieve heat flow and evaporation. With a natural circulation heater, particularly when the liquid is viscous, a Δt_{act} is required larger than the minimum which can be used with a forced (pump) circulation type of heater.

Case 3 in Fig. 8.1 shows approximate conditions for evaporation from lye to semi-crude glycerine in double effect using low pressure steam, a system which was once widely used.

Case 4 shows possible conditions for evaporation from lye to crude glycerine in triple effect using steam which condenses at 30 psig, 133°C.

Some general points related to these topics are :

(i) Boiling point elevation due to hydrostatic head varies with the design of the heaters and is assumed to be included in the figures for bp_{elev} used in Fig. 8.1. It does, however, reduce the effective temperature difference, even if the liquor is heated without boiling and only flashes when it arrives near to the liquor surface.

(ii) Boiling point elevation due to dissolved substances depends only on the composition of the liquor. It is high for crude glycerine; and this is a disadvantage of continuous operation in which the liquor in the effect from which the product leaves must be maintained at the product concentration. If an evaporator is operated semi-batch, that is the unit is fed with fresh liquor but the bulk of the product is not discharged unit its concentration reaches the desired level, the average boiling point elevation is lower and the average Δt_{act} higher, other things equal, than with continuous operation. At one time semi-batch operation to concentrate lye to semi-crude and, separately, semi-crude to crude glycerine was normal, but the usual advantages of continuous operation are now commonly considered to outweigh this disadvantage.

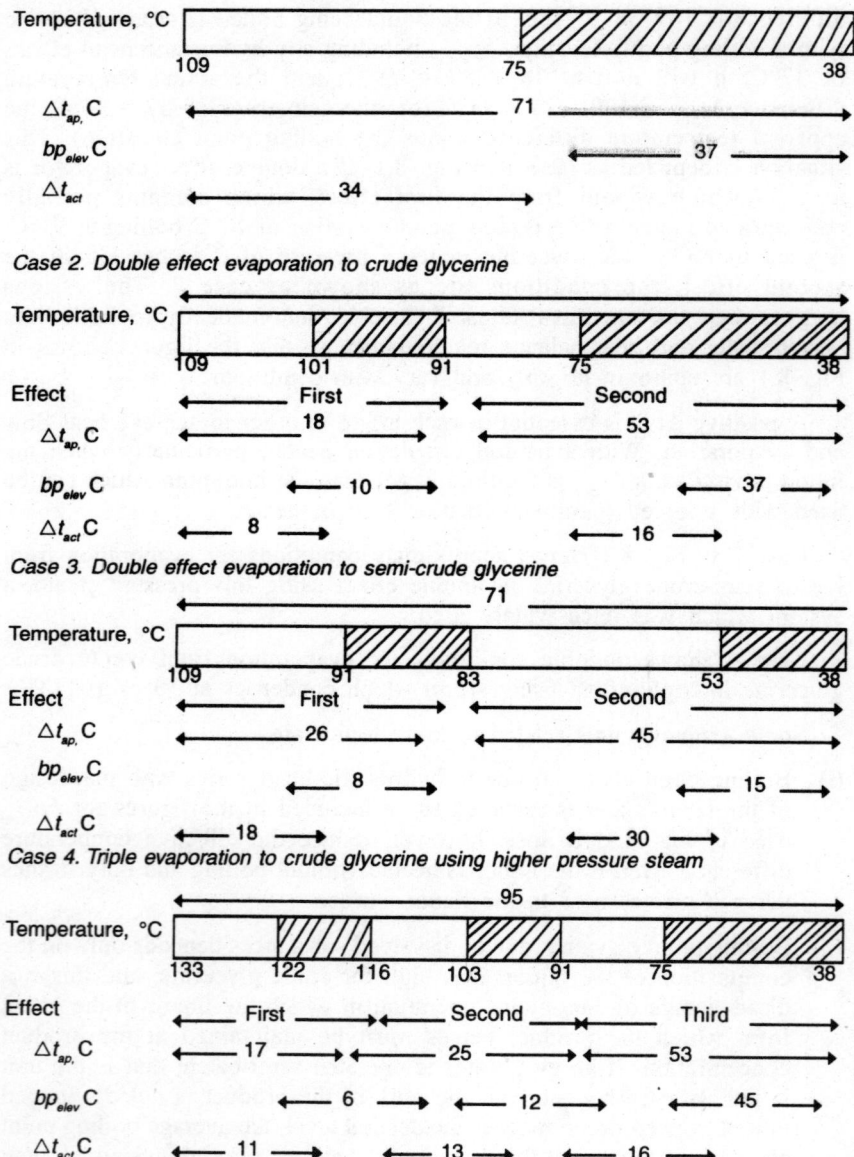

Fig. 8.1. Temperature conditions for single, double and triple effect evaporation to soap lye crude or semi-crude, glycerine.

(iii) Forced circulation heaters, now often used in all effects, are particularly valuable for effects in which crude glycerine concentration is reached, as they permit operation with a low Δt_{act}, even when the liquid is quite viscous.

(iv) The need for an adequate Δt_{act} in each effect limits the number of effects which are physically possible with given terminal conditions. In general, an increase in the number of effects leads to a reduction in steam consumption, but to an increase in capital cost. There is an economic optimum which may well use less than the maximum number of effects physically possible.

Double effect evaporators to concentrate lye to semi-crude were traditional for many years; and further concentration to crude glycerine was carried out in a separate single effect finisher, or in the second effect of the double effect unit operated either as a single effect using a connection directly to the steam supply, or as a double effect using water in the first effect. These plants used steam at only just above atmospheric pressure; and the supply commonly included exhaust from steam pumps and small steam engines. Increased emphasis on fuel economy led to the construction of at least one quadruple effect natural circulation evaporator to produce semi-crude glycerine from soap lyes. This was fed with steam at 90 psig and the pressure in the first effect heater rose to 30–50 psig depending upon conditions. Initially, this unit was operated semi-batch like its double effect predecessors, but it was found preferable to work it continuously. Triple effect evaporators have been, and are, also used for soap lyes and sweet waters. They now commonly use forced circulation and take the concentration to crude glycerine strength. The concentration of glycerol in the lye, or sweet water, to be evaporated is an important factor in the design of the plant.

Vapour recompression evaporation with an electrically driven mechanical compressor is only likely to be economical when the compression ratio (that is, the ratio of the pressure of the heating steam to that of the vapour it produces by evaporation) is low, which means a low Δt_{app}. A low Δt_{act} can be achieved by the use of a forced circulation heater with a generous area of heating surface, but δp_{elev} is inevitably high if the unit is to produce crude glycerine; and mechanical recompression is not suitable for this duty. Jet recompression, combined with multiple effect operation, is valuable; and is now widely used when steam at boiler pressure, rather than steam including low pressure exhaust, is to be the heating medium. Fig. 8.2 shows the main features of a triple effect system with jet recompression of part of the vapour leaving the first effect. Again, recompression is most effective when the pressure ratio is small, as it can be when the liquor concentration in the first effect is fairly low. When conditions permit satisfactory recompression, a useful improvement in thermal efficiency is achieved without increase in other costs and for only a very small additional capital cost for the jet.

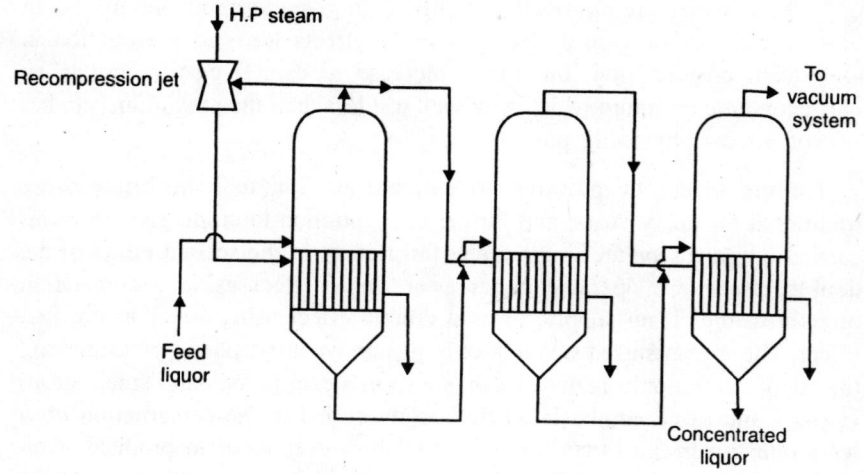

Fig. 8.2. Principle of triple effect evaporation with vapour recompression.

Parallel feed, that is separate feed to each effect, is suitable for the evaporation of brine to produce salt, but not for glycerine liquors in which the concentration of glycerol increases very substantially as the evaporation proceeds. Feed can be forward, that is 1 – 2 – 3 etc., the effects being numbered in the direction of heat flow; or backward which is the reverse direction, for example 3 – 2 – 1. The direction of feed often has an important influence upon steam consumption, depending upon the temperature of the feed liquor. If the feed is cold, it has to be heated to its boiling point under the pressure in the effect to which it goes. With forward feed, the temperature of the whole of the feed must be raised to the relatively high boiling point in the first effect by new steam and this sensible heat does not produce any vapour in that effect, although the heat in the partly concentrated liquor does produce some vapour by flashing when it is transferred to the second and subsequent effects which operate under lower pressures than the first effect. With backward feed, a cold liquor is heated in stages as the unevaporated portion is transferred from effect to effect. Only the final portion which reaches the first effect is heated with new steam, and the sensible heating in the second and subsequent effects is done by vapours from an earlier effect; that is by latent heat which has already been used one, or more, times in the evaporator. If the feed liquor is very hot, for example above the boiling point in the first effect, no sensible heat is required even with forward feed; and flash vapour is produced in each effect and forms part of the steam used in the next effect. With backward feed, the hot liquor flashes to the pressure in the last effect and the latent heat is not used again, at least in the evaporator. The residual liquor then has to be re-heated to the temperature in its next effect, as for cold feed. These considerations show that it is beneficial thermally to use forward feed with hot liquor and backward feed with cold liquor; at some intermediate temperature the factors balance and there is no difference between backward and forward feed. The thermal advantage of backward feeding with relatively cold feed liquor can nearly be achieved by the use of forward feed with interstage preheaters as indicated in Fig. 8.3. The feed is preheated stage by stage with vapour at as low a temperature as possible to make the maximum re-use of latent heat. A preheater cannot operate without a positive temperature difference in the required direction; and with modern vacuum equipment it is often not possible to use a preheater in the vapours from the last effect (number 1 in the diagram), but any heat which can be recovered from the final vapours is clear gain. Heat transferred to the feed in interstage preheaters is not available for use in the next effect, but the overall result is beneficial with cool feed for the reasons discussed in relation to forward and backward feed. Heat can also often be recovered from other streams leaving the effects, particularly the condensates. These condensates can be handled in several ways, bearing in mind the fact that condensate leaves the first effect heater at a higher temperature than that from the second effect, and so on :

(i) Condensates can be allowed to flash on reduction in pressure on transfer from one effect heater to the next, or in separate vessels. The flashed vapour is used as steam at a suitable point in the system, and the residual hot, or warm, water remains available. This condensate can be re-used as process water in the soapmaking or fat-splitting process, as well as a source of heat.

(ii) Unflashed, or flashed, condensate can be used in a heat-exchanger.

Other factors to be considered in the selection of a system of feeding are:

(i) With forward feed the most concentrated liquor is evaporated at the lowest temperature, so that its viscosity is higher than that of the fully concentrated liquor using backward feed. This adversely affects heat transfer, particularly with natural circulation heaters.

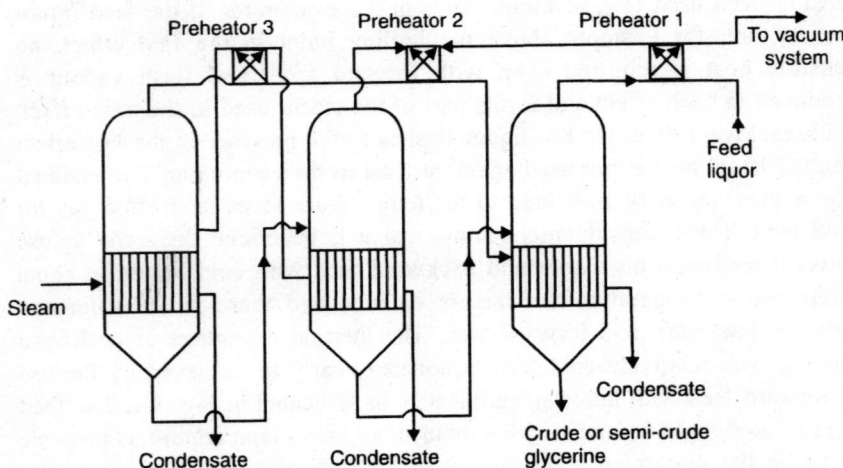

Fig. 8.3. Evaporation with forward feed and interstage preheaters.

(ii) More sensible heat leaves in the crude, or semi-crude, with backward than with forward feed; but some of this may be recoverable in heat exchangers.

(iii) Liquor can normally be transferred from effect to effect by pressure difference with forward feed, although a pump may be needed to introduce the feed liquor into the first effect. With backward feed it is normally possible to suck the feed liquor into the last effect, but pumps are needed for transfer from effect to effect, unless unusual levels are adopted which provide sufficient gravity head to more than offset the adverse pressure differences between the vapour spaces.

(iv) The salt crystals which separate during the evaporation of soap lyes are rather fine; and they tend to become finer with rise in the temperature at which they are separated and with increase in the glycerol content of the liquor. Fine crystals are more difficult to wash effectively than coarser ones. The same factors also tend to induce 'salting' of the heater tubes, a phenomenon discussed later. Salting can be a serious problem with natural circulation soap lye evaporators under some conditions and forward feed may then be advisable.

When the feed liquor is relatively cool it is usual to use forward feed with interstage preheaters. For special reasons, it is also possible to use other arrangements. For example, the quadruple effect evaporator mentioned earlier was for many years fed 1 – 2 – 4 – 3.

As mentioned earlier, and for the reasons just discussed in relation to methods of feeding, it is commonly a serious error to assume 1 kg of live steam will evaporate n kg of water in n-effect evaporator, unless the feed is exceptionally hot or a recompression jet is used. A good estimate of the performance to be expected from an evaporator, including its evaporative capacity, can be made with the aid of detailed heat balances over each stage of the process. Because of the dynamic equilibria maintained within a unit, the temperatures and pressures adjust themselves automatically to meet both heat balance and heat transfer criteria in each section of the plant. For a plant which does not exist, it may be necessary to make provisional assumptions as to intermediate temperatures, or likely heat transfer coefficients; and to make adjustments, if necessary, to reconcile the two sets of criteria. Such calculations are not difficult, but manual calculation can be tedious, particularly when several effects are involved. Plant manufacturers will provide estimates in response to serious enquiries, doubtless by the use of computer programmes, but it is sometimes useful for an operator to estimate what is likely to result from changes such as the inclusion of an additional heat exchanger, or a different feed arrangement. In making such estimates, it must be remembered that any change is likely to induce other changes to maintain the dynamic equilibria.

As a first approximation it is sometimes assumed that an n-effect evaporator costs n times as much as a single effect plant for the same duty, because the overall temperature difference is split between the effects with a consequent need for additional heating surface. This is also subject to considerable error. Estimates obtained some years ago from a well-known plant manufacturer for single and double effect units for the same, rather small, duty showed that the cost of the double effect was substantially less than twice that for the single effect. At least in part this was because of the larger final entrainment separator, condenser, and vacuum equipment needed for the single effect unit in which the whole of the water evaporated passes to the condenser.

The discussion of thermal efficiency has so far considered the evaporator alone, but it is sometimes worthwhile to involve other plant. Steam from another unit which is not at high enough pressure to be used as first effect steam may be able to be introduced between effects. Similarly, vapour can sometimes be bled between effects for use elsewhere. Different plant units commonly do not synchronise at all times, so that it is usually necessary to provide an alternative steam supply, probably through a reducing valve. Surplus heat from a high pressure fat splitting plant can often be used in the evaporation of the sweet water.

Types of Heating Unit

Long tube climbing film heaters, in which much vapourisation takes place in the tubes, are used for sweet waters, but are not suitable for soap lyes, because the tubes would block with salt. Traditionally, short tube natural circulation evaporators were usual for soap lyes. These have a bundle of vertical tubes (say 1–2 m long) arranged in a steam chest to give what is called a calandria. Commonly calandrias built into the main evaporator bodies, and provided with downcomers of various designs, such as a sector, a large central hole, or a series of tubes larger than the boiling tubes, are used. The liquor circulates up the small tubes and returns through the downcomers. It is also possible to have natural circulation evaporators in which the liquor rises through the vertical tubes in an external heater and returns via the liquor reservoir in the main body. In order to obtain the maximum rate of heat transfer, and hence of evaporation, it is necessary to work with a static level of liquid below the top of the tubes, so that some vapourisation takes place in the tubes giving an increased rate of circulation due to the vapour lift effect. With soap lyes care must be taken to avoid too low a static level, as otherwise salt will deposit in the tubes in effects in which the liquor is saturated with salt. Salting of the tubes can sometimes occur when an evaporator is being operated in a way which would normally be satisfactory; this is presumably due to some small changes in the composition of the purified lye. It is also sometimes possible to operate natural circulation evaporators with a static level well above the upper tube plate of the calandria, particularly when finishing batchwise to crude strength. In this case, the liquid is heated in the tubes to above its boiling point at the pressure in the vapour space and flashes as it rises to a region of lower hydrostatic pressure. It is usually necessary to avoid operation at some intermediate levels at which 'priming', or boiling over, is liable to occur.

Evaporator effects in which salt separates usually need to be washed out at regular intervals with water, or with other liquor which is not saturated with salt. When a plant is operated semi-batch, rather than continuously, the charge of concentrated crude, or semi-crude, containing some solid salt, is discharged from the last effect; and it is common practice to start the next batch by refilling with unsaturated lye, or even water, to dissolve deposited salt. If the tubes become blocked with salt, it is necessary to soak with water and to boil as long as is needed to clear them. Mechanical methods are sometimes used, but care is necessary to avoid damage and this procedure is best avoided if possible. Various insoluble scales can build up slowly in the tubes of a glycerine evaporator and these normally have to be removed mechanically. Scales can contain silica, but they are particularly common when lye, sweet water, or other liquor containing sulphate and calcium is introduced, for example, if the plant is washed out with hard water. Calcium carbonate scales are also encountered; these can sometimes be removed by careful use of acid. The modern tendency is to use external calandrias with forced circulation by pump. This system improves heat transfer, and so permits the use of less heating surface and/or a lower Δt_{act}. External heaters with forced circulation minimise the risk of salting and scaling; and an external heater can be arranged to permit mechanical descaling more readily than an internal calandria. As mentioned earlier, forced circulation is particularly useful for the final effect of an evaporator in which crude glycerine is produced.

Whatever type of heater is used, it is necessary for satisfactory heat transfer to remove effectively both condensate and non-condensible gases, mainly air. When the pressure in the steam chest is above atmospheric, ordinary steam traps can be used for the condensate; and air can be discharged, with a little steam, through a slightly opened vent valve, or cock. For steam chests at below atmospheric pressure, the air must be vented through a small connection to a vapour space at a sufficiently low pressure; and a pumping trap, or pump, used for the condensate. A convenient system for the condensate is shown in Fig. 8.4. The centrifugal pump runs continuously, but discharge of condensate is intermittent. The condensate is virtually at its boiling point at the pressure in the steam chest and can only be pumped when there is a sufficient head of liquid in the receiver to prevent vapourisation (cavitation) in the pump. The level in the receiver rises until the pump starts to discharge and then falls until flow virtually ceases; the cycle being repeated indefinitely.

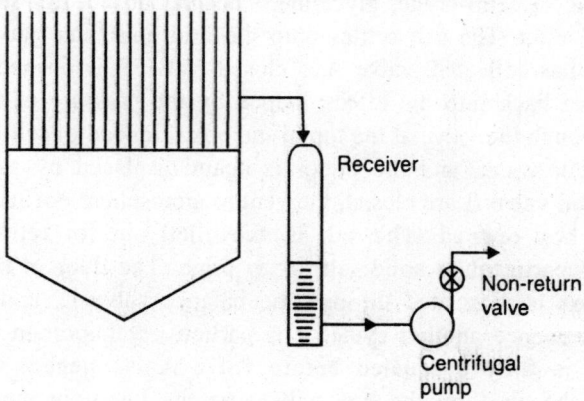

Fig. 8.4. Arrangement for discharge of condensate from a vessel at sub-atmospheric pressure.

Removal and Recovery of Salt in the Evaporation of Soap Lyes

All effects in which the lyes have become saturated, and solid salt separates, need a facility for the removal of the solid as soon as possible after it is formed and the crystals have grown to a reasonable size, otherwise the liquor may become unduly thick. Effects which do not normally produce salt may be fitted with simple vessels to collect sludge and any abnormal salt. Three systems are, or have been, used :

(i) Salt filters, or dry salt boxes.

(ii) Wet salt boxes.

(iii) Devices through which the effect contents are circulated, a thickened slurry is discharged, and substantially clear liquor is returned to the effect.

Dry salt boxes are messy, labour intensive, and do not permit adequate removal of glycerol from the salt to meet the requirements of efficient modern soapmaking. In the larger, and the more efficient, factories they were replaced many years ago, but they are still in use in some places. The other two systems remove the solid salt from the evaporator in the form of a slurry in a more, or less, concentrated glycerine liquor; and the separation process is completed in some form of centrifuge.

The operation of dry and wet salt boxes can be explained by reference to Fig. 8.5. In both types salt settles into the quiet zone of the effect body below the calandria and down into the salt box. A sufficiently deep cone and settling zone is important; and, if the plant is to be operated semi-batch, sufficient liquor volume is necessary in the last effect to hold an adequate batch of crude, or semi-crude, glycerine. Consider now a dry salt box as shown in Fig. 8.5a. The salt settles onto the grid and filter gauze. When sufficient salt has collected, valve A is closed, valve B is opened, and the liquor is drawn back into the effect, helped by the pressure of the steam introduced through the valve at the top of the box. The bed of salt is washed with lye, and/or water, and the liquor is again displaced by steam. The steam valve and valve B are closed, the vent to atmosphere opened, and the door on the box opened. The salt is shovelled out for return to the soapmaking department as solid salt, or as brine. The door of the box is closed, the box is evacuated through the balance valve C, and valve A opened to commence another cycle. It is particularly important to ensure that the box is fully evacuated before valve A is opened; otherwise expansion of the air from the box will cause the liquor in the effect to prime. It is also important to ensure that air cannot leak in through the valves; lubricated cocks have been found useful for these duties.

Wet salt boxes can be cylindrical drums, or vessels of any other convenient shape. They are operated in a manner similar to, but somewhat more simply than, the dry boxes. When sufficient salt has accumulated, as judged by inspection through sight glasses, or in accordance with a time schedule, valve A is closed, valve B and the vent are opened and the slurry is run out and/or blown out with steam after the vent has been reclosed. The box is evacuated via valve C before it is reconnected to the effect.

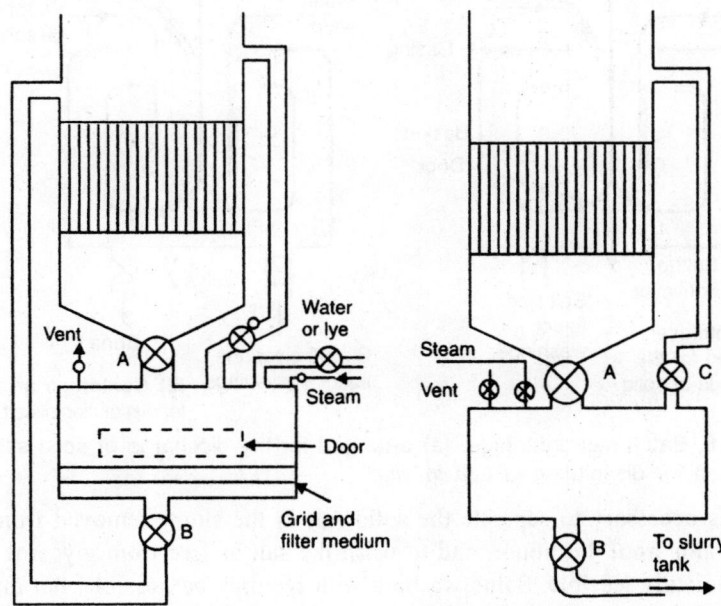

Fig. 8.5. The operation of salt boxes. (a) Dry salt box, or salt filter (b) Wet salt box.

Pairs of boxes, dry or wet, and used alternately, are common on effects in which large quantities of salt are produced. If only one box is fitted, salt accumulates in the cone of the effect whilst the box is disconnected. Discharge by continuous pumping from the base of the effect is arranged as shown in Fig. 8.9 later for a evaporator which uses a baffled settling vessel. Thickened slurry is discharged from the base of the vessel and the rest of the liquor is returned to the effect and/or transferred to the next effect. Sufficient pressure is maintained in the circulation loop by the pump and the valves in the return and transfer lines to enable the discharge of thickened slurry to be controlled by a valve without another pump. Hydroclones (liquid cyclones) instead of the baffled vessels have been found very satisfactory separating devices for systems of this kind. Care must be taken not to interfere with the circulation through the heater tubes if a continuous pumping system is introduced into a natural circulation evaporator. An attempt to do this many years ago produced a well thickened slurry, but led to unacceptable salting of the calandria tubes. Return of the circulated liquor to the base of the effect, with sufficient volume below the calandria, may prevent this difficulty. The system works well with forced circulation effects and is probably best modern practice.

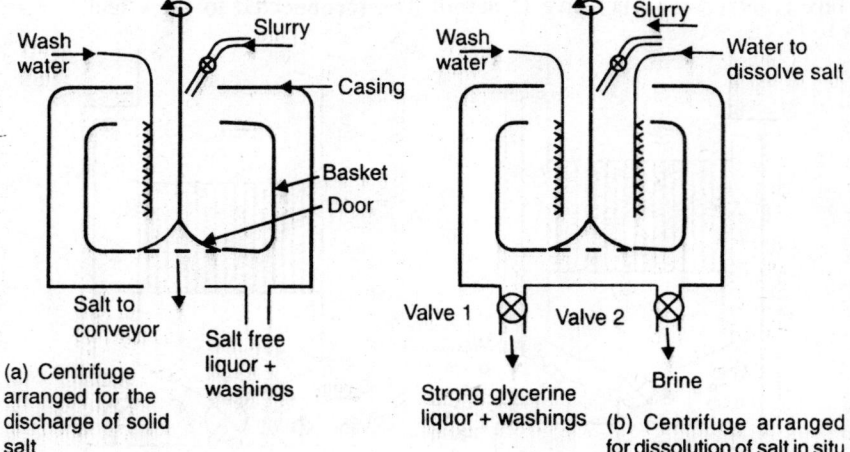

Fig. 8.6. Batch salt centrifuge, (a) arranged for the discharge of solid salt, (b) arranged for dissolution of salt *in situ*.

It is necessary to separate the solid salt in the slurry removed from the evaporator from the liquor, and to wash the salt as free from glycerol as is economically feasible. Some washing with lye may be possible, but mainly fresh water is used; and this water has to be evaporated, which sets a limit to the amount of washing which is worthwhile. Traditionally, basket centrifuges operated batchwise are used, and the salt is discharged as a crystalline powder. It is returned to the soapmaking department as such, or after solution in water, preferably condensate, to form a near saturated brine. Modern soapmaking plants use only brine; and systems have been developed in which the salt is dissolved in the centrifuge itself. Alternatively, the salt is dissolved in a separate unit. When sulphate is introduced into the system, by the use of aluminium sulphate or sulphuric acid, it is necessary to discard some recovered salt from time to time to prevent undue build up of sodium sulphate in the salt. The batch centrifuges used are normally of the suspended basket type; and some points regarding their use are discussed in relation to Fig. 8.6.

(i) Perforate basket machines are used; and the filter medium is a fine weave metal gauze backed by a coarse gauze to facilitate drainage. A high speed to give a high centrifugal force is desirable to remove the liquors from the solid salt as completely as possible.

(ii) The slurry should be fed with confidence in intermittent bursts having a high rate of flow. A 4" pipe with a control cock operated with a long handle is convenient. This rate of feed makes the slurry run up the basket wall to the top and gives a salt cake of fairly uniform thickness. Too high a feed rate, or unduly long bursts, cause slurry to overflow the rim of the basket and to carry solid salt over into the centrifuged liquor. Actually some salt always gets into the clear liquor and arrangements must be made to deal with it; as also from any vessels in which soap lye crude, or semi-crude, is held. Too low a feed rate gives a very uneven salt cake, with a large thickness at the base and a small one at the top; this is because most of the liquor has time to penetrate the cake before the slurry can rise far up the basket wall.

(iii) When most of the liquor has been removed, the salt cake is washed with a controlled quantity of water and spinning is continued to remove the wash water as completely as possible. Sometimes two short washes, separated by a period of spinning, are given. If a fixed perforated pipe is used for wash water, care must be taken to ensure that it does not interfere with the discharge of the salt and a spray on a flexible hose is an alternative. Timers to help the operatives to adhere to a cycle laid down by the management are useful. In order to reduce the glycerol content of the salt to the required level with a minimum use of wash water, careful control is essential, particularly of the time during which the water is sprayed.

(iv) With the system shown in Fig. 8.6(a), the machine is stopped, the door over the bottom outlet raised and held up on a hook, and the salt dug out to fall into a conveyor, or receptable, below. A narrow spade with the cutting edge cut at an angle is useful for this work. When lower speed, and less efficient, centrifuges were used, a rope placed round the bottom of the basket before the slurry was fed provided a simple means to dislodge the salt cake. The end of the rope had a hook which held it to the upper rim of the basket; and this end was pulled to bring down the salt at the end of the cycle. However, well washed and dried salt in modern machines is normally too hard for this technique to be used.

(v) With the system shown in Fig. 8.6(b), valve 1 is open and valve 2 closed during most of the cycle for the glycerine liquor and the washings to flow away. When the salt has been washed and dried, valve 1 is closed, valve 2 opened, and a controlled flow of water introduced to dissolve the salt. With proper adjustment of the conditions the salt can substantially all be dissolved to produce nearly saturated brine. Care is necessary to ensure that the casing is arranged to allow the two streams of liquid to drain with minimum contamination one with the other. Also that the valves close tightly, and that the operators control the valves and sprays with care. With proper precautions, this system can give very satisfactory results without the heavy manual labour required with system (a).

In large factories, the modern trend is to automatic centrifuges. Several types have been used, but some have not been very successful because either they could not handle fine glycerine salt satisfactorily, or because they need too much wash water to reduce the glycerol in salt to the required level. The best type is probably one in which the operation is very short cycle, automatic batch. The salt is discharged with a knife; and it is a useful refinement for the knife mechanism to be arranged so that the cut leaves a slightly different thickness of salt in consecutive cycles in order to minimise the formation of a 'glaze' of very fine broken crystals which may impair drainage.

The glycerol left in salt varies considerably with the system employed, the quantity of wash water used, and the care with which the operation is carried out. With centrifuges 0.4 – 0.6% glycerol on salt should be possible, but figures from 0.2–0.1% are quoted. With dry salt boxes, 1% is sometimes claimed, but 2–4% is probably more usual (corresponding figures calculated on saturated brine are about one quarter of those for salt).

Minimisation of Losses of Glycerol

During evaporation, liquor containing glycerol and salt can be lost by being carried out in liquid/solid form with the vapours and discharged in a condensate. The carry-over can be as droplets, foam, or, if a serious prime occurs, as bulk liquid. To minimise this loss :

(i) Adequate vapour space should be provided above the level of the boiling liquid.

(ii) Suitable separators, and external 'catchalls', usually cyclonic, should be included in the system. It is useful to provide sight-glasses in external separators, and/or the return lines, so that the plant operator can see immediately when any serious carry-over occurs and take remedial action.

(iii) The liquor levels in the effects should be controlled with care, automatically or manually.

(iv) The treatment of the lyes, or sweet water, should ensure that the liquors have no undue tendency to foam.

External catchalls are sometimes fitted only to the effect in which the concentration of glycerol is highest. Devices which measure roughly the electrical conductivity of the condensate from the evaporated vapours are very useful indicators of carry-over. They actually indicate the presence of salt, or other electrolyte, but this is relevant as salts and glycerol are carried over together. The main problem is to obtain a continuous flow of condensed vapour from the last effect when, as is common, a direct contact condenser is used in the vacuum system; and the condensate is mixed with cooling water. It is not essential to obtain a representative sample of the vapours plus any entrained liquor; all that is necessary is to get a sample which will indicate when any abnormal amount of salt is present in the vapours, and this can often be achieved by the use of a small partial condenser in the vapour line. For example, a ring carrying cold water placed above a collecting tray, in a vertical vapour main, as shown in Fig. 8.7, can be used. There is a continuous flow of condensate through the box containing the electrodes which are connected to a suitable low A/c voltage and a measuring instrument or, better, recorder. An audible alarm can be included to draw immediate attention to any substantial increase in conductivity.

Losses of glycerol during evaporation can occur in other ways, for example, if a leak develops in a calandria. The effects are normally left containing water during shut-downs (to dissolve deposited salt) and, if a leak is present, some salt solution can be drained from the steam chest of the calandria after the plant has stood for a period of time.

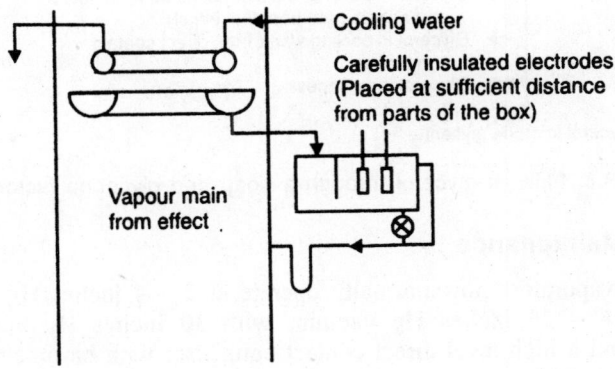

Fig. 8.7. Partial condenser for entrainment indicator.

Fig. 8.8 shows the flow of glycerol through a soap factory with glycerine recovery; and indicates the sources of loss. The diagram is drawn for a double-effect evaporator with direct contact condenser; and refers to closed coils in soap pans. Such coils are now obsolete, but they can be a serious source of loss if present. The diagram can readily be modified to suit other plant arrangements and similar ones prepared for the splitting route. It is useful to determine the recovery in each section of the factory, particularly if the overall loss is high, but, this is usually not done easily; and, for routine purposes, those with limited resources may find it best to concentrate on checks on the controllable losses, and the important glycerol in re-cycled salt. Care is necessary to ensure that truly average samples of materials such as muds are tested. Misleading samples of treatment mud can be taken from near the point at which water is introduced into the filter presses.

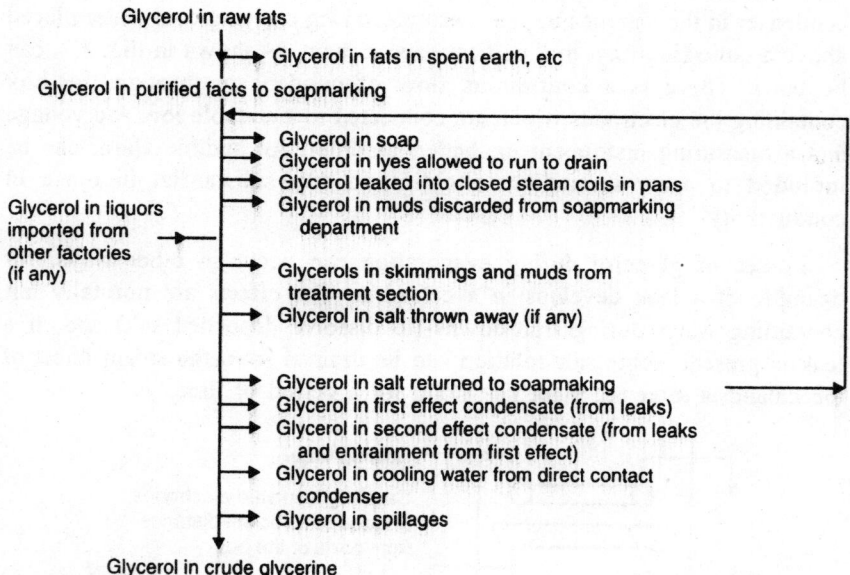

Fig. 8.8. Flow of glycerol through a soap and glycerine factory.

Vacuum Maintenance

Glycerine evaporators now normally operate at 2 – 4 inches Hg absolute pressure (28 – 26 inches Hg vacuum with 30 inches Hg barometric pressure); and a high-level direct contact condenser with barometric leg is commonly used. The non-condensible gas can be removed with a two-stage ejector system, with an intermediate condenser, or by a mechanical vacuum pump. Because of the level of pressure needed in a glycerine evaporator, no booster jets are normally installed. The direct contact main condenser can be replaced by a surface condenser if the condensate from the last effect is required, for example, for use with other condensates in the splitting section of a fatty acid plant.

Evaporator Control

Traditionally glycerine evaporators were controlled manually, but standard instrumental control techniques are now used to make operation increasingly automatic. Possible difficulties due to dissolved and solid salt must be remembered with soap lye evaporators. Also, the influence of vapour on the apparent density of boiling liquid in, for example, a natural circulation evaporator.

A reliable indication of the concentration of glycerol in the material being discharged from an effect producing crude glycerine is required. This is frequently based upon the relationship between boiling point elevation and concentration; in other words, between boiling point under a known pressure and concentration. A very simple way for a plant operator to control a manually operated finisher, or other unit producing crude glycerine, is to provide :

(i)   A barometer to measure atmospheric pressure (b inches Hg).

(ii) A mercury U-tube with an open end to the atmosphere and the other end connected to the vapour space in the effect. This gives the vacuum under which the liquor is boiling, x inches Hg; that is the difference between the atmospheric pressure and the pressure in the vapour space. The absolute pressure in that vapour space is thus $(b - x)$ inches Hg. Alternatively, gauges which measure this absolute pressure directly can be used.

(iii) A thermometer set to measure the temperature of the boiling liquid. Because of the effect of hydrostatic head, this temperature will vary somewhat with the quantity of liquor in the effect and the point at which the thermometer is placed.

Because of the influence of hydrostatic head, and of impurities in the crude glycerine being produced, it is necessary to calibrate the system by the analysis of samples taken in relation to various sets of readings. Roughly, a boiling temperature of about 77°C with an absolute pressure of 2 inches Hg indicates 83% glycerol in a soap lye crude glycerine. It has also been found that an extra 0.1 inch Hg pressure requires an extra 1°C in boiling liquor temperature. Hence, if 77°C is found to be correct for 2.0 inches Hg, 79°C will be about right for 2.2 inches Hg pressure, and 74°C for 1.7 inches Hg pressure.

Materials of Construction for Glycerine Treatment and Evaporation Plants

Traditionally, mild steel, wrought iron, or cast iron tanks and pipes have been used for soap lye treatment plants, with cast iron filter presses clothed with cotton cloths. Particularly for the early stages of treatment carried out under acidic conditions, the presence of chlorides and blowing with air give highly corrosive conditions; and the life of tanks is often short. The lining of such tanks with rubber, cement, or some acid resistant plastic coating, is highly desirable. Cast iron filter presses have a reasonably long life, although it is often useful to scrape the grooves in the plates from time to time. Expensive polypropylene filter plates are now sometimes used; and it may be desirable to assess the economics of using polyester cloths.

For many years, cast-iron was used for the bodies of soap lye evaporators, with arsenical copper tubes in the calandrias and heat exchangers. Such evaporators lasted a very long time; examples are known where the cast-iron bodies were very difficult to break-up after a life of about forty years, although they had been re-tubed 3 – 4 times during this period. These evaporators used exhaust steam, plus make-up, at about 4 psig in the main, and less in the actual calandrias. The adverse effect of higher temperatures, resulting from the use of higher pressure steam, is shown dramatically by experience with the quadruple-effect evaporator mentioned earlier. The arsenical copper tubes in the first effect calandria lasted only 1–2 years; and the rate of corrosion fell with reduction in temperature from first to last effect. Experiments showed that monel metal was a particularly good material for tubes, but that the less expensive cupronickels (including the cheapest 90:10 type) were adequate for the duty. The cast-iron bodies also behaved less well than expected, but this may have been due, at least in part, to the quality of war time cast iron. Special cast irons, such as Ni-resist, have also been used for the bodies of soap lye evaporators. Copper has been used successfully for the bodies of sweet water evaporators.

These traditional materials are still recommended by some plant manufacturers, but others alternatives, include mild steel. Sacrificial anodes of aluminium have sometimes been found beneficial in reducing the corrosion of mild steel, or cast-iron, evaporators.

The baskets of centrifuges used for salt separation have to withstand high stresses as well as corrosive conditions. Mild steel was normally used, but, with modern, more efficient, machines, failures, particularly at welds, are known. High nickel alloys (or, perhaps certain stainless steels), with the necessary mechanical properties as well as corrosion resistance, should normally be selected. Design to avoid high stresses is very important, whatever metal is selected; and rivetted, rather than welded, side seams may be advantageous.

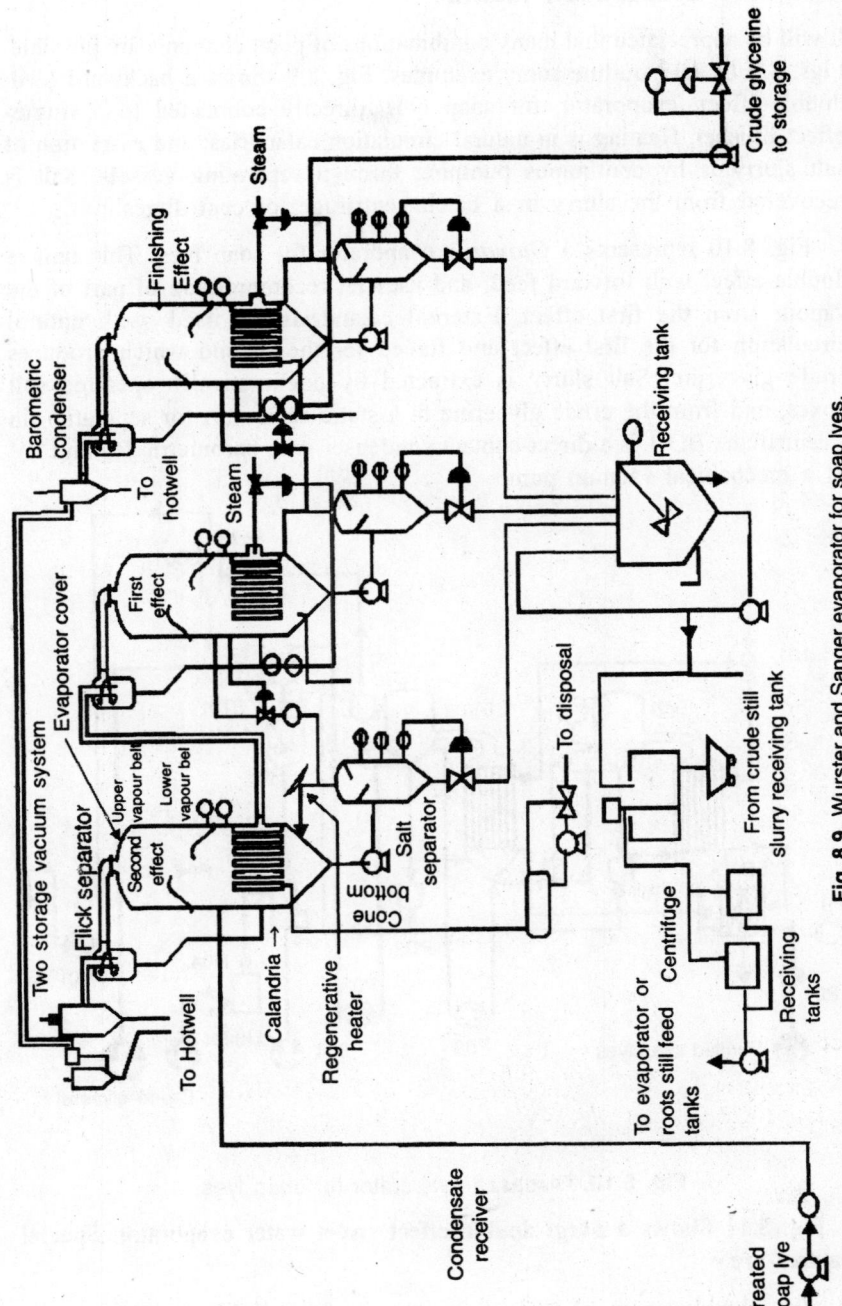

Fig. 8.9. Wurster and Sanger evaporator for soap lyes.

Complete Evaporation Plants

It will be appreciated that many combinations of plant elements are possible. Figs. 8.9 to 8.13 outline some examples. Fig. 8.9 shows a backward feed, double-effect, evaporator, for soap lyes, directly connected to a single-effect finisher. Heating is in natural circulation calandrias; and extraction of salt slurry is by continuous pumping through separating vessels. Salt is recovered from the slurry in a batch centrifuge, or centrifuges.

Fig. 8.10 represents a *Gianazza* evaporator for soap lyes. This unit is double-effect with forward feed; and uses jet recompression of part of the vapour from the first effect. External calandrias are used, with natural circulation for the first effect and forced for the second which produces crude glycerine. Salt slurry is extracted by pneumatically operated salt boxes, and from the crude glycerine in a static separator, for separation in a centrifuge. BO 1 is a direct-contact condenser with barometric leg and PV 01 a mechanical vacuum pump.

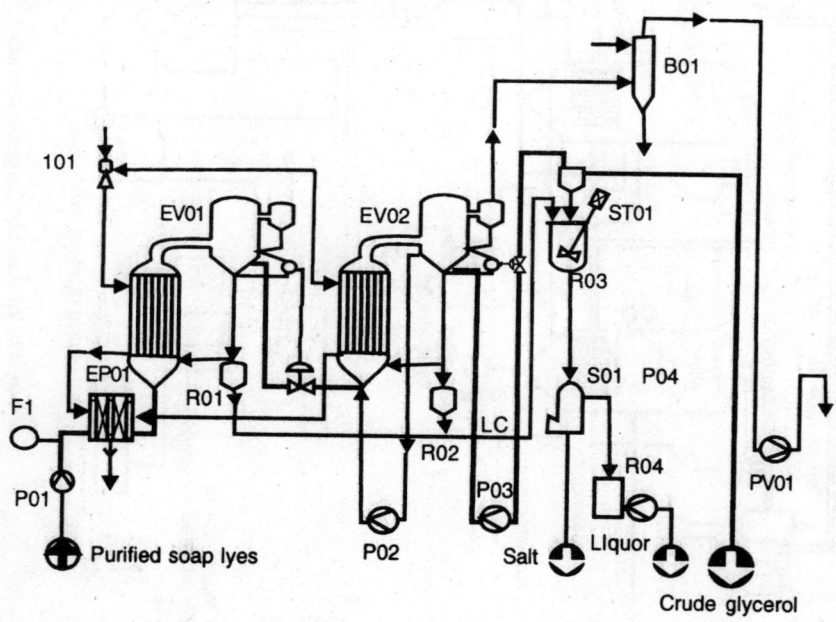

Fig. 8.10. Gianazza evaporator for soap lyes.

Fig. 8.11 Shows a *Lurgi* double effect sweet water evaporator. Special features are :

(i) Jet recompression of part of the vapour from the first effect.

(ii) Multi-compartment, long tube, climbing film calandrias, arranged so that the liquid flow is once through each compartment in series, with the vapour discharge from each into a single space. The last part of the second effect is separated from the rest and is fed with steam directly from the recompression jet; that is the final concentration is in single effect. As an example, Lurgi quotes concentration from sweet water at about 10% glycerol to 25% in the first effect, to 70% in the second effect, and to 88% in the final stage.

(iii) A surface condenser is used for the final vapours to increase the quantity of condensate available for re-use.

(iv) The feed is preheated by the first effect condensates (including that from the final stage).

Fig. 8.11 shows a Lurgi triple effect plant for sweet water which is generally similar to the previous example. No recompression jet is shown, but one could be used if the available heating steam is at a suitable pressure. The feed is preheated by condensate, followed by live steam. A separate finishing section, operating in double effect, is included in the third effect shell. Fig. 8.12 depicts a Wurster and Sanger quadruple effect evaporator for sweet waters. Feed is forward. External calandrias are used with natural circulation for the first three effects and forced circulation for the fourth effect.

TREATMENT OF CRUDE GLYCERINE TO REMOVE LOW MOLECULAR MASS FATTY ACIDS

The quality of soap lye crude glycerine can be improved by steaming it in an acid condition, so as to remove volatile fatty acids. This can be done by acidification before the later stages of evaporation, but, with normal equipment, corrosion problems are likely. The crude glycerine is acidified to about pH 3.5 with sulphuric, or hydrochloric acid. Sulphuric acid is cheaper than hydrochloric, but has the usual disadvantage that it adds sulphates to the system, although this is less important than when salt to be re-cycled can be affected. Both mineral acids have similar effects on corrosion, because substantial concentrations of chloride ions are already present. Type 316 stainless steel (16 – 18% Cn, 10 – 14% Ni plus 2 – 3% Mo) has been found to give a reasonable plant life, although some pitting occurs. The main design and operating problem is to adjust conditions so that a substantial part of the low molecular mass fatty acid is removed without the loss of too much glycerol. A continuous unit which has been used satisfactorily in shown diagrammatically in Fig. 8.12. The process can also be carried out batchwise.

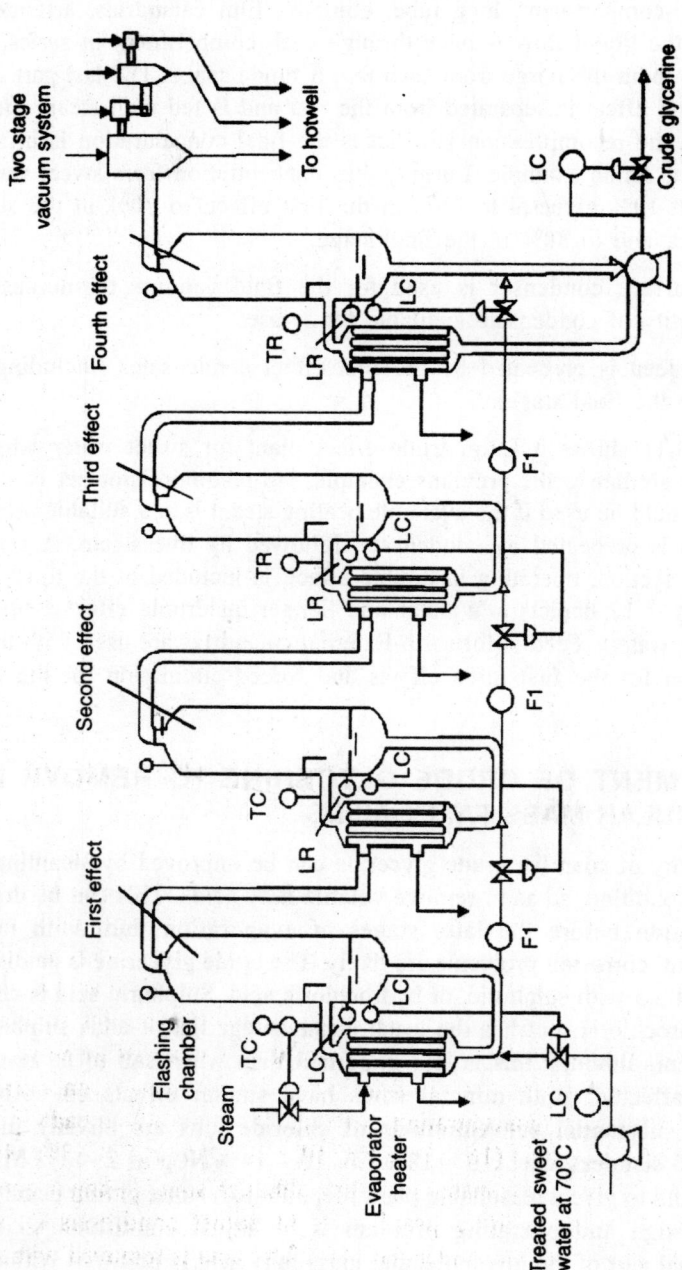

Fig. 8.11. Wurster and Sanger quadruple-effect sweet water evaporator.

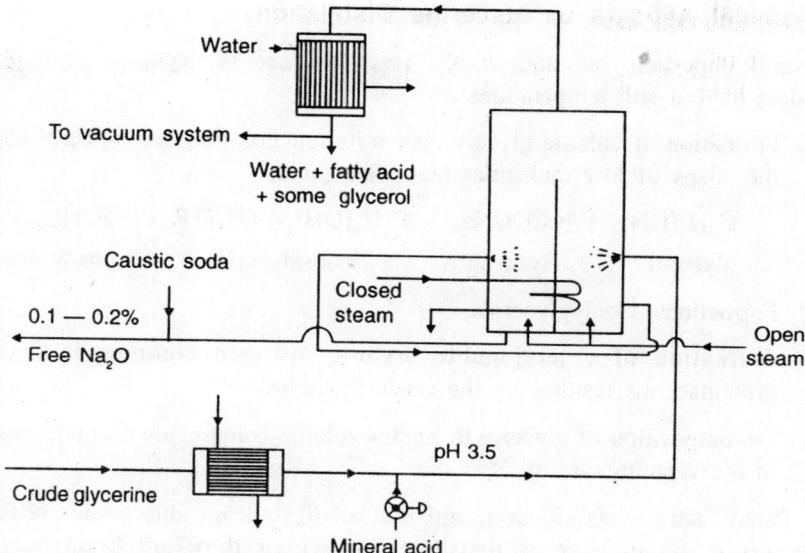

Fig. 8.12.   Removal of volatile fatty acids from soap lye crude glycerine.

DISTILLATION OF GLYCERINE

General

Under atmospheric pressure pure glycerol boils with decomposition at about 290°C and, as for fatty acids, it is necessary to use reduced pressure, and/or open, or carrier, steam to reduce the boiling temperature in glycerine stills. Principles of distillation discussed for fatty acids apply to glycerine distillation, but there are also important differences between the processes for fatty acids and for glycerine. In particular :

(i) The nature of the non-volatile residue, which from soap lye crude glycerine includes large proportions of salt and soapy matter.

(ii) The fatty acid vapours contain a considerable range of compounds of varying volatility, whereas in glycerine distillation the vapours are predominantly glycerol and water, although the relatively small concentrations of organic impurities are important.

(iii) Glycerol and water are completely miscible in the liquid phase, whereas fatty acids and water are not. Also glycerines do not solidify at temperatures likely to be encountered in still condensates.

Chemical Aspects of Glycerine Distillation

Several important, but undesirable, reactions tend to occur in glycerine liquors held at still temperatures :

(i) Formation of volatile glyceryl esters by reaction between glycerol and the soaps of low molecular mass fatty acids.

$$C_3H_5(OH)_3 + R\cdot COONa \rightarrow C_3H_5(OH)_2—O\cdot CO\cdot R + NaOH$$

glycerol   soap   mono-glyceride   caustic soda

(ii) Formation of polyglycerols.

(iii) Formation of objectionable organic nitrogen compounds from proteinaceous residues in the crude glycerine.

(iv) Decomposition of glycerol to highly volatile compounds normally lost in the vacuum system.

Esters, fatty acids (if any), and the volatile nitrogenous bodies effect adversely the quality of distillates in which they are condensed. Polyglycerols are substantially non-volatile under distillation conditions; but they are undesirable because they represent a loss of glycerol, and because they modify the residue in ways which can be adverse. The condensation reactions by which polyglycerols are formed are not reversible and the ether links they contain are not broken under still conditions. Careful materials balances determined for glycerine distillations normally show a small unaccounted for loss, which may be associated with radical decomposition of glycerol to gaseous products.

Ester formation is a balanced reaction which is suppressed by the presence of caustic soda in the still vapouriser; and polyglycerol formation is catalysed by alkalis, with caustic soda more potent than the milder alkalis sodium carbonate and soaps. It is, therefore, important to carry out the distillation with as little caustic soda in the still as will permit the production of distillates satisfactory as regards content of esters and fatty acids (measured by the saponification equivalent, but also likely to affect taste and odour). Physical conditions of distillation are very important. Increases in the time for which, and the temperature at which, glycerine is held in the still vapouriser both increase the formation of polyglycerols. There is an interaction between the two conditions, so that a very short time at a relatively high temperature may be less damaging than a longer time at a lower temperature. More severe physical conditions also increase the influence of alkaline catalysts, so that mild alkalis may not be very significant if the physical conditions are mild, but become important when they are more severe. It has also been shown that with a given crude glycerine the distillate is much lower in saponification equivalent when distillation is very rapid. Similar factors probably affect the formation of volatile nitrogeneous bodies. The introduction of a small quantity of water into the still in, or with, the crude glycerine may also help to suppress polyglycerol formation, even though the water is rapidly vapourised. Under some conditions, the use of open steam may also help to reduce ester formation. When distillates need to be re-distilled, it is important to saponify any esters in them before the second distillation. This can be done by agitating the hot distillate with air after addition of caustic soda; and a suitable quantity of caustic soda is needed in the still during re-distillation to suppress re-formation of esters.

Formation of polyglycerols is commonly a major source of loss of yield in glycerine distillation; and it may be useful to estimate the polyglycerol in the still residues, which are often called foots. Elaborate separation procedures have been developed for special investigations, but the polyglycerol content is usually deduced from a periodate and an acetin, or dichromate, test on the foots, or, better, from all three tests. As explained the periodate test does not respond to polyglycerols and, unless sugar or other unusual substances are present, records only glycerol. The acetin test includes polyglycerols, so that the difference between glycerol estimated by acetin and glycerol estimated by periodate is a measure of the polyglycerol content of the sample. To convert this difference to the glycerol from which the polyglycerols were formed, it must be multiplied by a factor which depends on the composition of the polyglycerols. If di-glycerol is assumed, the factor is $6/4 = 1.5$, and if tri-glycerol, it is $9/5 = 1.8$. To obtain the general relationship, when n moles of glycerol condense, $(n-1)$ ether links are formed with a loss of $(n-1) \times 2$ hydroxyl groups. Hence, $3n$ hydroxyl groups form a polyglycerol containing $3n-2(n-1)$ hydroxyls. In practice, a factor around 1.8 is probably reasonable. Similarly, the difference between glycerol percent by dichromate and glycerol percent by periodate is a measure of the polyglycerols present, and in this case no factor is required. Some stills give a very poor distillation yield when found to be operating at 60–70 mm Hg absolute pressure in the still instead of the 10–14 mm Hg for which it was designed. This raises the boiling liquor temperature to 170–180°C, substantial quantities of caustic soda had to be used, and the yield is only about 75%. A sample of liquid residue run out of the still, not a true average sample of the foots, tested 7% glycerol by periodate and 37% by bichromate. It, therefore, contained 7% glycerol and 30% polyglycerol calculated as glycerol; and showed qualitatively that major conversion of glycerol to polyglycerol had occurred and was a major cause of the very bad yield. In order to relate the polyglycerol found in foots to distillation loss due to this cause, it is necessary to test a true average sample of the residue and to find the quantity of this residue formed from a known quantity of crude glycerine by measurement, or by calculation from the salt contents of the two materials.

Vapourisation Principles

The vapour pressure of pure glycerol is as shown in the table.

Temperature °C	120	130	140	145	150	155	160	165	170	175
Vapour pressure mm Hg	0.75	1.35	2.45	3.2	4.3	5.5	7.4	9.2	11.5	14.7

Temperature °C	180	190	200
Vapour pressure mm Hg	18.5	29	45

The more relevant part of this information is shown graphically in Fig. 8.13.

Glycerol and water are miscible in the liquid phase and no water remains in the liquid in a glycerine still of steam distillation. That is $p_G + p_W = P_T$, where p_G is the vapour pressure of glycerol, p_W the vapour pressure of water, and p_T the total pressure in the still. The vapour pressure of glycerol in equilibrium with a still liquor is lower than that of pure glycerol at the same temperature, because of boiling point elevation due to dissolved substances. This elevation varies with the composition of the still liquor, which, in turn, varies with the composition of the crude glycerine fed and the degree of concentration which has been achieved. In the case of soap lye, crude, from which salt leaves solution as the distillation proceeds, a figure based upon experience of roughly 10–15°C is assumed for the main part of the distillation. That is to say that the vapour pressure of glycerol in equilibrium with a still liquor at 155°C is taken to be about that for pure glycerol at 140–145°C, or about 2.5–3.2 mm Hg.

Substituting 92, the molecular mass of glycerol, and including the vapourisation efficiency E, the equation derived for steam distillation can be re-written, lb of open steam/lb glycerol distilled = $(P_T - Ep_G) \times 18$ /$(Ep_G \times 92)$. Table 8.2 gives estimated usages of open steam for a range of conditions with the provisional assumption that E = 1. These estimates, which show a wide variation with conditions, refer only to the major part of the distillation run; considerably more open steam is used in the final stages when the residue is being worked down. Crude glycerines contain a small proportion of water; in some stills this is removed in a preliminary dehydrator, but in others it enters the main vapouriser, in which case it contributes to the open steam.

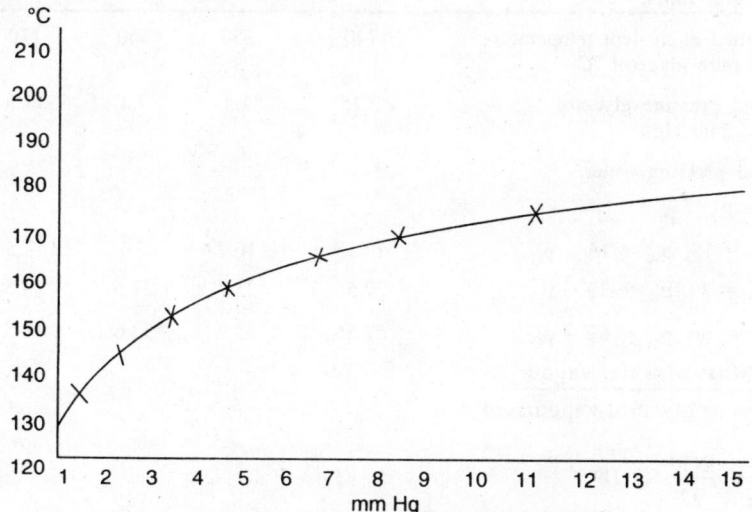

Fig. 8.13. Vapour pressure of pure glycerol.

In a vapouriser which uses open steam as well as reduced pressure, the distillation conditions are established by automatic adjustment to meet three sets of requirements simultaneously :

(i) Heat transfer from the closed steam, or other heating medium.

(ii) Heat balance, which determines the rate of vapourisation of the glycerol plus that of any liquid water fed to the main still.

(iii) The partial pressure relationship just discussed.

If, for example, the rate of heat transfer is increased, the additional heat will vapourise more glycerol. With a fixed rate of open steam supply this reduces the open steam/glycerol vapourised ratio, which causes the boiling liquor temperature to increase to the level required by the new ratio. Similarly, if the flow of open steam is increased, and the vapour passages and steam condensing equipment are such that the total pressure P_T is not raised significantly, the increase in the open steam/glycerol vapourised ratio results in a fall in the boiling liquor temperature.

So far it has been assumed that $E = 1$, that is that the open steam becomes saturated with glycerol as it passes through the glycerine liquor in the vapouriser.

Table 8.2. Open steam usage for glycerine distillation under various conditions.

Approximate temperature of still liquor °C	150–155	160–165	170–175	180–185
Assumed equivalent temperature of pure glycerol °C	140	150	160	170
Vapour pressure glycerol p_G, mm Hg	2.45	4.3	7.4	11.5
Partial pressure water,				
$p_W = P_T - p_G$				
If $P_T = 15$, $p_W = 15 - p_G$	12.55	10.7	7.6	3.5
$= 30$, $p_W = 30 - p_G$	27.55	25.7	22.6	18.5
$= 60$, $p_W = 60 - p_G$	57.55	55.7	52.6	48.5
$\dfrac{\text{Mass of water vapour}}{\text{Mass of glycerol vapourised}}$				
$\dfrac{(P_T - p_G)}{p_G \times 92} \times 18$				
If $P_T = 15$	1.00	0.49	0.20	0.06
$= 30$	2.20	1.17	0.60	0.31
$= 60$	4.60	2.53	1.39	0.83

The actual usage of open steam is much higher than that calculated with $E = 1$. It has been found that with allowance for boiling point elevation and excluding the work-down period, the calculations using $E = 1$ are in fair agreement with experience, for at least some stills. Analysis seems to provide some approximate support for this view as follows.

Assuming the soap lye crude used contained 83% glycerol and 7% water, and that the distillation yield was 95%, 1 lb of distillate came from crude glycerine containing $0.07/(0.83 \times 0.95) = 0.09$ lb of water.

For a Garrigue still, the findings are :

Heating steam in closed coils 175–200 psig (192–198°C)
Distillation temperature about 177°C.
Vapour leaving still 166–171°C.
Still pressure 40–50 mm Hg.
Open steam used during run 0.5 – 0.6 lb/lb glycerol.
Total open steam (including boil down and finishing of foots) 1–2 lb/lb glycerol.

From these figures it is assumed that the equivalent boiling temperature for pure glycerol was about 168°C, with a p_G of 11 mm Hg. With the mean operating pressure of 45 mm Hg, this gives (lb open steam)/(lb glycerol distilled) = $((45–11) \times 18)/(11 \times 92) = 0.60$. With allowance for the water in the crude glycerine fed, the ratio becomes $0.60 – 0.09 = 0.51$.

For an Ittner still, the findings are :

Distillation temperature 166°C
Still pressure 10–15 mm Hg
Open steam used 0.25 lb/lb glycerol.

It is assumed that the equivalent boiling temperature for pure glycerol was about 155°C, with $p_G = 5.5$ mm Hg. With the mean operating pressure of about 13 mm Hg, the ratio = $((13 – 5.5) \times 18)/(5.5 \times 92) = 0.27$, which, with allowance for water in the crude, becomes 0.18.

Water in crude glycerine obviously becomes of increasing importance as the total usage of open steam falls.

Vapourisation Equipment

Soap lye crude glycerine contains some 10% of salt and soapy matter; and the disposal of this non-volatile material influences the choice of vapourisation equipment which can be used, and makes it usual with many types to operate semi-batch, rather than continuously. This problem is much reduced with saponification crudes, particularly those made from high pressure split sweet waters which contain little inorganic matter. For this reason, saponification crudes are often considered superior to soap lye crude, but, especially when produced from low grade fats, they are liable to contain nitrogenous and other impurities which can lead to problems with the quality of the distillates made from them.

As a batch of soap lye crude glycerine is distilled semi-batchwise, with additional crude fed to replace the glycerol vapourised, the concentrations in the residue of sodium chloride, other inorganic materials, and MONG increase; and the viscosity of the residue rises. The recovery of as much glycerol from this residue as is economically worthwhile is a major problem; and several procedures are possible :

(i) Traditionally the main distillation was stopped whilst the residue was still liquid enough to flow out of the still. The residue was then diluted and purified as much as possible using treatments similar to those adopted for lyes or sweet waters. The purified liquor was re-evaporated to give 'foots crude', which was distilled separately to produce a low grade distillate. A variant was to return the still foots to the soap lyes before treatment. This was simple, but is not recommended because of the build-up of impurities in the system, despite some purification in the re-treatment.

(ii) Some plants remove residues from the main still in liquid form for the rest of the glycerol to be stripped in a separate foots still. This enables the main still to be operated more, or less, continuously, although it needs to be washed out from time to time. Recently the stills offered are of the type in which solid salt is separated in a centrifuge through which the liquor is passed on its way from the main still to the foots still. Apart from the cost of the additional equipment (offset by the additional capacity of the main still), the use of a foots still tends to increase the time during which the material is exposed to distillation temperatures. Separation of salt and other solids in a centrifuge helps the final stages of vapourisation, but exposure of hot liquor to air may entail some risk of oxidation of glycerol.

(iii) For many years it has often been found best to continue the main distillation until little more glycerol can be vapourised. With many still designs this involves a loss of glycerol in foots discarded to drain of at least about 2% of that fed. If a reasonably good quality soap lye crude glycerine has been distilled, and conditions are such that only a little polyglycerol is formed, the residue is so stiff that it cannot be run out of the still; and it has to be dissolved by being boiled with water. If the quality of the crude glycerine is poor, polyglycerol formation is high, or the final steaming has not been complete, some of the residue can be run out as liquid foots. When liquid foots are formed, investigation is desirable to ascertain the cause and, if possible, to arrange remedial action for future distillations.

In most stills the residue after as much glycerol as possible has been vapourised from a soap lye crude is a toffee-like material, but, as described later, it has been found recently that under some circumstances a dry, powdery, residue can be obtained from a once-through vapouriser.

When re-distillation of once distilled glycerine is used to produce chemically pure grade, a procedure which is no longer common, a liquid residue containing glycerol, polyglycerols, caustic soda, and soapy matter is left. From time to time, this needs to be transferred to a first distillation of crude glycerine.

Three major types of vapouriser are used in the distillation of crude glycerine:

(i) Pot stills, and variants, in which open and closed steam are used to provide natural circulation vapourisation.

(ii) Stills with external heaters and forced circulation.

(iii) Scraped wall, once through, vapourisers.

Types 1 and 2 can be used for saponification as well as soap lye crudes, but type 3 cannot because the necessary dry, powdery, residue is not produced, unless a liquid residue can be discharged satisfactorily.

Natural circulation pot vapourisers are used in Van Ruymbeke stills, a type which was very widely employed for many years; and in various other designs. In essence, they comprise pots fitted with closed heating coils (usually steam) and a perforated ring, or cross, near the bottom for the introduction of open steam. The Van Ruymbeke still operated at around 60 mm Hg; and hence required about 1–1.5 lb open steam/lb glycerol vapourised, with a boiling temperature around 170–175°C. Steam in the closed coils at 175–200 psig (condensing at 192–198°C) gives a reasonable temperature difference to induce heat flow and a good rate of distillation. One major installation with several Van Ruymbeke stills for the distillation of soap lye crude glycerine, plus two smaller stills for re-distillation to produce chemically pure grade, operated satisfactorily for many years with steam from Lancashire boilers at 150–160 psig (185–188°C) which gave a rather small temperature difference. Special arrangements to make the best use of the available steam are shown diagrammatically in Fig. 8.14. Features are :

(i) A good flow of steam through the coils, with efficient removal of condensate and air, was induced by use of surplus steam which was recovered from the 90 psig collector.

(ii) Superheated steam was used in the crude still coils. This is contrary to the usual view that superheat in the steam used in heating coils reduces rather than increases heat transfer, but this use of superheat was found to increase distillation rate, presumably because it increased the temperature of the upper parts of the coils which were in contact with froth, as well as more compact liquor. Saturated steam was found advisable for re-distillation, when the boiling point elevation was lower.

Classical Van Ruymbeke stills included a 'superheater' for the open steam. This comprised a tube and shell heat exchanger in which the flow of open steam towards the perforated distributor in the still was exposed to heat from full pressure steam. The superheater was presumably intended to avoid the introduction of liquid water into the hot glycerine liquor where it might cause primes. It can, however, be shown by theoretical analysis, using the Fanno line technique, that expansion of steam through a control valve and pipe is very different from that through a nozzle, because of the 're-heating' effect of pipe wall friction as the linear velocity becomes very high; and that liquid water is unlikely to be formed. Later on it was found that these superheaters were unnecessary, and they were removed without any ill effect.

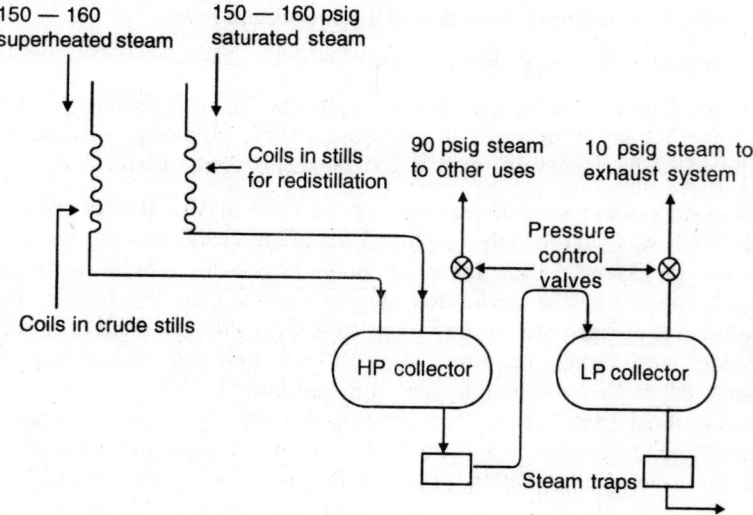

Fig. 8.14. Steam supply arrangements for a group of Van Ruymbeke stills.

In addition to its role as a carrier fluid, an important function of the open steam in pot-still vapourisers is to provide agitation, with proper circulation of the liquor over the closed coils without local overheating. Some plant manufacturers have introduced the open steam through jet-type vapour-lift devices to improve the circulation; for example up the centre of the pot and down over the annular heating coils. This is particularly useful when the still is designed to work at a relatively low absolute pressure and with a low open steam ratio.

It is important to ensure that flow passages through condensers and ducts are adequate to take the maximum requirement of open steam without undue pressure drop. Stills have been constructed in which this was not done, so that an adequate flow of open steam could not be used without a substantial increase in still pressure.

Several important modern stills use *external heaters and forced circulation* which permit operation at low pressures of about 2 – 4 mm Hg. Often no open steam is used during the main part of the distillation, but it is also possible to inject some open steam into the vessel to and from which the liquor is circulated; in which case a rather higher still pressure is usual. One early type was a Scott still for which the soap lye crude is dried before being fed into the still in which a pressure of about 2 mm Hg is maintained. This still is operated with a liquor temperature of 177°C, which is surprisingly high for a pressure of 2 mm Hg, even with no water in the crude glycerine as fed to the vapouriser. Still of this type requires a large addition of caustic soda to produce distillate with an acceptable saponification equivalent and that the yield was not good. The large amount of caustic soda used, and the polyglycerol formed, would tend to account for the high boiling temperature; and the high boiling temperature, plus the excessive caustic soda, for the poor yield.

Since that time very successful stills with vapourisers of this type have been developed. Additional stills of this type and similar vapourisers have been used. These stills operate at 3 − 4 mm Hg without open steam, but with a little water left in the crude glycerine. The boiling liquor temperature is around 155°C and the yield is superior to that with Van Ruymbeke stills. To maintain an absolute pressure in the Vapouriser of 3–4 mm Hg, the vapour system, spray arresters, ducts and condensers, must be designed to have very low vapour velocities, and hence low resistance to flow. As explained in Appendix 2, booster jets are used to compress the water vapour and non-condensible gas leaving the last glycerine condenser; and the capacity of the booster jet, or jets, is often the factor which limits the capacity of the still. This means that the water content of the crude glycerine should normally be kept low. The significance of the presence of a little water in the crude glycerine as fed to the vapouriser on polyglycerol and ester formation is not clear, but it may be important. To enable the distillation to be completed in the main vapouriser, without a foots still, closed and open steam coils are fitted in the vessel, together with a damper in the circulation pipe, as shown in Fig. 8.15. During the main part of the distillation the butterfly damper is open and the liquor is circulated at a high rate by the centrifugal, or axial flow, pump through the external heater. It is returned to the vessel through a tangential inlet onto a ledge round which it swirls; vapour flashes off and the residual liquor runs back into the reservoir to be recirculated with the new feed. The closed and open steam coils in the still vessel are not used during the main part of the distillation. When sufficient crude has been fed, and the liquor worked down until its viscosity becomes too high for continued operation in this way, the damper is closed and the liquor in the circulating system is transferred by the small pump to the main reservoir. The residue is then worked to dryness using the closed and open steam coils, as for a pot still. During this final stage the pressure in the still and the liquor temperature are higher than for the main run.

A more recent development is the use of a 'once-through' scraped wall vapouriser for the distillation of good quality soap lye crude glycerines. When conditions are right the glycerol is vapourised very rapidly and almost completely, without formation of polyglycerols, and a dry, powdery, residue is discharged from the base of the unit. In the Gianazza still (Fig. 8.16) the residue is collected in a large vessel from which it is said to be discharged once, or twice, a week. It is also possible to remove the residue continuously through an automatic air-lock system and screw conveyor.

The vapouriser is heated with steam at, say, 250 psig (208°C); or with hot oil, or other high temperature heat transfer medium. A low pressure of a few mm Hg is maintained as for other stills not using open steam. Because of the very short residence time it seems probable that the working pressure may be less critical than with other types of vapouriser. Because of its almost instantaneous vapourisation, this type of unit has great advantages when it works well. The yield can be exceptionally high, because virtually no polyglycerol is formed, and vapourisation of glycerol is nearly complete. The formation of esters and other undesirable compounds is also substantially lower than with other vapourisers using the same crude glycerine. On the other hand, it can only work if a dry, powdery residue is formed; patches of toffee-like material can cause major problems. This requires the unit to be kept in good mechanical condition and capable of uniform heat transfer. It also requires a good soap lye crude glycerine with a sufficiently high ratio of salt to MONG; at least 3.5 is said to be a useful criterion. It has also been found that particularly good crude glycerines, which produce a very dusty residue at the base of the vapouriser, can cause some difficulties due to carry over of dust, because there is little organic matter to bind the particles of residue. The dry residue produced by this type of vapouriser may be more readily disposed of than the rather foul smelling and polluting liquors obtained when other types of vapouriser are washed out.

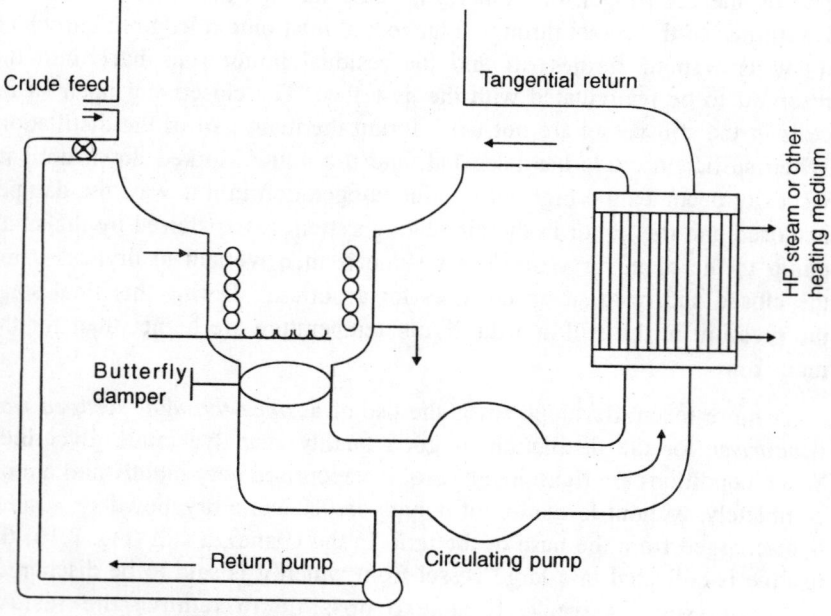

Fig. 8.15. Forced circulation vapouriser.

Spray Arresters

Traces of liquor, or solid residue, carried over in the vapours from the still can spoil a distillate, particularly the first (and best) when more than one fraction is collected. The strict limits for chlorides mentioned in the specifications discussed earlier make this particularly critical with soap lye crudes.

The main problem is to remove spray without excessive flow resistance, especially in stills which operate at very low pressures. Thin beds of rings having a large cross-section, so as to minimise linear velocity and hence resistance, are now usual. Such beds are often situated in the upper part of the still vessel; and they should be provided with water sprays to enable them to be washed when the still is cleaned. Older stills, such as the Van Ruymbeke, which operated without booster jets, commonly used catch-all vessels.

Particularly when foots are worked down to dryness and considerable boiling out with water is needed to remove the residue, carry over is liable to occur during the wash out. Some stills include a large valve in the vapour line which can be closed to isolate the vapouriser from the condensing system during the washout. Such valves can be valuable; but care must be taken to avoid constriction of the vapour passages and the possibility of in leakage of air.

Condensation, Condensers, Stripping Columns, Concentrators, and Deodorisers

Apart from any carry-over of droplets, the vapours leaving a glycerine still contain glycerol and water, plus small concentrations of esters, TMG, and other organic impurities. The proportion of water to glycerol depends upon the quantity of open steam used, including water in the crude glycerine as fed to the vapouriser. Because of the large difference in volatility of water and glycerol, separation of these two components by fractional condensation is relatively easy; and is the main way in which refined glycerines are produced. Fractionating columns with reflux and reboil are not usual; and are generally undesirable because of the pressure drop they introduce and the extra time for which the alkaline crude glycerine is kept at distillation temperature.

There is a pressure drop from the vapouriser through the system to the vacuum maintenance equipment, so that P_T falls on passage from section to section, but at any point $P_T = p_G + p_W + p_I + p_N$ where

p_G = partial pressure glycerol

p_W = partial pressure water

p_I = partial pressure organic impurities

and p_N = partial pressure non-condensible gas

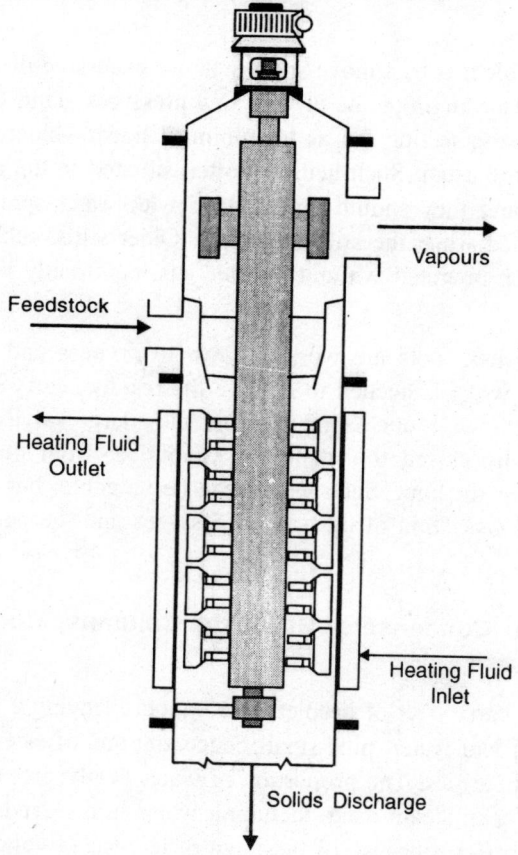

Fig. 8.16. Scraped wall vapouriser for Gianazza GDTF still.

As a condensate rich in glycerol separates, pG falls and the other partial pressures rise to maintain the total pressure.

Consider first only glycerol and water, and disregard changes in total pressure. The vapour-liquid equilibrium diagram is roughly as shown in Fig. 8.17. The shape of the curve mirrors the large difference in volatility of the components. If a superheated vapour containing 50% glycerol is cooled carefully, condensation will start at point a where the 50% vertical cuts the vapour line; and the drop of liquid which separates will, under equilibrium conditions, have composition b (that is it is nearly pure glycerol). As more heat is removed, further liquid will separate with compositions along the line bd; and the residual vapour composition, and temperature, will fall along the line ac. If the cooling in the first condenser stops at temperature t_2, the liquid condensed will have an average composition between b and d; and the remaining vapour composition c. Similarly, on further cooling to t_3, another fraction with average composition between d and f separates, leaving vapour of composition g. To remove practically all the glycerol from the vapour g, the temperature has to be lowered substantially more in a cold water condenser and the condensate contains a considerable percentage of water. The fall in pressure through a real still does not affect this qualitative discussion. The organic impurities are mainly more volatile than glycerol, but considerably less volatile than water. They, therefore, condense with the glycerol, but some fractionation occurs, so that a first fraction, bd, contains less impurities than a second df, and the highest concentration of impurities is in the final, relatively weak, fraction. In modern, high vacuum, stills practically all the glycerol remaining after the product condenses is condensed as an impure fraction containing, say, 60% glycerol before the water vapour and non-condensible gas pass on to the vacuum maintenance system. In some older stills, such as the traditional Van Ruymbeke, the condensers for vacuum maintenance were also the final condensers for weak glycerine which produced what was sometimes known as 'weaks' containing about 2–6% glycerol. Many years ago it found in some factories that much of the steam needed to evapourate the water from weaks could be saved by the introduction of a scrubbing tower as shown in Fig. 8.19 discussed later. This tower recovers nearly all the residual glycerol as an impure fraction containing about 60% glycerol, so that only a negligible quantity passes into the vacuum maintenance system.

Stills have been made which use several vapourisers, and in which the vapour, mainly water, left after the product has been condensed from the vapours from the first vapouriser is used as open steam in the next vapouriser; and so on through the system. This system is used in the Wood still.

So far equilibrium between vapour and condensate has been assumed. When surface condensers are used they should ideally have a large heat transfer area; and so work with a small temperature difference which gives condensation under near equilibrium conditions. If a small surface area and a large temperature difference are used to achieve the same heat flow, local overcooling can occur with less fractionation. This point can be illustrated by the extreme, fortunately not real, case in which a pocket of vapour is condensed completely by itself; the condensate would then have the same composition as the vapour. It is possible to arrange contact between liquid and vapour after the initial condensation. Interchange of components then occurs in a move towards equilibrium. Hence counter-current condensers may be beneficial; and stripping, or washing, columns can be used to improve distillate quality. In such columns, distillate flows downwards in counter-current to rising vapours, usually directly from the still, which vapours are superior in quality to others left after some condensate has been separated. Stripping columns are sometimes used to correct for quality deficiencies due to condensation with a large temperature difference.

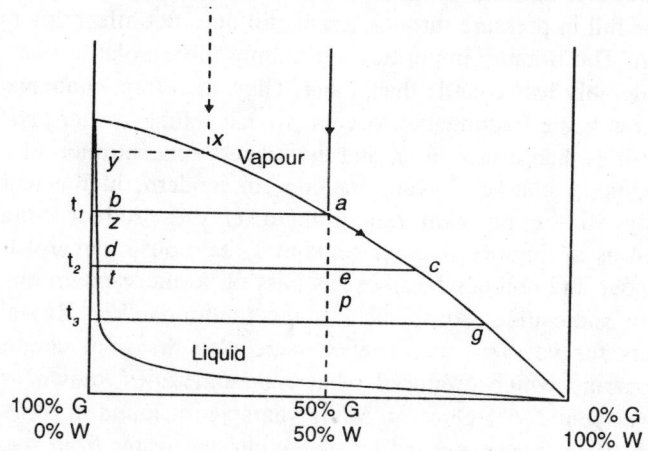

Fig. 8.17. Vapour/liquid equilibrium for glycerol and water.

Some stills, particularly small ones, have only one product condenser, but there are often advantages in having two. The best distillate, from the first condenser, is then available for the top quality product and the somewhat less good distillate from the second product condenser can be diverted to the less exacting grades. This may enable distillation to be carried out with less caustic soda in the vapouriser and so tend to a better yield. Fig. 8.17 shows that a little water condenses with the glycerol; and a comparison of routes *abdf* and *xyz* indicates that the amount is larger in stills using open steam than in modern high vacuum stills in which the vapours from the vapouriser contain little water. Concentration of the distillates is sometimes, but by no means always, needed to meet specification. This can be done in a separate evaporator, but stills in which concentration of distillate is needed may include vessels in which the distillate is subjected to heat from closed coils whilst under vacuum. Such vessels sometimes also use open steam, when they are known as deodorisers and are intended also to remove relatively volatile organic impurities which have condensed in the distillate. Older type stills, such as the Garrigue which produce large quantities of weaks, sometimes include built-in concentrators which also produce open steam for use in the vapouriser.

TMG is more volatile than glycerol and concentrates in the later fractions. The final, impure fraction from the cold water condenser is normally returned for redistillation, after treatment with caustic soda and possibly concentration; and if the crude glycerine being used contains a relatively high concentration of TMG this compound may build-up in the system to the extent that the distillates cannot meet the high glycerol clauses in some specifications, such as the 99% in the Indian standards, even if all water is eliminated. Unless customers can be found who do not mind the TMG, the best procedure is usually to collect the impure last fractions until sufficient is available for re-distillation alone, from which the cold condenser fraction is discarded with some loss of glycerol. If necessary, it may also be possible to push more TMG into the cold condenser fraction, plus additional glycerol, by an increase in the temperature of the vapour as it leaves the last product condenser. It is much more satisfactory for the producer of the crude glycerine to avoid formation of TMG, which, as stated earlier, also indicates a loss of glycerol by decomposition.

The condensers used in glycerine stills fall into three main types, although variants of each are possible :

(i) Air cooled cylinders.

(ii) Surface condensers, using a hot cooling medium for the main product condenser, or condensers.

(iii) Direct contact condensers.

Air cooled cylinders are now obsolete, but they were used in Van Ruymbeke and the even earlier direct-fired stills. The vapours from the vapouriser pass through a lagged catchall and then through a series of large unlagged cylinders. Some 8–12 cylinders are usual and they can be arranged in line, or in tiers. Heat is lost by natural convection and radiation from the surface of the cylinders. Distillation has to be continued for several hours after the commencement of a cycle to bring the system up to the temperatures necessary to produce satisfactory condensate. During this period the liquid condensed is returned to the still; and this helps to remove impurities carried over into the cylinders during the washing out of the foots. When the return to still is closed, condensate collects in the cylinders, but, with some designs, 'drop-cylinders' are provided for removal of condensate during the run. The following temperatures for the sequence of cylinders in the condensing system of a Van Ruymbeke still are quoted as under :

Cylinder	1	2	3	4	5	6	7	8	9	10	11	12
Temperature °C	197	195	191	185	178	170	160	149	122	106	90	72

Distillation was said to commence at 199 – 203°C, using closed coils fed with steam at 225 psig (203°C). The distillation, and early condensing cylinder, temperatures seem rather high, but the figures illustrate :

(i) The favourable temperatures for near equilibrium condensation.

(ii) The way the condensing temperature falls slowly at first and then at an accelerating rate as the vapours are denuded of glycerol.

Objections to air condensers are the space required and the lack of possible control. However, with some designs, particularly the linear arrangement of cylinders, the condensates from the individual cylinders can be combined to suit the operator's requirements. For example, the best fraction might normally be taken from cylinders 1–5; but if a larger quantity of slightly lower quality is needed, a change might be made to cylinders 1–7.

Surface condensers using cooling media at relatively high temperatures and condensers of this type have been, and are, used in many more recent stills. The main features of one version of the Garrigue still are shown in Fig. 8.20, discussed later. The following temperatures for the vapours at various points in a Garrigue condensing system can be quoted as under :

	First		Second	Cold water	Vacuum
Vapouriser→	Superheater→	condenser→	condenser→	condenser→	system
166–171°C		149–154°C	121–124°C	60°C	29°C

Figures for the condensates from a modern still working at 3 mm Hg without open steam, other than from water in the crude glycerine, are :

	First	Second	Cold water	Vacuum
Vapouriser→	condenser→	condenser→	condenser →	system
	↓	↓	↓	
	130°C	75°C	15°C	

The actual temperatures needed at the various stages depend upon the distillation conditions and the quality requirements, but it will be evident that when two product condensers are used a cooling medium temperature above 100°C is normally required for the first condenser in order to achieve the appropriate cooling of the vapours with a modest temperature difference. In different designs this is obtained by :

(i) Condensers in which heat is removed by water boiling at a pressure, and temperature, controlled by conditions in an auxiliary steam condenser.

(ii) Circulation of hot water under pressure.

(iii) Circulation of a cooling liquid, such as glycerine at a suitable concentration, with a boiling point higher than that of water.

For the second product condenser, hot water is normally adequate. A cold water condenser separates most of the remaining glycerol as an impure fraction.

Direct contact towers containing a mesh, or other low pressure drop packing can be used as condensers by the circulation of suitably cooled glycerine condensate as shown diagrammatically in Fig. 8.18. Advantages of this system are :

(i) The condenser also functions as a stripping column in which the quality of the distillate is improved by counter-current contact with the best available vapours. A separate stripping section for the glycerine leaving the unit can, if required, be installed below the condensing section round which condensate is circulated.

(ii) It is cheaper than a large surface condenser.

(iii) Control is simple; and considerable flexibility can be achieved.

The problem is to achieve adequate contact between liquid and vapour without excessive pressure drop. This may make the system more suitable for use with vapourisers which do not require very low absolute pressures of 2 – 4 mm Hg, such as those operating at 10 – 15 mm Hg with a little open steam, and those with once through, scraped wall vapourisers.

Points in the design of a direct contact condenser include :

(i) The quantity of heat to be removed from the circulated glycerine is controlled by the kg/hr of vapour to be condensed. The higher the flow of circulated glycerine, the smaller the decrease in temperature required to achieve this heat removal. The extent to which the vapours need to be cooled is controlled by the vapour-liquid equilibrium discussed earlier. The temperature of the circulated liquid, which rises as it passes down the condenser, must be sufficiently below that of the vapour at all points in order to induce the required heat flow, but should be kept as near as possible to that required for equilibrium.

(ii) The diameter of the column must be selected so that the vapour velocity is sufficiently low to give an acceptable pressure drop with an adequate depth of packing; but also so that the liquid flow is enough to wet the packing properly without any risk of flooding.

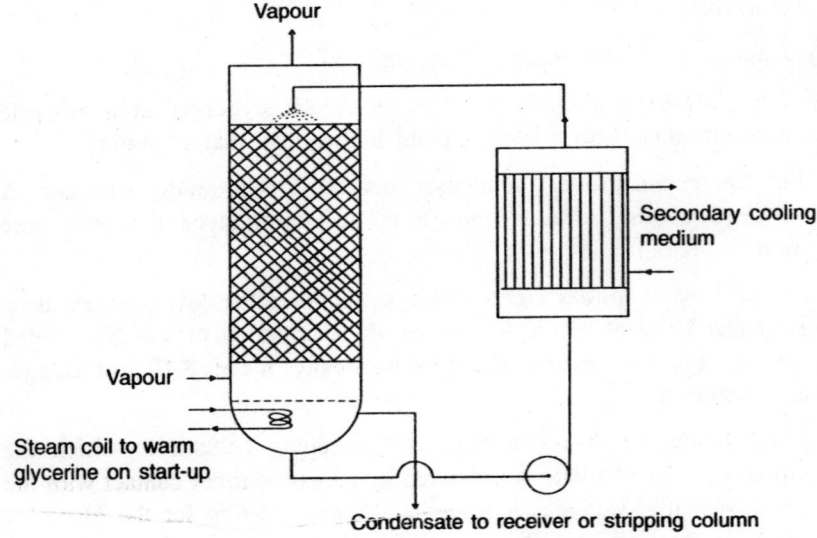

Fig. 8.18. Direct contact condenser.

Materials of Construction for Glycerine Stills

The older stills, such as the Van Ruymbeke, were normally made of mild steel, with only a few parts, such as the perforated open steam distributor, and the valves, of more resistant metal, such as phosphor bronze. Some discolouration of distillate occurred. For the production of chemically pure grade by redistillation in mild steel equipment it is advisable whenever possible to use a separate still for the second distillation. Without the need to boil the foots from a distillation of crude glycerine with water, the condensing cylinders in the second still retain a film of concentrated glycerine which minimises contamination.

Most modern stills use stainless steel, or aluminium, for all parts in contact with the vapours and distillates. Some designers consider it worthwhile also to make the vapouriser of stainless steel.

Complete Stills

The various possible elements considered in previous sections can be selected and combined in many ways to form complete glycerine stills; and many designs have been used and described. Some plant manufacturers have used widely differing designs at various times, so that mention of a plant manufacturer does not necessarily indicate the nature of a still. In this section, only two stills of particular historical importance, plus some designs from recent manufacturers are outlined.

Fig. 8.19 shows one form of *Van Ruymbeke still*. The vapouriser is a pot with closed and open steam coils. Traditional stills did not include a scrubbing tower; and the weaks from the cold water condensers had to be evaporated to recover glycerol. The inclusion of a packed scrubbing tower, fed with a small quantity of cold water sufficient to reduce the temperature of the vapours leaving the tower to an appropriate level, recovers practically all the glycerol leaving the air cylinders, the weaks are then discarded to sewer.

Fig. 8.20 shows one version of the *Garrigue still* which was widely used. The vapouriser is similar to that of the Van Ruymbeke still. Surface condensers are used to produce product by fractional condensation. Weaks, sometimes called sweet waters, are concentrated in a built-in single effect evaporator which provides the open steam for the main vapouriser. This open steam is superheated in a heat exchanger which also serves as the first product condenser. A steam heated concentrator is included to bring the third product fraction up to the strength required for dynamite ('high gravity') grade.

The *Gianazza* HPS still which is very similar to the *Frazer Scott*, glycerine is dehydrated before it is fed to the forced circulation vapouriser. Salt slurry, which separates as the crude glycerine loses water in the dehydrator, is returned to the crude glycerine department. Condensation is in surface condensers which use a circulated fluid cooled in auxiliary coolers. The *Gianazza GDTF 'Turbotherm'* still includes a scraped-wall, once-through, vapouriser, otherwise it is similar to the Gianzza type HPS.

In recent process developed by *Lurgi*, the *Lurgi* still preheated crude glycerine is vapourised in a vessel with one, or more, legs, each equipped with closed heating elements over which the liquor rises by the vapour lift principle induced by open steam injected at the base of the leg. With this steam also acting as carrier fluid, vapourisation is at about 10 mm Hg total pressure. This vapouriser is similar in principle to that used in Lurgi fatty acid stills. In some large units the multi-compartment vapouriser distils soap lye crude continuously, a liquid residue being withdrawn to be worked down in a separate foots still. Distillate from the foots still is redistilled. Other Lurgi stills of this type are operated on the semi-batch principle. The main condenser is of the direct contact type using glycerine cooled slightly in a heat exchanger in which it boils water. In addition to the stripping which occurs in the condenser, the product leaves through a separate

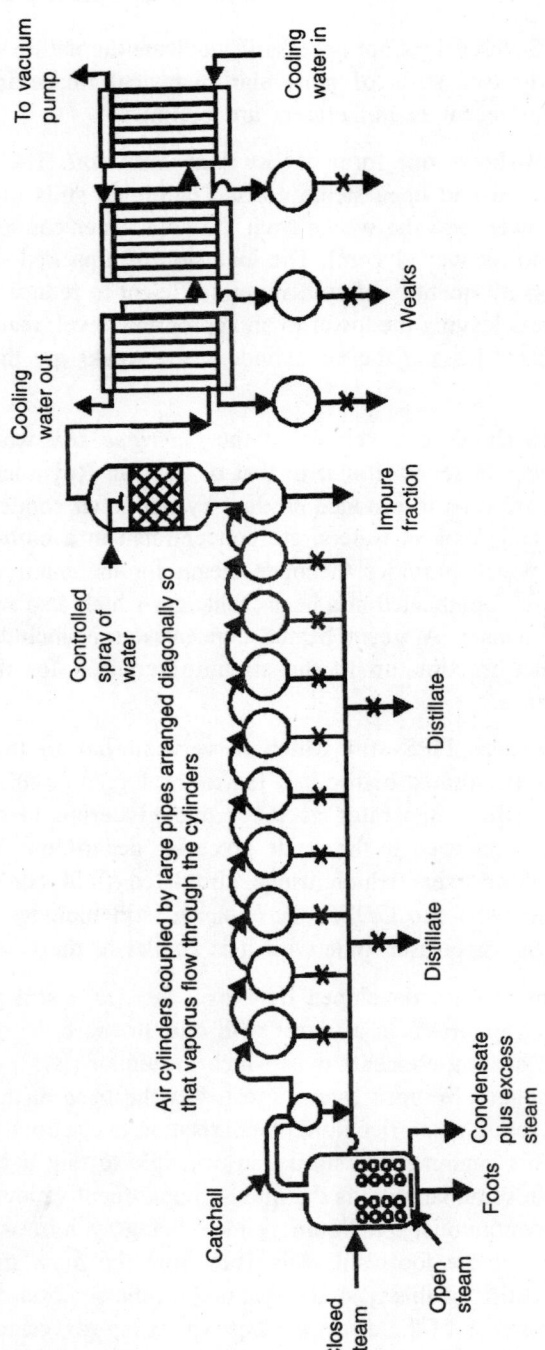

Fig. 8.19. Van Ruymbeke still.

stripping section of the tower in which it contacts the vapours from the vapouriser. Practically all the glycerol left in the vapours leaving the main condenser is separated as a 90–94% glycerol fraction in a cold glycerine direct contact condenser. This fraction, which needs to be reprocessed, amounts to 6–10% of the total glycerol fed.

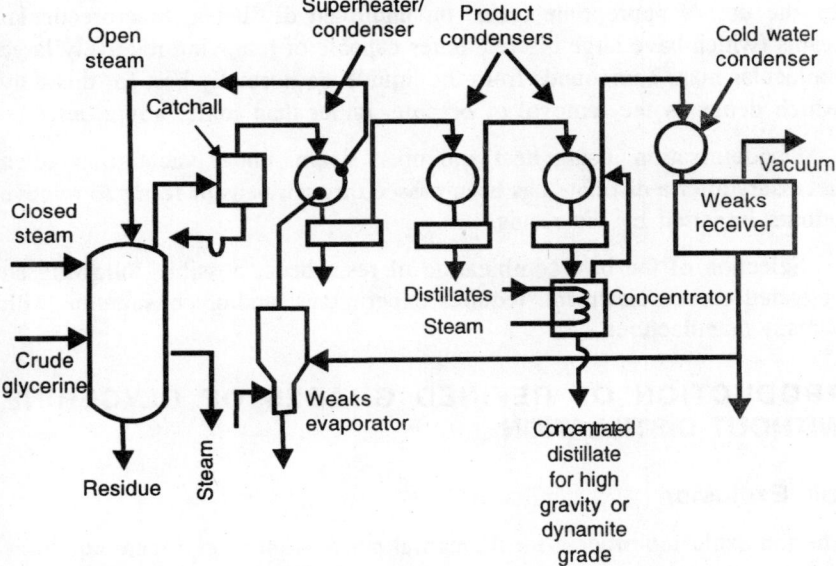

Fig. 8.20. Garrigue still.

BLEACHING AND THE TREATMENT OF DISTILLATES WITH ION-EXCHANGE RESINS TO UPGRADE QUALITY

Traditionally, distilled glycerines are bleached batchwise by an addition of 0.2–2.0% of activated carbon. The mixture at about 80°C is stirred, held for 1–2 hours, and filtered through a press. The carbon retains a high proportion of glycerol even after it has been steamed; and it may be worthwhile to slurry the used carbon with condensed water, and to re-filter, wash, steam and/or air blow in the press to recover weak glycerol. It is also possible to re-activate the washed carbon by stoving it in lidded crucibles in an oven (although to a lower level of activity than that of new activated carbon). This recovery process is not necessarily economic. Some bleaching can also be achieved by pumping distillates through the carbon cake in a filter press. It is possible to bleach glycerines in continuous columns which contain beds of granular carbon.

A relatively recent process is to improve the quality of distillates by the use of ion-exchange resins instead of, or after, activated carbon. At one time it was thought necessary to dilute the distillate to about 80% glycerol, and to reconcentrate by evaporation of water after treatment with the resins, but it has since been found that useful purification can be achieved by the use of appropriate resins on undiluted distillates. Macrorecticular resins (which have large discrete pores capable of removing relatively large molecular mass compounds from the liquid) are normally best for this duty which demands the removal of organic, rather than ionic, impurities.

Deodourisation, using heat and open steam, under vacuum, is often necessary after a distillate has been passed through beds of resins to remove odours imparted by the resins.

Selection of the best combination of resin beds, possibly following an activated carbon treatment, requires experiments, and/or consultation with a resin manufacturer.

PRODUCTION OF REFINED GRADES OF GLYCERINE WITHOUT DISTILLATION

Ion Exclusion

The ion exclusion process for the partial purification of glycerine liquors is based on the fact that in a bed of granular resin (50–100 mesh) of a suitable type, such an 8% cross-linked medium porosity resin consisting of sulphonated styrene-divinyl benzene co-polymer fed with an impure glycerine solution, there are three phases; the resin, the liquor within the resin granules (called the resin liquor), and the surrounding liquor (called the interstitial liquor). Most low molecular mass substances of nonionic, or slightly ionic, nature (such as glycerol) diffuse freely in and out of the resin liquor and the interstitial liquor. Ionic substances, such as sodium chloride and sodium sulphate, are repelled by the ionic charges of the resin, and exist at a much lower concentration in the resin liquor than in the interstitial liquor. Thus when a mixture of ionic and non-ionic solutes is passed through a resin column and is followed by a water rinse, ionised solutes are excluded from the resin granules and leave the column first. The nonionic material is held back and appears in the rinse water. By the use of properly balanced flows of impure glycerin and of rinse water a partial purification can be achieved, as shown in Fig. 8.21. Purification can be completed by ion-exchange. It is said that Indian Pharmacopoeia grade can be produced from saponification crude glycerine without any distillation.

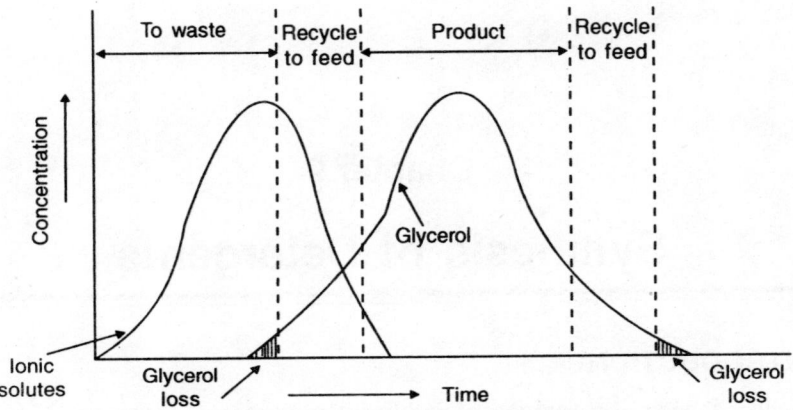

Fig. 8.21. Partial purification of glycerine liquors by ion-exclusion.

Ion Exchange

Glycerine liquors can be purified by the use of ion exchange resins in a process similar to that used for the de-mineralisation of water. The cations, Na^+, Ca^{2+} and Mg^{2+} are exchanged for H^+ in a bed of cation resin (regenerated with hydrochloric or sulphuric acid); and the anions Cl^- and SO_4^{2-} (or their acids) are removed by an anion resin (regenerated with caustic soda). Organic impurities are also removed. Because of the quantities of resins and of regenerating chemicals needed, the process is impracticable for soap lyes, but it was at one time considered at least marginally economical for soap lye crude glycerine diluted to 25 – 30% glycerol (much of the salt having been removed in the evaporation of purified lyes to crude glycerine). The weak base anion resins are most economical for the removal of the strong acids, hydrochloric and sulphuric, but they need to be 'backed' by a strong base anion resin to remove weak organic acids.

Reverse Osmosis Using Semi-permeable Membranes

Experiments have shown that weak glycerol liquors can be concentrated some ten-fold by forcing water through a suitable cellulose acetate membrane under a pressure of 1500 psig (103 bar), with a recovery of 96% in the concentrate. This is an application of reverse osmosis in which preferential sorbtion of water by the membrane, as well as flow through capillary pores, is involved. This use of reverse osmosis might, perhaps, provide a basis for the recovery of glycerol from liquors which would otherwise be too weak to be processed.

Chapter 9

Synthesis of Detergents

INTRODUCTION

In this chapter descriptions of industrial processes for the synthesis of detergents are discussed. Sulphonation with sulphuric acid or oleum of various free sulphurtrioxide strength plays an important role in synthesis. The advantages by using sulphur trioxide for sulphonation is the additional asset that its oxygen content is low compared with vapourised liquid sulphur trioxide, where air acts as a carrier gas. The synthesis of these materials can be done with relatively simple equipment a glass-lined or stainless steel reactor equipped with a dual purpose heating/cooling jacket or stainless steel dissolving tank.

RAW MATERIALS FOR ANIONIC SYNTHETIC DETERGENTS

Fatty Alcohols, Natural and Synthetic

The process used for the reduction of the glyceride to the alcohol (it also works on esters) iss reduction by metallic sodium in the presence of a reducing (hydrogen donating) lower alcohol.

The reaction can be written :

$$RCOOR' + 4Na + 2R''OH \rightarrow RCH_2ONa + R'ONa + 2R''ONa$$

and the sodium alcoholates are then hydrolysed with water and fractionated. The reducing alcohols can be butanol, cyclohexanol or 4-methyl-2-pentanol. The use of metallic sodium requires that all the reactants be completely anhydrous.

The reduction is only on the ester linkage and all double bonds originally present remain intact.

With the development of high-pressure hydrogenation in the organic process industry and its application to the 'oil-hardening' process for natural oils, attention is directed to the conversion of fats and oils by this means. Catalysts and process conditions are rapidly developed and the sodium reduction process fell into disfavour.

High-pressure hydrogenation was originally carried in batch processing units. More recently, however, continuous plants have been introduced. Such plants may use fats, fatty acids or preferably methyl esters. Pressure is high, between 30 and 32 MPa, and the temperature is between 300 and 320°C. Earlier, somewhat lower pressures but even higher temperatures had been used. However, for product quality it is better to use a higher pressure and a lower temperature for the reaction. In this way residence time is reduced to a minimum and the temperature not higher than 320°C maximum. The catalyst is usually copper chromite, sometimes promoted by cadmium or cerium compounds. The catalyst is well dispersed in the fatty acids or methyl ester, and dosed by dosing pump into the reaction system proportionally to the required hydrogen. Sometimes it is recommended to pre-mix the catalyst with a fraction of the fatty alcohol obtained in the process. As can be seen from the flow diagram (Fig. 9.1) part of the spent catalyst is recycled and fresh catalyst added.

Typical process data are as follows :

For 1000 kg tallow fatty alcohol from distilled tallow fatty acids :
    1060 kg distilled fatty acids

    5–8 kg catalyst

    hydrogen 270 m^3

    cooling water 50 m^3

    electricity 350 kWh

    yield 94 per cent calculated on the fatty acids

The figures will vary slightly for coconut fatty acids or their methyl ester owing to their lower molecular weight. They would require somewhat more hydrogen and also slightly more electricity, steam and cooling water. Approximately 25–30 per cent more for coconut fatty acid, some 20 per cent more for coconut methyl ester. The yield would be 92.5 per cent calculated on distilled coconut fatty acids.

During the hydrogenation process, the unsaturated fatty acids are transformed into saturated fatty acids of the same molecular weight, and, of course, the amount of hydrogen consumed will be correspondingly higher for unsaturated fatty acids than for saturated fatty acids or their methyl esters.

By altering the process conditions somewhat and using a selective catalyst ($Al_2O_3/CdO/Cr_2O_3$) the double bonds present can be retained in the final alcohol. The chemical reaction for high-pressure reduction of fats and oils, or 'hydrogenolysis', as it is often called, is as follows :

$$\underset{\displaystyle R-\overset{\displaystyle O}{\overset{\|}{C}}-OH}{} + 2H_2 \rightarrow RCH_2OH + H_2O$$

The equation and heat of reaction for fatty acid esters of lower alcohols and triglyceride hydrogenation. (The heat of reaction is for the hydrogenation of the COOR group only and not for the transformation of unsaturated double bounds in case unsaturated feedstock is used) are given as under :

$$R_1COOR_2 + 2H_2 \rightarrow R_1CH_2OH + R_2OH$$
$$\Delta H_R \approx -500 \text{ kJ}$$

$$\begin{array}{l} CH_2OCOR \\ | \\ CHOCOR \\ | \\ CH_2OCOR \end{array} + 8H_2 \rightarrow \begin{array}{l} CH_3 \\ | \\ CHOH \\ | \\ CH_3 \end{array} + 3RCH_2OH + 2H_2O$$
$$\Delta H_R \approx -105 \text{ kJ}$$

Virtually all plants for the production of alcohols by high-pressure hydrogenation use as starting materials the methyl esters. These have now become an important intermediate in the industry for at least two further processes.

Fig. 9.1 shows a detailed flow sheet for high-pressure hydrogenation. During past, the plants using both OXO and Ziegler chemistry came on stream to manufacture synthetic alcohols. These were based on petroleum and with the low price of crude oil at the time, synthetically produced alcohols quickly surpassed natural alcohols in volume.

In essence, the OXO process uses an olefin, carbon monoxide and hydrogen as the raw materials to form an aldehyde with one carbon more than was originally present in the olefin thus :

$$R-CH = CH_2 + CO + H_2 \xrightarrow[\text{high pressure}]{\text{catalyst}} \begin{array}{l} R-CH_2-CH_2-CHO \\ \text{or} \\ R-CH_2-CH-CHO \\ | \\ CH_3 \end{array}$$

and this aldehyde is then reduced to an alcohol.

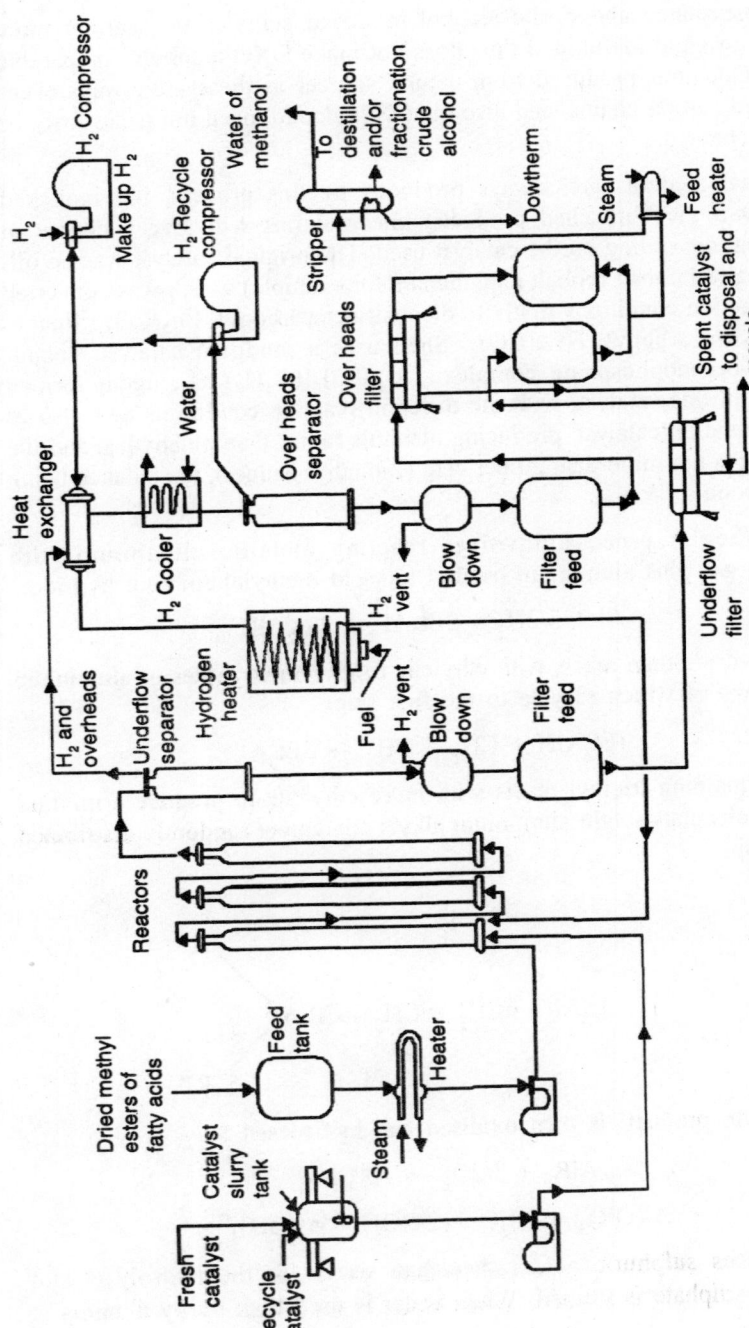

Fig. 9.1. Flow chart for high-pressure hydrogenolysis to obtain fatty alcohols from natural fats.

As mentioned above, the alcohol produced contains one carbon more than the original olefin, and thus does not make OXO alcohols comparable with the alcohols produced from natural sources as these are always even-numbered carbon chains, and also the OXO alcohol need not necessarily be straight-chained.

Normal alcohols are always produced by this process, but branched olefins will give branched products, linear olefins will give a degree of branching, depending on the catalyst used. The original catalyst was an oil-soluble salt of cobalt (cobalt naphthenate for example) which was converted in the reaction conditions firstly to dicobalt octacarbonyl, $Co_2(CO)_8$, then to cobalt hydrocarbonyl, $HCo(CO)_4$. Shell uses a modified catalyst, cobalt-carbonyl-organophosphate complex, $Co_2(CO)_6[(C_4H_9)_2P]_2$, again formed *in situ*, and this catalyst, with the different reaction conditions acts also as a hydrogenation catalyst, producing alcohols rather than aldehydes, and the branching is not more than 20 per cent (2-methyl isomer), the balance being linear alcohols.

The Ziegler process, involves reacting metallic aluminium with hydrogen gas plus aluminium triethyl to yield diethylaluminium hydride:

$$Al + 3/2H_2 + 2Et_3Al \rightarrow 3Et_2AlH$$

The hydride then reacts with ethylene to give three moles of aluminium triethyl, two of which recycle to the first step:

$$3Et_2AlH + CH_2 = CH_2 \rightarrow 3Et_3Al$$

The remaining triethyl reacts with more ethylene to produce a mixture of high-molecular-weight aluminium alkyls containing randomly distributed alkyl groups :

$$Et_3Al + nCH_2 = CH_2 \rightarrow R.Al\begin{array}{c} R' \\ \diagdown \\ R'' \end{array}$$

This 'growth product' is then oxidised and hydrolysed :

$$AlR_3 + 3/2O_2 \rightarrow Al(OR)_3$$

$$Al(OR)_3 + 3H_2O \rightarrow 3ROH + Al(OH)_3$$

Ethyl uses sulphuric acid rather than water for the hydrolysis and aluminium sulphate is formed. When water is used high purity alumina is obtained.

The alcohols produced can range from C_6 to higher than C_{20}, depending on the process conditions. They are normally grouped round a peak following a Poison distribution. Ethyl has modified the conditions to produce 'controlled peaking'. It is surmised that the company makes use of the fact that aluminium alkyls can crack at high temperatures forming dialkyl aluminium and an alpha olefin. These olefins are separated into high and low boiling fractions, the lower fraction is sent upstream for further growth and the higher fraction to another reactor for 'transalkylation', both reactors contacting the incoming streams of olefin with aluminium trialkyl. A narrower chain distribution of the final alcohol mixture is thus obtained. Table 9.1 shows a comparison of the distribution of alcohols produced by the normal Ziegler type plant and the Ethyl modification.

Table 9.1. Comparison of alcohol distribution from two types of Ziegler production units.

Carbon length	Normal distribution	Ethyl controlled peaking
6	9.6	1.4
8	16.9	3.2
10	20.7	7.7
12	19.4	,34.5
14	15.1	26.3
16	9.8	16.7
18	5.3	8.9
20	3.2	1.3

Reaction conditions can be controlled to shift the peak higher or lower, and the C_{14} alcohol (not very abundant in nature) can be produced at will. The alcohols in the C_{12} to C_{20} range are the important ones for the industry, the lower fractions have other uses—solvents, plasticisers, etc. Very seldom is a pure alcohol used as a detergent intermediate, blends are quite sufficient. It has been found that a blend of a (relatively) low-molecular-weight alcohol sulphate or ethoxylate with a higher molecular weight one can have synergistic effects.

In the East European countries petroleum is oxidised to fatty acids and then reduced to alcohols. These alcohols are straight-chain random secondary, not suitable for sulphonation but can be ethoxylated to non-ionic detergents.

The fatty alcohols, either natural or fully synthetic, are used for three main synthetic detergent processes.

(i)   Alcohol sulphates

$$CH_3(CH_2)_nCH_2OSO_3^-Na^+$$

(ii)   Alcohol ethoxylates

$$CH_3(CH_2)_nCH_2O(CH_2CH_2O)_mCH_2CH_2OH$$

(iii)   Alcohol ether sulphates

$$CH_3(CH_2)_nCH_2O(CH_2CH_2O)_mCH_2CH_2OSO_3^-Na^+$$

Tables 9.2, 9.3, 9.4 and 9.5 give the specifications of commercial fatty alcohols produced by high-pressure hydrogenation of natural fats and oils, and of selected Ziegler alcohols.

Table 9.2. Properties of typical commercial fatty alcohols derived from oils and fats.

Name	Number of C-atoms	Iodine number	Solidification point °C	Boiling range °C/ 101 kPa	Hydroxyl number
Lauryl alcohol wide-range type	C_{10}—C_{18}	Less than 0.5	17–21	220–320	275–85
Lauryl alcohol	C_{12}—C_{16}	Less than 0.5	18–22	240–320	280–5
Lauryl alcohol c.80% C_{12}	C_{12}—C_{14}	Less than 0.5	17–23	255–85	283–93
Myristyl alcohol c. 95% C_{14}	C_{14}	Less than 0.5	36–38	280–95	255–62
Cetyl alcohol c. 95% C_{16}	C_{16}	Less than 0.5	46–49	316–30	225–35
Stearyl alcohol c. 95% C_{18}	C_{18}	Less than 0.5	55–57	340–55	203–10
Tallow fatty alcohol	C_{14}—C_{18}	Less than 0.5	48–52	120–90	210–20
Oleyl-cetyl alcohol mixture	Mainly C_{18} saturated and unsaturated	45–120*	4–35*	310–65	200–20

* The more unsaturated the mixture, the higher the iodine value and the lower the solidification point.

The alpha olefins are reacted with dry hydrogen bromide in the presence of peroxides or ultra-violet light to give the primary alkyl bromides, which are then converted to an ester through metal-halide catalysed reaction with organic acid. The ester is then hydrolysed with superheated steam to yield the corresponding primary alcohols. The process has the advantage that anhydrous hydrogen bromide produced in the esterification and the organic acid can both be recycled. The Ziegler process consumes expensive compounds such as tri-butyl aluminium and boron hydride, while the OXO process for the production of secondary alcohols uses a cobalt catalyst over which olefins are passed in the presence of carbon monoxide and hydrogen.

Higher alcohols made by the Ziegler and modified OXO processes have one characteristic which make them particularly useful in detergent manufacture: both anionics and non-ionics produced from them are easily degraded by the bacteria flora in normal sewage treatment plants.

Table 9.3. Properties of pure fatty alcohols.

Name of alcohol	Formula	Mol. weight	Melting-point °C	Boiling-point/ Boiling range °C	Hydroxyl number
Decanol	$C_{10}H_{21}OH$	158	7	231 (101.1 kPa)	355
Undecanol	$C_{11}H_{23}OH$	172	14	131 (1.98 kPa)	326
Dodecanol (lauryl alcohol)	$C_{12}H_{25}OH$	186	24	135–7 (1.33 kPa)	301
Tridecanol	$C_{13}G_{27}OH$	200	30	155 (1.73 kPa)	280
Tetradecanol (myristic alcohol)	$C_{14}H_{22}OH$	214	38	159–61 (1.33 kPa)	262
Pentadecanol	$C_{15}H_{31}OH$	228	44		246
Hexadecanol (palmitic alcohol or cetyl alcohol)	$C_{16}H_{31}OH$	242	49	179–82 (1.59 kPa)	232
Heptadecanol	$C_{17}H_{35}OH$	256	54		219
Octadecanol (stearyl alcohol)	$C_{18}H_{37}OH$	270	58	202 (1.33 kPa)	208
Octadecanol (oleyl alcohol)	$C_{18}H_{35}OH$	268	15–16	177–83 (0.40 kPa)	209

Olefins

Olefins are required as intermediates for three kinds of detergent materials, alpha olefins for the OXO process and for sulphonation, and internal olefins as alkylation reagents for alkylbenzene and alkyl phenol.

Alpha olefins

Cracking of hydrocarbons to produce lighter fractions has been practised for many decades, both by catalytic and thermal processes. For the production of alpha olefins thermal cracking is preferred. The main reaction can be shown :

$$C_nH_{2n+2} \rightarrow C_nH_{2n'} + C_{n''}H_{2n''+2}$$

where $n' + n'' = n$ and n'' is a low integer.

Table 9.4. Typical composition and properties of selected alfol alcohol blends.

	Alfol 1014 CDC	Alfol 1214	Alfol 1412	Alfol 1216	Alfol 1218 DCBA	Alfol 1618
Individual alcohol content %						
C_{10}	31	0.6	0.6	0.3	0.7	–
C_{12}	37	55	38	64	39	–
C_{14}	31	44	60	24	30	1.0
C_{16}	0.8	0.9	1.9	11	18	58
C_{18}	–	–	–	tr.	11	39
C_{20}	–	–	–	–	0.7	2.4
Colour APHA max	0	30	20	30	30	40
Density 22/22°C	0.834	0.836	0.839	0.840	0.840	0.814*
Freezing point °C	5	22	23	20	21	46
Average molecular weight	184	198	203	198	211	253
Flash point PMCC °C	121	129	132	129	135	163
Hydroxyl number	302	277	272	276	260	215
Iodine number	0.07	0.1	0.1	0.1	0.2	0.9

* Density at 60/4°C.

A linear paraffin will produce a linear olefin. Several side reactions are possible but a highly refined paraffin wax will yield 80–90 per cent alpha olefins with a small amount of branched olefins, some internal and diolefins.

The refined wax is pre-heated, mixed with steam (to avoid coking) and fed to a cracking furnace at a temperature of about 550°C under pressure two to four times atmospheric. A low residence time is necessary (about 10 seconds), this maximises alpha olefin production, and cracking conversion is kept low (20–40 per cent per pass).

Table 9.5. Typical composition and properties of selected epal blends.

	Epal 12/85	Epal 1214	Epal 1218	Epal 1416	Epal 1418	Epal 1618
Individual alcohol content %						
C_{10}	0.1	–	0.3	–	–	–
C_{12}	86	66	48	0.3	0.7	–
C_{14}	13	27	20	62	35	1.5
C_{16}	0.01	7	17	36	40	47
C_{18}	–	–	14	1.4	23	50
C_{20}	–	–	0.3	–	0.8	2
Colour APHA max	5	5	5	5	5	5
Density	0.831*	0.831*	0.834*	0.825†	0.825†	0.819‡
Iodine value	0.02	0.02	0.34	0.2	0.4	1.0
Hydroxyl number	297	283	266	250	235	219
Average molecular weight	189	198	211	224	239	256
Flash point PMCC °C	137	137	125	143	149	202
Freezing point °C	20	22	25	37	42	46

* At 25°C.
† At 40°C.
‡ At 50°C.

The vapour exiting from the furnace is quenched with water and the uncondensed vapour is separated from the liquid in a flash tower. These vapours are then fractionated to remove the (unwanted) lighter fraction and distilled into the various commercial cuts.

The quenched liquid is recycled to the cracker. The light fraction contains ethylene, propylene and butylene, and needs to be disposed of economically, possibly with a link to a lower olefin plant.

The specifications for linear alpha olefins of wax origin indicate that the branched alpha olefin content is fractional, internal olefins can be from 5 to 10 per cent, di-olefins upto 4 per cent, saturates upto 4 per cent with fractional amounts of aromatics. If it is desired to sulphonate wax-cracked olefins, the exact composition should be considered as some of the extraneous constituents can adversely affect the quality of the sulphonate. The olefins will of course have both even and odd numbered carbon chains.

As mentioned in describing the Ethyl modification of the Ziegler process for the manufacture of alcohols, olefins can also be produced by this growth reaction. After the aluminium alkyl has grown to the required length, the temperature is raised and a low-molecular-weight olefin (ethylene) is added. The ethylene replaces an alkyl from the aluminium trialkyl, yielding an alpha olefin.

Olefins produced by this method are of even numbered carbon lengths. Table 9.6 indicates the typical values for detergent grade olefins produced by ethyl. It will be noted that for the higher molecular weight products the amount of branched terminal olefins is high. Despite this the sulphonated alpha olefin produced from the olefin biodegrades at a faster rate than LAS.

Table 9.6. Typical composition and properties of selected ethyl alpha olefins.

	Tetra decene-1	Hexa decene-1	Octa decene-1	C12-14	C14-16	C16-18
Carbon no. %						
C_{10}	–	–	–	0.3	–	–
C_{12}	2	–	–	62	1	–
C_{14}	96	2	–	36	65	1
C_{16}	2	97.5	18	1.2	33	55
C_{18}	–	0.5	81	–	1	36
C_{20}	–	–	1	–	–	8
Mono-olefin %	99.6	99.1	98.9	99.6	99.6	99.2
Paraffin %	0.4	0.9	1.1	0.4	0.4	0.8
Olefin isomers mol %						
Linear terminal	81	72	59	87	76	63
Branched terminal	14	21	35	9	19	29
Linear internal	5	7	6	4	5	8
Colour APHA	– – – – – – – – All less than 5 – – – – – – – – –					
Density 20°C	0.760	0.780	0.788	0.764	0.776	0.787
Flash point °C	107	128	143	81	113	135
Boiling range °C						
IBP	245	276	298	216	245	285
FBP	250	283	316	250	279	316

Linear Internal Olefins

Internal olefins are one of the preferred feedstocks for the alkylation of benzene as these are random in nature. It has been pointed out that the presence of 2- and 3-phenyl alkanes impairs the detersive effects of the final sulphonate, whereas by increasing the 5-, 6- and 7- isomers the detergency is increased. Internal olefins can also theoretically be used as feedstock for OXO alcohol production but here the alpha type is preferred.

Kerosene is the normal base material but the term is very loose as refineries produce to physical rather than to chemical constants. The chemical characteristics depend on the crude being used, but a kerosene with a median boiling point of 220°C will contain upto 25 per cent normal paraffins of C_{11} to C_{16} chain length, with aromatics, iso-paraffins and cyclic naphthenes. If substantially free of olefins, this kerosene is suitable for processing. The absence of olefins is necessary as the dehydrogenation step will convert to any olefin present to the undesirable di-olefin. The feedstock needs to be hydrotreated to remove any impurities which would impair the catalyst used in subsequent operations. Depending on the molecular weight required of the final linear alkylate, the hydrotreated kerosene is first fractionated to either C_{10}–C_{13} or C_{11}–C_{14} fractions or any other ranges the market calls for.

Normal paraffins are then separated by selective adsorption processes:

Urea has the property of adsorbing to itself straight-chain hydrocarbons with at least 7 carbon atoms in the molecule. It is suggested that the mechanism is that the urea molecules wrap themselves round the hydrocarbon molecule in a hexagonal spiral and these spirals form channels which can accommodate straight-chain molecules but not the branched ones. In practice urea can adsorb some 30 per cent of its own weight of hydrocarbon.

An activator is necessary for this action to take place. Suitable activators are methanol and the lower ketones.

The procedure basically is to add the kerosene to the solution of the urea in the activator with stirring. The hydrocarbon/urea complex crystallises (the reaction is not instantaneous), when it is filtered, washed with a short-chain paraffin solvent (less than 7 carbons) and the n-paraffin is released from the complex by hot water.

Variations of this basic method are the use of continuous processes and decomposition of the complex by non-aqueous solvents. Many patents have been granted for these variations.

These normal paraffins can now be used for alkylation (via chlorination) or be dehydrogenated to internal olefins, also as the alkylation reagent. Dehydrogenation of the paraffin to an internal olefin is an endothermic, catalytic process.

In brief, the process involves passing the vapourised feedstock, together with recycle, through a fixed bed reactor where only part of the paraffin is dehydrogenated to linear mono-olefins. Hydrogen is produced as a by-product, and the olefin/paraffin mixture is separated from light materials by fractionation. The olefins are separated from the unreacted paraffins by selective solvents, the paraffin being fed back to recycle. If the olefin is to be used for alkylation, this final separation need not be too thorough as the paraffin can act as a diluent in the reaction and can easily be separated after the alkylation, then recycled. Alkylation units therefore normally start with an olefin unit, if not with a linear paraffin unit.

The catalyst is the important point in this operation and new and efficient ones have been developed capable of working for long periods without the need for regeneration.

Normal Paraffins

For the production of linear internal olefins, normal paraffins are required. When paraffins are required as feedstock for the production of alkane sulphonates, the aromatic content needs to be brought down to below 50 ppm, otherwise undesirable side reactions will occur in the sulphoxidation process.

The removal of aromatics can be accomplished by two methods, hydrogenation, when they are converted to cycloparaffins, which remain in the liquid and do not increase the n-paraffin content, or by sulphonation and removal as the sulphonic acid.

Alkyl Benzene

Among the many different synthetic detergents the alkyl aryl sulphonates are the most important. There are many important reasons for this. Not only are the alkyl aryl sulphonates outstanding in their properties as detergents, but they are also based on raw materials which are easier to obtain and cheaper than those on which the other types of detergents are based.

The main source of alkyl aryl sulphonates is the petroleum industry, but before going into details about raw materials and methods of manufacture, it will be convenient to deal with their chemical structure.

As the name implies, these products are based on aromatic compounds combined with an aliphatic chain bound to the aromatic nucleus. Thus, dodecyl benzene, which is the most important type of alkyl aryl condensate, has the following structural formula :

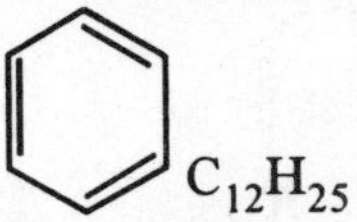

The nature of the propylene polymerisation is such that the tetramer (also the trimer when used) has a branched chain. When the problems with biodegradation arose the cause was quickly pinpointed to this branched chain structure and it was proved that linear alkyl benzene sulphonates (LAS) are biodegradable. The propylene tetramer type is still manufactured for use in countries which have not as yet legislated against it and for specialised purposes such as agricultural emulsifiers, where the problems of re-formulating, both administrative and technical, are great.

The chemistry of the alkylation is represented by the equations :

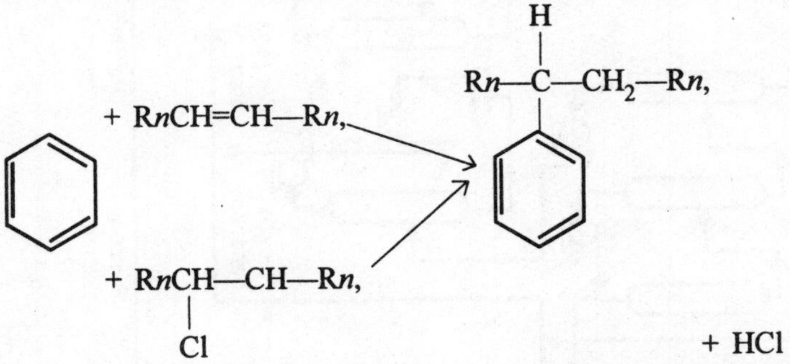

Certain side reactions can and do take place; dimerisation of the olefin, alkylation of this dimer, di-alkylation of the benzene, possibly di-phenylation of the olefin or chlorparaffin and cyclo-alkylation of the benzene. Thus in the reactor together with the alkylbenzene, we can find di-phenyl alkane, di-alkylbenzene, di(alkylbenzene), all in the ortho- or para-positions:

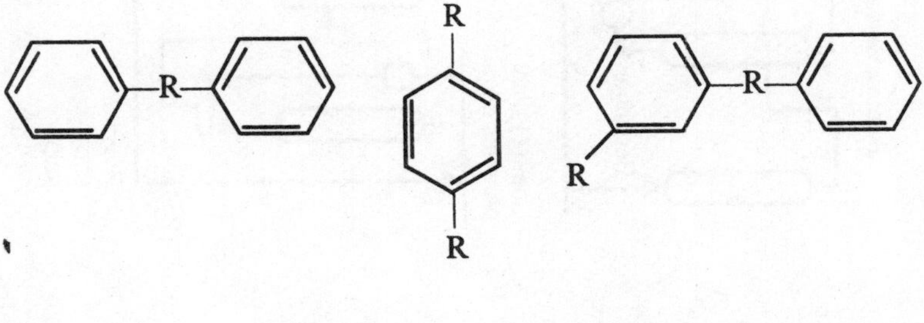

and also alkylbenzene with a side chain consisting of the dimer of the olefin. Although we have mentioned mainly the reactions of an olefin, the 'heavy alkylate' from chlorparaffin alkylation is surprisingly similar to that of the olefin based product.

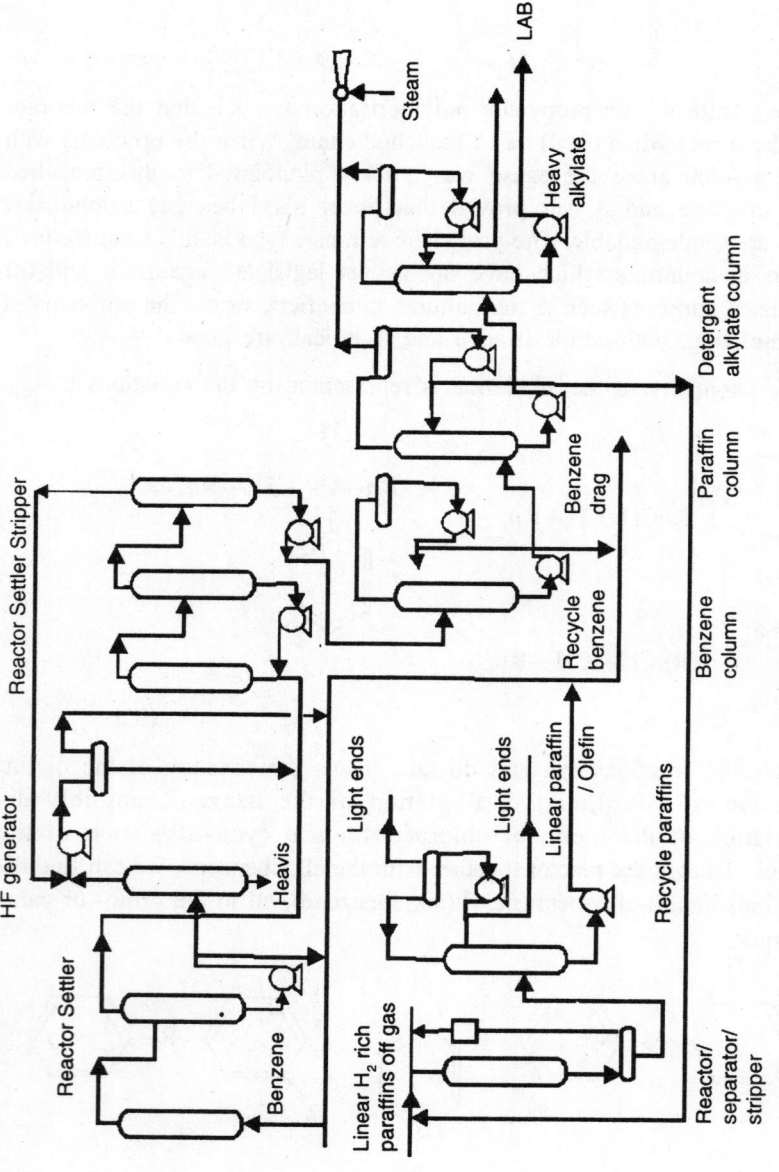

Fig. 9.2. Production of linear alkyl benzene schematic flow (UOP process).

The typical properties of linear alkyl benzene are given in Table 9.7.

Table 9.7. Typical properties of linear alkyl benzene (LAB).

Bromine no.	0.02
Saybolt colour	+ 30
Alkyl benzene content	97.4%
Doctor test	Negative
Unsulphonatable matter	1.0%
Water	0.1%
Specific gravity 15°C	0.8612
Refractive index $n_D 20$	1.4837
Flash point	138°C
Average molecular weight	240
Distillation range °C	
IBP	281
50 vol %	290
EP	309
2-Phenyl isomer	20%
Biodegradability (ASTM D-2667)	> 95%

The paraffinic C_{12-14} petroleum fraction) the HCl is recovered during the dehydrochlorination process, in the Rheinpreussen process HCl from the monochlorinated paraffinic petroleum fraction (or a synthetic paraffin fraction from the Fischer-Tropsch process) is recovered during alkylation. The idea of using pure aluminium rather than the hygroscopic and corrosive aluminium trichloride is rather attractive and has become fairly universal in the industry when chlorparaffins are used.

The material balance for the conditions of reaction is given as :

Raw materials to produce 1000 kg linear alkyl benzene :

	kg
n-paraffin	76
Benzene	359
Chlorine gas	368

By-products produced :

Heavy alkylate	81
HCl	365

Chemicals used :

Aluminium powder	1
Flake caustic soda	5
Caustic soda solution (50 per cent)	74

Methyl Esters

Methyl esters are used in the industry as an intermediate for three purposes, high-pressure hydrogenation of fats to produce fatty alcohols, production of 'superamides' and as the raw material for sulphonation to produce the sulphonated methyl esters. Each of these has its own requirements as to quality, etc. In addition, with the present low price of methanol, methyl esters are being mooted as the raw material for soap production.

For converting a fat or oil into a methyl esters, certain pretreatments are necessary. No matter what the end use, the fat needs to be de-gummed to remove both gums and other extraneous matter. This is usually achieved by contacting the fat with slightly diluted sulphuric acid. The water, acid, coagulated gums and dirt are then separated by settling. If the colour of the fat is poor it is then treated with an absorbent earth and filtered.

For sulphonation the fat needs to be hydrogenated or to have an intrinsic low iodine value. The hydrogenation is of the normal fat hardening kind, not as described in the production of alcohols, i.e., using a nickel catalyst at moderate pressure. For the preparation of alkanolamides the fat is not normally hydrogenated unless special properties are sought.

Thirdly, the free fatty acid in the oil must be reduced to less than 0.4 acid value as the presence of the fatty acids will react with and impair the performance of the catalyst. The free fatty acid reduction can be achieved by one of the conventional methods used in the production of edible oils, using a caustic wash or steam distillation under vacuum.

An alternative method of reducing free fatty acid for methylester production is by pre-esterification of the ffa with methanol thus transforming the ffa into methylesters prior to the transesterification process.

Pre-esterification may either be carried out in a simple batch process or, alternatively, in a continuous system under pressure.

Distillation of excess methanol required for esterification, is a simple operation; methanol for esterification of fatty acids needs not to be completely anhydrous, as for the transesterification process, and a simpler rectification may be carried out independently from the recirculation system in the transesterification unit.

The classical ester-interchange alcoholysis reaction occurs when pre-refined or pre-esterified oils and fats are reacted with anhydrous methanol in the presence of an alkaline catalyst. In order to obtain maximum conversion into methylesters, the following conditions have to be maintained: the fat and oil must be degummed and deacidified to reduce the acid value below 0.4. Methanol must be anhydrous (max. 0.3 water).

The catalyst of choice is sodium methylate. The catalyst originally used is dry NaOH or KOH, dissolved in methanol.

During extensive laboratory and pilot work, a conversion rate as good as with sodium methylate, when a conversion rate or at least 95 per cent could easily be obtained, provided the sodium methylate was fresh and the methanol practically anhydrous.

Sodium methylate powder should however immediately be dissolved in methanol as a stock solution, otherwise it very quickly absorbs water and even CO_2 from the air, thereby reducing its efficiency as a catalyst.

Two alternatives exist to separate the glycerine layer set free after completion of the interchange reaction. One is the classical method, namely to separate it from the reaction mixture in the form of crude glycerine containing the soap formed by the catalyst, some methylesters and methanol.

This glycerine layer then undergoes a prerefining step: distilling off methanol, splitting the soap with acid and separating the fatty acids and some methylester on top of a glycerine solution slightly diluted with water to give a prerefined glycerine of about 70 per cent concentration, which can of course easily be brought upto a higher concentration.

The alternative method to separate glycerine is as follows: after the transesterification reaction is terminated, excess anhydrous methanol is distilled, and recovered for recirculation from the total reaction mixture containing methylester, glycerine, soap formed by the catalyst and some small percentage of unconverted triglyceride, at about 70°C. A low temperature is recommended in order to prevent any reverse reaction of methylester with glycerine. The last traces of methanol are then stripped off preferably under reduced pressure.

When all the methanol has been removed from the reaction mixture, about 3 per cent water containing sufficient acid to neutralise any free sodium methylate and to split the soap formed, is added.

Glycerine separates rapidly from the methylester in the form of an about 70 per cent glycerine solution. The sodium salt content of the glycerine solution, either sodium sulphate or chloride, is less than the salt content of crude glycerine obtained from soap spent-lye.

Organic matter, too, is less. The methylester may be washed free from last traces of glycerine and salt by again adding about 3 per cent water. This rinse is again made up with acid and used for the next process batch.

Obviously, when using this system for glycerine recovery, the methylester contains a small percentage of free fatty acid (ffa) from the split soap. However, when the methylester is used for alkanolamide production or for soap, the presence of ffa does not represent a disadvantage. Only when the methylester has to undergo distillation and the distilled methylester is marketed as a commodity, the acid value has to be reduced to a low level below 1.

This can easily be accomplished by conventional alkali or carbonate neutralisation, a process which is not much more difficult, than washing the methyl ester free from glycerine and soap in the conventional process.

Alternatively, it was found that, on neutralising the ffa with monoethanolamine prior to distillation, no foam problems occur and no highly viscous sticky bottom fraction is obtained. The modified method of separating glycerine is especially attractive, when transesterification is carried out continuously. Fig. 9.3 shows a batch system for transesterification.

The components, fats and oils, sodium methylate stock solution in methanol, and methanol are introduced, by automatic batching, into the reactor. The reaction mixture is kept boiling under reflux at about 70°C for about two hours, when conversion of triglyceride into methyl ester reaches 95 per cent for tallow and even 97 per cent for coconut or palmkernel oil. Subsequent processing steps have already been described.

In the continuous process, Fig. 9.4 which can also be carried out under pressure, residence time is considerably reduced; the components are dosed by triple-head dosing pump, the fat or oil passes through a preheater, so as to give a temperature of the total mixture, catalyst stock solution, methanol and fat of about 70°C in the static mixer.

From there the reaction mixture enters the reactor. The temperature is above 70°C when the alcoholysis is carried out at atmospheric pressure, and higher when carried out under pressure.

Correlation between temperature and pressure is, e.g., 3.4 bar at 100°C, 10.4 bar at 140°C. From a practical constructional and operational point of view, operation at 5–7 bar at a temperature around 120°C would be preferable.

From the reaction column, the reaction mixture enters the distillation column for distilling off the excess anhydrous methanol for recirculation via the surge tank.

Splitting off the soap, separation of glycerine and washing are generally performed discontinuously. It may be mentioned that the batch and the

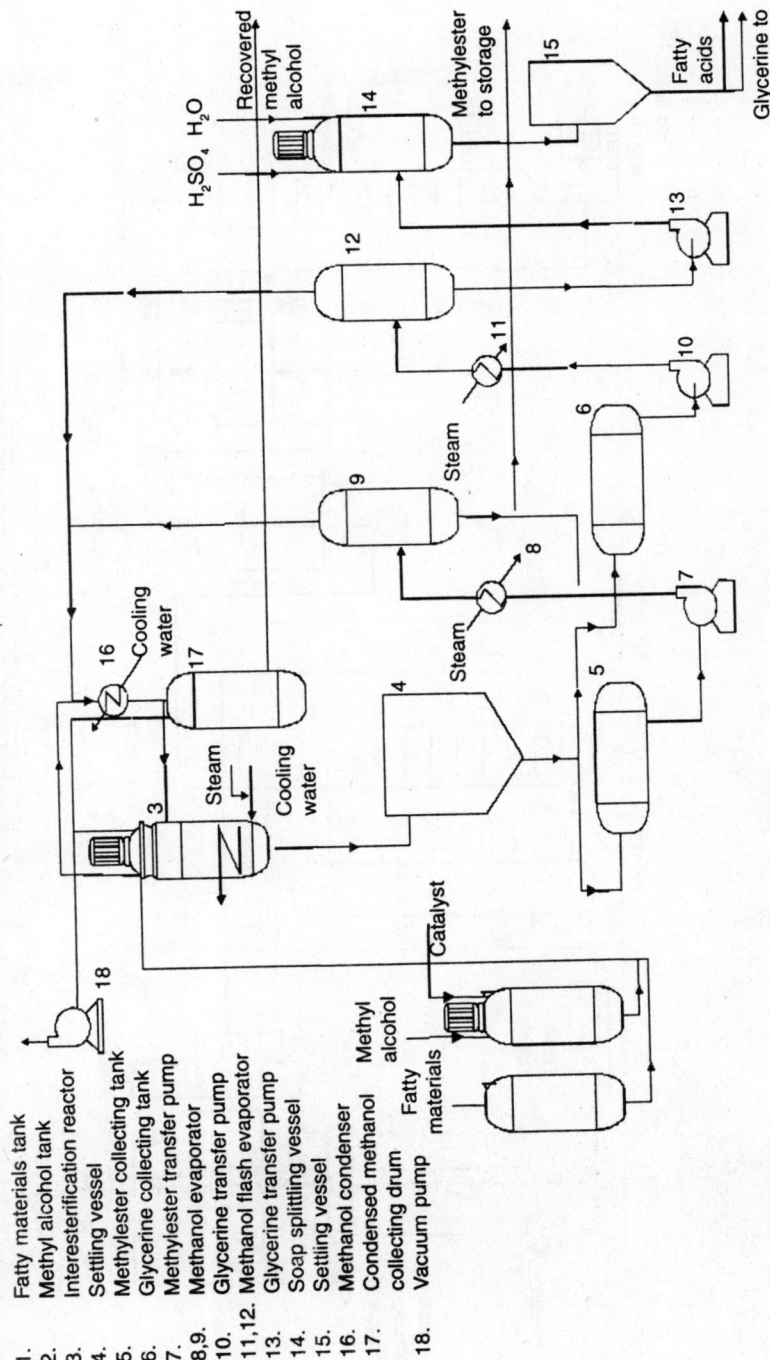

1.   Fatty materials tank
2.   Methyl alcohol tank
3.   Interesterification reactor
4.   Settling vessel
5.   Methylester collecting tank
6.   Glycerine collecting tank
7.   Methylester transfer pump
8,9. Methanol evaporator
10.   Glycerine transfer pump
11,12. Methanol flash evaporator
13.   Glycerine transfer pump
14.   Soap splitting vessel
15.   Settling vessel
16.   Methanol condenser
17.   Condensed methanol collecting drum
18.   Vacuum pump

Fig. 9.3. Ester interchange process—batch process.

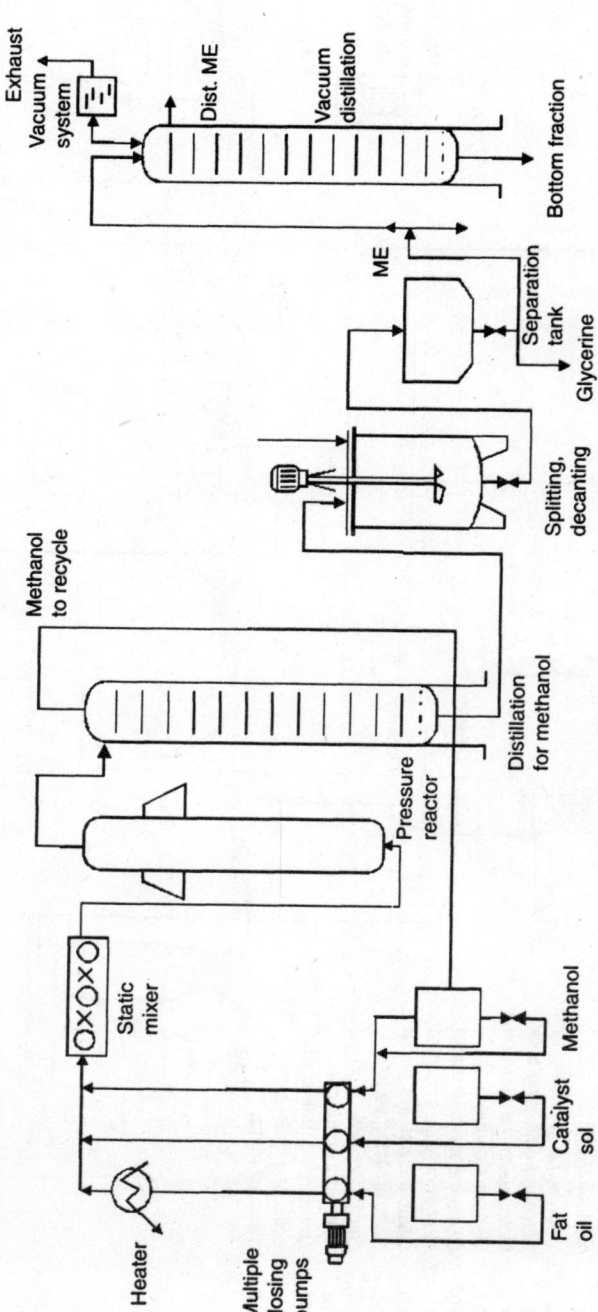

Fig. 9.4. Methyl ester production—continuous system.

continuous transesterification processes should be able to use the two alternate methods separating the glycerine.

SULPHONATION OF DETERGENT RAW MATERIALS

Sulphonation and sulphation are the most important processes for the production of anionic detergents, which, as we have seen, are by far the most important synthetic detergents. In recent years sulphonation processes have been developed which have revolutionised the detergent industry. Sulphonation and sulphation with acids, such as sulphuric acid, oleum of various concentrations, and chlorosulphonic acid, is now being replaced by methods which use gaseous SO_3, obtained either by vapourisation of liquid stabilised SO_3 (sulphuric acid anhydride) or directly as converter gas from a contact sulphuric acid plant. However, sulphonation with acids is still widely practised, so that a description of the more conventional acid-sulphonation processes will also be given in this chapter.

The Inorganic Raw Materials for Sulphonation

A precondition for properly carrying out sulphonation is the knowledge of the inorganic raw materials. We will therefore give a short description of the most important inorganic acid raw materials used in detergent sulphonation processes. (In this context we consider liquid SO_3—i.e., sulphuric acid anhydride—as an acid, too. We will also give the properties of sulphur since this is the starting material for the most modern sulphonation process.)

Sulphuric acid and oleum (fuming sulphuric acid)

The oldest, and still widely used, acid in the detergent industry is sulphuric acid and oleum of various strengths. The chemical formula of sulphuric acid is H_2SO_4, molecular weight 98.08. For sulphonation processes either the so-called monohydrate of 98/99 per cent acid strength is used, or oleum, i.e., sulphuric acid with various percentages of free SO_3 dissolved. The physical properties of more dilute sulphuric acid are of interest for the practical man in the detergent industry, as he might well encounter dilute sulphuric acid in the form of so-called 'spent acid' and in acid cleaners.

The tables show that, especially at a percentage between 97 and 100 per cent sulphuric acid, there is a curious irregularity in density. The specific gravity of an acid of 97 and 98 per cent is identical, and the density of an acid of 99 and 100 per cent strength is practically the same as that of an acid of 94–95 per cent strength. This fact is of great practical importance, because it makes determination of acid strength by simple hydrometer reading impossible within this range of concentration, which unfortunately is the most important concentration range for acid sulphonation processes. There are other physical control methods for the determination of acid strength in the range of 93–100 per cent, such as the measurement of conductivity and 'heat of dilution'. Incidentally, these physical methods have been worked out for the control of SO_2 absorption and air-drying in the classical contact sulphuric acid production processes.

Table 9.8. Properties of sulphuric acid.

Degrees baume	Specific gravity 60°F	% sulphuric acid H_2SO_4	Weight of 1 Imp gal in lb av	Normal boiling-point °C	Freezing (melting)-point °C
0	1.0000	0.00	10.011	100.0	0.00
1	1.0069	1.02	10.079	100.1	− 0.39
2	1.0140	2.08	10.150	100.3	− 0.83
3	1.0211	3.13	10.222	100.4	− 1.28
4	1.0284	4.21	10.295	100.6	− 1.72
5	1.0357	5.28	10.369	100.8	− 2.22
6	1.0432	6.37	10.443	100.9	− 2.77
7	1.0507	7.45	10.517	101.1	− 3.44
8	1.0584	8.55	10.595	101.3	− 4.11
9	1.0662	9.66	10.673	101.5	− 4.72
10	1.0741	10.77	10.752	101.7	− 5.39
11	1.0821	11.89	10.832	101.9	− 6.11
12	1.0902	13.01	10.915	102.2	− 7.00
13	1.0985	14.13	10.996	102.4	− 7.94
14	1.1069	15.25	11.082	102.7	− 8.72
15	1.1154	16.38	11.167	102.9	− 10.00
16	1.1240	17.53	11.251	103.2	− 11.11
17	1.1328	18.71	11.340	103.6	− 12.28
18	1.1417	19.89	11.430	103.9	− 13.61
19	1.1508	21.07	11.521	104.3	− 15.33
20	1.1600	22.25	11.612	104.7	− 17.22

(*contd.*)

Degrees baume	Specific gravity 60°F	% sulphuric acid H_2SO_4	Weight of 1 Imp gal in lb av	Normal boiling-point °C	Freezing (melting)-point °C
21	1.1694	23.43	11.707	105.1	− 19.17
22	1.1789	24.61	11.802	105.4	− 21.39
23	1.1885	25.81	11.898	105.8	− 23.94
24	1.1983	27.03	11.996	106.3	− 26.67
25	1.2083	28.28	12.096	106.8	− 29.94
26	1.2185	29.53	12.198	107.3	− 33.33
27	1.2288	30.79	12.301	107.9	− 37.56
28	1.2393	32.05	12.407	108.6	− 42.39
29	1.2500	33.33	12.513	109.2	− 48.50
30	1.2609	34.63	12.622	110.0	− 56.39
31	1.2719	35.93	12.732	110.9	− 61.50
32	1.2832	37.26	12.845	111.8	− 59.56
33	1.2946	38.58	12.959	112.7	− 57.78
34	1.3063	39.92	13.077	113.6	− 56.27
35	1.3182	41.27	13.197	114.6	− 54.72
36	1.3303	42.63	13.317	115.7	− 42.50
37	1.3426	43.99	13.441	116.9	− 49.44
38	1.3551	45.35	13.565	118.2	− 45.88
39	1.3679	46.72	13.694	119.7	− 42.22
40	1.3810	48.10	13.819	121.2	− 39.00
41	1.3942	49.47	13.952	158.3	− 36.10
42	1.4078	50.87	14.087	124.6	− 33.70
43	1.4216	52.26	14.226	126.4	− 31.90
44	1.4356	53.66	14.365	128.6	− 30.40
45	1.4500	55.07	14.510	130.8	− 29.40
46	1.4646	56.48	14.656	133.1	− 28.50
47	1.4796	57.90	14.806	135.8	− 28.40
48	1.4948	59.32	14.958	138.4	− 28.80
49	1.5104	60.75	15.114	141.4	− 30.00
50	1.5263	62.18	15.274	144.4	− 31.90
51	1.5426	63.66	15.437	147.8	− 34.20
52	1.5591	65.13	15.601	151.1	− 36.80
53	1.5761	66.63	15.772	155.0	− 38.20

(*Contd.*)

Degrees baume	Specific gravity 60°F	% sulphuric acid H_2SO_4	Weight of 1 Imp gal in lb av	Normal boiling-point °C	Freezing (melting)-point °C
54	1.5934	68.13	15.944	159.2	− 39.70
55	1.6111	69.65	16.121	163.6	− 42.80
56	1.6292	71.17	16.302	168.3	− 40.40
57	1.6477	72.75	16.488	174.1	− 39.70
58	1.6667	74.36	16.678	179.7	− 33.60
59	1.6860	75.99	16.872	186.3	− 22.80
60	1.7059	77.67	17.071	193.1	− 11.40
61	1.7262	79.43	17.273	200.6	− 1.50
62	1.7470	81.30	17.482	208.6	− 4.20
63	1.7683	83.34	17.695	218.8	7.60
64	1.7901	85.66	17.914	230.7	7.10
64¼	1.7957	86.33	17.969	234.2	6.10
64½	1.8012	87.04	18.024	238.1	4.40
64¾	1.8068	87.81	18.080	243.4	2.20
65	1.8125	88.65	18.138	247.2	− 0.40
65¼	1.8182	89.55	18.194	252.2	− 4.20
65½	1.8239	90.60	18.252	258.9	− 9.40
65¾	1.8297	91.80	18.310	268.3	− 16.40
66	1.8354	93.19	18.366	279.4	− 29.40

Table 9.9. Properties of Oleum. (Free SO_3 total SO_3 and equivalent H_2SO_4).

% free sulphur trioxide SO_3	% actual sulphuric acid H_2SO_4	% total sulphur trioxide SO_3	% equivalent sulphuric acid H_2SO_4	% free sulphur trioxide SO_3	% actual sulphuric acid H_2SO_4	% total sulphur trioxide SO_3	% equivalent sulphuric acid H_2SO_4
0	100	81.63	100.00	21	79	85.49	104.73
1	99	81.82	100.23	22	78	85.67	104.95
2	98	82.00	100.45	23	77	85.86	105.18
3	97	82.18	100.68	24	76	86.04	105.40
4	96	82.37	100.90	25	75	86.22	105.63
5	95	82.55	101.13	26	74	86.41	105.85
6	94	82.73	101.35	27	73	86.59	106.08
7	93	82.92	101.58	28	72	86.78	106.30
8	92	83.10	101.80	29	71	86.96	106.53
9	91	83.29	102.03	30	70	87.14	106.75

(Contd.)

% free sulphur trioxide SO$_3$	% actual sulphuric acid H$_2$SO$_4$	% total sulphur trioxide SO$_3$	% equivalent sulphuric acid H$_2$SO$_4$	% free sulphur trioxide SO$_3$	% actual sulphuric acid H$_2$SO$_4$	% total sulphur trioxide SO$_3$	% equivalent sulphuric acid H$_2$SO$_4$
10	90	83.47	102.25	31	69	87.33	106.98
11	89	83.65	102.48	32	68	87.51	107.20
12	88	83.84	102.70	33	67	87.69	107.43
13	87	84.02	102.93	34	66	87.88	107.65
14	86	84.20	103.15	35	65	88.06	107.88
15	85	84.39	103.38	36	64	88.24	108.10
16	84	84.57	103.60	37	63	88.43	108.33
17	83	84.75	103.83	38	62	88.61	108.55
18	82	84.94	104.05	39	61	88.80	108.78
19	81	85.12	104.28	40	60	88.98	109.00
20	80	85.31	104.50	41	59	89.16	109.23
42	58	89.35	109.45	72	28	94.86	116.20
43	57	89.53	109.68	73	27	95.04	116.43
44	56	89.71	109.90	74	26	95.22	116.65
45	55	89.90	110.13	75	25	95.41	116.88
46	54	90.08	110.35	76	24	95.59	117.10
47	53	90.27	110.58	77	23	95.78	117.33
48	52	90.45	110.80	78	22	95.96	117.55
49	51	90.63	111.03	79	21	96.14	117.78
50	50	90.82	111.25	80	20	96.33	118.00
51	49	91.00	111.48	81	19	96.51	118.23
52	48	91.18	111.70	82	18	96.69	118.45
53	47	91.37	111.93	83	17	96.88	118.68
54	46	91.55	112.15	84	16	97.06	118.90
55	45	91.73	112.38	85	15	97.25	119.13
56	44	91.92	112.60	86	14	97.43	119.35
57	43	92.10	112.83	87	13	97.61	119.58
58	42	92.29	113.05	88	12	97.79	119.80
59	41	92.47	113.28	89	11	97.98	120.03
60	40	92.65	113.50	90	10	98.16	120.25
61	39	92.84	113.73	91	9	98.35	120.48
62	38	93.02	113.95	92	8	98.53	120.70
63	37	93.20	114.18	93	7	98.71	120.93
64	36	93.39	114.40	94	6	98.90	121.15
65	35	93.57	114.63	95	5	99.08	121.38

Fig. 9.5 showing the freezing point of concentrated sulphuric acid and oleum, again demonstrates another irregularity of great practical importance in handling oleum at various strengths, especially during the colder months. For sulphonation of detergent raw materials, mainly alkyl benzene, grades of oleum with 10–22 per cent free SO_2 are used. It is thus of interest to note that it is within this range of free SO_3 that the greatest irregularity in the freezing temperature occurs. 100 per cent H_2SO_4 has a relatively high freezing point, whereas an oleum of 15–20 per cent has a relatively low freezing point. Oleum with 60 per cent free SO_3 and a rather low freezing point serves as a source of SO_3 in some detergent sulphonation processes.

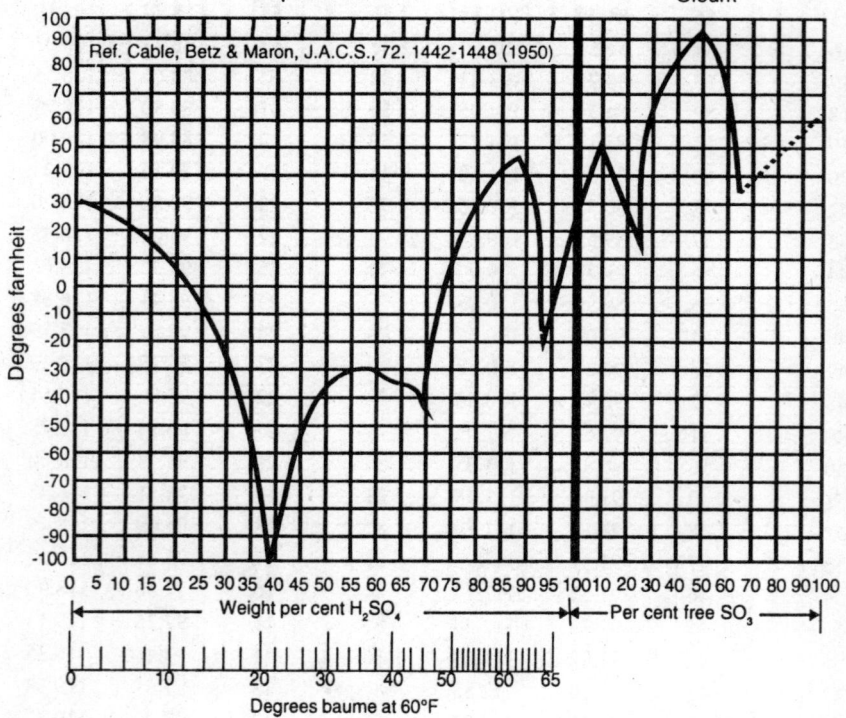

Fig. 9.5. Freezing points of sulphuric acid and oleum.

We emphasize, however, that in many countries explicit regulations exist for the safe handling, storage, etc., of sulphuric acid and oleum. These regulations should be carefully studied and followed by anybody starting a sulphonation plant.

Storage areas should have facilities for drainage and for the washing down of spills with water. Major spills of acid should not be discharged into a sewer, or washed into a sewer or river until they have been neutralised. Where convenient, the use of crushed limestone as a foundation under storage tank areas is considered good practice since it both provides good natural drainage and neutralises the spills before they are discharged into the sewer.

Storage tanks should be equipped with vents of such size as to maintain the tanks at atmospheric pressure. Metal catwalks should be provided for working on top of tanks.

Drums should be stored with the plugs up. The plugs should be loosened upon receipt and every week thereafter, or more frequently depending upon the weather, to release the internal pressure created by the evolution of hydrogen.

Although sulphuric acid is not inflammable, it should not be stored near organic materials, nitrates, carbides, chlorates and metal powders. Contact of high concentrations of sulphuric acid with these materials may cause ignition.

Sulphuric acid in drums, tank cars and metal storage tanks evolves hydrogen. Therefore open lights, flames and spark-producing tools should not be permitted near such containers.

Heating of storage tanks or storage areas is necessary for certain strengths of acid which freeze at common winter temperatures.

Sulphuric acid, like any other corrosive liquid, is dangerous when improperly handled. However, if suitable precautionary steps are taken, proper handling procedures followed, and adequate protective equipment worn, there is relatively little danger in working with this chemical.

The storage of sulphuric acid in bulk storage tanks, and piping it directly to process, is by far the easiest and safest way to handle the acid. This not only results in a considerable saving of space, but eliminates the hazards involved in lifting, transporting and discharging of drums and carboys. With properly designed and maintained equipment, handling hazards are reduced to a minimum.

Before being moved, drums should be inspected for loose plugs, for leakage and for bulging. To remove drum plugs, a pipe wrench or a plug wrench with a long handle should be used. The plug should be turned very slowly for one full turn, and accumulated pressure should then be permitted to vent itself. After the hydrogen is released, the plug may be further loosened and removed.

Drums of sulphuric acid should be emptied by gravity, using a faucet or safety siphon. A pump may also be used. Air pressure should never be used. Since the drums may contain hydrogen gas, they should never be struck with a spark-producing tool.

Employees should wear face shields and rubber gloves when opening drums.

For all practical purposes the dilution of sulphuric acid and/or oleum with water should not be carried out in a detergent plant. There is a strong exothermic reaction, and as a general rule one should never add water to acid. Dilution of acid is a process step during sulphonation, and will be described later in this chapter. (The elimination of spillages has already been mentioned.)

The handling of sulphuric acid and/or oleum as well as of 'spent acid' presents a corrosion problem, too.

Liquid sulphur trioxide (SO$_3$)

Appearance and colour:	Clear to slightly turbid liquid.
	30–50 APHA
Assay:	99.5 per cent SO$_3$
Boiling point:	44.8°C
Freezing point:	16.8°C
Density at 20°C:	1.922 g/ml
Heat of reaction with water:	1.096 kJ/g

Vapour pressure (mm of mercury)	
25°C	265
50°C	950
75°C	3000

Sulphur trioxide in the vapour form always exists as the monomer, but when cooled below its boiling point, in the presence of minute traces of moisture or even on standing for any length of time, it can polymerise to liquid and solid polymers. Thus liquid SO$_3$ can only be transported or used in its stabilised form.

Stabilised liquid sulphur trioxide is chemically the gamma form of SO$_3$. The alpha and beta forms are solids having a higher melting point. Since transformation between these three forms of sulphur trioxide can result in formation of the solid varieties, stabilised liquid SO$_3$ (gamma) which is protected against the formation of high melting-point solids by the addition of stabilisers such as borates or sulphonic acids, is the preferred sulph(on)ating agent.

Gamma form of sulphur trioxide (Trimer)

In the detergent industry, liquid SO_3 is not used directly in its liquid form as a sulphonation agent, but mainly in its vapourised form strongly diluted with dry inert air. It must be mentioned that liquid SO_3 is also used as a liquid sulphonation agent diluted with undercooled liquid SO_2.

Sulphur trioxide from Oleum Stripping

In areas where liquid SO_3 is not readily available (there are stringent restrictions on the marine transport of this type of material), and where for some reason or another a sulphur burning plant is not to be established, SO_3 can be obtained by the stripping of oleum —60 per cent oleum is preferred although 25 per cent can be used.

The method is to distil the oleum, condense the evolved SO_3 gas into a storage vessel and add a stabiliser. Continuous evolution of SO_3 to feed into a reactor cannot be employed because as the SO_3 is removed from the oleum, the rate of evolution will fall, and a constant supply of SO_3 can never be maintained.

The liquid SO_3 thus formed needs to be treated as described above, i.e., to be vapourised prior to feeding into the reactor(s). When the oleum concentration falls to 20 per cent, distillation is discontinued. This 20 per cent oleum has to be disposed of, usually by returning to the sulphuric acid manufacturer for enrichment.

Another possibility is to blow dried, heated air through the liquid, possibly in a heated column, thus obtaining directly, diluted SO_3 gas for sulphonation. To maintain an accurate and constant gas flow, the concentration of SO_3 in the evolved gas should be monitored with a feedback to control the air supply.

As can be seen, problems of corrosion and handling will be enormous, and the plant is in essence two plants, one for vapourisation and the other for treatment of the liquid SO_3. These factors should influence the potential manufacturer to install a sulphur burning plant.

Sulphur

As already mentioned, sulphur is an important raw material for the most modern sulphonation method, which applies converter gas for direct sulphonation of detergent raw materials. It is this process which will be described in much detail later in this chapter. For this process elementary, technically sulphur of high purity (atom wt 32.06, specific gravity 2.03) is required. Three types of sulphur are most commonly used. Crude, run-of-mine 'bright' 99.5 per cent pure sulphur, free of arsenic, selenium and tellurium. Refined sulphur, which is an elemental sulphur produced by distilling crude sulphur, of not less than 99.8 per cent purity. Free-burning; available in lumps or cast sticks. A more recently available variety is the so-called 'gas-sulphur' which is obtained from natural gas or petroleum-refining processes. These types of sulphur solidify at about 114.5°C.

For practical purposes, data on the viscosity of liquid sulphur are important. As will be seen later, an exact pumping of liquid sulphur by dosing pump is the basis of the modern sulphonation process with converter gas. The pumping is best carried out at the lowest viscosity, i.e., 150–155°C (302–311°F). In this connection it must be mentioned that carbonaceous and hydrocarbon impurities tend to cause 'sticking valves' in the dosing pump.

The advantage of sulphur as basic inorganic raw material for sulphonation is obvious indeed. It is safe to handle, practically non-toxic and non-corrosive to the skin. (Irritation of ocular conjunctivate and the mucous membranes of the respiratory organs may occur.) What has to be taken into account, however, is its inflammability. Even so, for all practical purposes of sulphonation, solid sulphur would not have to be ground in a mill, which would represent a certain ignition hazard. Storage of solid sulphur should exclude humidity and ignition hazards. Solid sulphur has to be molten in steam-jacketed vessels.

Chlorosulphonic Acid

This acid is still widely used in the detergent industry for the sulphonation of fatty alcohols, ethylene oxide condensation products, etc. For alkyl benzene it cannot be used, as in the presence of an aromatic ring structure sulpho-chlorination would occur, whereas in all other sulphation procedures using chlorosulphonic acid free hydrochloric acid (HCl) in gaseous form is evolved during the sulphation reaction and has to be absorbed in water, to give liquid HCl of various strengths.

Physical and chemical properties

Chemical formula:	$ClSO_2OH$
Molecular weight:	116.53
Colour:	Colourless
MP:	$- 80.0°C$
BP:	151.5°C at 101.75 kPa
Solubility:	Decomposes in water, decomposes in alcohol, insoluble in CS_2
Index of refraction nD:	1.437 at 14°C
Specific gravity:	1.787 at 25°C
Specific heat:	1.18 J/g, °C, 15°C to 80°C
Heat of vapourisation:	406 kJ/kg
Heat of formation (from elements):	597.1 kJ/mol
Viscosity:	2.5 mPa s at 30°C
Inflammability:	Non-flammable, but may cause ignition of combustible materials on contact. In contact with metals hydrogen gas is formed which can be very explosive in certain mixtures with air.

The physical data listed summarise the bulk of information available on chlorosulphonic acid and are adequate for many engineering calculations. Additional data on a good commercial grade acid are :

Appearance:	Pale yellow liquid
Assay $ClSO_3H$:	99.4 per cent
Iron—(Fe):	< 5.0 ppm
Free SO_3:	0.40 per cent

For handling chlorosulphonic acid it is advisable again to ask the supplier for detailed safety data, and also about materials of construction, etc.

Sulphamic Acid

HSO_3NH_2 (also called amido-sulphonic acid) is occasionally used commercially for specialised sulphonations. It is non-corrosive and the plant required is relatively simple. Alcohol and phenol ethoxylates have been sulphated by sulphamic acid, although results with linear alcohols have not been successful.

Compared to other methods of sulphation, this process is rather expensive. Materials sulphated with sulphamic acid are neutralised naturally to the ammonium salt :

$$R\text{—}O\text{—}CH_2\text{—}CH_2\text{—}OH +$$

$$HSO_3NH_2 \longrightarrow R\text{—}O\text{—}CH_2\text{—}CH_2\text{—}O\text{—}SO_3NH_4$$

In general, the fact that the material is neutralised with ammonia is usually no detriment, but if the final formulation has a pH of over 8, a smell of ammonia will develop. The ammonium cation can then be replaced by adding the stoichiometric amount of caustic soda solution and heating to 100°C, when the ammonia will be distilled off.

SULPHONATION WITH SULPHURIC ACID AND/OR OLEUM

Sulphonation with sulphuric acid and/or oleum of various free SO_3-strengths is still widely carried out in the detergent industry. The sulphonation with acid is mainly used for the sulphonation of alkylate, either ABS or LABS. The procedure of sulphonation is described in detail in publications issued by the producers of alkylate.

The chemical reaction is as follows :

alkylate

As is to be seen, water is set free by the reaction. The presence of water retards the sulphonation reaction, or even prevents it altogether. This is the reason why either a very large surplus of H_2SO_4 or a lesser surplus of oleum ($H_2SO_4 + SO_3$) is required to bring the reaction to completion.

Because alkyl benzene is produced to precise specifications, the detergent manufacturer can work out sulphonation procedures for detergent alkylates from various sources, without changing process conditions which, once established, have become standard procedures for an established sulphonation set-up. This does not mean, however, that there are no differences in quality among detergent alkylates of various provenances. Another important point must be mentioned: that sulphonation procedures vary to a certain extent depending on whether straight-chain alkylate or PT alkylate is used. In the following we will give a description of a batch sulphonation process of alkylate.

For sulphonation, a glass-lined (or stainless) jacketed vessel with an anchor type agitator of c 120 rpm is suitable. During sulphonation, optimum temperature is maintained by cooling with water through the jacket. In hot climates it may be necessary to augment the cooling effect by a refrigeration system and/or a side-arm heat exchanger through which the reaction mass is recirculated by means of a gear pump. The pipes through which the reaction mass is circulated should be of stainless (type 316) steel. It is our opinion that in all cases where the batch size is above 1000 kg a side-arm heat exchanger should be installed.

For acid sulphonation the following process variables are decisive for the quality of the finished product :

1. Acid strength (and, of course, purity of acid, colour, etc.).
2. Ratio alkylate/acid.
3. Sulphonation temperature and time of acid addition.
4. Digestion time and temperature.

After the actual sulphonation step the separation of spent acid (i.e., the greater part of surplus acid) has to be accomplished. Sulphonation without subsequent spent-acid separation is rarely carried out, because in this case all the surplus acid has to be neutralised with caustic soda, transforming it to sodium sulphate. This is rather uneconomical and, owing to the high reaction temperature for neutralising considerable amounts of free sulphuric acid, special precautions are required to keep neutralisation temperatures under control. We will therefore describe a sulphonation process with subsequent spent-acid separation.

Table 9.10 shows the variables in sulphonation for various acid strengths.

This table should serve as a guide. In practice, some variations will be necessary, e.g., when sulphonating with 98 per cent acid, it might be advisable to warm the alkylate to c 40°C, in order to start the reaction smoothly. Otherwise, when starting at too low a temperature, the reaction might not begin until a very large amount of acid is added, and then it might be quite sudden, with an excessive amount of heat which might cause the reaction to get out of control. Anyway, it is a good rule to follow. Reaction should start immediately with the addition of acid, so that the time of acid addition and cooling together keep the reaction well within the range of optimum conditions. Naturally the reaction temperature also has an effect on the ratio alkylate/acid. Saving on acid generally means a sulphonation temperature at the upper limit. Furthermore it may often be preferable not to aim at too complete a degree of sulphonation, and to be satisfied with, say 1.2–1.5 per cent of unsulphonated oil calculated on 100 per cent active matter, rather than 0.8 per cent, which would require more acid, higher temperature, longer ageing time, etc. This would result in a product of poorer colour and odour properties—on the whole a disadvantage that might not be warranted by a decrease of unsulphonated content from 1.5 to 0.8 per cent. With straight-chain alkylate, the reaction conditions are very much the same as for PT alkylate; however, the reaction temperatures should generally be about 5°C lower at the end of the reaction than those given for PT alkylate.

Table 9.10. Acid sulphonation conditions.

Acid strength	Ratio alkylate/acid	Reaction temperature (°C)	Digestion time (°C)
20–22% SO$_3$ oleum	1:1.1 (min. 1:1)	30° until half of acid addition 45° at the end	2 h at 40–45°
12–14% SO$_3$ oleum	1:1.2 (min. 1:1.1)	45° half, 55° end	2 h at 50–55°
100% H$_2$SO$_4$	1:1.5	50° half, 55° end	2 h at 55–60°
98% H$_2$SO$_4$	1:1.65 (min. 1:1.55)	45° half, 55° end	2 h at 50–55°

Now to the separation of spent acid. As a rule of thumb, 10 per cent of water calculated on the total of alkylate plus acid are added to obtain a good spent-acid separation. The water is added as it is, or in the form of crushed ice (not too large lumps of ice, in order to prevent damage to the sulphonation equipment). Under no circumstances should the temperature rise above 60°C, when all the water has been added. Separation takes about six to eight hours, and the resulting spent acid has a strength of 70–80 per cent. In case the sulphonic acid is to be used for the production of liquid detergents, where freedom of electrolytes is of great importance, a modified 'double-wash' spent-acid separation is recommended.

As previously mentioned, sulphuric acid and/or oleum sulphonation is still used in the detergent industry, mainly for the sulphonation of alkyl benzenes, PT benzene, or straight-chain alkylates. It is not suitable for the production of fatty alcohol sulphates, or for the sulphation of fatty alcohol and/or alkyl phenol ethoxylates.

SULPHONATION WITH SULPHUR TRIOXIDE

The greatest advance in sulphonation technique in recent years is the use of SO_3 as sulphonating agent, made possible by the commercial availability of this material. Whereas sulphuric acid and/or oleum is mainly used for the sulphonation of alkyl benzene, and chlorosulphonic acid is only suitable for detergent raw materials where an OH group is present, SO_3 is suitable for practically all types of detergent raw materials.

The reaction with SO_3 takes place without formation of water (see above). The following equation clearly shows this :

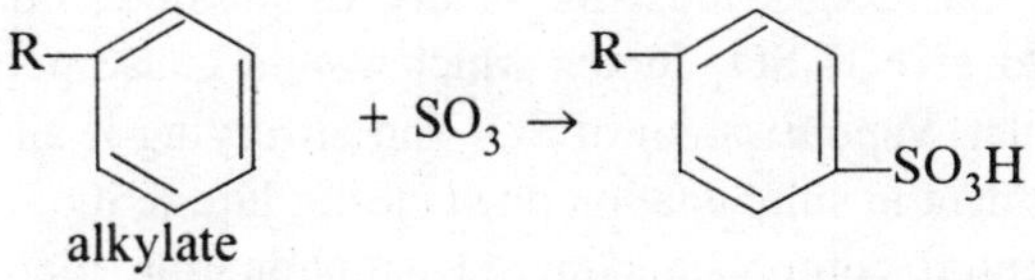

alkylate

One hydrogen wanders from the aromatic ring to form SO_3H, sulphonic acid.

As SO_3 is a vigorous sulphonating and dehydrating medium, on occasion certain side reactions take place :

(a) Formation of anhydride

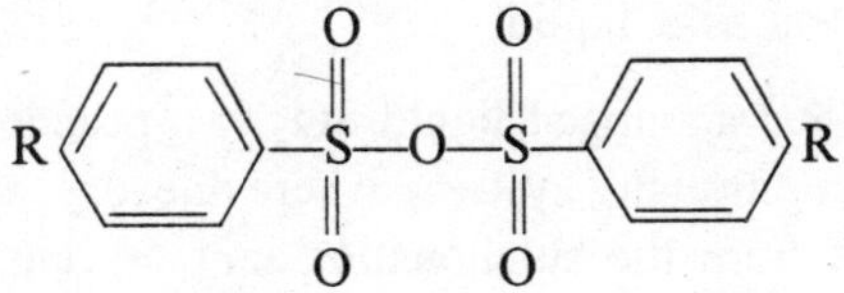

(b) Formation of sulphone (not to be confused with sultone)

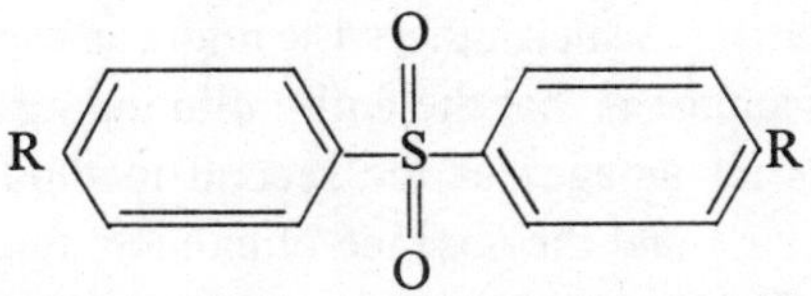

In practice, either or both of these materials can be formed in small quantities, usually less than 1 per cent of the total mass.

Various methods of hydrolysing these two products have been patented, all consisting of adding small quantities of water, dilute alkali, aromatic solvents or alcohols to destroy either the anhydride or the sulphone.

It can be seen that the anhydride hydrolyses to two molecules of sulphonic acid, whereas the sulphone produces one molecule of sulphonic acid and one of alkyl benzene on hydrolysis.

For fatty alcohols the reaction is as follows :

$$RCH_2OH + SO_3 \rightarrow RCH_2OSO_3H$$

Again, a hydrogen wanders to the SO_3 to form the acid sulphate which in the case of fatty alcohols has to be neutralised immediately to prevent hydrolysis.

The most widely accepted method of using liquid SO_3 as sulphonating agent is to employ it in its vapourised form diluted with inert carrier gas, mostly dry air, to give a dilute SO_3 gas stream of 7–12 per cent (vol) SO_3. Naturally the carrier gas must be as dry as possible, otherwise the SO_3 would react to give H_2SO_4 fumes which would cause poor quality of the finished product. Vapourisation of SO_3 and air-drying is an essential part of process equipment in sulphonation plants using liquid SO_3. (Air-drying units are also essential components in SO_3-sulphonation plants starting with elementary sulphur. Producers of liquid SO_3 generally give advice on the proper type of vapouriser to be used. Usually, the liquid SO_3 is fed by dosing pump (e.g., Teflon-lined pumps) into the electrically heated vapouriser, so that not much liquid SO_3 enters at a time, and vapourisation is aided by the stream of dry air passing through the vapouriser. Often the equipment is designed as a film vapouriser, so that at the lower end, the 'sump', the stabiliser collects as a liquid.

For the air-dilution of the vapourised liquid SO_3 two principal methods may be used : (a) a 'once-through' system, where the dry air used for dilution of the SO_3 passes from the sulphonation unit to vent; and (b) a 'closed' system, where the air leaving the sulphonation unit is recircled to the SO_3 vapouriser. The advantage of the first method is that no droplets and fumes recycle to the vapouriser, which otherwise might cause serious colour deterioration. The disadvantage is that the entire dilution air must be dried all the time. The obvious advantages of the second method are the elimination of continuous air drying and the absence of exhaust fumes from the vent. But, in order to be efficient and to prevent colour deterioration, an extremely efficient mist-collector must be inserted between the vent from the sulphonation unit and the vapouriser. In a closed system, nitrogen may be used as the carrier gas, which otherwise would be too expensive, but has the advantage of being completely inert.

The actual sulphonation technique with vapourised liquid SO_3 is the same as with converter gas from a sulphur-burning unit. However, sulphonation with liquid SO_3 has the advantage that such a plant may be operated on a daily shift basis, whereas a plant based on sulphur burning and conversion to SO_3 is only economical when operated continuously, or at least on a weekly basis. Some sulphonation plants using dilute SO_3 vapour obtain the SO_3 by stripping it from 60 per cent SO_3 oleum.

This method of obtaining vapourised sulphur trioxide diluted to about 7–10 per cent in dry air may seem to be unduly complicated in view of the fact that the converter gas which enters the sulphur trioxide absorber in any contact sulphuric acid plant has the same percentage of gas as one wishes to obtain by vapourisation of the liquid. However, upto now, there have been good reasons why converter gas has not been so widely used as a direct sulphonating agent. For instance, it may happen that the industry wanting to use converter gas from a sulphuric acid plant is not situated near a contact plant. Furthermore, it is not always economically advisable for the industry producing sulphonated materials to become dependent on a sulphuric acid plant. Technically, it is not a simple matter to use 'split-off' converter gas from a sulphuric acid plant, mainly because of the difficulty of measuring exactly the amount of sulphur trioxide taken from the cooler to be used in the sulphonation process.

To control the actual amount of sulphur trioxide, one must know the volume of converter gas at constant temperature and the concentration within a very narrow range. This does not appear to be too difficult, but in practice it is not a simple matter. For many sulphonation processes the amount of sulphur trioxide used must be accurately measured to within ±1 per cent, as any greater variations will lead to either over- or under-sulphonated material. In the case of detergent alkylate, this would mean a product with too much unsulphonated matter or with an unacceptable dark colour. With fatty alcohols it is even more dangerous, as inaccurate addition of sulphur trioxide might even lead to a reversal of the sulphonation reaction.

However, methods for controlling the degree of sulphonation have been worked out, which automatically record the amount of unsulphonated material, and by feedback adjust the flow of alkylate to the sulphonation plant. This method of control might overcome the difficulties in working with 'split-off' converter gas.

It is only with those sulphonation processes where accuracy is less important that the use of 'split-off' converter gas is a relatively easy matter. Thus, white oil and highly refined kerosene can be produced by treating the appropriate petroleum fraction with converter gas either in a continuous or discontinuous process.

The possibility of designing a plant to produce sulphur trioxide on a small scale, where all the sulphur trioxide from the converter is cooled and used for the sulphonation of detergent raw material. To do this, certain conditions had to be fulfilled. First of all, to ensure a constant dosing of sulphur into the system, a metering pump was used. Then, in order to have the sulphur transformed into a constant amount of dry sulphur trioxide, the air used for burning the sulphur in the sulphur-burner had to be completely dry, as otherwise sulphuric acid would be formed by the reaction with the water present. Drying of air was therefore carried out in a silica-gel drying unit, which brought the humidity down to about 0.5 g/m^3 air. If a still higher degree of drying is required, it is possible to carry out a kind of 'threshold drying' by having the silica-gel-dried air pass through a molecular sieve system, which brings the final humidity down to as little as 0.05 g/m^3 air.

A more efficient method to reach a very low dew point is by pre-cooling the air to about 5°C, condensing and then removing the bulk of the moisture present. The silica gel, operating at a lower temperature works better and the final dew point of the air can be −30°C or even −60°C.

A simple method of controlling the efficiency of the drying unit is to use the Draeger system of gas detection. The method is to draw by means of a hand pump, a constant volume of air through the appropriate Draeger tube and the concentration of moisture is read off. For more exact determinations special instrumental methods are available.

The same principle can be applied to the measurement of the SO$_2$ content of the exhaust gas leaving the sulphonators. Determination of SO$_2$ gives an indication of the original degree of convertion and also is important from the point of view of air pollution.

As already mentioned, the sulphur fed into the sulphur-burner in a converter-gas plant must also be measured very exactly. Furthermore, it was found in practice that the sulphur-burner used should preferentially be so designed that the sulphur is not sprayed under pressure through a nozzle into the burner, but flows through a relatively large orifice. Such a design is less liable to irregularities because of the clogging of the nozzle, and there is also the advantage that the pump is not working against a high pressure.

The sulphur used in small converter plants should be of high quality.

SULPHONATION PROCESSES

In addition to the advantages previously mentioned, the use of sulphur trioxide for sulphonation, generated as described, has the additional asset that its oxygen content is low compared with vapourised liquid sulphur trioxide, where air acts as the carrier gas. In the production of the trioxide direct from sulphur, about half of the oxygen in the air is used for conversion. This reduction in the amount of oxygen in the sulphur trioxide gas stream eliminates side effects often encountered with vapourised gas. Furthermore, the traces of sulphur dioxide in converter gas are also beneficial in this respect.

Sulphonation was carried out semi-continuously in two small sulphonating vessels of about 250 litres capacity. The unit consumes 1000 kg sulphur/24 hour, and this produces 2400 kg of sulphur trioxide (as 7–8 volume per cent converter gas), i.e., enough for about 7500 kg alkyl benzene/24 hour.

Some 175 kg (200 litres) of alkyl benzene are charged into one of two sulphonation vessels and sulphur trioxide admitted. Sulphonation requires about 60 min and, while the reaction is proceeding in one vessel, ageing is taking place in the other. After ageing, which takes about 15 min, a small amount of water or NaOH solution is added, and the alkyl benzene sulphonic acid is either run to storage or immediately converted to the neutralised AB-sulphonate. The small amount of water (about 2–3 per cent in the sulphonate mixture) helps to prevent 'anhydride-sulphonate' formation and keeps the colour of the AB-sulphonic acid stable, even on prolonged storage in mild steel drums or tanks. By introducing a time control and automatic-operated valves, etc., the semi-continuous sulphonation system can be converted to a continuous process.

The Sulphurex process Fig. 9.6 represents a further development. Here both the sulphur and the detergent raw material are metered into the system in the requisite proportions.

After being cooled, the SO_3 gas stream passes into the sulphonators at about 50°C. These consist of two or more reactors—different in number and size depending on the capacity of the plant. The SO_3 gas stream enters the reactors in parallel, with the major portion entering the first reactor and the minimum portion entering the last reactor. This proportioning of the SO_3 gas into the various reactors is done automatically and requires no attention from the operator.

The alkyl benzene (or any other material to be sulphonated or sulphated) essentially passes through a series of reactors in cascade, but because of the 'free oil' controller, a small amount is proportioned into the last reactor, since it is this quantity which is varied according to the completion of reaction.

Special high-speed turbine mixers disperse the SO_3 gas and the alkyl benzene with a high degree of efficiency, which prevents and local over-heating of the reactants. Inside the reactor are precisely located cooling coils and baffles which immediately remove the large amount of heat released during the reaction.

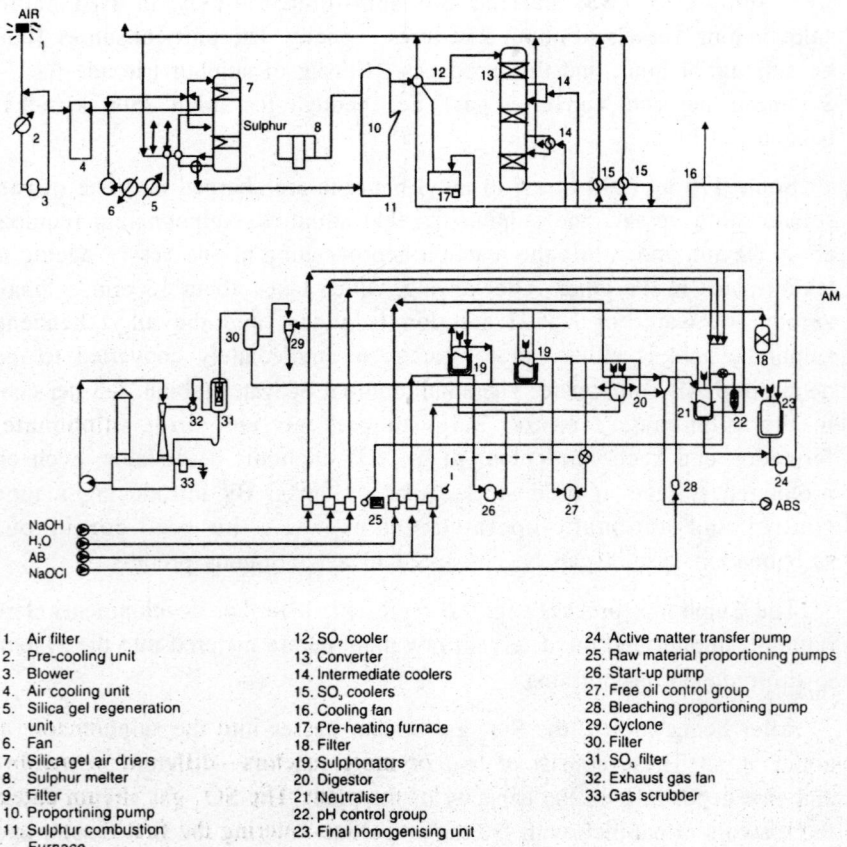

1. Air filter	12. SO_3 cooler	24. Active matter transfer pump
2. Pre-cooling unit	13. Converter	25. Raw material proportioning pumps
3. Blower	14. Intermediate coolers	26. Start-up pump
4. Air cooling unit	15. SO_3 coolers	27. Free oil control group
5. Silica gel regeneration unit	16. Cooling fan	28. Bleaching proportioning pump
6. Fan	17. Pre-heating furnace	29. Cyclone
7. Silica gel air driers	18. Filter	30. Filter
8. Sulphur melter	19. Sulphonators	31. SO_3 filter
9. Filter	20. Digestor	32. Exhaust gas fan
10. Proportining pump	21. Neutraliser	33. Gas scrubber
11. Sulphur combustion Furnace	22. pH control group	
	23. Final homogenising unit	

Fig. 9.6. Flow diagram of Ballestra sulphurex process.

The temperatures normally used for various raw materials are :

(a) for the usual types of alkyl benzene (branched or straight-chained) :
Sulphonators : 50°C
Digestion: 55°C
Neutralisation : 45°C

(b) for lauryl alcohol :
Sulphonators : 35°C
Neutralisation : 45°C

(c) for tallow alcohol :
Sulphonators : 55°C
Neutralisation : 50°C

(d) for C_{14}–C_{18} Ziegler alcohols :
Sulphonators : 50°C
Neutralisation : 45°C

After the sulphonators, the main stream of sulphonic acid passes on to the digestion vessel, where it is held, to allow the reaction to go to completion. A side portion of this stream is directed to a patented device which automatically analyses the sulphonic acid and determines its 'free oil' content. If the free oil is increasing, this control unit automatically cuts down on the quantity of alkyl benzene being fed to the last reactor, and conversely, if the free oil is going too low, the unit automatically increases the flow of alkyl benzene to the last reactor.

The gases from the reactors are essentially free of SO_3. However, the plant is equipped with a scrubbing system for removing any SO_3 gas that may be in the exit stream, and for scrubbing out the SO_2, so that the exhaust gas is free of these undesirable components as it passes to the atmosphere.

The scrubbing solution containing mainly sodium sulphite may be used directly as a slurry component in an adjacent spray-drying unit (sulphite has no detrimental effect on optical brighteners) or better still, it could be oxidised by a special air injection system to sodium sulphate, which again may safely be used as a slurry component, or in case no spray-drying unit is near the sulphonation unit, be discharged to the sewer. The scrubbing system is efficient enough to absorb the SO_2 so that the exhaust gas exits with as little as 5 ppm SO_2.

After the digestion vessel the sulphonic acid stream can be passed into another small reactor where water or NaOH solution can be added if desired. This unit is to take care of any anhydride formation that may occur.

Of course, no water or NaOH solution is used during the sulphation of the fatty alcohols. Because of the combination of reactors it is very easy to perform co-sulphonation–sulphation reactions and produce a high-quality product.

The sulphonic acid then passes on to the neutraliser, where caustic soda and water are continuously added to give a white paste containing the desired percentage of active ingredient. An automatic pH control unit controls the addition of caustic by sending a signal to the controller which varies the stroke of the proportioning pump of the neutralising agent by means of a servomotor.

The neutralised paste overflows into a storage tank, where it is continuously homogenised. From here it is pumped to any further processing operations, such as spray-drying, liquid production, etc.

It is important to note that after the sulphonators no transfer pumps are required. The reactants and acid flow from vessel to vessel by gravity. Operation and control of this plant are extremely precise and yet simple. One operator runs the entire plant—sulphur-burning, conversion, and sulphonation.

Neutralisation of sulph(on)ated products can be an important operation, particularly for materials other than alkyl benzene sulphonates made by SO_3 sulphonation. If oleum sulphonation is used, the sulphonic acid contains both free sulphuric acid and water, rendering it corrosive. Unless special storage conditions are observed the acid material will both corrode iron and darken on standing. Neither of these restrictions applies to SO_3 sulphonated alkyl benzenes, they can be stored and transported in mild steel with no deleterious effects. For all that even the SO_3 sulphonated variety is often neutralised immediately after sulphonation.

Alcohol sulphates and alcohol ether sulphates need to be neutralised immediately, otherwise, in acid medium, the sulphuric ester will split to its original components (the reverse of the sulphation reaction). Olefin sulphonates need to be hydrolysed in addition to neutralisation. Sulphonated methyl esters need special treatment to prevent (or minimise) ester saponification in an alkaline medium.

For all the above reasons, the system of double step neutralisation has been developed recently has been developed, which eliminates the local concentrations of acid or water (or alkali), preventing hydrolysis of the ester group, or the formation of highly viscous hydrated lumps which dissolve slowly, causing pH drift. If hydrolysis were to occur, free sulphuric acid and free oil are formed. Free oil increases the unsulphonated portion of the finished paste, free sulphuric acid requires considerably more neutralising agent which can increase both the costs and the amount of inorganic salts present. Increase of free acid (even for a short period) can cause corrosion problems to the equipment and this is enhanced if, as is sometimes the case, bleaching agents are added during the neutralisation step. The plant is also so designed that the temperature of the paste is kept within reasonable limits.

To overcome all these problems, the Neutrex system neutralises 90 per cent of the incoming material in the first of two reactors, equipped with specially designed low speed impellers, which mix and homogenise. The heat of neutralisation is removed in this vessel. The remaining 10 per cent is fed into the second (larger) mixer and this mixes with the incoming neutralised paste from the first. The neutralising base which can be any of the alkaline materials commonly used is accurately dosed with linked pH control. The mass in the second neutraliser is kept at a temperature such that it can be easily transported further.

This unit can be coupled to any SO_3 sulph(on)ation system, and by using this system the hitherto difficult to produce 70 per cent alcohol ether sulphates can easily be made.

Comparison of oleum and sulphurex processes

	Oleum process	Sulphurex processes
Sulphonating agent	Oleum	Sulphur
—characteristics	Corrosive liquid	Inert powder
—supply	Tanks, drums	In bulk (powder)
—storage	Tanks, drums	In bulk (powder)
Sulphonic acid purification	Washing and separation	No purification required
By-products	Dark 70% spent sulphuric acid	No by-products
Products storage	Lead-lined tanks	Mild steel tanks

For the production of 100 kg of 100 per cent active material from 72.5 kg alkyl benzene, the acid sulphonation process requires 75 kg of oleum, while the Sulphurex process consumes 10.5 kg of sulphur.

Only one man per shift is needed to operate the 'switch-over' and the Sulphurex processes. A sulphur-trioxide absorption tower is not essential for warming up and cooling down the unit. While the unit is being brought into equilibrium, the converter gas of irregular composition may be absorbed, for example, in alkyl benzene.

To use the converter-gas plant for the sulphonation of other compounds, such as toluene, xylene, etc., some modifications are necessary. Thus, for sulphonating low-boiling aromatic compounds, special cooling condensers have to be provided, to prevent undue losses by carrying over of unsulphonated product. By special modification of the sulphonation process on the basis of the 'dominant bath' system for low-boiling aromatics, a less expensive condensing system is obtained.

It is also possible to co-sulphonate toluene or xylene in conjunction with detergent alkylate, to give the proper proportion of hydrotrope or anti-caking agent in the finished detergent.

Production of toluene sulphonates with oleum or SO_3 is rather costly, so the price of separately bought toluene sulphonate is higher than the price of AB-S. Since only a small percentage of toluene sulphonate is used (5–10 per cent calculated on AB-S used), it would be worth while to co-sulphonate toluene with the alkylate. The main difficulty encountered here is the low boiling point and high volatility of toluene. There is danger of 'carry-over' of the toluene with the inert gas stream leaving the system through the exhaust, before the toluene is sulphonated completely in the system. This leads to losses, unpleasant exhaust problems, and danger of ignition; problems which can be overcome in the laboratory by adding condensers and scrubbers to the sulphonation unit, but which are costly and difficult to set up on a plant scale. A different approach is used. The toluene is fed into the sulphonation system (in this case a 'switch-over' system) via the side-arm heat exchanger. Once the alkylate is sulphonated, toluene is fed into the suction line of the circulation pump for the side-arm heat exchanger in a ratio corresponding to the feed of SO_3 entering the system. The toluene is thus pre-mixed with the AB-sulphonic acid in the circulation pump and heat exchanger, and then further mixed into the 'dominant bath' of the AB-S in the sulphonation system. No losses occur, and the toluene can be properly co-sulphonated. (The same principle can, with modifications, be applied to the fully continuous system.) If an automatic control system is not available, or if it is desired to control the system from time to time, the unsulphonated toluene can be controlled by a quick and simple technique: a sample of the AB-sulphonic acid is with-drawn and heated under an infra-red drying balance. Weight losses during five minutes give the amount of unsulphonated toluene, since neither AB-sulphonic acid nor toluene-sulphonic acid is volatile enough to disturb this quick test. Of course the same technique applied to co-sulphonation of cumene or xylene.

Practical test runs on a large scale have been made in a Sulphurex plant in which the standard Sulphurex set-up was used. Sulphonation in the first vessels was carried to about 85–90 per cent, and toluene was fed by the second dosing pump into the last sulphonation vessel, where the practically fully sulphonated AB-S (or for that matter fatty alcohol sulphate) was used as the 'dominant bath' for the toluene to be co-sulphonated. No special mixing pump was required in this case. Adjustment of the automatic instrument for sulphonation control can easily be made.

Recently there has been an increased interest in α-sulpho-fatty-acid methyl ester. They form useful components for many types of powder detergents, and are also used as components for syndet bars and syndet/ soap combinations. They can be produced by sulphonating methyl esters of coco fatty acids, stearic and palmitic fatty acids. It is important for the quality and colour of the sulphonated end product, that the methyl esters should have a low iodine value, i.e., below 0.5.

With the introduction of large-scale SO_3 sulphonation which, as we have stated, revolutionised the industry, considerable work has been done on improving the process both to produce higher quality materials and to adapt the procedure to allow of the sulphonation of the hitherto hard to sulphonate hydrophobes.

Sulphonation units for detergent production based on the falling film reactor are now being almost universally supplied to the exclusion of all other types, as these have proved to be able to cope with practically all the base materials except of course paraffin sulphonates.

Basically these units use a film of the material to be sulphated/ sulphonated, injected on to the wall(s) of a vertical tube or series of tubes (bundle type). The SO_3 is fed concurrently into the centre of the space. The reaction takes place on the surface of the tube(s) and the heat of reaction is dissipated through the walls of the conduit.

As the reaction is almost instantaneous, and as the film is flowing downwards continuously, and as 70–80 per cent of the SO_3 is consumed in the first third of the reactor, the sulph(on)ated material does not remain in contact with a large concentration of SO_3 for any length of time.

The advantages of the falling film reactor are :

1. Contact time of already sulphonated material with SO_3 is minimal.

2. Low residence time, thus acid sensitive materials (alcohol, alcohol-ether sulphates) can be transported rapidly to the neutralising section.

3. Relative flow rates and dilution of the SO_3 can be adjusted so that olefins, which react very vigorously, are contacted with dilute SO_3 for a relatively short time, on the one hand, and slow reacting materials, such as methyl esters, can have their residence time increased, on the other hand.

The problems in designing the reactor are :

1. Maintaining a constant uniform film of the organic feed on the walls.

2. Metering almost stoichiometric quantities of organic feed and accurately diluted SO_3.

3. Dissipating the heat of reaction but not at such a rate that the already sulph(on)ated material becomes a viscous mass and ceases to flow.

4. Controlling the velocity of the gas stream so that the organic feed is not carried over as a mist or the lower boiling components are not volatilised.

5. Having and maintaining a uniform small diameter (or aperture) as this increases the contact between the SO_3 and the feed.

The Ballestra Multitube Film Reactor consists of a vertical bundle of long, small-daimeter tubes.

The liquid film, which forms inside the tubes, comes into contact with the gaseous reactant. The length and diameter of each tube is so dimensioned as to provide the optimum product quality performance.

The most interesting feature of the Ballestra reactor is the simple way in which the exact reactants ratio is maintained in all the tubes.

The automatic adjustment of the reactants in each of the tubes is the result of the geometric design of the reactor itself and is completely independent of the actual number of tubes. It is a simple matter, therefore, to scale up the reactor to meet any production capacity requirement. All that is required is to increase the number of reaction tubes.

Fig. 9.7 illustrates the principle of the reactor. The SO_3 gas enters at the top of the reactor and then flows, without restriction, into each of the parallel-mounted reactor tubes. The liquid is uniformly distributed, by means of a simple concentric slot, over the inner walls of each tube.

The concentric slot permits the same amount of liquid to be fed to each tube. A difference of about 20 per cent between each tube is permissible, and the correct amount of gas is automatically fed to each tube, due to the aforementioned geometric design of the reactor.

The special patented feature of the reactor is that the required molar ratio between the reactants in each tube is self-adjusting and, furtheremore, is close to the preset ratio of the total liquid to the total gas flow rates in each of the tubes. The total gaseous reactant is fed to the reactor at the proper rate and at the proper concentration. The total liquid reactant is pumped to the reactor by means of the proportioning pump.

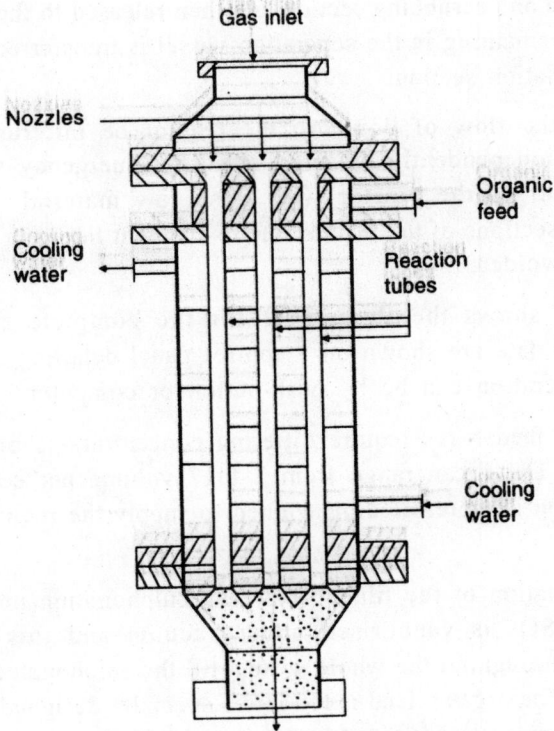

Fig. 9.7. Ballestra multitube film reactor.

The flow of gas and liquid inside the reactor is parallel, with the liquid phase forming the film on the inner surface of each tube and the gas flowing through the tubes at very high speed. The outsides of the tubes, the shell side, are provided with cooling to ensure correct reaction temperature. The reaction takes place at a very high rate, with about 70 to 80 per cent of conversion occurring in the first third of the reactor.

Each individual tube, being fed at a particular average rate of liquid flow which is different from the next, has a correspondingly different film thickness and viscosity profile. This contributes to the self-adjustment of the gas-to-liquid mol ratios to the optimum average values, since the flow of gas in each tube then is directly proportional to the flow of liquid itself. The sulphonated liquid and exhaust gas leaving the reactor enter the gas separation vessel.

The separated exhaust gas is sent to the cleaning, electrostatic precipitation and scrubbing section and then released to the atmosphere. The liquid remaining in the separation vessel is transferred by pump to the neutralisation section.

Should the flow of liquid to the reactor be interrupted for any reason, an independently powered stand-by emergency system takes over and purges the reactor with fresh raw material. In this way charring of sections of the reactor due to stagnant liquid in the presence of SO_3 is avoided.

Fig. 9.8 shows the flow sheet for the complete sulphonation process. All data are shown on a control panel detailing the complete process. Operation can be by push-button or computer.

Different feedstocks require differing concentrations of SO_3 in the gas stream. These can range from 2 to 7 volume per cent. The air-drying section is therefore dimensioned to supply the requisite dilution of dry air.

A combination of the film reactor and sulphonation under vacuum and liquid SO_3 is vapourised under vacuum and this vacuum is maintained throughout the whole system till the sulphonated material is discharged. The organic feed is fed into a specially designed egg-shaped reactor by a rotor to form a thin film on the internal wall of the egg.

These patentees claim that all the feedstocks normally used as detergent raw material intermediates can be sulph(on)ated by this unit. Advantages are that no air-drying or gas-liquid separation is necessary, considerably reducing both capital costs and space required.

Film reactor plants are being used to sulphonate all types of conventional detergent materials. They are, however, not suitable for co-sulphonating volatile materials, such as toluene, xylene, or cumene. They have found great success in the sulphonation of α-olefines which are playing an increasing part in detergent formulations and which hitherto had required very special methods of sulphonation.

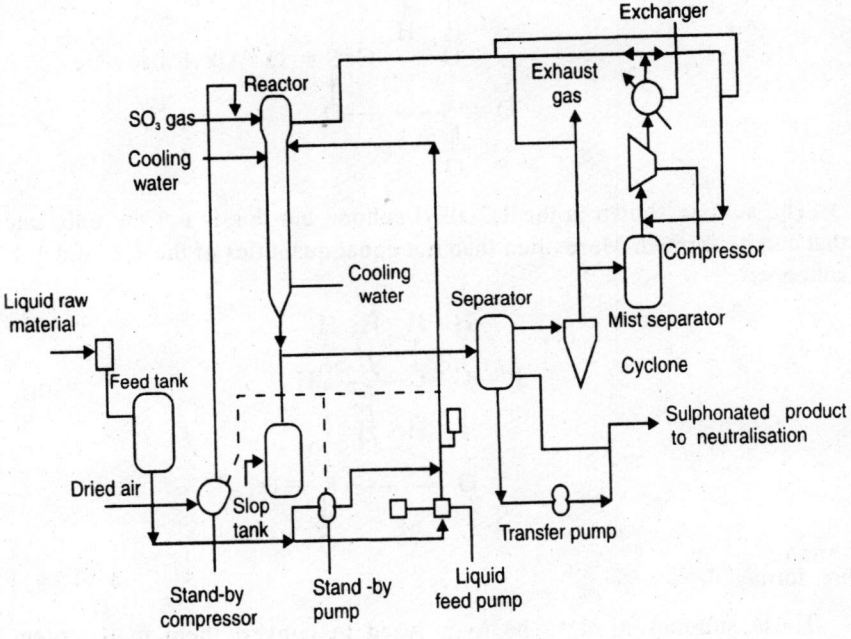

Fig. 9.8. Ballestra's film sulphonation system.

Alpha Olefin Sulphonates (AOS)

The sulphonation of α-olefins produces in approximately equimolar proportions :

(a)

$$R-\underset{\underset{H}{|}}{\overset{\overset{H}{|}}{C}}=\underset{\underset{H}{|}}{\overset{\overset{H}{|}}{C}}-SO_3H$$

Alkene sulphonate

(b)

Alkyl sultone

with minute amounts of

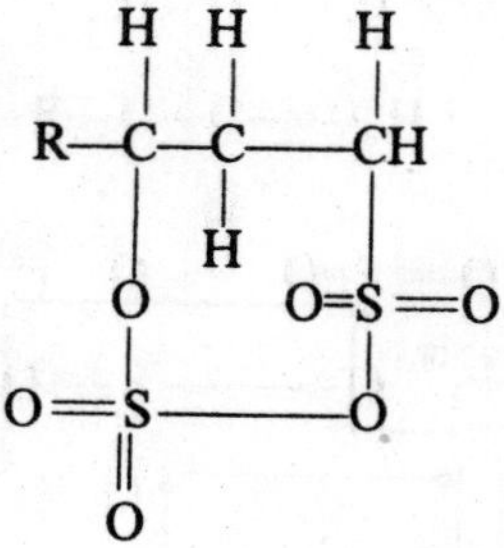

$$
\begin{array}{ccc}
\overset{\displaystyle H}{\underset{\displaystyle |}{}} & \overset{\displaystyle H}{\underset{\displaystyle |}{}} & \overset{\displaystyle H}{\underset{\displaystyle |}{}} \\
R\!-\!C\!-\!C & & CH \\
\end{array}
$$

R—C—C——CH   (with H below second C)

O   O=S=O   Alkyl disultone

O=S——O

‖
O

The sultone shown is the 1,2-alkyl sultone but this is not the only one that can be formed. More often than not equal quantities of the 1,2- and 1,3-sultones:

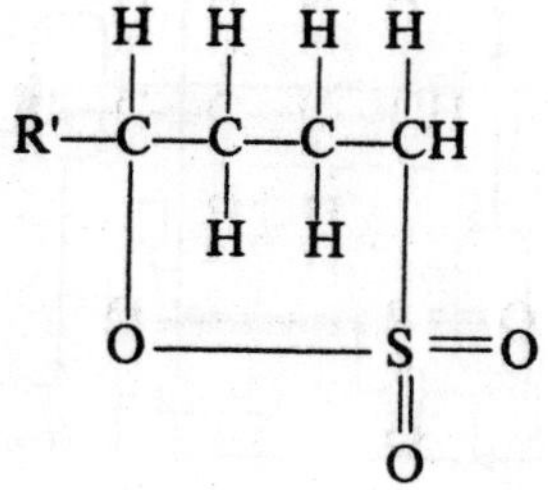

H H H H
| | | |
R'—C—C—C—CH
      | |
      H H

O————S=O
       ‖
       O

are formed.

These sultones need to be hydrolysed to convert them to detergent active materials. This hydrolysis is done in an alkaline medium, when the ring opens to form the sodium hydroxy sulphonate :

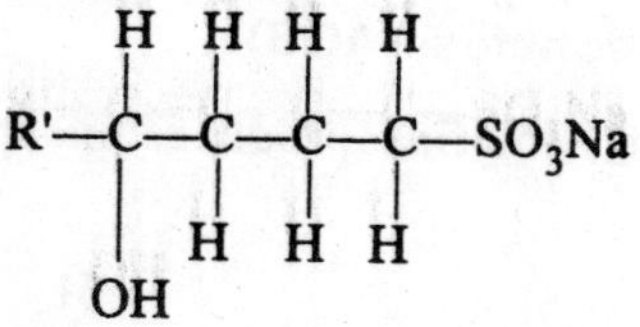

H H H H
| | | |
R'—C—C—C—C—SO$_3$Na
| | | |
OH H H H

β-Hydroxy sulphonate is an undesirable product of this hydrolysis as it is not soluble in water, giving a hazy solution. Fortunately the 1,2-sultone from which it is formed is thermodynamically unstable, and will isomerise to the 1,3- or 1,4-sultone on standing. This isomerisation takes only a matter of minutes so that sulphonation/neutralisation/hydrolysis conditions need to be arranged so that there is a 'hold' period between the sulphonation on the one hand and the neutralisation and hydrolysis on the other.

Formation of disultones is dependent on sulphonating conditions, excess of SO_3 and temperature, and should be kept as low as possible. The disultone on hydrolysis forms

R'—C—C—C—C—SO₃Na   Alkane sulphato-sulphonate

As stated the α-olefin plants are almost invariably of the falling film type and must have neutralisation/hydrolysis units coupled to them.

α-Olefin sulphonates have not been accepted as universally as their initial proponents hoped, mainly because the chemistry is somewhat different from normal sulphonation.

Table 9.11 shows, among other items, the colour of unbleached detergent materials sulphonated in a F film reactor. It can be seen that the colour of all the materials, including the olefin sulphonate, is such that no bleaching is required.

Methyl Ester Sulphonates

The basic chemistry of the sulphonation of methyl esters of fatty acids has been touched on as has their method of preparation. The chemistry is not completely elucidated but appears to be the rapid formation of the complex, which can be termed a 'sulpho-anhydride' and this rearranges (relatively) slowly to the sulphonic acid.

It is evident that sulphonation is best done in a film reactor, requires an excess of SO_3, a relatively high temperature and a holding period to allow the intermediate complex to rearrange to the true sulphonic acid. Colour of the sulphonate will be high but this is easily bleached, either on the acid material using peroxide, or on the neutralisate using hypochlorite.

Neutralisation needs to be done with care, the pH, even localised, should not rise above 8.5, otherwise saponification of the ester might occur. The Neutrex double step neutralisation unit is therefore an ideal · solution to the after-treatment of the sulphonic acid.

Table 9.11. Production characteristics of materials sulphonated in a Ballestra Sulphurex F film reactor.

Material	Neutralised material			On 100% material		Klett colour*
	Am %	Uns %	Na$_2$SO$_4$	Uns %	Na$_2$SO$_4$	
LABS	-	-	-	0.8	0.9†	25
Lauryl alcohol sulphate sodium salt natural or synthetic	35	0.3	0.5	0.9	1.4	7
Lauryl ether sulphate 3 mol EO- sodium salt	30	0.3	0.4	1.0	1.3	15
Alcohol C$_{16-18}$ sodium salt	25	0.65	0.4	2.6	1.6	60
α-Olefin sulphonate sodium salt	40	0.72	0.56	1.8	1.4	70

* On 5 per cent AM solution, 40 mm cell, # 42 filter, Unbleached products.
† Free sulphuric acid.

Heavy Alkylate Sulphonates

The heavy alkylate produced as the 'bottoms' in the alkylation of benzene is sulphonated by either oleum or SO$_3$ to produce synthetic petroleum sulphonates, not to be confused with paraffin sulphonates. Natural petroleum sulphonates are produced as by-products in the production of white oils, and the 'synthetic' variety is almost identical functionally, although not chemically.

As already mentioned, the heavy alkylates are complex mixtures but all the constituents of the mixture contain an aromatic ring, coupled to alkyl, alkyl–aromatic, other aromatic or cyclic nuclei in the ortho- or para-positions. Sulphonation of alkyl benzene is almost invariably at the para-position, so if the para- position is blocked, sulphonation in the ortho-position is not feasible due to steric hindrance.

Heavy alkylates can therefore be only partially sulphonated and commercial products normally contain 60–70 per cent sulphonated matter.

Sulphonation is best achieved by SO$_3$ plus a non-reactive paraffinic solvent (boiling range *c* 210–230°C) to facilitate the separation of any oil-insoluble acid sludge. The unsulphonated and unsulphonateable portion acts as a solvent for the neutralised material allowing neutralisation to be done with concentrated caustic soda or alkaline earth hydroxide slurries. The small amount of water introduced and produced in the neutralisation process can either be left in the material or distilled off if the specifications require.

The sodium salts of the heavy alkylate sulphonates are used as emulsifiers, wetting agents and dry-cleaning additives.

The alkaline earth salts (calcium, barium and magnesium) are used in lubricants and greases and their 'over-based' varieties as rust preventatives. Over-basing is achieved by adding surplus hydroxide or carbonate to the neutralise in such a way that the base remains in colloidal suspension.

Sulphonation with Chlorosulphonic Acid

Chlorosulphonic acid is widely used for the sulphonation of fatty alcohols, fatty alcohol-ethoxylates, alkylphenol-ethoxylates, and related detergent raw materials with OH groups available for the attachment of an SO_3H group. The reaction of lauryl alcohol with chloro-sulphonic acid illustrates the chemistry involved :

$$C_{12}H_{25}OH + ClSO_3H = C_{12}H_{25}O—SO_3H + HCl$$

The hydrogen chloride must be absorbed to give a 30 per cent HCl solution as a by-product. One can well consider the HCl in the formula of chlorosulphonic acid as a kind of 'carrier gas' present as liquid bound to SO_3, and set free during the sulphonation reaction. Whereas sulphonation with SO_3 within an inert carrier gas consists of a gas/liquid reaction, sulphonation with chlorosulphonic acid is a liquid/liquid reaction, with liberation of HCl.

Sulphonation with chlorosulphonic acid requires special corrosion-proof equipment, either glass-lined steel or all-glass, the latter especially suitable for small-batch-size sulphonation. The HCl-absorber, too, is built either of glass-lined steel or is all-glass. Some firms use glass-lined steel reactors and all-glass HCl-absorbers. A typical batch sulphonation process of lauryl alcohol is carried out as follows :

250 kg lauryl alcohol are charged into the glass-lined reaction vessel. 156 kg chlorosulphonic acid (the calculated stoichiometric amount of the lauryl alcohol used) are added slowly with cooling. The reaction temperature is kept at an optimum of 25°C (not exceeding 30°C). 47 kg HCl gas are set free and have to be absorbed to 30 per cent HCl acid solution. (Some small amounts of HCl remain dissolved in the acid sulphonate and form a neutral chloride with the neutralisation agent.) If NaOH is used as the neutralisation agent, c 85 kg NaOH are required. Generally, the resulting 359 kg of acid lauryl-alcohol-sulphate are run into a crushed ice/water mixture c 1000 kg,

while the NaOH is run into the neutralisation vessel (as 255 kg 38° Be NaOH), keeping a small excess of NaOH, so that the mixture is always at a slightly alkaline pH. This prevents splitting off of the sulphuric acid by hydrolysis of the O—SO₃H group, which would rapidly occur at an acid pH in the presence of water. Neutralisation should be carried out as quickly as possible after the sulphonation process step; otherwise hydrolysis will occur, even in the absence of water.

The slurry (c 35 per cent active matter) is adjusted to a pH of 7.7–8.0, and is then stable on storage. Good mixing during neutralisation is very important to prevent agglomeration of acid sulphonate, which again would lead to hydrolysis and a high unsulphonated alcohol content in the finished paste. Any bleaching required is carried out with sodium hypo-chlorite at a near neutral pH. Neutralisation with ethanolamines is also often practised, and this method, and neutralisation with ammonia, is, in fact, less critical: the mixture remains liquid and agglomeration of acid sulphonate does not occur so frequently. The heat of neutralisation is more easily dissipated than in the case of caustic soda neutralisation.

In sulphating ethers (and to a lesser extent alcohols) with chloro-sulphonic acid, the HCl liberated in the initial stages of the reaction dissolves exothermically in the reaction mass, and in the second stage is displaced from solution together with the fresh portion of HCl being formed. The evolution of this double quantity of gas in the second stage of the reaction causes troublesome foam.

This foam formation, in continuous process particularly, can be over-come by arranging the flow of the liberated HCl on the one hand, and the incoming alcohol or ether on the other hand in such a manner that the evolved HCl is allowed to flow upwards through an absorption tower, down which the incoming alcohol or ether is allowed to trickle. This saturates the unsulphated material with HCl and does not allow any further HCl to dissolve during the reaction process. To a large extent this eliminates foaming in the case of alcohols but when ethers are being sulphated, in addition to the above arrangement it is necessary to dissolve in the unsulphated ether about 40 ppm of an anhydrous, oil-soluble, silicone antifoam. With these precautions, although a certain amount of foaming takes place, the foam formation remains within reasonable limits.

It is interesting to note that in the sulphation of ethers with SO₃, the same troublesome foam is encountered due to the large excess of air or other inert gas. Addition of silicone will prevent undue foam formation. In SO₃ film sulphonation systems this trouble does not occur.

Commercial specifications of various anionic detergents produced by the several sulphonation methods are given below, where the first comparison made is between commercial sodium alkyl benzene sulphonate pastes produced by either oleum or sulphuric acid sulphonation with spent-acid separation and those produced by SO_3 sulphonation (either with vapourised SO_3 or directly with converter gas).

Anionic detergents	By acid sulponation	By SO_3 sulphonation
	%	%
Sodium alkyl benzene sulphonate	50	50
Unsulphonated free oil	0.5	0.5
Sodium sulphate	7.5	0.8
Water to make	100	100
Colour	White	White

Typical specifications for converter-gas-derived AB-sulphonic acid are as follows :

	Dodecyl benzene sulphonic acid	Tridecyl alkylate sulphonic acid†
Active detergent matter calculated as the sodium salt	100% ± 1%	100% ± 1%
—calculated as sulphonic acid with molecular weight 325	94% ± 1%	94%± 1%
Unsulphonated free oil (calculated on 100% am)	1.7% max	1.7% max
Acid number	182–7 mg KOH/g	174–81 mg KOH/g

† Specifications for sulphonic acid produced with converter gas from 'straight-chain' alkylates have principally the same data, depending on the molecular weight of the alkylate, generally either in the range of PT alkylate or of tridecyl benzene. Heat of neutralisation with NaOH is about 45 cal/g.

Not all sulph(on)ated materials are made using either oleum or SO_3. Sulphamic acid can be used for special applications, sulphosuccinates and sulphosuccinamates are made using sulphite addition and alkanes are 'sulphonated' using SO_2 and oxygen.

Sulphation by Means of Sulphamic Acid

A method of sulphating alcohols and ethoxylates is by the use of sulphamic acid. The acid is quite expensive so the final cost of the finished product will be high compared to conventional methods of sulphation, but against this must be balanced the fact that no large investment in plant is needed.

The plant for sulphamation is relatively simple; all that is required is a closed stainless steel reactor with stirring, heating and cooling arrangements and a coil for the introduction of a nitrogen purge.

The reaction of sulphamic acid with an—OH group is :

$$—CH_2OH + H_2NSO_3H \longrightarrow —CH_2—O—SO_3H + NH_3$$
$$—CH_2—O—SO_3H + NH_3 \longrightarrow —CH_2—O—SO_3NH_4$$

Thus the ammonia liberated immediately neutralises the sulphuric ester produced and an ammonium salt is formed. Sulphamic acid is particularly useful in sulphonating alkyl phenol ethers as in this case the aromatic ring is not attacked, whereas with SO_3 and chlorosulphonic acid upto 20 per cent ring sulphonation takes place.

It is not considered that ring sulphation detracts from the quality of the finished product, but the two types of sulphates (the ring has a sulphonate group) will be different. If 20 per cent ring sulphonation occurs, an equivalent amount of the terminal—OH will remain unsulphated (assuming stoichiometric quantities are used). This molecule will therefore be somewhat more, hydrophilic than the non-ring sulphonated material. Recentally a process has been developed in which all types of alcohols and alcohol ethoxylates can be sulphated using this acid.

The procedure is relatively simple but varies for different materials. In principle the material being sulphated is charged into the reactor, the stirrer started and 1.05 mol of sulphamic acid, which should be finely divided, is fed in rapidly. The rate of stirring and rate of feed of the sulphamic acid should be such that the powder is well dispersed throughout the mass of the material. When all the sulphamic acid has been added, the lid of the reactor is closed and nitrogen is passed through the vessel to purge all the air. Simultaneously the reactor is heated to 110°C rapidly and kept at 110–115°C for 1½ hours. After this period the reaction mass is cooled rapidly to 70°C and dropped into water or a mixture of water and alcohol, the amounts depending on the final concentration required and the viscosity of the finished sulphate. pH is then adjusted with ammonia to 6.5 to 7.

Sulphosuccinates and Sulphosuccinamates

Synthesis of these materials can be done with relatively simple equipment, a glass-lined or stainless steel reactor equipped with a dual purpose heating/cooling jacket and a stainless steel dissolving tank.

The choice of hydrophobe is wide, it can be any of the long-chain alcohols (natural or synthetic), long-chain primary amines (if a succinamate is to be made) or an alkanolamide or a mixture of any of these.

For this reason hard and fast rules as to quantities cannot be given but it is essential for the manufacture that the (average) molecular weight be known.

For alcohols and alkanolamides, the molecular weight can be obtained from suppliers' specifications or it can be calculated from the hydroxyl number by the formula :

$$\text{Molecular weight of monoalcohol} = \frac{56,100}{\text{Hydroxyl number}}$$

The hydroxyl number if not available can be determined by standard procedures on the mixture. If an alkanolamide is being analysed for hydroxyl number, the natural alkalinity present will affect the result but as this alkalinity can also be a factor in the coupling to the maleic anhydride, no correction need be made.

For amines the hydroxyl procedure will also give a result for molecular (or equivalent) weight but a simpler method is to determine the base number by titrating in alcoholic solution with acid using bromcresol green indicator.

The hydrophobe, which needs to be practically anhydrous, is charged into the reactor and heated to 60–70°C. Heating is now stopped and maleic anhydride in the proportion of 1.04 moles of the anhydride calculated on the hydrophobe, is added slowly. Maleic anhydride is normally supplied as lenses or flakes and the heat required to melt roughly balances the heat of reaction of the esterification. The rate of addition of the anhydride is adjusted so that the temperature remains within the range 60–70°C.

Concurrently with this the sodium sulphite or bisulphite is dissolved in water to make an approximately 25 per cent solution. The sulphite should be freshly prepared and free of thiosulphite. If monoethanolamine sulphite is to be used, prepare a solution of 8 per cent monoethanolamine in water and bubble in approximately half its weight of SO_2, for the sulphite, and an equal weight for the bisulphite. Completeness of the reaction can be checked by alkalinity and iodine titrations.

In every case the amount of sulphite salt should be the molar equivalent of the amount of maleic anhydride used. For determining these figures we give the molecular weights of the reactants :

Maleic anhydride	91.1
Maleic acid	134.1
Na_2SO_3	126.1
$Na_2SO_3.7H_2O$	252.2
$Na_2S_2O_5$ (the anhydrous form of sodium bisulphite)	190.1 (equivalent weight 95.05)
Monoethanolamine sulphite	187.0
Monoethanolamine bisulphite	126.6

When all the maleic anhydride has been added, maintain a temperature of 60–70°C for 1 hour and check acidity. Figures for acidity should be set as a standard based on laboratory work, because if the half-ester is being produced this also gives an acid reaction in addition to the unreacted maleic anhydride. For the di-ester only maleic acid will react if no half-ester is formed concurrently.

Once the reaction is found to be complete, the maleic ester (or amide) is transferred by gravity or pump to the solution tank containing the sulphite solution. Stirring is continued for 1 hour without further heating till the residual sulphite content has fallen to below 0.5 per cent (checked by iodimetric titration). Surplus sulphite is then destroyed by adding peroxide. This procedure produces a paste of 40–45 per cent active matter.

Alkane (Paraffin) Sulphonates

This group of anionic detergents, initially obtained by sulpho-chlorination under ultra-violet irradiation. The reaction involved is as follows :

$$R_1 \cdot CH_2 \cdot R_2 + SO_2 + Cl_2 \rightarrow R_1 \cdot CHR_2 + HCl$$
$$\underset{SO_2Cl}{|}$$

The source of the paraffins being the coal hydrogenation process. This process is carried out in special reactors under the influence of ultra-violet rays. The next step is saponification of the chlorosulphonate by treatment with NaOH :

$$R_1 . CHR_2 + 2NaOH \rightarrow R_1 . CHR_2 + NaCl + H_2O$$
$$\underset{SO_2Cl}{|} \qquad\qquad \underset{SO_3Na}{|}$$

In brief, water and n-paraffin (which needs to be specially purified, are fed into a photochemical reactor equipped with lamps emitting ultra-violet radiation of wavelength between 3300 and 3600 Å. Sulphur dioxide and oxygen are introduced with vigorous agitation.

In another process sulphur is burnt in oxygen to produce the correct ratio of SO_2 to O_2 for sulphoxidation. Separation of the sulphoxidised paraffin from by-product dilute sulphuric acid is performed by extraction with hexanol.

The paraffin for the ATO process is n-paraffin from the appropriate petroleum fraction separated from the iso-paraffins by molecular sieve. It has to be practically free of aromatics the presence of which would severely inhibit the sulphoxidation reaction.

Pressure and temperature in the reactor are close to ambient. As the reaction proceeds, a portion of the reaction mass is withdrawn to be replaced by fresh or recycled reactants.

The outflow is passed to a decanter and consists of sulphonic acid, n-paraffins, dissolved SO_2, water and sulphuric acid. This is separated into two phases, n-paraffins which are recycled to the top of the reactor, and a sulphonic/sulphuric acid mixture.

The aqueous phase is degassed and passed to the solvent extraction unit, where it is mixed with an alcohol of at least five carbon atoms. This again separates into two phases, sulphonic acid/solvent and a sulphuric acid solution in water of about 25 per cent acid.

The sulphuric acid solution can be either neutralised to sodium sulphate (for captive use) or after concentration sent to a sulphuric acid/fertiliser plant.

The organic phase is neutralised with caustic soda and contains sodium alkane sulphonate, solvent, water and unreacted paraffin. The water and solvent are stripped off under vacuum and finally the unreacted paraffin is distilled off and recycled.

A flow sheet for the complete process is shown in Fig. 9.9.

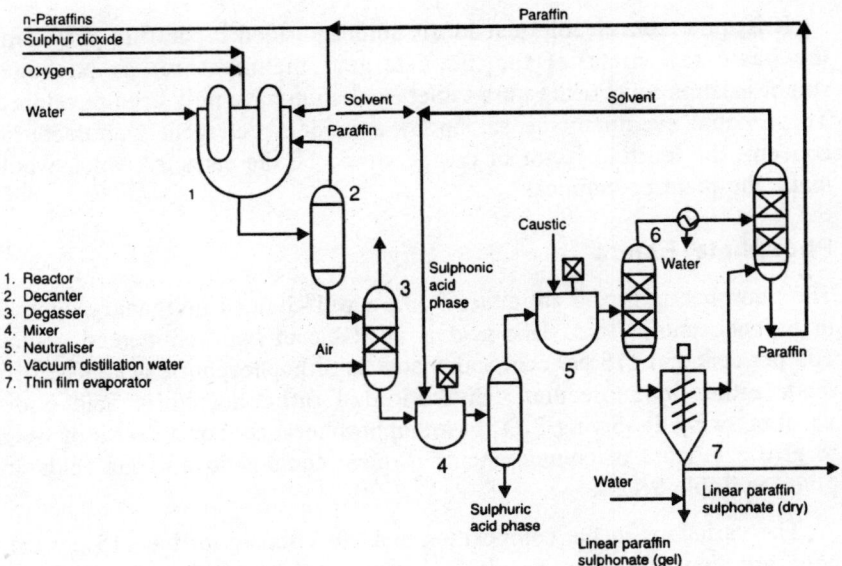

Fig. 9.9. Process for making linear paraffin sulphonate detergent.

To produce 1000 kg active matter the plant uses the following raw materials :

n-Paraffins	675 kg
SO_2	410 kg
Oxygen	110 kg
Caustic soda (100%)	145 kg
Process water	1865 kg
and utilities :	
Cooling water	310 m³
Steam	2000 kg
Power	0.65 k WH
Fuel	215,000 kcal

As far as the detergent processor is concerned it would be better if future development would lead to a change in the raw material for sulphonation (either straight-chain alkylates or synthetic fatty alcohols) rather than in the production of ready-sulphonated material supplied by giant concerns, which leaves nothing to the detergent industry but the last stages of processing: spray-drying and compounding. Producers of primary raw materials for anionic detergents have recently begun to give special attention to this requirement, and are now supplying manufacturers with new raw materials rather than finished products.

It is quite conceivable that in the future purified n-paraffins may form the basic raw material for the detergent manufacturer as packaged sulphoxidation units using ultra-violet irradiation have now been developed. These could eventually be set up by a modern detergent manufacturing concern; the limiting factor of course would be the capacity which would make the plant economical.

Phosphate Esters

The phosphating media are either P_2O_5, a well-defined inorganic chemical, or polyphosphoric acid. This acid is available in two commercial grades, 105 per cent and 115 per cent, calculated as orthophosphoric acid. They are made either by molecular dehydration of orthophosphoric acid under vacuum, or by dissolving P_2O_5 in orthophosphoric acid or a dearth of water to give a mixture of complex acids. Typical compositions of the acids are given in Table 9.12.

The variations in the composition and the viscosity of the 115 per cent acid are dependent on the P_2O_5 content which differs for different manufacturers. Handling of this polyphosphoric acid at ambient temperatures is difficult and the containers should be heated over 50°C for efficient flowing or pumping. It is also to be noted that these acids are hygroscopic. Absorption of water tends to hydrolyse the complex acids to orthophosphoric acid rapidly.

Table 9.12. Composition of polyphosphoric acids.

	105% acid	115% acid
Orthophosphoric acid (H_3PO_4)	49–54	3–5
Pyrophosphoric acid (H_4PO_7)	41–42	9–16
Triphosphoric acid ($H_5P_3O_{10}$)	5–8	10–17
Tetraphosphoric acid ($H_6P_4O_{13}$)	0–1	11–16
Higher acids	–	46–67
Total P_2O_5%	76	83.5–84.5
Density g/cm³	1.92	2.06
Viscosity at 25°C Pa s	0.8	30–60
at 100°C Pa s	0.035	0.5–1.7

The favoured acid for phosphating is the 115 per cent acid but for very special reactions the 105 per cent acid suffices.

NON-IONIC DETERGENTS

Ethylene and Propylene Derivatives

The most important non-ionic detergents are those obtained by reacting various hydrophobic groups with ethylene oxide or propylene oxide. It is essential that any compounds to be condensed, with either ethylene oxide or propylene oxide, should contain a suitably reactive group, and Table 9.13 shows the principal types of reaction involved.

Before describing the various groups of compounds for condensing, it will be useful to outline the processes for making ethylene oxide and propylene oxide. Ethylene oxide is produced, either by reacting ethylene with hypochlorite followed by hydrolysis with calcium hydroxide or by direct oxidation of ethylene with oxygen. It is essentially a two-stage process, in which ethylene is first chlorohydrinated and then hydrolysed with calcium hydroxide. The reactions may be represented by the following equations :

$$CH_2 : CH_2 + HOCl \longrightarrow CH_2ClCH_2OH$$

$$2CH_2ClCH_2OH + Ca(OH)_2 \longrightarrow 2O . CH_2 . CH_2 + CaCl_2 + 2H_2O$$

Ethylene dichloride, CH_2ClCH_2Cl and beta-beta'-dichloroethyl ether, $O(C_2H_4Cl)_2$ are produced by side reactions.

Table 9.13. Preparation of non-ionic surfactants from ethylene oxide.

Hydrophobic portion	Product

n and n' = the number of ethylene molecules

This process is presently of historical interest only.

In the newer process, ethylene reacts with oxygen or air at temperatures between 220 and 300°C over a silver catalyst at elevated pressure.

The reaction involved is as follows :

A competing reaction which can take place is :

$$CH_2 = CH_2 + 3O_2 \rightarrow 2CO_2 + 2H_2O$$

This obviously reduces the yield but modern technology has settled on reaction parameters to minimise this reaction. In modern plants the yield is 70–72 per cent, thus 100 kg ethylene oxide will produce approximately 100 kg ethylene oxide.

Initially the conditions were that a mixture of 3–5 volume per cent ethylene in air was fed into a tubular reactor with a fixed bed supported silver catalyst at a pressure of 10–20 bar with a temperature of 220–280°C. Selectivity is highly sensitive to temperature, dropping as temperature rises. The reaction is highly exothermic and temperature control is maintained by heat exchangers with a circulating heat transfer agent.

The gas leaving the reactor is cooled by another heat exchanger, and scrubbed by water to retain the ethylene oxide produced. The waste gas, containing unreacted ethylene is then split, part is fed back as recycle to the reactor and part is passed through a second reactor for further reaction. Ethylene oxide is then stripped from the water solution and condensed.

Many inhibitors have been suggested to lower the competing carbon dioxide reaction, the one in most common use is ethylene dichloride.

Modern plants, instead of using air as the source of oxygen use pure oxygen, which eliminates the necessity of purging the system of inert and possibly reactive components.

Propylene oxide is obtained from propylene generally via its chlorohydrin by reacting it with hypochlorite similar to the Wurtz reaction described for ethylene oxide. A recent process for the direct oxidation of propylene to propylene oxide has been developed.

Further developments are the production by oxidation of propylene by hydrogen peroxide via perpropionic acid and still another by the epoxidation of propylene with peracetic acid.

Tables 9.14 and 9.15 give selected physical properties of pure ethylene oxide and propylene oxide. The commercially produced compounds are very pure and have virtually the same properties.

Table 9.14. Physical properties of pure ethylene oxide.

Auto-ignition temperature	571°C
Boiling-point (101.1 kPa)	10.7°C
Explosive limits in air by vol %	3–100%
Flash-point (open cup)	< –17.8°C
Heat of vapourisation (101.1 kPa)	569.4 J/g
Molecular weight	44.05
Refractive index (8.4°C)	1.3599
Solubility in water	Complete
Specific gravity 20/20°C	0.8711
Specific heat, liquid (0°C)	0.44
Vapour pressure (20°C)	146.0 kPa
Viscosity (0°C)	0.32 mPa s
Density (20°C)	870 kg/m^2

Table 9.15. Physical properties of pure propylene oxide.

Boiling-point (101.1 kPa)	34.1°C
Explosive limits in air by vol %	2.1–38.5%
Flash-point (open cup)	– 37.2°C
Freezing-point	–104°C
Heat of vapourisation (101.1 kPa)	372.6 kJ/kg
Molecular weight	58.08
Refractive index, n_D (20°C)	1.3657
Solubility in water, weight % (20°C)	12.5
Specific gravity 20/20°C	0.8305
Specific heat, kJ kg^{-1} K^{-1} (20°C)	2.1304
Vapour pressure (20°C)	60.5 kPa
Viscosity (20°C)	0.33 mPa s
Density (20°C)	8.29 kg/m^3

The handling and storage of ethylene oxide and propylene oxide, as well as their reactions, must be carried out with the utmost caution. When starting a plant where ethylene and/or propylene oxide is used it is advisable to contact the supplier for details of safety precautions, and to follow these instructions conscientiously. An outline of these precautions will be given later, when we describe ethylene oxide reactions.

We will now describe some of the compounds that constitute the hydrophobic portion of surface active ethylene oxide compounds. We will confine ourselves to three main groups : (a) fatty acids; (b) fatty alcohols; and (c) alkyl phenols.

Fatty acids react with ethylene oxide to form esters. The same product can theoretically be formed by the esterification of the fatty acid with the appropriate polyethylene glycol, but in practice some di-ester is formed.

Table 9.16 shows the properties of fatty acids (representative commercial products) suitable for reaction with either ethylene oxide or propylene oxide, and also for reaction with alkylolamines. The properties of fatty alcohols, natural and synthetic, have already been described.

Alkyl phenol ethylene oxide condensation products still comprise a substantial proportion of the total world production of non-ionic detergents. This in spite of the fact that the alkyl phenol ethoxylates, obtained from polymerised propylene or butylene, represent a similar sewage problem to that caused by anionic detergents derived from PT benzene; less serious, however, owing to their low foam property.

The most important alkyl phenols are C_8 to C_{12} phenols. They are obtained by alkylation of phenol with the corresponding olefins. The olefins are the dimeric or trimeric polymers of propylene or butylene or their mixture.

Table 9.16. Properties of fatty acids.

	Titre °C	Acid value	Saponification value	Iodine value	Unsaponifiable %
Lauric acid 94/96%	39–41	279–84	279–84	max 0.5	0.5
Lauric acid 98/100%	43–43.6	279–82	279–82	max 0.2	0.2
Dist. stripped coconut oil fatty acid	25–28	250–60	252–62	8–14	0.5
Dist. stripped palm-kernel oil fatty acid	23–27	250–60	250–60	16–22	1.0
Myristic acid 94/96%	50–53	244–54	245–55	max 1.0	0.5
Myristic acid 98/100%	51–54	244–54	245–55	max 1.0	0.5
Palmitic acid 94/96%	60–61	215–25	215–25	max 2.0	0.5
Palmitic acid 98/100%	61–62	219–24	219–24	max 2.0	0.5
Stearic acid 94/96%	64–66	195–8	196–9	1–3	1.5
Stearic acid 98/100%	66–68	195–9	196–200	max 2.0	1.5
Stearine single pressed	52–53	207–11	207–11	8–10	0.5
Stearine double pressed	53–54	208–11	208–11	5–8	0.5
Stearine triple pressed	54–56	207–11	207–11	1.5–5	max 0.5
Dist. tallow fatty acid	39–43	204–8	205–9	53–57	max 1.5
Oleine light coloured	7–10	186–204	188–206	85–92	2–6
Dist. palm oil fatty acid	42–46	207–13	209–15	44–54	max 1.5
Dist. refined tall oil light coloured	–	188–93	190–5	155–65	1.5–2.2
Tall oil refined	–	163–8	165–70	153–67	7–8

Table 9.17 gives the properties of three of the most important alkyl phenols for the production of surface active ethylene oxide condensates.

Before discussing laboratory and industrial ethoxylation processes, a general remark must be made on the hydrophobe/hydrophile balance of ethylene oxide and propylene oxide condensates: the higher the molecular weight of the hydrophobe component, the more moles of ethylene oxide (or propylene oxide) are needed to produce water-soluble condensates. Conversely, the lower the molecular weight, the smaller the number of moles required to obtain a water-soluble product. For the purpose of producing non-ionic detergents, it is generally desirable that the products be easily soluble in water. For the production of non-ionic emulsifying agents, on the other hand, a certain affinity to the oil phase is preferable. Thus, to condense the hydrophobe component, fewer moles of ethylene oxide will be needed.

As has been mentioned, the production of ethylene and propylene oxide condensates requires a great deal of experience and it is thus advisable to study the reaction of ethoxylation on a laboratory scale. We give details of a laboratory synthesis of alcohol ethoxylates and alkyl phenol ethoxylates.

Quantities of about a pound are readily made in the laboratory by the following procedure : A 1-litre four-necked flask, fitted with a water-cooled stirrer, condenser, thermometer, and gas-dispersion thimble is connected through two 500 ml safety bottles to a graduate containing white oil and a manifold connected to low-pressure nitrogen and a tank of ethylene oxide (Fig. 9.10). 200 g (1.00 mol) of tridecyl alcohol is charged to the flask followed by 0.4 to 0.5 g of powdered sodium hydroxide. The flask, thermometer, gas disperser, and contents are weighed. A slow stream of nitrogen is started through the alcohol and the temperature raised to 150–160°C by means of a glass-insulated heating mantle. Ethylene oxide gas is admitted to the manifold and allowed to escape momentarily to the atmosphere through the three-way stopcock. The three-way stopcock is then turned to vent the nitrogen to the atmosphere and the ethylene oxide is simultaneously introduced into the flask. The operation should be carried out in a well-ventilated hood and away from naked flames.

Absorption takes place immediately, with a rise in temperature. The ethylene oxide rate is increased to the point where an occasional bubble of ethylene oxide escapes through the white oil. The temperature is permitted to rise to 180–220°C, and is maintained within this range until the required amount of ethylene oxide has been absorbed. For trial purposes, between 8.5 and 10 moles of ethylene oxide per mole of alcohol is suitable. This is equivalent to an absorption of 374–440 g of ethylene oxide. Approximately three to six hours are required to absorb the above quantities.

In shutting down, the ethylene oxide is vented to the air, and the nitrogen is turned into the flask and a slow stream maintained until all of the ethylene oxide has been purged from the system. The flask and contents are weighed to determine the ethylene oxide absorption.

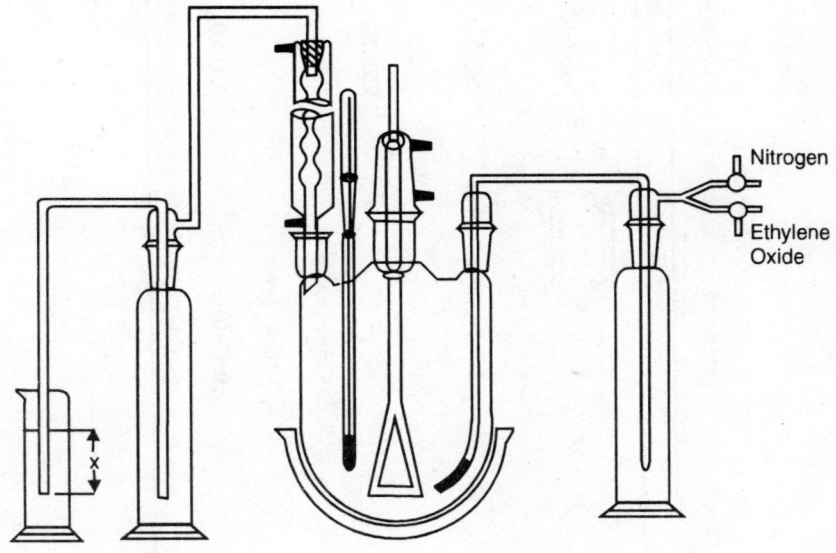

x= Hydrostatic head equivalent to 2.5 Cm Hg

Fig. 9.10. Laboratory ethoxylation plant.

When the desired quantity of ethylene oxide has been absorbed, the free alkali is neutralised with either 30 per cent sulphuric acid or glacial acetic acid to a phenolphthalein end point (external indicator). Ten grams of decolourising carbon are then added and the mixture stirred and heated at 100°C for 15 minutes. The mixture is filtered through a Büchner funnel using a small amount of a filter-aid on the paper as a precoat in order to ensure a bright filtrate. The small amount of water resulting from neutralisation can be permitted to remain, or may be stripped from the product by heating under reduced pressure.

Another example is for the ethoxylation of 1 mole C_{12} lauryl alcohol (either natural or synthetic) with 4 moles of ethylene oxide :

Table 9.17. Properties of three important alkyl phenols for the production of surface active ethylene oxide condensates.

	Octylphenyl	Nonylphenol	Dodecylphenyl
Formula	C_8H_{17} —OH	C_9H_{19} —OH	$C_{12}H_{25}$ —OH
Molecular weight (theoretical)	206.2	220.2	262.2
Molecular weight of commercial products	200–7	221–4	262–7
Appearance	Yellowish crystalline mass	Colourless to yellowish oil	Yellowish oil
Density 30°/4°	–	0.93–0.95	0.92–0.94
Distillation range at 101.1 kPa	280–95°C	290–315°C	330–35°C

Lauryl alcohol C_{12}	186 g (1 mol)
Ethylene oxide	176 g (4 mol)
NaOH powdered	0.6 g

Proceed as described for tridecyl alcohol ethoxylate, including the same purification method.

Finally an example for the ethoxylation of nonylphenol with 9.5 moles ethylene oxide :

Nonylphenol	215 g (1 mol)
Ethylene oxide	418 g (9.5 mol)
NaOH powdered	0.7 g

The same procedure is used.

This account of a laboratory ethoxylation synthesis makes it easier to understand the description of an industrial batch ethoxylation plant.

An ethylene oxide kettle should have provisions for maintaining the desired reaction temperatures. Heating-cooling coils and a jacket are usually suitable. The kettle should be designed for predetermined pressures. A pressure-relief valve and rupture disc should be provided. A suitable vent should be supplied which will vent to a safe area through a flame arrester. The speed with which ethylene oxide reacts with other materials depends on intimate mixing; therefore, care should be exercised in the choice of the agitation system. Stainless steel or glass linings are recommended if the quality, especially colour, of the reaction product is important. Carbon steel is suitable for other purposes. Acetylide-forming metals such as copper, silver, magnesium and their alloys should never be used in direct contact with ethylene oxide.

Heating of the charge can be accomplished with steam or some other heat-transfer medium. Cooling water should be available to remove heat of reaction, for emergency cooling, and for routine cooling of the finished product preparatory to removing it from the kettle. Precautions should be taken to avoid overcooling the reaction mass and thus allowing the accumulation of unreacted ethylene oxide in the kettle. With an excess of unreacted ethylene oxide in the vessel, the reaction could proceed with explosive violence. The inert gas, preferably nitrogen or methane, used to purge the system should be free of impurities such as oxygen, ammonia, hydrogen sulphide and acetylene. Prior to purging with an inert gas, vacuum equipment must be used to remove most of the air. Suitable facilities should be provided as required for introducing the charge materials, cooling water and inert gas. Automatic shut-off of ethylene oxide feed should be available for activation by : (1) drop in temperature below that at which the reaction rate is enough to consume the ethylene oxide being introduced; or (2) rise in pressure or temperature above that which has been determined as safe for the design of the equipment.

An intermediate ethylene oxide charge tank, located between the reaction kettle and ethylene oxide storage tank, should be available to serve as a work tank and to prevent the backflow of catalyst from the reactor to the ethylene oxide storage vessel. The work tank might also be used as a volumetric or weighing tank. Ethylene oxide transfer lines to process equipment should contain at least one check valve. Adequate storage facilities should be available for the other reactants and the product. Processing equipment and transfer lines should be thoroughly cleaned, leak tested and purged free of air before being placed in service. The equipment and transfer lines should be electrically bonded and grounded. Explosion-proof motors and spark-proof tools should be used.

The catalyst and oxide accepter are charged to the kettle. The charge is purged free of oxygen and brought to the temperature at which the reaction may be initiated. Ethylene oxide is cautiously added to the kettle. A temperature rise will indicate that the reaction has begun, at which point the ethylene oxide feed rate may be increased to the desired level, but never to a rate greater than that at which it is being consumed. The build-up of unreacted ethylene oxide in the vessel must be carefully avoided. Most ethylene oxide reactions take place in the temperature range of 120–200°C and in the pressure range of 205–410 kPa.

In general, ethylene oxide reactions are exothermic in nature, giving up about 80 kJ/mol of ethylene oxide reacted. This heat of reaction must be removed. Overcooling may result in stopping the reaction and allow accumulation of unreacted ethylene oxide in the vessel. The reaction may then resume and proceed with explosive violence.

After the predetermined amount of ethylene oxide has been added, flow to the kettle is stopped. The temperature of the kettle is held for a time to allow the last of the ethylene oxide to react, as indicated by a gradual drop in pressure. When the pressure is steady, the product is cooled, treated where necessary and removed from the kettle. Minimal moisture content of reactants is essential, otherwise side reaction with the formation of polyethylene glycol occurs.

Completely enclosed explosion-proof magnetic mixers, without a stuffing-box, are very much to be recommend as agitators for the reactor.

Furthermore, it is our experience that the safety of the set-up is greatly increased by a side-arm heat exchanger. Reaction time and reagent flow should be similar to the laboratory procedure, so that the process on a plant scale is, in fact, an upscaled laboratory process. Many ethylene oxide (and propylene oxide) condensation products and the processes for their production are subject to patents in various countries. Thus it is always advisable to study the patent situation carefully, before starting with the production of ethylene oxide and/or propylene oxide condensation products.

It must be mentioned that the condensation reaction may be carried out in a continuous 'run-through' pressure tube reactor. Such a system employs, e.g., an 'in line' mixing system (without moving parts). The reactants are fed in proper proportion into the tube reactor by means of dosing pumps. Heat control—and thus control of reaction—is, in fact, easier than for a batch reactor system. The quantity of reaction mixture in the continuous reactor is relatively small; and the safety of such a system is to a certain extent easier to guarantee than for a batch system. When using a continuous system, the NaOH catalyst is pre-mixed with the alcohol or alkyl phenol respectively, to be condensed with ethylene oxide or propylene oxide. The refining step with acid and active carbon can be carried out after the continuous condensation reaction is accomplished, either in a subsequent batch refining and filtering system or on a continuous basis.

In a commercial plant for the production of ethylene oxide adducts the production of 2500 kg of C_{16}—C_{18} alcohol natural or synthetic with 25 mol ethylene oxide condensed requires :

C_{16}—C_{18} alcohol	1250 kg
Ethylene oxide	1250 kg
Steam	150 kg
Nitrogen	5–6m^3
Cooling water at 15–20°C	16 m^3
Electricity	45 kW

Operation of the plant is as follows (refer to Fig. 9.11).

A preset amount of material to be reacted with ethylene oxide and/or propylene oxide by the RP programming system together with the catalyst is fed into the process unit through line 1. (The amount of catalyst is generally less than that usually introduced in a conventional ethoxylation plant, and this has a beneficial effect on the colour and purity of the end product.) After the starting material and catalyst have been introduced, the unit is set under vacuum, heating begins through F 1, and the recirculation through circulation pump PC 1 is set into operation. The mixture flows again into the upper part of SA 1, the gas–liquid contactor, and the desirable temperature is maintained by F 2. During this phase of operation the material is de-acrated and de-humidified. This operation is carried out in a very short period of time. Low residual moisture is essential to prevent any undesirable side reactions, e.g., the formation of polyglycols and irregularity in the formation of the ethylene oxide and/or propylene oxide chain. Quick nitrogen injection and venting is then carried out; some nitrogen pressure is left in the system, and the actual condensation reaction starts by feeding the preset (RP programming system) amount of ethylene and/or propylene oxide into the pre-treated starting material and catalyst. Recirculation through circulation pump PC 1 into the gas–liquid SA 1 is carried out and the heat of reaction removed by the cooling heat exchanger F 2. In some cases it is desirable to pass the cooling medium, either water or a diathermic fluid, through a refrigeration system in a closed circuit.

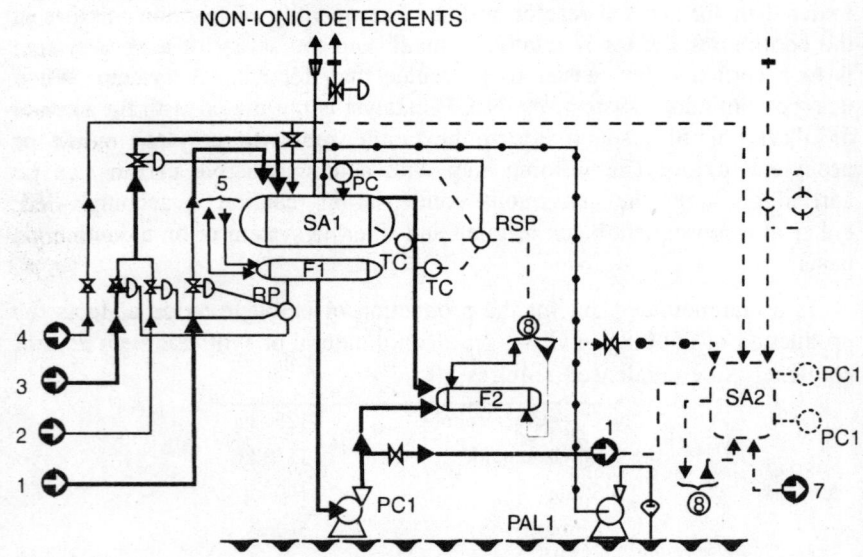

1. Basic raw material (eg alkylphenol or fatty alcohol)
2. Ethylene oxide from storage
3. Propylene oxide from storage
4. Nitrogen for purging and pressurising
5. Heat transfer fluid for reaction start-up
6. Cooling circuit
7. End product discharge
8. Cooling-water for final cooling of product
SA 1. Special 'Pressindustria' gas/liquid contactor

RSP. Automatic reaction rate governing system.
RP. Automatic programming system (formula preset system)
SA 2. Cooler (optional) for reaction product
PAL 1. Vacuum pump
PC 1. Process circulation pump
F2. Heat-exchanger for reaction heat control
F 1. Heater for reaction start-up

Fig. 9.11. Pressindustria plant for the manufacture of ethylene oxide adducts.

An automatic control system regulates the flow of ethylene oxide and/ or propylene oxide in relation to the temperature of the reaction mixture. This feedback system guarantees a constant reaction rate and constant temperature which accounts for uniform quality of the end product. The end product is cooled down by the F 2 cooling system and by the optional additional cooling vessel SA 2. No bleaching of the end product is required.

In small pilot plant for the continuous production of ethylene oxide the reactor is in two parts, the first reactor working at a temperature between 170 and 240°C giving a condensate of molecular ratio 1:1 to 1:6, and the balance of the ethylene oxide being added in the second reactor at a temperature of 240–360°C. As a catalyst 0.1 to 1.5 per cent sodium as sodium methylate is used. Pressure is between 5 and 10 MPa and the residence time between 8 and 150 seconds. Use of continuous type reactors involving both small charges and contact time can obviously lessen considerably the danger inherent in this type of reaction.

In the last twenty years or so the block polymers have assumed increasing importance. Their special qualities are the variable hydrophobe/ hydrophile properties, and the fact that most of them are produced as solid flakes, which makes them easy to incorporate, e.g., into powder detergent compounds.

Polyoxyethylene compounds are water-soluble—no matter how high their molecular weight. For many years it was generally assumed that polyoxypropylene compounds would also be water-soluble. However, at a molecular weight of about 800–900, polyoxypropylene glycols become essentially water-insoluble. As the molecular weight of this chain increases it becomes more hydrophobic.

If, however, water-soluble polyoxyethylene groups are added to both ends of the water-insoluble polyoxypropylene chain, a complete series of new block polymers with highly desirable surface active properties can be obtained.

These block polymers are prepared by adding propylene oxide to the two hydroxyl groups of a propylene glycol nucleus. The resulting hydrophobic base can be made to any controlled length varying from 800 to many thousands in molecular weight.

By adding ethylene oxide to both ends of this hydrophobic base, it is possible to put polyoxyethylene hydrophilic groups on the ends of the molecule. These hydrophilic groups are controlled in length to constitute anywhere from 10 to 80 per cent of the final molecule. The simplified structure can be represented as :

$$HO(CH_2-CH_2-O)_a(CH_2-CH_2-O)_b(CH_2-CH_2-O)_cH$$
$$\underset{CH_3}{|}$$

If the hydrophobe portion of a normal non-ionic is calculated it will be found to be of the order of 30–40 per cent of the finished product. Block polymers are produced by (usually) propoxylating and then ethoxylating on to an initiator. This initiator is seldom 5 per cent, usually ½–1 per cent of the final product. It can be seen that in batch process it will be extremely difficult to start the process with such a small amount of material in the reactor. It is common practice to produce a 'precursor', of the initiator with a definite portion of (say) propylene oxide. This precursor is then stored till required and the requisite amount(s) of ethylene and/or propylene oxides are added.

The newest development in the non-ionic detergent field is the extra-low-foam products. As has been mentioned, ethylene oxide condensates, especially with a short ethylene oxide chain (4–5 mol), produce poor foam but for all that the amount of foam generated in certain operations can still be a hindrance.

Modified non-ionics, usually made by coupling an alkyl group to the terminal —OH, are now available and are used in household powders for foam sensitive machines and more particularly for dishwashing machine powders which also contain active chlorine, as the blocking of the terminal —OH reduces the tendency for degradation of the material producing the active chlorine.

Block polymers specially designed to give no (not low) foam are available and these are being used successfully in automatic dishwashing machine detergents, both liquid and powdered.

Fatty Acid Alkanolamides

This important class of non-ionic detergents is derived by condensing alkanolamine with fatty acids. These are the so-called fatty acid alkanolamides.

Alkanolamines combine readily with long-chain fatty acids, such as oleic and stearic acids, to give neutral alkanolamine soaps.

$$HOCHRCH_2NH_2 + C_{17}H_{35}COOH \rightarrow HOCHRCH_2NH_2 . HOOCC_{17}H_{35}$$

This neutralisation reaction takes place at room temperature. The products are waxy, non-crystalline materials which have widespread commercial application as emulsifiers.

At elevated temperatures (140–160°C) n-alkylol amides are the chief products formed by reacting mono- and dialkanolamines with fatty acids in a 1:1 ratio.

$$HOCHRCH_2NH_2 + R'COOH \rightarrow HOCHRCH_2NH\overset{\overset{\displaystyle O}{\|}}{-C}-R' + H_2O$$

At the same time, significant quantities of amine esters and amide esters are formed by side reactions involving the hydroxyl moiety.

$$HOCHRCH_2NH_2 + R'COOH \rightarrow H_2NCH_2CHROOCR' + H_2O$$

$$HOCHRCH_2NH\overset{\overset{\displaystyle O}{\|}}{-C}-R' + R''COOH \rightarrow R''COOCHRCH_2NH-$$

$$\overset{\overset{\displaystyle O}{\|}}{-C}-R' + H_2O$$

In addition, when dialkanolamines are the starting materials, small amounts of amine diesters and amide diesters result.

$$(HOCHRCH_2)_2NH + 2R'COOH \rightarrow HN(CH_2CHROOCR')_2 + 2H_2O$$

$$(HOCHRCH_2)_2N\overset{\overset{\displaystyle O}{\|}}{-C}-R' + 2R''COOH \rightarrow (R''COOCHRCH_2)_2-$$

$$\overset{\overset{\displaystyle O}{\|}}{N-C}-R' + 2H_2O$$

When diethanolamine is the starting dialkanolamine, some morpholine and some piperazine derivatives are also obtained.

Reaction of dialkanolamines with fatty acids in a 2:1 ratio at 140–160°C gives a second major type of alkanolamide. These products, in contrast to the 1:1 alkanolamides, display aqueous solubility. In composition, they are complex mixtures which comprise n-alkylol amides, amine esters and di-esters, plus a considerable percentage of unreacted dialkanolamine. This latter constituent chiefly accounts for the aqueous solubility of the 2:1 dialkanolamide.

Both the 1:1 and the 2:1 alkanolamides are of commercial importance as detergents and detergent additives.

An improved process comprised mixing an ester of an aromatic or aliphatic carboxylic acid with an alkanolamine, adding an alkali metal alkoxide catalyst and heating the mixture to 100°C at atmospheric or above atmospheric pressures.

A continuous process for making fatty alkanolamides in a thin-film reactor. The reaction involved condensation of a methyl ester of a fatty acid with a mono- or dialkanolamine, in the presence of an alkali metal, alkali metal alkoxide, or alkali metal amide catalyst. A short contact time in the reactor produces a high purity alkanolamide.

More recently another process for making high purity alkanolamides has been developed. A fatty acid is first reacted with an excess of alkanolamine, forming amine and amide esters in addition to the intended unsubstituted alkanolamide. In a second step involving an alkali metal catalyst, the amine and amide esters are converted to the unsubstituted alkanolamide.

Also a process of making high-purity alkanolamides by condensing a dialkanolamine with a fatty triglyceride, then adding phosphoric acid to remove the excess amine and most of the glycerine by-product.

There are two types of alkanolamide products. The first is the Kritchevsky-type liquid product, made by reacting an alkanolamine with a fatty acid or fatty acid derivative at elevated temperatures in a 2:1 ratio. Such a product contains 60–70 per cent alkanolamide, plus some amine esters and di-esters, amide esters and di-esters, and piperazine derivatives that are formed by side reactions. In addition there is significant unreacted alkanolamine. This excess alkanolamine renders the Kritchevsky-type alkanolamides water soluble.

The second type of alkanolamide is the so-called 'super' amide, prepared by reacting an alkanolamine and a fatty acid ester in a 1:1 ratio. These are generally solid products which have an alkanolamide content above 90 per cent. Some of the same by-products formed in preparing 2:1 alkanolamide are likewise formed in preparing super amides, but in much smaller quantities. For this reason, and because they contain only relatively small amounts of free alkanolamine, super amides have poor water-solubility. They are therefore always used in conjunction with a small amount of anionic or non-ionic surfactant which acts as a solubiliser, converting an aqueous alkanolamide dispersion into a viscous, clear solution. The isopropanolamine-based alkanolamides do however show an increased solubility in water over the ethanolamine types.

The starting material for the super amides are the methyl esters of fatty acids.

For the Kritchevsky type of fatty acid diethanolamide no catalyst is needed. The high surplus of diethanolamine (DEA) makes this unnecessary.

One mol lauric acid is reacted with 2 mol DEA in an electrically heated stainless steel reactor fitted with an agitator and a sparger-pipe for the introduction of nitrogen as purging gas. The reaction is carried out at 160°C for 5 h. (Purge nitrogen is introduced when the reaction mass has been heated to about 120°C). The reaction may be controlled by having a water trap and cooler fitted to the reactor, so that the amount of water (1 mol) liberated during the reaction may be measured. Of course, any water present in the DEA must be taken into account. In the same manner other fatty acid diethanolamides can be produced.

'Superamides' are generally prepared by reacting methyl esters of fatty acids with diethanolamine, using sodium methylate as a catalyst.

Recently it has been found :

1. The optimum mol ratio of DEA to methyl ester is 1.1:1.

2. The optimum temperature is 105°C if the reaction is carried out at atmospheric pressure.

3. There is very little difference in the catalystic effect of sodium methylate between 0.15 and 0.25 per cent, calculated on the batch as a whole and under atmospheric pressure.

4. The undesirable amine and amide esters are not stable in sodium methylate catalysed reactions and if formed will revert to the amide.

All the above parameters yielded an amide content verging on 90 per cent. To produce over 90 per cent it was found that a reduced pressure (4 kPa) and temperature (60°C) with 0.75 per cent catalyst were necessary.

The temperatures are all above the boiling point of methanol which is liberated by the reaction. The plant needs to be equipped with condensing units to recover the methanol for recycle.

From infra-red studies of the superamide it is found that on allowing the material to stand for a month or so at room temperature, there was a considerable drop in ester content with a corresponding rise in the amide, despite the fact that superamides are solid at room temperature.

To produce high quality superamides it is suggested to hasten this post-reaction by keeping the finished amide at 50°C for 3–4 hours when the bulk of the esters will have reverted to the amide.

It has been found that in producing monoethanolamides, a surplus of monoethanolamine is useful. This reduces reaction time, and improves the quality and odour of the end product. The surplus amine is stripped at the end of the reaction under reduced pressure. In general, operating under reduced pressure, especially towards the end of the condensation, was found to be of advantage.

In Fig. 9.12 shows the commercial production of alkanolamides, using methyl esters. This process produces the 'superamides'. If the older types are to be produced, the methyl ester is simply replaced by the equivalent fatty acid or neutral fat.

The monoethanolamides and monoisopropanolamides of fatty acids are very easy to manufacture. The production process consists in simply heating stoichiometric amounts of fatty acids or neutral fats with monoalkanolamine at about 160°C for 2–3 hours.

If neutral fats (glycerides) are used, the glycerine liberated will remain in the final product, lowering its setting point. This will also add to the emolient effect of the finished formulation.

The monoethanolamides and monoisopropanolamides are waxlike substances, practically insoluble in water, but solubilised by another hydrophilic anionic or non-ionic detergent. Table 9.18 gives the setting points and main uses of various monoethanolamides and monoisopropanolamides. These types of detergents are generally marketed in the form of flakes produced by running the molten products over chilling rolls fitted with doctor blades for scraping off the flakes.

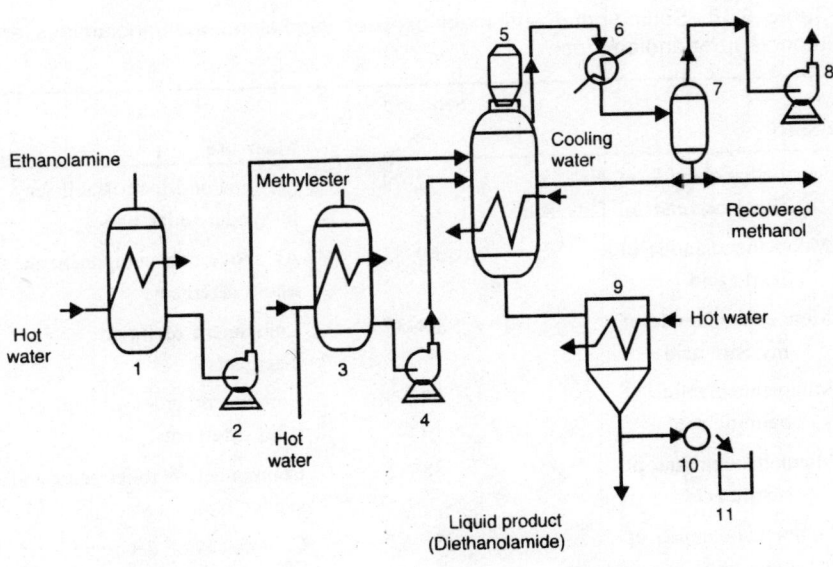

Liquid product
(Diethanolamide)

1. Ethanolamine tank
2. Ethanolamine charging pump
3. Methylester tank
4. Methylester charging pump
5. Reactor (monoethanolamide)
6. Condenser

7. Vacuum drum
8. Vacuum pump
9. Surge tank
10. Cooled flaking drum
11. Flaked product

Fig. 9.12. Plant for the production of normal and super alkylolamides.

It will be noted from the table that the isopropanolamides show a lower setting point than the monoethanolamides of the same fatty acids.

CATIONIC DETERGENTS

A very good description of the principle of manufacture and the use of these detergents is given as under :

Cationic Surface Active Agents

Cationic surface active agents contain a long-chain cation which is responsible for their surface active properties. They are marketed in solid form or as pastes or in aqueous solution. Important examples include :

Amine acetates $[R.NH_3]O_2C.CH_3(R = 8$–18 carbon atoms$)$.

Alkyl trimethyl ammonium chlorides $[R.N(CH_3)_3]Cl$ $(R = 8$–18 carbon atoms$)$.

Table 9.18. Setting-point and main uses of various monoethanolamides and monoisopropanolamides.

	Setting-point °C	Main use
Monoethanolamide of distilled coconut oil fatty acids	72–74	Perfume and foam stabiliser in syndet toilet bars
Monoethanolamide of lauric acid	80–82	As above and component of liquid detergents
Monoethanolamide of myristic acid	88–89	Component of liquid detergents
Monoethanolamide of palmitic acid	90–93	Component of powdered and solid detergents
Monoethanolamide of stearic acid	88–92	Component in toilet soap and syndet bars
Monoethanolamide of oleic acid	59–62	Component of detergent pastes
Monoisopropanolamide of distilled coconut oil fatty acid	46–50	Increases viscosity and foam stability of liquid and paste detergents
Monoisopropanolamide of lauric acid	54–58	Perfume stabiliser, 'feel improver' in syndet toilet bars.

Dialkyl dimethyl ammonium chlorides [$R_2N(CH_3)_2$]Cl (R = 8–18 carbon atoms)

Alkyl pyridinium chlorides and bromides, e.g., (R = 12–18 carbon atoms)

$$\left[\bigcirc N—R \right] Cl$$

Lauryl dimethyl benzyl ammonium chloride

$$\left[R.N(CH_3)_2—CH_2— \bigcirc \right] Cl$$

Amine acetates are produced by neutralising fatty amines with acetic acid, and are water-soluble. Quaternary ammonium compounds are normally prepared by one of four methods, viz, exhaustive alkylation of a primary or secondary fatty amine, alkylation of a low-molecular-weight tertiary amine with a fatty alkyl halide, alkylation of a tertiary fatty amine or by treating a tertiary amine or its salt with an epoxide.

In the detergent industry the main use of cationic detergents is for fabric softeners. For this purpose quaternary ammonium compounds derived from hydrogenated tallow are used; distearyl dimethyl ammonium chloride is preferred in Europe while in the USA distearyl dimethyl ammonium sulphate is also used. Although called 'distearyl' the tallow molecule contains an appreciable quantity of palmitic (C_{16}) acid and also, depending on the degree of hydrogenation, a proportion of oleyl amine (as the quaternary compound) might also be present.

Starting materials are either the fatty acid or alcohol, both of which are converted to the secondary (sometimes primary) amine.

If fatty acids are used they are reacted with ammonia successively to the ammonium salt of the fatty acid, dehydrated to the amide and again dehydrated to the nitrile :

$$R—COOH \rightarrow R—COONH_4 \rightarrow R—CONH_2 \rightarrow R—C \equiv N$$

This nitrile is then reduced (hydrogenated) to form as required the primary or secondary amine :

$$R—C \equiv N + 2H_2 \rightarrow R—CH_2NH_2$$

or

$$2R—C \equiv N + 4H_2 \rightarrow (R—CH_2)_2NH + NH_3$$

If an alcohol is to be the starting material, it is reacted with ammonia gas in the presence of a catalyst, when the nitrogen atom replaces the reactive hydrogen. The reaction for the preparation of a secondary amine is

$$2ROH + NH_3 \rightarrow R_2NH + 2H_2O$$

The primary or secondary amine can then be methylated with formaldehyde to the tertiary amine, an example of the reaction being

$$2R_2NH + 3HCHO \rightarrow 2R_2CH_3N + CO_2 + H_2O$$

or the primary or secondary amines undergo exhaustive methylation with methyl chloride according to the equations :

$$RH_2N + 3CH_3Cl \longrightarrow R(CH_3)_3N^+Cl^- + 2HCl$$

$$R_2HN + 2CH_3Cl \longrightarrow R_2(CH_3)_2N+Cl^- + HCl$$

In both instances HCl is formed and as the reaction cannot go to completion in an acid medium, it is necessary to provide a scavenger for the acid produced. Industrially sodium or potassium carbonates are used in a polar medium such as water or alcohol and at a temperature between 60 and 95°C under slight pressure.

Tertiary amines can also be methylated as above, but they are often reacted with dimethyl sulphate :

$$2R_3N + (CH_3O)_2SO_2 \rightarrow 2R_3CH_3N^+SO_4^=$$

forming the quaternary ammonium sulphate.

The same reaction can be employed on secondary amines to add two methyl groups, producing for example a di-alkyl dimethyl ammonium sulphate.

Cationic surface active agents have wetting, foaming, and emulsifying properties. They are not, however, good detergents, being substantive to solids. Most of their applications are in fields where non-ionic detergents cannot be used and depend on this substantiveness to solid surfaces, whilst others arise from their excellent germicidal properties (all have high phenol coefficients). Amine acetates are used to eliminate the normal static electrical charges on resins and plastics, as collectors and frothers for a variety of minerals, e.g., mica, phosphates, lead ores, etc., as emulsifying agents, as bactericides, and for other purposes. Alkyl trimethyl and dialkyl dimethyl ammonium chlorides are used as antistatic agents for plastics, as softening agents for textiles, as germicides, as flotation agents for the separation of certain minerals from low-grade ores, and for other purposes. The dialkyl dimethyl compounds are in addition used for emulsifying oils into water, as corrosion inhibitors, and as mould inhibitors. Cetyl tri-methylammonium bromide is used as a skin-sterilising agent, for cleansing wounds, and as a hospital disinfectant. Alkyl pyridinium chlorides and bromides and alkyl dimethyl benzyl ammonium chlorides are used in the textile industry for a variety of purposes (e.g., as antistatic and lubricating aids, in dyeing processes, and for stripping vat and azo dyes, in finishing processes, as germicidal and mildew-retarding agents); in the leather industry as bactericides and dye fixatives; as sterilising agents for plant and equipment in food production and catering, brewing, and allied industries; in medicine; in the paper industry to control moulds and slime and as assistants in dyeing.

Amphoteric detergents

Amphoteric detergents are speciality detergents, which find application mainly in the field of cosmetics.

Chapter 10

Manufacture of Finished Detergents

INTRODUCTION

Any substance that reduces the surface tension of water; specifically, a surface-active agent which concentrates at oil-water interfaces, exerts emulsifying action, and thus aids in removing soils. The older and still widely used types are the common sodium soaps of fatty acids, which are relatively weak. The much stronger synthetic detergents are classed as anionic, cationic, or nonionic, depending on their mode of chemical action. The latter functions by a hydrogen bonding mechanism. The most widely used group comprises linear alkyl sulphonates (LAS), often aided by "builders". LAS are preferable to alkyl benzene sulphonates (ABS), because they are readily decomposed by microorganisms (biodegradable). LAS are straight-chain compounds having ten or more carbon atoms in the chain; the branched-chains characteristic of ABS resist decomposition; these have been largely replaced by LAS because of water pollution.

POWDER DETERGENTS

The bulk of detergent materials are eventually converted into powders by one process or another. The problem in the production of powders is that all active detergent materials, with only a few exceptions, are not in themselves solids. It is, therefore, necessary to combine them with the builders and filling materials in such a way that the finished powder does not tend to cake or lump and remains dry to the touch. For each detergent raw material there is a limit of active material that can be incorporated into a powder, and the method of production of the powder also limits its active matter.

The principal methods of producing powders are :

(a) Absorption of a liquid detergent on to inorganic salts.

(b) Simultaneous absorption and neutralisation of a sulphonic acid by soda ash.

(c) Dry mixing of previously dried concentrated detergents with the other powder ingredients.

(d) Spray-drying.

(e) A combination of spray-drying with one of the other above alternatives.

(f) Drum-drying.

Simple Absorption

This method is the most limited in application, and the amount of active material that it is possible to incorporate into the finished powder depends on the physical form of the active ingredient. If the detergent is available in the form of a water solution, containing not more than 40 per cent active matter (i.e., it is of relatively low viscosity), then 8 per cent active matter calculated on the final powder can easily be incorporated; and with special manipulation, upto 12 per cent can be achieved. If the detergent is in a more concentrated form, for example, a non-ionic detergent, which is usually available in 100 per cent concentration, 15 per cent active matter is usually the practical limit.

The plant required for this method is any one of the conventional bladed powder mixers available, such as ribbon mixers, plough mixers, incorporators, screw mixers. In recent years several sophisticated powder mixers have been developed and these are finding more and more applications in the absorption process of making detergent powders.

Manufacture of powders by adsorption with a high percentage of active matter depends on the physical structure of the surfactant to be absorbed. A solid non-ionic is obviously no problem at all. In the other extreme, low ethylene oxide non-ionics can cause stickiness. It has been found that a nonyl phenol with 12 EO gives better flow characteristics to the finished powder than does one with only 10 EO.

The physical characteristics or structure of the builders have a marked effect on adsorption. One has to distinguish between the water absorption capacity of the builder, i.e., the water of crystallisation and the adsorptive capacity for organic matter, the surfactant.

In these methods, anhydrous inorganic salts, which can be hydrated, are used, and on the addition of the detergent solution the salts are partially hydrated and the active matter is dispersed with this water of crystallisation.

It is essential that the detergent solution be of very low viscosity and that it contain more water than active matter. However, if low concentrations of active matter are to be incorporated, this condition need not necessarily be fulfilled. In this case, the detergent is merely adsorbed on to the surface of the inorganic material. To obtain the higher concentrations of active matter, it is necessary for the formula to contain a large proportion of soda ash, as it can theoretically absorb as water of crystallisation 170 per cent of its weight of water. This figure is impossible to achieve in practice, as the organic detergent material does not allow the full 10 molecules of water of crystallisation to be formed. The same effect can be achieved by the use of borax either in the anhydrous or pentahydrate forms. Borax is, of course, more expensive than soda ash but has two added advantages in that it is a better builder and can also to a certain extent eliminate stickiness.

As mentioned above, if a 40 per cent active matter solution is used, 8 per cent active matter can easily be incorporated. If, however, the vessel is steam-jacketed, or if the mixing blades are hollow and can be steam-heated, powders containing upto 11 per cent active matter can be produced.

The method of producing powders is to charge the mixer with the dry ingredients, start the mixing, and to pump (or pour) the solution through a jet into the powder slowly while mixing. The rate of addition of the liquid must be adjusted so that the liquid is absorbed as it reaches the surface of the powder; if not, hard crystalline lumps will be formed.

After all the liquid has been added, mixing is continued for at least 15 minutes, and the powder is discharged on to a concrete floor to age for 12–24 hours. If the total amount of liquid is small (not more than 5 per cent) this ageing can be dispensed with, but in general the ageing serves the purpose of allowing the crystals to form and cool. If one of the more sophisticated mixers, mentioned above (especially one with a disintegrating effect) is used, this ageing process and the grinding can often be dispensed with.

By these methods a free-flowing powder, free from lumps, should be obtained; but if any of the original dry ingredients contain an appreciable amount of lumps, it will be necessary to pass the powder through a hammer mill after ageing. This grinding can be obviated by charging the powders into the mixer through a 60-mesh sieve. In this case, the lumps are screened out and can be themselves be milled to recover for use in future batches.

Because this process permits only a limited amount of active matter to be incorporated into the powder, it yields materials with only limited application for household use. However, the process can be used for industrial cleaners where the active-matter requirement is low.

To illustrate this process, we give below one typical formulation which can be used for the washing of stone floors :

Formula 1

Floor-washing Compound

Soda ash	77
Sodium tripolyphosphate	5
Sodium metasilicate pentahydrate	5
ABS-Na (40 per cent active matter)	12
Pine oil	1

The soda ash, phosphate and metasilicate are charged into the mixer, the mixing started and the ABS-Na is added slowly with stirring. After all the ABS-Na has been added, the mixing is continued for 15 minutes and the pine oil added. Mixing is continued for a minute or two after the pine oil has been incorporated and the powder discharged on to the floor to age.

Combined Absorption and Neutralisation

The method of neutralisation and absorption is more versatile than that of simple absorption. The process depends on the utilisation of an unneutralised alkyl benzene sulphonic acid, and neutralising it with soda ash which has previously been dry mixed with all the other dry ingredients of the formula. It is immaterial whether this sulphonic acid is of the 90 per cent (oleum or acid sulphonation) or 100 per cent (SO_3 sulphonation) type. If the manufacturer is doing his own sulphonation, it is obviously more economical to use his sulphonic acid without any intermediate manipulation, and if the manufacturer buys his raw detergent material, the use of a substantially anhydrous material will obviously save freight costs on water or other inert filling materials. It should be pointed out that if material other than 100 per cent sulphonic acid is being used, precautions against iron pick-up must be taken in storage and transport.

By this method there is obviously no lower limit to the amount of active material that can be incorporated into the powder, and a product such as Formula 1 can be manufactured with ease. The upper limit of active matter is generally 20 per cent, but it is possible to achieve as much as 24 per cent. Another advantage of this method is that powders incorporating solvents (solvent detergents) can be manufactured (in this case, however, milling of the powder should be dispensed with). If an attempt were made to manufacture solvent detergents by spray-drying, the solvent would be carried off in the air stream and could also constitute a fire and explosion hazard.

The plant required is the same as that mentioned in the section on absorption, such as plough mixers, ribbon mixers, incorporators and screw mixers, but in this case no heating arrangements are required. Although, theoretically, the powder produced by this process should not need to be ground, in practice a better appearance is obtained if the finished powder is passed through a hammer mill.

Systems combining spray-drying with other forms of preparation, are ancillary units and can also of course be used by themselves for powder production. With all of these specially designed mixers, ageing and grinding can usually be dispensed with. If perborate is to be added, the powder after manufacture can be discharged into a silo, fed on to a conveyor belt on to which perborate (and enzymes if required) is directly charged, and thence to a further simple drum-type mixer.

The method of manufacture is as simple as that for the absorption process. The dry materials are charged into the mixing vessel and the mixer started. The sulphonic acid is now poured in slowly while the mixer is running. If 90 per cent sulphonic acid is being used, the acid immediately reacts with the soda ash present to form the sodium salt, and the sulphonic acid should be run in at such a rate that no large unabsorbed excess of the acid is allowed to form. If 100 per cent sulphonic acid, which is more inert, is being used, it will not react so easily with the soda ash, and it can be added fairly fast and the mixer will merely disperse the acid. When the sulphonic acid has been well dispersed, i.e., when there is a uniform brownish-blue discoloration of the powder, 2 per cent of water, based on the final weight of powder, is added. This water renders the sulphonic acid reactive and almost immediately the powder will assume a light-yellow colour without any dark lumps of sulphonic acid. The sodium silicate can now be added, followed by the optical brightener, if the formulation calls for it.

In this reaction of alkyl benzene sulphonic acid with soda ash, no carbon dioxide is liberated, as in every case the formulation is arranged so that at least double the amount of soda ash is present over the stoichiometric amount and the reaction proceeds only as far as :

$$C_6H_4C_{12}H_{25}SO_3H + Na_2CO_3 = C_6H_4C_{12}H_{25}SO_3Na + NaHCO_3$$

Therefore, only one sodium atom of the two in the soda ash is utilised to form the sodium salt of the sulphonic acid and the soda ash is converted to sodium bicarbonate. If this were not the case, the evolution of carbon dioxide would cause the mixture to overflow the vessel. For this reason, when the 100 per cent type of sulphonic acid is used, it is necessary to disperse the acid well, so that no local concentrations are formed before the water is added.

An alternative method has been developed for neutralising sulphonic acid, by using anhydrous sodium metasilicate. The reaction can be written:

$$C_6H_4C_{12}H_{25}SO_3H + 2Na_2SiO_3 \longrightarrow C_6H_4C_{12}H_{25}SO_3Na + Na_2O:2SiO_2 + H_2O$$

the metasilicate has neutralised the sulphonic acid and been converted to a colloidal type of silicate, obviating the use of energy intensive liquid or dry forms. The above equation converts the silicate to the 1:2 ratio. As already mentioned, the preferred form can be the ratio 1:2:4. To obtain this form of the colloidal silicate, 3.2 parts of sulphonic acid are reacted with 1 part of anhydrous metasilicate to give a ratio of sodium sulphonate to colloidal silicate of 1:7.

Once the reaction is completed, which in practice should not take more than 30 minutes for a 300–500 kg batch, the powder is discharged on to a concrete floor and allowed to age overnight. After the ageing period, the powder is fed into a hammer mill to break up any lumps and is then ready for packing.

If one of the sophisticated mixers described later, is being used the ageing and grinding can of course be dispensed with.

Powders made by this process will have a cream to light-yellow colour, the exact shade depending on the colour of the sulphonic acid used. If a whiter powder is required, this can be achieved by bleaching it with sodium hypochlorite. When 100 per cent sulphonic acid is used, a 10 per cent solution of sodium hypochlorite is employed instead of the water, and when 90 per cent sulphonic is being used, the sodium hypochlorite is added thus:

The mixer is charged with all the powdered ingredients, except the optical brightening agent, and the mixing started. As soon as the powders are uniformly dispersed, the sulphonic acid (either 90 per cent or 100 per cent) is added. When the sulphonic acid has been well mixed in, the sodium hypochlorite solution, calculated at the rate of 2 per cent on the final weight of powder, is added. This will help to disperse the sulphonic acid, trigger the reaction with the soda ash and simultaneously bleach the powder. The silicate of soda is added after the bleaching is completed.

When sodium hypochlorite is used for bleaching of powders, the optical brightening agents should not be added to the mixing vessel immediately after the neutralisation, as the hypochlorite will attack the dyestuff. If perborate is to be added to the powder, the optical brightener can be incorporated simultaneously with the perborate. If the perborate is not to be added, after the reaction of the sodium hypochlorite has ended, the powder can be qualitatively checked for active chlorine. If this test is negative, the optical brightener can then safely be added. If this test is positive, 50 g of an 'anti-chlor' (sodium thiosulphate, sodium sulphite or sodium bi-sulphite)/ 300 kg batch of powder are added, the mixing continued for 3–5 minutes and then the optical brightener can be added with safety. The qualitative test for chlorine could even be dispensed with and the anti-chlor added as a routine, but one must stress that the anti-chlor can only be added after the powder has been bleached to the required shade, as otherwise it will interrupt the bleaching reaction.

Sodium perborate, if called for in the formulation, cannot be added prior to the ageing or grinding (unless one of the modern powder mixers is used) whether sodium hypochlorite is used or not. If sodium hypochlorite is not used, the perborate will start decomposing when the powder is damp, prior to ageing, if it is introduced initially into the mixing vessel, or it will start decomposing in the hammer mill as a result of the heat of friction in the grinding process. If sodium hypochlorite is used, the hypochlorite and perborate (peroxide) ions are mutually incompatible (so much so that perborate can be used as a rather expensive form of anti-chlor if nothing else is available). Sodium perborate can only be added after the ageing and grinding processes by reintroducing the powder into the mixer and adding the requisite amount of perborate. It is preferable to make use of a separate simple powder mixer for this operation.

The active ingredient by this process need not necessarily be only alkyl benzene sulphonic acid, but this can be mixed with any other active material desired to give special properties. These other materials can be non-ionics of the ethylene oxide condensate type, alkylolamides (the preferred types being lauric acid monoethanolamide or monoisopropanolamide) and even fatty acids to form a soap *in situ*. If the material to be mixed with the sulphonic acid is a solid at room temperature, it should be warmed about 10°C above its melting-point, then added to the sulphonic acid immediately before the addition to the powders. By this method the reaction of the sulphonic acid with the soda ash causes the fatty acid, if used, to be saponified completely and eliminates the need for the presaponification of fats or fatty acids prior to their incorporation into powders of this sort. To obtain this 'auto-saponification' the fatty acid should be of the distilled grade with a maximum of 1½ per cent neutral oil (triglyceride) or unsaponifiable matter present. If neutral oil is present it will not be saponified under these conditions and will tend to make the powder sticky if present in too large quantities.

If solvent detergent powders are to be manufactured, the solvent must be chosen with care for the particular operation involved.

Solvents incorporated in detergent powders made by dry neutralisation should have a boiling range of 125–260°C, such as deodorized kerosine. The finished product should be marketed in metal or other containers non-permeable to vapour, or in high-density polyethylene bags which may be used as liners for jute or paper sacks.

If the solvent is mixed with the alkyl benzene sulphonic acid prior to neutralisation with builders, the process should preferably be carried out so as to dispense with grinding after mixing. During processing, the mixture must be protected against exposure to electric sparks or flame.

The solvent may be incorporated in the detergent powder by two different methods. In the first, the alkyl benzene sulphonic acid is mixed with the solvent and this mixture is added to the builders. (Fatty acids are not recommended for incorporation in solvent/detergent combinations, because the foaming power of such products is too low.) A solvent must be chosen which does not react with alkyl benzene sulphonic acid. Pine oil, for instance, does react and can be used only if diluted with non-reacting solvents to the point where little or no reaction with alkyl benzene sulphonic acid occurs.

As an alternative, the detergent mixture can be neutralised dry as described earlier and the solvent added separately to the finished detergent mixture. A low percentage of alkyl benzene sulphonic acid should be used in this procedure, otherwise the final product may not feel dry enough. A modification of this method calls for pre-mixing of the solvent with colloidal silica. In this case, a non-ionic detergent must be added to the solvent prior to premixing with the silica. The non-ionic renders the silica/solvent combination more readily dispersive in the wash liquor, thus increasing its efficiency.

The first method can dispense with the use of the rather expensive air-floated silica either entirely or in part.

If pine oil is added to alkyl benzene sulphonates, it acts as a synergist towards wetting speed. This effect can be demonstrated and measured by the Draves test method. First observed by one of the authors in combinations of pine oil with liquid alkyl benzene sulphonic acid and sulphated fatty alcohols, this synergism is evident also in combinations with powders based on alkaline builders and alkyl benzene sulphonic acid.

As mentioned above, if solvents are to be incorporated into a powder, grinding must be dispensed with as the current of air through the mill will carry with it some of the solvent, which is both a loss and a fire hazard. This therefore calls for a type of mixer.

One of these types uses the fluid bed principle and again for the same reasons, this cannot be considered for solvent/detergent powders.

It is often required to dye powders to a colour other than white. This can be done by dissolving a water-soluble dye in the water naturally being added to the powder or to disperse with the powders finely divided lakes or pigments, e.g., Monastral blue. In general, if a light-coloured sulphonic acid is used, it will not be necessary to bleach the powder prior to dyeing, so that no complication about the use of sodium hypochlorite need arise. However, if the combination of the natural colour of the sulphonic acid and the colour of the dye or pigment produces a 'muddy' appearance, it is necessary to bleach the powder before adding the colouring matter, to give a brighter colour. Inorganic pigments are, in general, not affected by sodium hypochlorite, but lakes can be, so again it is necessary to treat the lake in the same way as the optical brightener when sodium hypochlorite is used.

Some powders, both spray-dried and non-spray-dried appear on the market with coloured spots dispersed throughout the mass. Elaborate arrangements are available for 'dotting' the powder but a simple one is to charge into any dry-powder mixer anhydrous metasilicate or borax and to add to this, while mixing, a water solution of the dye or dispersion of the pigment, the amount to be of the order of 0.5 per cent of the inorganic salt. Mixing takes only a minute or two. This dyed powder is stored until required and is added in the same way as, or together with, perborate to the finished powder, the rate of addition to be between 0.1 and 0.5 per cent of the finished powder.

Perfume is generally added to household powders. If sodium hypochlorite is not used, the desired amount of perfume can be sprayed into the powder towards the end of the mixing process. Sodium hypochlorite can destroy perfume material; so, if the powder is being bleached, it is advisable to add the perfume in the same manner as the optical brightener.

On occasion it is required to manufacture powders containing more than the practical limit of 20 per cent active matter. This is achieved by the use of highly absorptive special silica. This silica will tend to adsorb on to its surface all the stickiness that a high active matter will cause. The powder is manufactured in the normal way as described above without the addition of the silica and when completed prior to its being discharged for ageing 1–2 per cent silica is added. The actual amount is dependent on the actual formulation and can be as low as $\frac{1}{2}$ per cent and should rarely exceed 2 per cent.

To illustrate the above processes we detail below some formulations and procedures using both 100 per cent and 90 per cent alkyl benzene sulphonic acids. It should be pointed out that 100 per cent sulphonic acid never contains 100 per cent of the acid. The term 100 per cent means 100 parts of the sulphonic acid when neutralised to the sodium salt will yield between 99 and 101 parts of active matter. Similarly, for 90 per cent sulphonic acid, 100 parts of the acid will yield 90 parts of active matter after neutralisation to the sodium salt. These two figures, however, are very variable. One hundred per cent sulphonic acid made by the sulphonation of alkyl benzene with SO_3 gas does not usually deviate seriously from the practical limits, but for acid or oleum sulphonated material, unless special precautions are taken with the separation of the spent acid, the available active matter very rarely exceeds 90 per cent.

This process can be used to manufacture a cheap, heavy-duty household powder :

Formula 2

Heavy-duty household washing powder

	100% sulphonic acid	90% sulphonic acid
Charge and mix together :		
Light soda ash	58	58
CMC (66% active)	2	2
Sodium tripolyphosphate	15	15
then add with mixing:		
Alkyl benzene sulphonic acid (ABS)	18	20
followed by:		
Water and	2	–
Sodium silicate (40% solution)	5	5

The powder is now discharged on to the floor to age and on the following day is passed through a hammer mill. In practice, several batches can be prepared on the first day and all of these can be combined prior to milling. After milling, the powder is weighed into the same (or another) mixer and sodium perborate and the optical brightening agent are added in the following porportions :

Milled powder	89.9
Sodium perborate	10.0
Optical brightening agent	0.1

This is mixed only for a few minutes and is then ready for packing.

If linear alkyl benzene sulphonic acid is used, the addition of toluene sulphonate and/or commercial very light density silica will obviate stickiness.

For a whiter powder than that produced by this method the water is replaced by sodium hypochlorite solution containing 10 per cent available chlorine. The formula then becomes :

Formula 3

White household heavy-duty washing powder

	100% sulphonic acid	90% sulphonic acid
Light soda ash	58	57
CMC	2	2
Sodium tripolyphosphate	15	15
ABS	18	20
Sodium hypochlorite solution	2	2
Sodium silicate (40% solution)	5	–
Sodium silicate (54% solution)	–	4

It will be noted that, where the 90 per cent ABS is used, it is necessary to use a more concentrated silicate solution, because extra liquid cannot be added.

After milling, the powder is treated in the same way as the powder from Formula 2.

Recently special disintegrating high-speed mixers, fitted with special rotating knives have been successfully introduced by the authors to produce these dry neutralised powders in one single step, dispensing with ageing and grinding. It is even possible to add perborate immediately after the neutralisation action has been completed, even when hypochlorite has been used. If hypochlorite is being used, after the reaction has been completed add any suitable anti-chlor or 1 per cent sodium perborate. This will inactivate any residual hypochlorite. The desired percentage of perborate is now added, and mixing is carried out with the mixing blades only, not using the knives, and at a lower speed than during the neutralisation process. A variable speed motor, or a system of gears, is therefore advisable. This procedure diminishes loss of perborate activity.

If a relatively high amount of alkyl benzene sulphonic acid or fatty acids is being used, the mixer should be fitted with a water-cooling jacket to dissipate the heat of reaction.

Dry Mixing of Powders

Concentrated spray-dried or drum-dried detergent powders, containing at least 40 per cent active matter, often more than 60 per cent, are available. They are merely blended with the other required ingredients, in a powder mixer, dry blender or similar equipment.

If it is desired to preserve as far as possible the characteristics of a spray-dried powder, spray-dried sodium tripolyphosphate and silicate may be used in addition to the spray-dried detergent concentrate. In this way more than half of the ingredients are already of the spray-dried type. When mixing the ingredients care must be taken to avoid (as far as possible) physical breakage of the beads; a tube mixer or a slow rotating drum mixer similar to the mixer used for adding perborate and enzymes to spray-dried powder beads is suitable for this purpose.

As this is a dry-mixing process, it is not advisable to incorporate silicate solutions into the powder. Finely divided partially hydrated colloidal silicates are available or sodium metasilicate can be used if the high alkalinity will not adversely affect the final properties of the powder. Alternatively, some concentrated detergent powders are available with sodium silicate already included in the beads.

Spray-drying of Powders

The vast majority of household detergent powders are manufactured by the spray-drying process, because spray-dried powders have many advantages over the other types. These advantages can be summarised:

(a) The formulation is not limited. Relatively high amounts of active matter can be incorporated; soda ash is not an essential ingredient, and the moisture and bulk density can be varied at will (within definite limits).

(b) The powders present a pleasing appearance and being light have more sales appeal.

(c) Spray-dried powders are dustless and free-flowing and do not tend to lump. For normal formulations no special inner liners are required in packaging.

(d) Because it consists of hollow beads, with a large surface area, the powder dissolves instantly when added to water. This is important when powders are used in machines, where the wash cycle can be as low as 4 minutes. If the powder takes time to dissolve, valuable washing time is lost. Also, in tub-washing, where no mechanical agitation takes place, portions of powders other than spray-dried powders, can still be undissolved even when the operation has been completed.

(e) Heat-sensitive materials can, within limits, be handled in a spray-drier.

A spray-drier involves large capital outlay, but the results will pay for this investment. Many types are available, both of the jet and of the disc type, and also using both countercurrent and concurrent airflow patterns.

Most, although not all, spray-driers for detergent powders are of the countercurrent airflow, jet-spray type, which produce a large bead with a minimum of dust and a medium bulk density.

Spray-drying techniques depend to a great extent on the type of equipment being used. Spray-driers are equipped with various methods of producing the feed (slurry), which may be continuous (which has many advantages) or batch which allows for easier control of the constituents, since a batch can be prepared, its constituents checked and adjusted if necessary, and then passed for spray-drying. This involves considerable handling and work and also requires at least two batch-preparation tanks, from one of which the slurry is being pumped into the tower while the fresh batch is being prepared in the other. The size of each vessel should be large enough to hold sufficient material to spray during the time it takes for another batch to be prepared and checked in the second one.

If the vessels are comparatively small, this involves frequent preparation of batches during a shift. If the vessel is large, the slurry must be kept warm and stirred during the waiting period (which also includes the time that is taken to pump the slurry into the tower). This waiting period can cause hydration and/or hydrolysis of the phosphate present and may turn the CMC solution into a gel.

With continuous feed, the hold-up time is small and the above complications are largely avoided. More careful control is necessary on the setting of dosing units, but labour is reduced to a minimum. On the whole, the advantages of continuous feeding greatly outweigh the advantages of batch preparation. The source of active matter fed into the spray-drier can be any one of hundreds of alternatives: sulphonates as the acid, or as the already neutralised paste (a concentrated powder can also be used, but this is highly uneconomical, besides causing complications due to the inorganic filler naturally present); sulphates and non-ionics (although the possibilities of adding non-ionics to the powder prior to spray-drying are somewhat limited because of 'pluming' and stickiness, and even combinations of the above ingredients and, if desired, admixtures with soap.

It is obvious that if a sulphonic acid is used as the basic material it must be neutralised, either continuously or batch-wise, and it is essential to do this before the acid comes into contact with the rest of the ingredients of the slurry. Otherwise, insoluble silica may be precipitated; the polyphosphate can easily be hydrolysed to orthophosphate; and the optical brightener may also be affected adversely. In continuous operations, the sulphonic acid and caustic soda should be fed into a neutralising vessel with all the water required for the slurry. This sodium sulphonate paste is then fed into the slurry preparation vessel, where the rest of the ingredients are added. To produce a white powder, sodium hypochlorite can also be added at the neutralisation stage, at such a rate that it is completely consumed in this vessel and not allowed to come into contact with the optical brightening agents in the slurry-preparation vessel.

On batch processes it is best to start with the water, caustic soda, sodium hypochlorite (if required), and then the sulphonic acid. After the pH has been adjusted the rest of the ingredients can be added.

For certain formulations, mixtures of two or more active ingredients are used. If a continuous process of slurry preparation is being used and if the dosing arrangements are such that each constituent can be fed separately, there is no problem. Similarly, in batch preparation no difficulties should arise.

However, if, as is sometimes the case, automatic dosing arrangements are such that only one unit can provide the active ingredient and two or more materials are to be used, various pre-mixing arrangements need to be made.

If, as is often the case, a mixture of soap and alkyl benzene sulphonate is used as the active ingredient, the soap can be fed in as a ready-made powder with the other powder ingredients. This again, however, is un-economical, as it requires previous work to make the soap into a powder. A more usual and efficient arrangement is to pre-mix fatty acids with the alkyl benzene sulphonic acid prior to neutralisation, and to feed this mixture into the neutralisation vessel in the same way as alkyl benzene sulphonic acid alone. The intimate mixture of fatty acid and the sulphonic acid allows instantaneous neutralisation (saponification) in the water solution without any heating. If this pre-mix is used, and if the fatty acid contains any proportion of unsaturated fatty acids (for example, the oleic acid normally present in tallow fatty acids), a reaction between the sulphonic acid and the unsaturated fatty acid can take place. This produces an insoluble material (possibly a sulphone) which can be adsorbed on to the cloth during the washing process, causing an unpleasant odour. The reaction can be inhibited by the addition of 5 per cent water to the sulphonic-fatty acid mixture. However, although alkyl benzene sulphonic acid in the 100 per cent form is normally inert against metals, if water is added, the acid will become reactive. For this reason this sulphonic acid-fatty acid-water mixture should be made and stored in a stainless steel or plastic-coated vessel.

The fatty acids and the alkyl benzene sulphonic acid can of course be introduced into the neutralisation system in two parallel streams. The heat of neutralisation of the sulphonic acid is sufficient to saponify (neutralise) the distilled fatty acids as well.

If a mixture of long-chain fatty alcohol sulphates and alkyl benzene sulphonates is being used as the active ingredient, the alcohol sulphates are available only as the neutralised salts; and if these are mixed with sulphonic acids (or any acids) the sulphate will tend to decompose at the low pH to produce free insoluble long-chain fatty alcohols. If both the alkyl benzene sulphonate and the long-chain alcohol sulphate sources are the neutralised pastes, they can, of course, be pre-mixed, but if alkyl benzene sulphonic acid is to be used on its own the long-chain fatty alcohol sulphate can only be added to the slurry after the alkyl benzene sulphonic acid has been neutralised.

When alkyl benzene sulphonate and a non-ionic are to be used, there is in general no objection to pre-mixing the non-ionic with the alkyl benzene sulphonic acid or sulphonate. It is advisable to carry out small-scale trials before committing large quantities of material to production, because, if the non-ionic is a fatty acid or amine ethylene oxide condensate (manufacturers of non-ionic detergents do not always disclose the composition), there is again danger of acid hydrolysis of the non-ionic. It should be noted, however, that the bulk of non-ionics are either long-chain alcohol or alkyl phenol ethylene oxide condensates, which are perfectly stable under these conditions.

The CMC should be granular or powdered, so that it is easily dispersed and not necessarily dissolved. In fact, it is not desirable that it should dissolve at all, since this might produce thixotropic gels which cause trouble in pumping.

The incorporation of sodium tripolyphosphate poses a special problem. Sodium tripolyphosphate exists as the anhydrous material in two crystalline (or one crystalline and one meta-crystalline) forms, normally known as Phase (or Type) I and Phase II. Phase I is produced when the calcination temperature is of the order of 650°C and Phase II at a temperature in the neighbourhood of 400°C. It is obvious that very rarely can only one or other of these two types be produced, the resultant is usually a mixture. When dissolved in water both give the identical material $Na_5P_3O_{10}$, or when hydrated form the crystal $Na_5P_3O_{10}.6H_2O$. However, Phase I hydrates very rapidly with the evolution of ten times more heat of hydration than Phase II, which also takes a considerably longer time to hydrate. In figures the heat of hydration for Type I is 13,000 calories and for Type II 1041 calories. (For this reason Phase II is sometimes described as 'slow hydrating'.) Spray-drier feed systems, particularly the continuous system, take this property into account in their design. No hard and fast rules can be given as to which type is best suited for a particular spray-drier, but it should be mentioned that when this property of the phosphate was first observed all spray-drying plants demanded 100 per cent Phase II, because Phase I, when coming into contact with water, hydrated rapidly on the outside surface and formed lumps, and did not allow the inside of the lump to be dispersed. These lumps, at the worst, clogged filters, and at the best, were retained by the strainers and reduced the amount of phosphates fed to the jets. Nowadays it is recognised that for some designs of spray-drier feed systems, a certain portion of Phase I is desirable and some detergent manufacturers are specifying 20 and even 40 per cent Phase I content in their sodium tripolyphosphate.

A more recent tendency is to use pre-hydrated sodium tripolyphosphate which has now become available on the European market.

Another problem in the incorporation of sodium tripolyphosphate into the detergent slurries prior to spray-drying is that this material when dehydrated from solution or from its hexahydrate at a temperature of the order of 100°C can lose only five of the six molecules of crystallisation. To 'remove' all the water, at temperatures of 100°C the last molecule of crystallisation re-enters the molecule to form a mixture of pyrophosphate and orthophosphate according to the formula :

$$Na_5P_3O_{10} + H_2O = Na_4P_2O_7 + NaH_2PO_4$$

Thus, if the polyphosphate dissolves in the water of the slurry, or if it is hydrated at all, when the slurry is fed into the tower and all the water driven off, no polyphosphate will appear in the finished powder, but instead there will be a mixture of pyrophosphate and orthophosphate.

To overcome this difficulty, the slurry preparation should be organised so that the amount of phosphate which enters into solution is minimal. This can be achieved if the solids concentration in the slurry is not allowed to drop below 60 per cent. The phosphate in solution can be dehydrated, without any danger of decomposition, to its hexahydrate (or even mono-hydrate, but this is sailing close to the wind). Drying through the tower should thus be arranged so that the powder does not appear as 'bone-dry', but contains sufficient residual moisture to allow the phosphate to exist in the hydrated form. The other inorganic constituents generally present in powders either do not form crystalline hydrates or, if they do, the water of hydration can be driven off at a much lower temperature. Sodium sulphate, for example, has a transition point from the decahydrate to the anhydrous material at 32.4°C. In figures, the finished powder should therefore retain moisture at the ratio of at least 30 per cent of the sodium tripolyphosphate present. Keeping the slurry solids concentration high has an added advantage. Spray-driers are designed to evaporate water (they are normally rated by 'water evaporative capacity'). Concentrated slurries are therefore more economical because more dried powder is discharged per unit time. In general, also, concentrated slurries yield less friable beads with a smaller proportion of dust (i.e., a lower particle spread).

By virtue of their innate composition, detergent slurries can entrain air. This air entrainment can play havoc with a smooth flow of a slurry through the high pressure lines. De-aeration devices are available as optional additions to spray-driers.

All the above factors and possibly many more due to local conditions, need to be taken into account in designing a slurry unit, whether batch or continuous. Spray-drier manufacturers almost invariably offer slurry preparation units as adjuncts to their towers. One such continuous system is shown in Fig. 10.1. The old method of weighing into a rotating drum has been improved by the use of load cells with constant and automating taring through a memory, thus each portion is dosed by difference.

When standardising a new formulation it is advisable to take a small constant volume vessel of, say, 1 litre to determine the specific gravity of the slurry with a normal desired solids content and to check this specific gravity in the plant from time to time during the actual running operation.

Many household washing powders contain sodium perborate or one of the substitutes for perborate. The introduction of these persalts into the slurry is impossible, as the peroxide will decompose, so this builder is added after the spray tower. Most spray-driers have elaborate perborate dosing arrangements for providing a constant and uniform flow of perborate into the exit stream of the finished powder. This system does not, however, take into account the fact that the bulk density of the perborate is approximately double that of a spray-dried detergent powder and this difference can cause segregation of the perborate in the package; and more particularly, if pneumatic arrangements are used to move the powder, the two constituents will travel at different speeds.

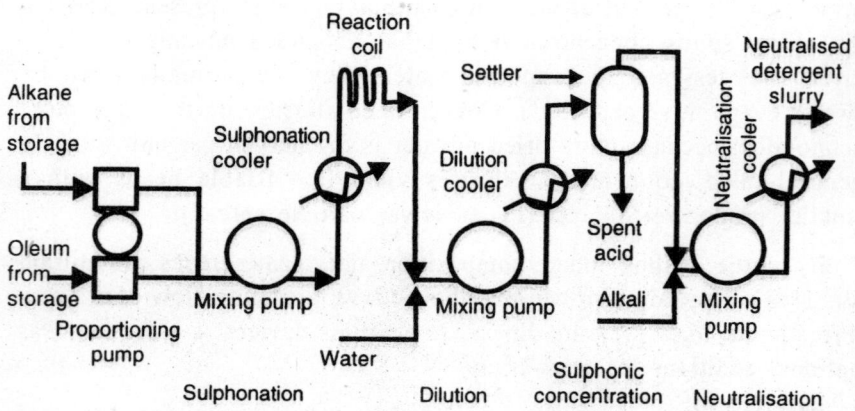

Fig. 10.1. Ballestra continuous slurry preparation system.

This problem has been overcome by feeding the perborate into a position near the base of the tower where the powder is already dry but still plastic and the perborate will adhere to the individual granules of detergent powder.

The addition of enzymes to powders poses the same problem as that for perborates. The enzyme cannot be added to the slurry, prior to spray-drying, nor can it be added to the base of the tower when the powder is hot. In addition, enzymes are added in quantities of the order of 1/2–2 per cent of the powder so the problem of mixing and segregation is more involved. Modifications of the physical state of the enzymes are being made to overcome some of these problems and the enzymes are best fed continuously into the powder stream generally via a continuously revolving drum mixer.

The bulk of household powders are perfumed. Perfume can be added in two ways, either as a powder (concentrated powdered perfumes are available for this purpose) in which case the powder is treated like perborate or enzymes, or as a fine spray on the stream of powder passing over a baffle, or by spraying into the rotating drum mixer used for the addition of perborate and/or enzymes to the beads coming from the tower. To spray the perfume it is sometimes advisable that it be diluted with a volatile solvent. This allows better dispersion of the perfume over the powder stream and prevents local build-up of the perfume oil, which may cause lumping of the powder.

The choice of the perfume for detergent powders is important and the following points should be borne in mind :

(a) The type of perfume. Perfumes have both an immediate and residual odour and both need to be considered.

(b) Whether the contents of the powder will adversely affect the perfume on storage.

(c) The volatility of the perfume under the particular conditions of packaging and storage.

(d) Whether the perfume will leave a residual smell on the washing. This is not always a requirement but account must be taken of this possibility.

In the case of perfumes, as part of the sales appeal of a powder depends on this, it is best to collaborate with the manufacturer of the perfume and not to buy the cheapest type available.

The actual manipulations involved in operating a spray-drier are usually laid down by the manufacturer of the plant and it is difficult to give generalised directions for all the possible types of spray-driers in use. However, some guides as to quality control of spray-dried powders are worth while recording :

An acceptable powder is characterised by :

1.   Good colour properties.
2.   Desirable particle size and spread.
3.   Correct bulk density.
4.   Correct residual water content.
5.   Absence of stickiness.
6.   Uniform composition and appearance.

Some of these characteristics are interrelated. For example, if the colour is bad owing to scorching the powder will also be bone dry with a minimal residual water content. If the residual moisture is excessive the powder might become sticky.

Colour

Bad colour may be due to off-colour ingredients in the slurry, such as excess iron in one of the inorganic powders or dark-coloured active matter. If discoloration is caused by some inorganic component of the slurry very little can be done about it. However, if an organic colouring matter is at fault, this can frequently be corrected by bleaching of the slurry with sodium hypochlorite or hydrogen peroxide solutions.

In the presence of optical brighteners, bleaching agents must be used judiciously. Not all optical brighteners are stable to oxidising agents. It is therefore necessary to arrange the bleaching of the dark-coloured constituents before the optical brighteners are added and to add these only after the oxidising agents have been effectively destroyed. This holds good for both continuous (Fig. 10.1) and batch-wise slurry preparation. A point to remember is that, due to the optical properties of fine opaque particles, a spray-dried powder always looks whiter than the appearance of the slurry would lead one to believe. This is particularly true of slurries containing optical brighteners.

The risk of scorching is greater in a countercurrent plant because the powder reaches the hottest point in the tower when it is practically dry. In a concurrent operation the powder is exposed to maximum heat when it contains most moisture.

Scorching may be counteracted by adding to the volume (i.e., the weight) of inlet air, which causes a drop in the temperature of the incoming air without reducing the amount of heat. In operating a spray-drier it is well to remember that the work is not being done by the temperature of the inlet air, but by the quantity of heat, i.e., the product of specific heat, mass, and temperature. The volume of the heat is therefore roughly proportional to the product of temperature and mass; roughly because the specific heat varies with temperature.

Any considerable build-up of matter on the walls of the spray-tower is a hazard to the quality of the final product. Such accumulations consist of material in contact with the hot air over extended periods of time and must be scorched. Furthermore, such build-up reduces the effective diameter of the tower and increases the tendency of the particles to impinge on the walls. Eventually the accumulation becomes too heavy for the detergent-metal bond and large chunks of material will break off and may temporarily overload the conveyor system. If the installation includes a screening system, the 'fall out' from the walls will be diverted to one side. But such overdried material is brittle enough to be broken up by the conveyors and the screens and yellow and brownish fragments (not beads) will find their way into the white beaded powder. Unfortunately, this build-up is often due to basic faults in tower design and in many instances little can be done to remedy the situation. However, finer atomisation will yield particles with less momentum which, in turn, will ensure that the particles are dried before they reach the walls. Being dry they will have less tendency to stick.

If build-up cannot be avoided, the walls of the tower need to be cleaned at frequent intervals, either by scraping by hand or mechanically or by a jet of water. Spray-driers are often equipped with an auxiliary system for continuous removal of build-up. Bowen spray drier for soaps, detergents and other coarse-particle products are shown in Fig. 10.2. Another method is to fit to spray-driers an air broom, which prevents build-up. The air broom is a pipe fitted with nozzles which rotates slowly around the chamber. Compressed air blown through the nozzles removes any material adhering to the walls and also cools them slightly.

Sometimes the powder is grey when leaving the tower. This condition can be traced to soot from a poorly adjusted burner, being absorbed into the bead. This problem is more acute in a concurrent drier where the soot comes in contact with the jet of product when it is still wet and able to absorb it. In a countercurrent drier most of the soot is exhausted by the air removal equipment and the fines are then discoloured. Adjustment of the furnace will cure this trouble at once. The most common cause of soot formation is a dearth of primary air in the burner, which causes incomplete combustion. Usually a slight cut in the fuel supply will give complete combustion which in turn will yield a higher temperature using less fuel.

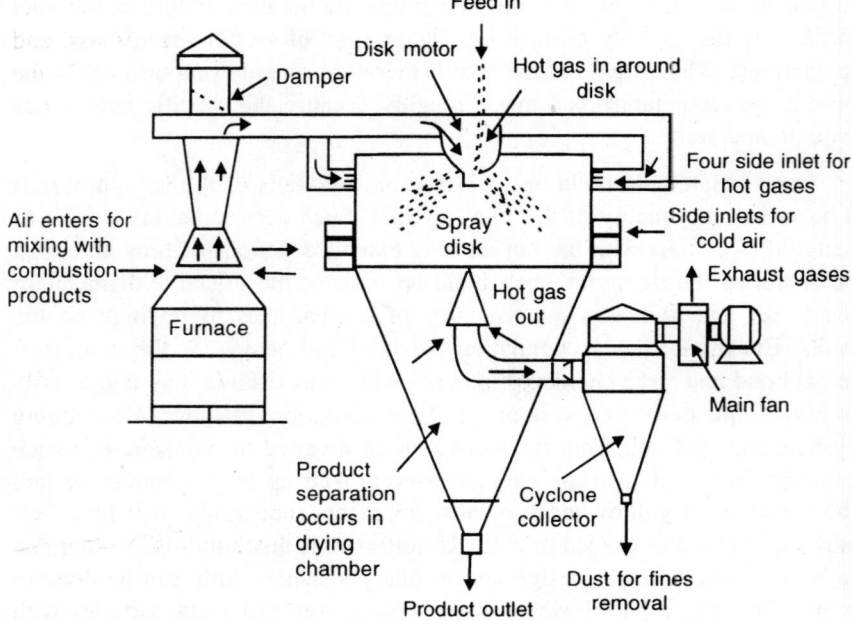

Fig. 10.2. Bowen spray drier for soaps, detergents and other coarse-particle products.

Particle size and spread

Particle size of the powder is dependent on the particle size of the atomised slurry which in turn in inversely proportional to the pressure applied at the jet or to the speed of the atomizing disc and approximately proportional to the viscosity of the slurry. One of the possible sources of abnormal particle spread, all other factors being equal, is a faulty jet or a disc with slurry encrusted on it. (This can also cause build-up on the walls.) A jet can appear to be working satisfactorily but if the orifice has been eroded away from the circular, it will atomize unevenly. For this reason jets should be stripped, cleaned and examined at regular intervals.

A narrow spread in particle size is desirable from the viewpoint of product appearance. Furthermore, if the spread is wide it may range from large particles to dust. The housewife, the ultimate consumer of the product, is particularly sensitive to dust. Finally, a wide spread will add to the percentage of fines produced by the process. In most instances fines can be considered nothing but a necessary evil to be recycled or disposed of—usually uneconomically.

However, where bulk density of a powder is too low, this may be remedied by a slight widening of the particle size range, since the small particles will tend to fill the voids between the large beads.

Bulk density

Bulk density is the key factor in the economics involved in the production of a spray-dried washing powder. Containers are of standard size and shape. It is not practical to sell a half-empty carton if the powder is too heavy and if it is too light the carton will not accommodate the requisite weight. Furthermore, since use directions are always given by volume, insufficient active material will be used by the housewife if the powder is too light and the product will appear inefficient.

The bulk density is dependent on the 'hollowness' of the bead assuming all the beads to be of the same size.

The mass of a hollow sphere is expressed by the formula

$$M = \frac{4}{3} \, (r_1^3 - r_2^3) D \pi$$

where M = mass

r_1 = external radius of sphere

r_2 = internal radius of sphere

D = absolute density of the material of the bead.

The weight of each sphere is therefore proportional to the difference between the cubes of the external and internal radii of the sphere.

The bulk density of each bead is given by the formula

$$BD = \frac{(r_1^3 - r_2^3)D}{r_1^3}$$

and therefore the closer r_2 approaches to r_1 the lower the bulk density. In other words, the thinner the wall the smaller the bulk density.

These calculations do not yield the bulk density of the final product which is a function of absolute bulk density and particle size spread and an inverse function of the mean diameter of the beads. However, the thicker the walls of the beads the greater will be the bulk density. This would suggest that lower feed concentration will give thinner walls and lower bulk density, and conversely, higher concentration will result in higher bulk density. This relation does hold good for solutions. For slurries, however, any change in slurry concentration affects the amount of material in solution and therefore the density and viscosity of the slurry which in turn play an important part in atomisation.

Rather contradictory reports on the effects of changes in slurry concentration have appeared in the literature. In actual practice when spraying solutions a drop in bulk density will result from a decrease in solids content of the slurry. Conversely, in spraying slurries containing powders in suspension, a cut in solids content will result in finer atomisation, smaller particles, and increased bulk density (within limits). This phenomenon is due to reduced slurry viscosity and density. Consider a slurry to be a uniform mixture of a solution and finely divided powders. When this slurry is atomised each single droplet will consist of a drop of solution with particles of a powder dispersed throughout the volume of the sphere. The solution will be dried into a hollow sphere and the powder which was present originally will be fused into the wall of the sphere. The powder, having not entered into solution, will in no way change its physical shape, whereas the liquid will give a very thin-walled sphere. The ratio of undissolved powder to solids in true solution will materially affect the bulk density of the finished product. By altering slurry concentration the percentage of undissolved matter is changed and thereby the amount of solid (undried) matter fused into the spheres is also altered. A drastic cut in this undissolved solids content will cause the slurry to approach a true solution. In this case lowering the solids content of the feed will lower the bulk density of the dried material.

Feed temperature has a similar though converse effect on slurry characteristics. Elevated temperatures bring a drop in viscosity and density and an increase in the amount of material in true solution (within limits). Thus an increase in feed temperature gives lower bulk densities. However, increasing the feed temperature causes a drop in Δt, which in turn will cause an increase in bulk density. (Δt is the difference in temperature of the atomised liquid and the ambient temperature. If this difference is large it causes 'puffing' of the bead.)

Actual hollowness of the beads depends on the rate of evaporation of water from the droplets. If the exterior surface of the droplet is dried rapidly to a semi-solid state, it will form a membrane which prevents further escape of moisture. The water retained inside the globule will inflate this sphere into a balloon which will eventually rupture. The size of the balloon depends on the rate of evaporation of the water inside the sphere.

The rate of heat transfer is proportional to the temperature differential between the atomised liquid and the surroundings (Δt). Narrowing of this differential slows and reduces evaporation and thereby reduces puffing of each individual particle. Bulk density is therefore increased. To accelerate the rate of evaporation, the rate of heat transfer must be stepped up by increasing Δt. This in turn will cause a drop in bulk density.

Adjustment of the inlet air temperature will give any desired changes in Δt in a concurrent tower. In a countercurrent installation, adjustment of the inlet air temperature will also change Δt but less so.

A temperature reading even more pertinent to Δt can be taken at the jet. It is general practice to observe the inlet air temperature in the inlet duct immediately prior to its entry into the tower, and the outlet air temperature in the outlet duct at some point between the tower and the exhaust fan. A very important further point of reference is in a plane cutting through the atomisation tower about 60 cm (2 ft) below the plane of the jet. Temperature readings at this point will give an accurate idea of any changes in Δt, and are more sensitive than the inlet air temperature because some heat from the inlet air is dissipated in warming up the dried powder and the walls of the chamber.

A change in Δt does not necessarily mean a change in fuel supply to the furnace.

The temperature of the inlet air can be modified by altering the volume of fuel fed to the furnace. In this case the amount of heat from the furnace is changed, which changes the drying conditions. As an alternative, the volume of air entering the tower may be modified while the amount of fuel remains constant. This does not change the quantity of heat but has a material effect on Δt.

In an installation permitting infinite variations in the rate of feed (in a spinning disc atomiser or by the use of a return flow jet) changes in the feed rate can be used to modify Δt. For instance, by stepping up the rate of feed Δt is increased. Here the total evaporation of water is greater but the amount evaporated from each individual particle is smaller. Product moisture content and therefore density are increased.

By virtue of the design Δt must be smaller in a countercurrent spray-drier than in a concurrent installation. Therefore, as a general rule countercurrent spray-drying will yield heavier powders than concurrent operations.

$$x = \frac{dk\,e^{0.705v}}{2P^{0.375}\sin\dfrac{1}{2}\sigma}$$

where x = mass median drop size
d = orifice diameter
k = constant
v = kinetic viscosity
P = nozzle pressure
σ = spray cone angle

shows the drop diameter to be an inverse function of pressure. The greater the pressure on the nozzle (or the faster the revolutions of a disc) the smaller the drop size. The same amount of material dispersed into more and smaller drops means a larger surface area giving more evaporation and lower residual moisture. Increased evaporation means a lower Δt. This in turn results in less and slower evaporation per droplet and therefore less puffing and greater density. Increasing the pressure on a jet or the revolutions of a disc will increase bulk density.

As will be noted from the above equation, the viscosity of the slurry plays a very important part in the atomisation. The viscosity of the slurry is dependent on the actual formulation (i.e., the types of ingredients), the pH, the degree of hydration of the phosphate, the concentration of solid matter, the amount of material in true solution and in suspension and the temperature. It is suggested that continuous viscosity metering be done with a dynamic viscometer to control any change in slurry structure. It may even be arranged with an automatic alarm if the viscosity changes radically from pre-set conditions.

The amount of residual water retained in the powder is another factor affecting density which is discussed below.

The actual formulation of a product plays an important part in bulk density. Generally, the higher the percentage of active ingredients the lower the bulk density, and the higher the alkaline builder content the higher the bulk density. Sodium sulphate rather surprisingly tends to reduce density. Bulk density is increased by sodium chloride.

Residual moisture

For economic reasons it may be necessary to retain a certain amount of residual water in the powder. As the powder must of necessity be free-flowing, this water can only be present as water of crystallisation and not as surface moisture. Account should also be taken of the fact that certain inorganic salts which have more than one physical form tend to assume the stable form. For example, washing soda crystals $Na_2CO_3.10H_2O$ will effloresce down to $Na_2CO_3.H_2O$, whereas anhydrous sodium carbonate will take up moisture from the atmosphere to its stable form $Na_2CO_3.H_2O$. Therefore if a formula containing soda ash is spray-dried to be bone dry, the soda ash will tend to absorb moisture up to 17 per cent of its weight. This phenomenon occurs with most inorganic salts used in detergents but it is difficult to give hard and fast rules because, in virtually every case, the problem may be complicated by double and co-crystals. Generally a spray-dried powder containing a fair proportion of inorganic salts can be expected to pick up moisture on storage. Table 10.1 lists some crystalline inorganic salts used as fillers or builders in detergent powders and the percentage of water absorbed by the anhydrous salts to form the stable molecule.

Table 10.1. Percentage of water absorbed by anhydrous salts to form crystals.

	% water absorbed by anhydrous salts	Temperature of instability of crystal
$Na_2B_3O_7.5H_2O$	44	–
$Na_2B_4O_7.10H_2O$	89	60
$Na_2CO_3.H_2O$	17	100
$Na_2CO_3.7H_2O$	105	32
$Na_2CO_3.10H_2O$	170	32
$Na_2CO_3NaHCO_3.2H_2O$	17	85
$Na_2HPO_4.7H_2O$	79	–
$Na_3PO_4.10H_2O$	110	–
$Na_3PO_4.12H_2O$	130	100
$Na_4P_2O_7.10H_2O$	67	93
$Na_5P_3O_{10}.6H_2O$	29	–
$Na_2SiO_3.5H_2O$	74	–
$Na_2SiO_3.9H_2O$	133	100
$Na_2SO_4.7H_2O$	89	–
$Na_2SO_4.10H_2O$	126	32.4

Storage conditions determine the rate at which water of crystallisation is taken up. If the powder is packed immediately in 'airtight' containers, the rate of absorption will be comparatively slow. (No container commonly used for packaging powders is truly air-tight). Storage in an atmosphere of fairly high relative humidity will result in fast moisture take-up and a tendency to agglomeration. However, the absorption will be slower the closer the moisture content of the powder is adjusted to that of the stable form. If a powder must be stored in ambient relative humidity exceeding 60 per cent, the risk of lump formation can be minimised by discharging the powder from the spray-tower with at least three-quarters of the stable moisture content. It is economically unsound to dry a powder to 1 per cent moisture content when the powder must revert to, say, 7 per cent. The operation would be a lot more efficient if the powder were dried to 7 per cent in the first place.

The stable moisture content can be determined only by prolonged storage tests. These call for small samples of the powder to be kept in desiccators designed to expose the samples to varying degrees of relative humidity.

Temperature of the outlet air and product residence time in the tower are the two factors used to control moisture content of the finished powder.

Contrary to general belief the outlet air temperature need not be above 100°C to obtain a desirable product. Spray-driers can work efficiently at temperatures well below 100°C. However, for a bone dry powder the temperature must be in the vicinity of 100°C or near the boiling point of water at the altitude of the particular plant.

To keep water of crystallisation in the finished powder the outlet air temperature must be low and residence in the tower must be extended to permit crystallisation (which has both a time and heat factor) to take place. Some spray-driers are designed to discharge the powder while slightly wet and then to transport it through a pneumatic conveyor or air lift to the packing department. This conveyor system serves a useful purpose by cooling the product and effecting crystallisation (for example $Na_2SO_4.10H_2O$ exists only below 32.4°C; to form the crystal the wet powder must be cooled below this temperature). However, the extra cooling step is not essential. A dry free-flowing powder of 10–15 per cent moisture content can be obtained from the tower without this aid. The amount of water of crystallisation retainable in a powder is determined by the type and quantity of the inorganic salts incorporated in the particular formulation.

Spray-drier manufacturers and operators sometimes do not pay sufficient attention to residence time in the tower. Disadvantages attendant upon excessive time in the chamber have been discussed above. But there are disadvantages also resulting from too short a time in the tower. It is not enough to expose the spray to a quantity of heat sufficient to dry it. The spray must be in contact with the hot air long enough to permit all necessary heat to be transferred. In a countercurrent installation operating at a low outlet temperature the temperature at the point of atomisation will be only a few degrees higher than that of the slurry. Therefore Δt will be small and the rate of heat transfer slow. The residence time in the tower must then be increased. This is a very important point.

Air pressures in spray-drier towers are usually balanced to range between 0 and 5 Pa vacuum water gauge. This is accomplished by means of dampers on either the inlet air duct or exhaust duct, or both. The rate of exhaustion of air plus vapour is adjusted to be either equal to or slightly greater than the rate of ingress of hot air into the chamber. Positive pressure inside the tower is not advisable sine it may cause blowing out of dust. The vacuum in the chamber can be raised, however, quite easily to 125 Pa water gauge provided that the tower is sealed off at its base with a rotary valve or other sealing-discharging device. The increased vacuum will retard the rate of fall of the dried particles and thus extend the residence in the tower. At the same time the higher vacuum will tend to add to the amount of fines produced, the actual increase being dependent on the particle spread of the powder.

Thus if the powder exists slightly wet, this may be remedied by increasing the vacuum in the tower (by opening the outlet air duct slightly). By the same token, if more residual water is required, a drop in outlet temperature (by increasing the rate of feed or decreasing inlet heat) and a higher vacuum in the tower will have the desired effect. Operating variables are tabulated in Table 10.2.

The following range of parameters shows typical tower operation conditions :

Slurry concentration	60–70 per cent
Hot air inlet temperature	300–450°C
Average air velocity	0.3–0.6 m/second
Exhaust air outlet temperature	90–95°C
Powder temperature at outlet	60–70°C
Powder bulk density at outlet	200–400 g/litre
Retained moisture in powder at outlet	2–15 per cent

Stickiness

Stickness is a phenomenon difficult to define but absolutely unacceptable in a spray-dried powder. It may be caused by poor drying or faulty atomisation leaving large particles incompletely dried or by some ingredient or lack thereof in the formulation. Addition of sodium silicate will somewhat improve a slightly sticky powder. However, certain materials tend, *per se*, to yield a sticky product. In particular the linear alkyl benzene sulphonates yield stickier products than the branched chain materials. This stickiness is usually easily overcome by the addition of one of the hydrotropes, particularly sodium toluene sulphonate. Certain alkylates, for all that and depending on their isomerism, do, when sulphonated, yield powders which remain sticky even when toluene sulphonate is added. This can be overcome by the addition of small percentages of fumed silica or calcium silicate. Magnesium silicate, added either as such or, more commonly, formed *in situ* by the reaction of magnesium sulphate (added as a stabilising agent for perborate) and sodium silicate, also adds to the free flowing properties of problematic powders.

Table 10.2. Effect of variation of operating conditions on powder characteristics.

	Particle spread		Particle size		Bulk density		Water content	
	Incr	Decr	Incr	Decr	Incr	Decr	Incr	Decr
Orifice diameter	+	−	+	−				
Pressure	−	+	−	+	+	+		
Slurry concentration (solids)	+	−	+	−	−	+*		
Active matter			+	−	−	+		
Alkalis			−	+	+	−		
Viscosity			+	−				
Rate of feed			+	−	+	−	+	−
Feed temperature					+	−†		
Δ*t*						−	+†	
Outlet air temperature					−	+	−	‡

+ Sign indicates increase of the function in the left-hand column, and — sign the converse.
* Within limits.
† These two factors tend to neutralise each other.
‡ In conjunction with increased residence time in the chamber.

Product uniformity

After careful adjustment of all processing variables one may attain all desired characteristics in the product, but this is not enough. For uniform product quality, optimum conditions must be established, maintained and repeated from run to run. Trial runs need to be made to establish the best processing conditions for each individual product and these must be recorded and repeated for each production run. Accurate measuring instruments located at all vital points in the spray-drying installation are absolutely essential. Readings should be made and recorded at frequent intervals or better still the instruments should be of the continuously recording type. A recording instrument will predict a trend which can only be derived from ordinary indicators by very frequent readings and recording of results.

Separation of powder

Separation of the powder from the exhaust air is an integral part of the function performed by a spray-drying installation. Factors pertaining to the sometimes third separation step is performed by external cyclone separators and/or bag filters or by wet scrubbing. The manufacturer is faced with a problem of collecting the resulting fines and using them.

Wet scrubbing

Even the best of cyclones or bag filters do not completely trap all the solid particles entrained in the exhaust air, particularly those of micron size. In addition due to intermittent faulty operation of the tower, the separation system can become overloaded.

The detergent manufacturer has a moral responsibility to ensure that he does not contaminate or pollute the air he discharges, this apart from any legislation which has been or will be enacted.

As dry separation systems are known to pass, at best, a few per cent of the entrained particles, recourse may be made to wet scrubbing, when the solid material is scrubbed out by water. Wet scrubbers can bring the concentration of solid particles in the outgoing air down to 15–25 ppm.

One such scrubber is shown in Fig. 10.3. The water is circulated through the unit until the concentration reaches a predetermined amount, when the solution is pumped to the slurry preparation.

To avoid troublesome foam formation, particularly when the active content is high, a salt or preferably a silicate solution rather than water can be used. If the fuel used for hot air generation contains an appreciable amount of sulphur, caustic soda can also be injected to neutralise acidity.

Use of fines

In some plants detergent fines can be used in the manufacture of scouring powders or other related products. In that case the fines can be collected in bags at regular intervals from the cyclones or filter bags. If no other use can be made of the fines they must be reintroduced into the slurry. Where the slurry is made up continuously this reintroduction should be continuous also. In plants where the slurry is made up in batches, reintroduction of fines simply calls for transporting them from the collecting points to the slurry vessels.

Where a wet scrubber is used the reintroduction of fines is easiest. Water is pumped through the scrubber to form a solution of fines and this solution is recirculated continuously until it reaches a predetermined concentration when it is pumped to the slurry vessel where the primary water must be adjusted accordingly.

1.NaCl solution vessel
2. NaOH constant level vessel
3. Scrubber
4. Collection vessel
5. Recycle pump
6. Transfer pump

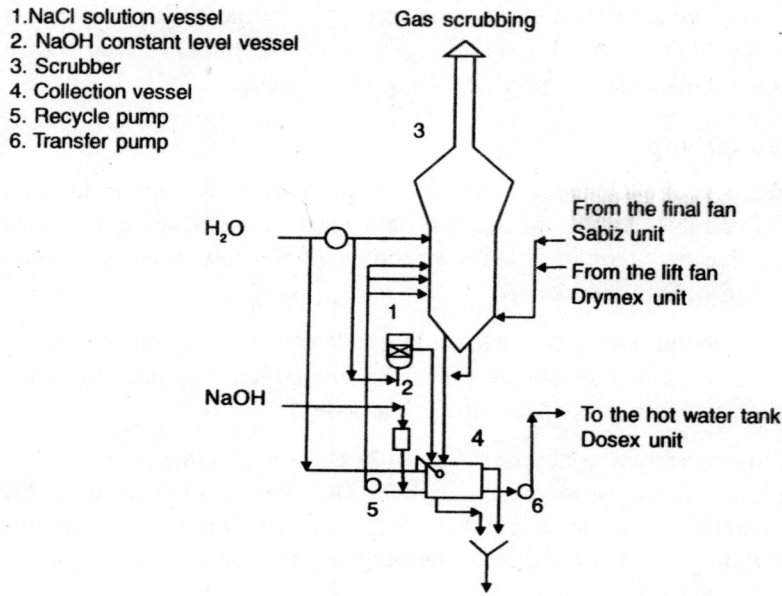

Fig. 10.3. Ballestra wet scrubber system.

Another method (Ballestra system) used is to blow the fines continuously to the top of the tower and to reintroduce them concurrently with the atomised slurry (Fig. 10.2). As they come into contact with the wet atomised slurry they tend to agglomerate with the atomised particles and in this way all the problems of handling fines are disposed of.

There is one prerequisite for the automatic reintroduction of fines into either the slurry or the tower: there must be no soot whatsoever and careful attention must be paid to the accurate adjustment of the furnace. The slightest trace of carbon in the fines will discolour a large amount of powder. Freedom from soot is not easily attained since some of the factors involved are beyond the operator's control. A temporary stoppage of electric current, for instance, can cause soot. An outside use for fines is therefore usually preferable to their reintroduction, if freedom from soot cannot be guaranteed. In a scouring powder for example, a trace of off-colour powder will not affect the appearance or performance of the final product but it is axiomatic that white linen cannot be washed with a powder containing soot.

Combination of Spray-dried and Dry-mixed Powders

The introduction of linear alkyl benzene led to problems of stickiness which were rapidly overcome. Nowadays more and more powders have certain amounts of non-ionic detergents incorporated into them and this is again leading to problems of stickiness. The use of non-ionic detergents in spray-dried powders causes difficulties as soon as the percentage (depending on the type) of the non-ionic reaches a certain figure. The difficulties encountered are stickiness of the powder, clogging of cyclones, build-up on walls of the tower, 'pluming', etc. Thus the development of chemically advanced powders can be hindered by technical problems.

Some of the above factors can be overcome by mechanical means such as wet scrubbers and automatic cleaning rings; but for all that, some of the newer additives are heat sensitive and in using certain non-ionic detergents the solution from a wet scrubber will become black due to scorching of the material.

A method used to overcome the problem was to dose powdered non-ionics on to the spray-dried powder. These are available in various grades, often with 25 or more ethylene oxide units which lower the biodegradability, and alter the hydrophile-hydrophobe balance, thus reducing the efficiency of the non-ionic as a detergent. In addition, technical problems were encountered when the non-ionic powder was mixed with a spray-dried powder while the latter was still hot.

All of the above problems have been overcome by the use of a combination of dry-mixing and spray-drying techniques. The principle is basically that the alkyl benzene sulphonic acid and the soap (fatty acid) portions of the active matter are spray-dried in the normal way with phosphates, CMC, etc. The non-ionic component, ethanolamide, and other difficult-to-spray constituents are adsorbed on to a portion of the inorganic filler, with or without the use of colloidal or fumed silica in a mixer.

Ballestra 'combex' system

This system consists of a special mixing/dispersing unit in which a powder containing ingredients that are difficult to spray-dry is produced. It has two advantages in that it can be a fully integrated system linked to a spray-drier or can be run separately to make 'heavy powders' and it may be adjusted to produce powders of a granular or fine structure.

This process consists of a dispersion system which distributes the active matter components on to the detergent builders. This very homogeneous dispersion of the surfactants on to the builders is accomplished in a mixer which has specially designed plough type horizontal mixing blades rotating at high speed, and passing very close to the mixer walls. This design guarantees very efficient blending of all the components. In addition, ultra high speed rotating desintegrators are inserted to prevent any lump formation.

In contrast to 'spray-mixing', 'fluid bed', and 'agglomeration' processes, fixation of the active matter components is accomplished by mechanical disintegration of the solid builder material on to which the active matter components are added by means of a rather simple dispersion system. In most spray-mixing systems special high absorption types of builders are used, whereas in a system based on mechanical dispersion and powerful disintegration a much wider variety of builders in normal powder form may be used. The mixing-disintegration unit is of a relatively small size in relation to its output. Loading, discharging and residence times are short. While most spray-mixing systems are designed to obtain powders approaching the structure of spray-dried beads with relatively low density, the system described here aims at a powder with relatively high density, but with a structure to make it adhere strongly to the surface of beads coming from a spray tower.

The mixing unit may be used either for the incorporation of high percentages of non-ionics or for the incorporation of sulphonic acid and/or distilled fatty acids, with or without non-ionics. In this case the LAS and fatty acids are neutralised immediately on contact with the builders. Fatty acids are transformed in a single step operation into soap without leaving any unreacted fatty acids. Certain processing details are to be observed to guarantee the complete neutralisation of LAS and the complete saponification of distilled fatty acids.

This 'dry neutralisation and saponification' is accomplished by using a small percentage of an alkaline aqueous phase (representing only a small fraction of the stoichiometric amount of NaOH) as a kind of accelerator or 'trigger' for the reaction. The aqueous phase, generally only 2–4 per cent on the total mix, is added at the same time, but not pre-mixed with the surfactant components. To give a general indication of the efficiency of the mixing system: only about 3–4 minutes are required for even distribution of non-reacting non-ionic components on to the builder mix, and about 4–6 minutes for 'reacting' components, i.e., AB sulphonic acid and/or distilled fatty acids. Automatic charge and discharge takes approximately an additional minute or two, depending on the size of the mixing system. The process can be run completely automatically with weighing and dosing similar to continuous slurry preparation.

Depending on the bulk density of the builders used, the amount of the surfactants added, the bulk density of this powder from the first phase of the process is in the range of 0.6–0.9. The product from this process may be used as such for detergent powders, e.g., washing powders for commercial laundries, metal cleaning detergents, etc. The powder is less dusty than otherwise produced 'heavy powders'. The process may be carried out in such a manner that powders, either with a granular or a finer structure, are produced. The process is flexible enough to produce powders in a wide range of particle size.

The dry-mixed powder can also be fed continuously on to the powder being discharged from the spray drier while it is still hot and the two streams of powder are passed through a slow-acting baffle mixer, a slowly rotating cylinder mixer or by weighing belts. From this mixer the powder is taken to storage or packing.

The entire system is completely co-ordinated and features automatic controls. Being a closed system, there are no problems with dust. Proportions of builders and liquid components, as well as those of the final mixture, are set on the central control panel and may be changed quickly; the density of the final mixtures of 'heavy' powder and beads from the spray-drier is determined more by the low density of the beads than by the high density of 'heavy' powder. Table 10.3 gives some examples of the density of various mixtures of 'heavy' powder with spray-dried beads having a density of 0.3.

Table 10.3. Density of mixed powders made by the combex system.

Density of 'heavy' powder	Density of mixture (proportion of heavy to light powder with a density of 0.3)		
	50:50	40:60	30:70
0.9	0.52	0.43	0.38
0.8	0.45	0.41	0.37
0.7	0.42	0.40	0.36
0.6	0.40	0.39	0.35

The approximate expected density of the resultant mixtures may be calculated as follows :

$$\text{density of mixture} = \frac{100}{\dfrac{a}{d'} + \dfrac{b}{d''}}$$

where a = percentage of heavy powder

b = percentage by weight of beads

d' = density of heavy powder

d'' = density of beads

This does, of course, give only an approximation of the expected density, but may serve as an indication.

Using this method the detergent chemist has a great amount of latitude in formulation and the output of the spray-tower is enhanced.

In certain countries of the world there is a swing away from light density spray-dried powders to medium density powders. Two of the factors influencing this swing are high packaging costs and disposal problems with packages. This combination system can be arranged to produce powders of intermediate bulk density, regardless of whether non-ionic detergents are used, by producing a powder and mixing it with a spray-dried powder in a proportion to give the desired formulation and density. This again will vastly increase the production of powder without the necessity of increasing the capacity of the spray-drying plant.

In further development is the alkaline powders are fluidised in a cooled air stream and the liquid components are sprayed on to the fluidised powders by special nozzles. Intimate mixing allows of the production of *expanded granules* by the explosive release of CO_2 from the reaction.

These expanded granules are agglomerated, cooled and aged in a second stage by the use of another turbo-reactor, with or without the addition of further liquid ingredients, from additional nozzles.

If no neutralisation occurs (when only non-ionics are used), agglomeration and granulation can still be achieved by spraying through four different nozzles, using agglomerating agents.

Perfume is added to the cooled powder by a further nozzle.

Patterson–Kelley systems

In this system V-shaped blender (Fig. 10.4) for both powder mixing and liquid/solids blending and agglomeration is used. As can be seen from the drawing, this method achieves a precision blend through a divergent flow and by the intermeshing action when two inclined cylinders combine their flow. Material is rotated close to the axis, reducing power requirement. As the blender rotates, liquid is sprayed into the material through a high-speed liquid dispersion bar located concentric to the trunnion axis. The blades on the bar aerate the material to increase the speed and thoroughness of the blend. Liquid is dispersed through disc apertures in a controlled pattern and extended to all solids throughout the blend.

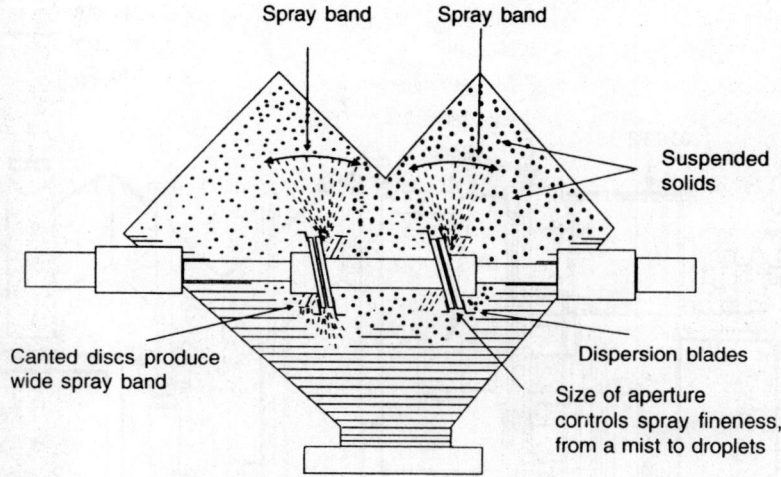

Fig. 10.4. Patterson–Kelley V-shaped blender.

The blender is used for manufacture of detergents by combined neutralisation/absorption processes, for the hydration of sodium tripolyphosphate, and if colloidal sodium silicate solution is added, agglomeration can be achieved.

An improvement of this system is their Cross-Flow Blender, where the two legs of the V are of unequal length, giving unequal displacement in the powder mass in each revolution of the shell, producing an axial exchange from each leg, thus faster blending.

The above two blenders are of the batch type, the solid material needs to be added while the blender is stationary and the machine needs to be stopped for discharge. There is no cooling so if perborate, enzymes and perfume are to be added, this can only be done after ageing and cooling, when the powder needs to be reintroduced into the same or another blender.

A development of this principle of uneven flow is the P–K zig-zag continuous blender shown in Fig. 10.5. The blender combines the action of a rotating, eccentric drum with multiple recycling to produce uniform blends of both solid/solids and liquid/solids materials. The zig-zag completes most blends within two minutes. Two liquids can be added simultaneously.

As the blender revolves, the blended material leaves the drum and flows into the legs, where recycling begins. At each half turn, part of the powder moves forward, part of it backwards within the legs. Splitting, tumbling and merging of the material bring particles into contact with each other to complete the blending.

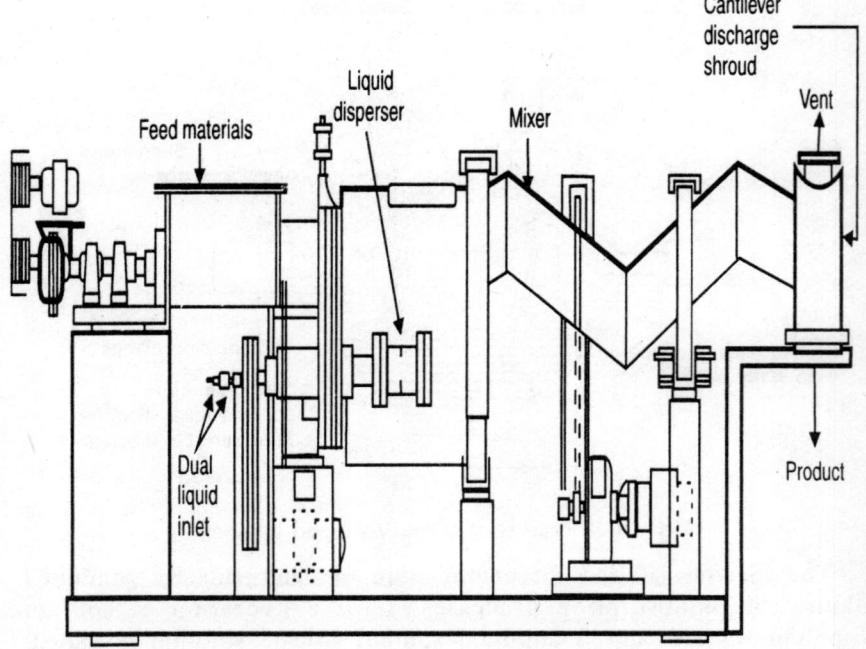

Fig. 10.5. Patterson–Kelley zig-zag continuous blender.

The blender typically recycles thirty parts of material for every part it discharges. Recycling serves as an averaging device to level off short-term feed variations. With each revolution, the machine discharges a uniform quantity of material of constant volume and weight.

This blender can also be combined with a spray-drying plant to increase capacity. STP is hydrated in a zig-zag blender (as shown in Fig. 10.6) mounted in series with a spray tower. The hydrated STP is then passed through a conditioner/cooler. The first part of the conditioner is a fluidised bed which both ages and cools the hydrate, the second part brings the power down to 30°C.

On passing out of the conditioner the material is split into two parts, the smaller part is fed to the crutchers to become part of the spray-dried matrix, while the larger part moves on to the second stage.

The second stage is a second zig-zag blender which is used to combine the spray-dried matrix, the STP hexahydrate and the non-ionic. Perborate, perfume and enzymes can be added at this stage. The non-ionic is dispersed through a liquid dispersion bar, designed for gentle but thorough mixing so

as not to destroy the bead structure of the matrix. The non-ionic acts as
a binder to form a soft agglomerate and to prevent segregation. Bulk
densities of 410–430 g/litre are obtained.

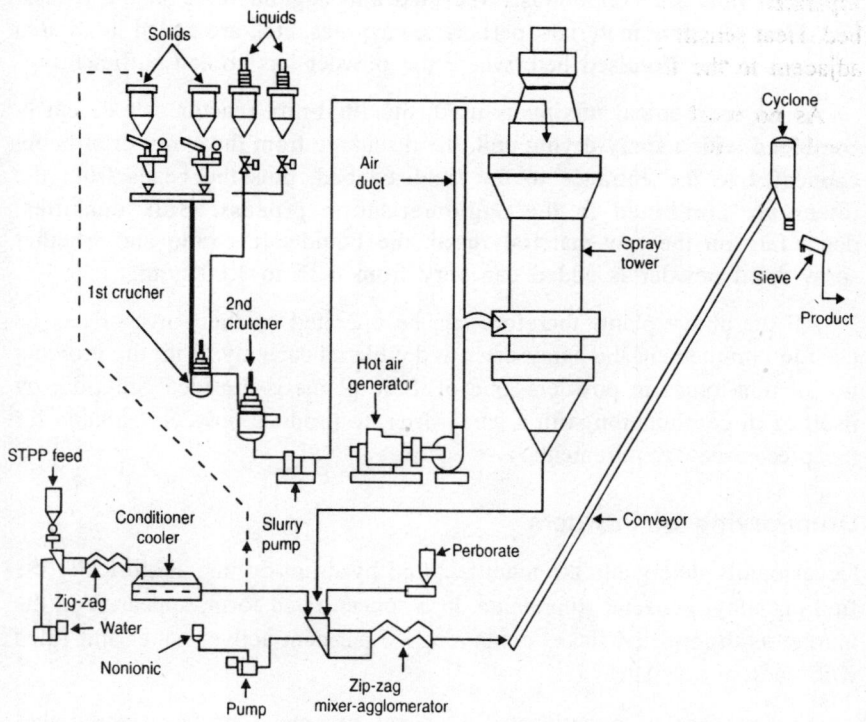

Fig. 10.6. Patterson–Kelley system for increasing spray-drier capacity.

Anhydro system

In Anhydro fluid-mix system, illustrated in Fig. 10.7, of the mixing,
neutralisation and agglomeration take place in a fluid bed reactor. Powders
are fed continuously from the day silos one of the other powdered
ingredients so that they can be measured efficiently in the volemetric
measuring device.

Materials discharged from the silos, after being measured by the
volumetric measuring devices at the base of the silos, are transported to the
fluid bed by belt conveyors. Liquids are fed into the reactor by metering
pump(s) and atomised by compressed air directly on to the bed. The fluid
bed reactor consists of a mixing chamber with a fluidised area, fines
separating area and a perforated plate for air distribution. The air supply is

divided into a number of compartments each with a guiding vane to allow differential fluidisation to take place. The reactor bed outlet is connected to a series of cyclones to separate fines from the exhaust air. The fines separated thus are continuously recycled and agglomerated in the reactor bed. Heat sensitive materials, perborate, enzymes, etc., are added in an area adjacent to the fluidised bed, where the powder has cooled sufficiently.

As no mechanical mixing is used, the fluid-mix reactor can be easily combined with a spray-drying unit, the discharge from the spray-drier being connected to the entrance to the fluidised bed, thus the beads from the tower are combined in the agglomerisation process. Bulk densities, depending on the raw materials used, the liquid/solids ratio and whether spray-dried powder is added can vary from 0.35 to 1.00 g/ml.

All the above plants therefore can be operated on their own or can be used to complement the spray-drier to double its capacity. With the growing use of non-ionics in powders, one of these plants is the ideal solution, by itself or in combination with a spray-drier, to produce powders suitable for the present-day requirements.

Drum-drying of Powders

Occasionally detergents are manufactured by drum-drying. Historically the first dry alkyl benzene sulphonate, in a concentrated form, appeared on the market as drum-dried flakes containing 40 per cent active matter and filled with sodium sulphate.

This process of manufacture does not commend itself, as the flakes produced are not very attractive and have a relatively small surface area. Compared to spray-dried powders, and even to well-made powders from the simultaneous absorption and neutralisation process, drum-dried powders are slow to dissolve in water.

The operation of a drum-drier is expensive (compared with a spray-drier) and there is always the risk of scorching the finished product. Furthermore, unless the drums are made of stainless steel, the detergent slurry will corrode them quickly.

Drum-driers have one advantage over spray-driers in that they can be started and stopped in a relatively short time, allowing for the possibility of short runs.

For the above reasons drum-driers are not normally used for the manufacture of built powders, but when a highly concentrated powder of alkyl benzene sulphonate or fatty alcohol sulphate needs to be made in relatively short runs, drum-drying has a definite economic advantage over spray-drying.

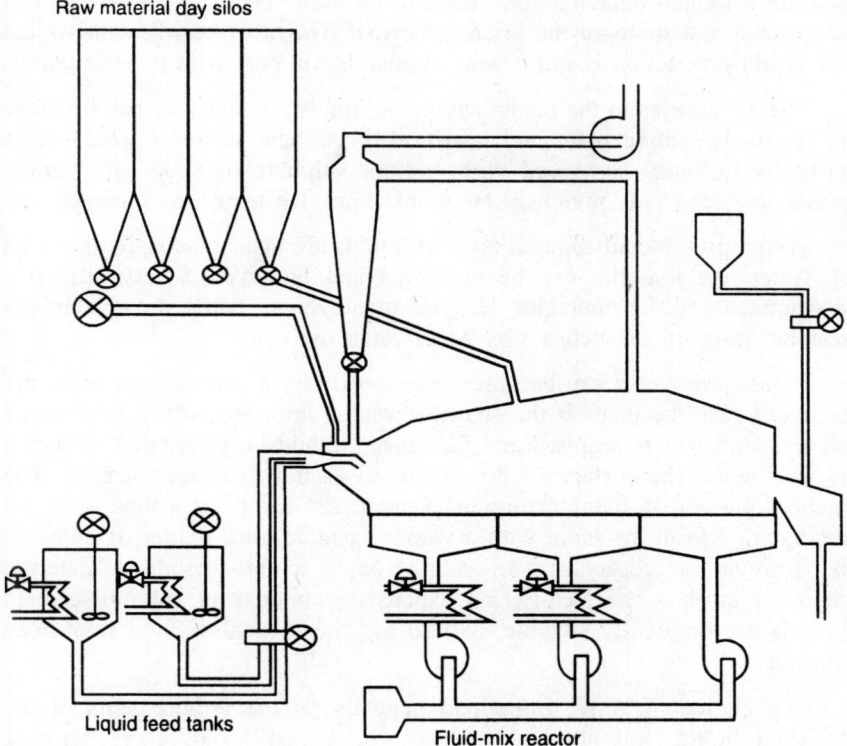

Raw material day silos

Liquid feed tanks

Fluid-mix reactor

Fig. 10.7. Anhydro fluid-mix process.

LIQUID DETERGENTS

The very first detergents sold in large quantities for household use were merely simple solutions of anionic detergent in water of active concentrations varying from 5 to 20 per cent. They rapidly became standard household requisites for dishwashing and for the washing of fine articles of clothing such as wool and silk. Of recent years, these light-duty detergents have become more sophisticated and liquids (or lotions) have now appeared with builders for heavy-duty washing, both by hand, and more particularly, in household washing machines.

At first glance, it might be considered that liquids are cheaper in raw material costs than powders, as the filler (or diluent) is water rather than soda ash or sodium sulphate. This might hold true for light-duty liquids, but where the heavy-duty product is to be considered, it is by no means axiomatic. For heavy-duty liquids hydrotropes, potassium rather than sodium salts and ethanolamines need to be used. These in no way add to detergency but increase the price somewhat. However, to offset this plant for liquid production is in no way comparable in cost with powder plants.

The advantages to the manufacturer of liquids are that they can be made in relatively simple plant and also that the diluent is water which costs virtually nothing, compared with sodium sulphate or soda ash, which, although cheap, are obviously far more expensive to use as diluents.

To the user, the advantages are that liquids are instantaneously dispersed in water, the material can be perfumed and be given a very attractive appearance and the container—a glass or polythene bottle, for example—can be designed in such a way as to catch the eye.

Liquid detergents can be made from a variety of starting materials, but in every case the plant is the same. A vessel equipped with a slow-speed stirrer is all that is required and the stirrer should be positioned so that it is well under the surface of the liquid, so as not to cause foaming. The authors know of a manufacturer who processes 4.5 m^3 at a time of liquid detergent, stirring by hand with a wooden paddle, in a matter of hours. It is, however, necessary that the vessel be of a non-corrodible material. Stainless steel is satisfactory, but expensive; concrete or asbestos-cement vessels are eminently suitable, and so are those of glass-fibre reinforced plastics.

The choice of active ingredient depends on the requirements of the finished liquid, but here particularly, the versatility of alkyl benzene sulphonic acid made by sulphonating with SO_3 is of particular value. Other anionic raw materials are available as the already neutralised salt, with or without the presence, whether incidental or not, of various non-detergent materials. Alkyl benzene sulphonic acid based on oleum sulphonation, when neutralised, will also contain a fairly large proportion of inorganic sulphate ions. Alkyl benzene sulphonic acid sulphonated with SO_3 and sold as the 100 per cent material, contains a minimum of free sulphuric acid and thus gives the manufacturer the greatest scope in formulation.

This 100 per cent material can be neutralised by any base to give various types of liquids, from low viscosity to gels. The solubility of the alkyl benzene sulphonic acid salt of the alkali metals is opposite to that of the regular order, in that the most soluble of the salts is that of lithium, followed by sodium, with potassium the least soluble. When manufacturing heavy-duty liquid detergents, this must be taken into account if potassium silicate and pyrophosphate are to be used, since the potassium ion of the inorganic filler can precipitate the sulphonic acid ion unless the formulation is carefully controlled.

Table 10.4 gives the physical properties of 10 per cent solutions of dodecyl benzene sulphonic acid, neutralised by various bases. In every case 100 g of the sulphonic acid was neutralised to pH 7 by the particular base and the final solution made up to 1 litre.

The figures in Table 10.4 will vary with the type of ABS used, as small amounts of free sulphuric acid in the original material will greatly influence the behaviour of the neutralised salt. Furthermore, these figures were given for propylene tetramer DDBS. Linear DDBS in general has a lower viscosity, both for the alkali and ethanolamine and ethylamine salts, but these figures give a relative idea of the differences according to the base used.

If partial neutralisation of the sulphonic acid is to be done with caustic soda (or potash) as is possible when using the DDBS produced from SO_3, a slight amount of ferrous iron present in the caustic soda can cause unpleasant complications in the final product. (The same holds good for sulphonic acid produced by oleum or sulphuric acid sulphonation and where iron pick-up is possible.) Traces of the ferrous salt of the sulphonic acid will in time precipitate as a black sludge. This can be eliminated by oxidising the ferrous to ferric ions, and is easily accomplished by the use of sodium hypochlorite thus:

All the water necessary for the formulation, together with the caustic soda solution, is run into the vessel. Then 1 per cent of a 10 per cent solution of sodium hypochlorite is added and the solution mixed. The sulphonic acid is now added and mixing continued until all the sulphonic acid has been dissolved, and the solution allowed to stand for at least 20 min. longer to permit the oxidation process to go to completion.

The ethanolamine or other organic basic which is to be used for the completion of the neutralisation is now added, together with all the other additives.

The sodium hypochlorite will have been partially destroyed in the initial oxidation process; then, when the sulphonic acid is added, the excess hypochlorite will completely bleach all the colouring matter if any is present in the sulphonic acid. After this process a small residue of active chlorine will still be present, but this need not be eliminated as it will react with the organic amines to produce a chloramine, which is unstable. The amines will thus reduce the balance of the active chlorine left and render it innocuous. (It is for this reason that sodium hypochlorite cannot be used to bleach amine-containing solutions.)

Table 10.4. Physical properties of ABS solutions neutralised by various bases.

Base	Appearance of sol at 20°C	Viscosity at 25°C (Pa s)	Surface tension of 0.1% active material N/m	Foam height of 0.1% active material solution (Ross Miles)	
				Initial cm	After 5 min. cm.
Ammonia	Separates into 2 phases	–	0.0310	17.5	17
NaOH	Clear liq	0.0045	0.0300	18	17
KOH	Gel	0.0285	0.0310	16	15
LiOH	Clear liq	0.0024	0.0310	17	16
Monoethanol-amine	Clear liq	0.0285	0.0310	16.5	15
Diethanolamine	Clear liq	0.0870	0.0300	17.5	16.5
Triethanolamine	Clear liq	0.0635	0.0310	18	17
Morpholine	Gel	0.4350	0.0300	12	11.5
Monoisopropanol-amine	Gel	0.3100	0.0310	15	15
Di-isopropanol-amine	Clear liq	0.3450	0.0300	15	14
Tri-isopropanol-amine	Clear liq	0.0910	0.0310	15	14.5
Mixed Isopropanolamine	Clear liq	0.3750	0.0305	16	15
Diglycolamine	Clear liq	0.1180	0.0310	16	15
Diethyl-ethanolamine	Clear liq	0.3050	0.0320	15	14
Triethylamine	Liquid*	0.1350	0.0320	14	13
Trimethylamine	Liquid*	0.3100	0.0360	13	12
Cyclohexylamine	Liquid*	0.0240	0.0310	6.5	6
Iso-butylamine	Gel*	0.7000	0.0340†	5†	5†
n-Butylamine	Cloudy liq*	0.0078	0.0310	5	4
sec-Butylamine	Gel*	0.4450	0.0305	5	4
ter-Butylamine	Gel*	0.2250	0.0310	5	4
Diethylamine	Gel*	0.5250	0.0310	12	11
Dimethylamine	Gel* separates	0.1250	0.310	12.5	12

* With characteristic smell.   † Not completely soluble.

If for any reason it is not possible to use sodium hypochlorite for this oxidation process, the same effect can be achieved by the use of hydrogen peroxide, either as such, or in the form of sodium perborate. It is general practice to add to detergents based on alkyl benzene sulphonic acid one of the alkylolamides as a foam stabiliser, thickening agent and skin-protecting agent. No special procedures are involved. Opacifying agents, especially for solutions based on linear alkyl benzene sulphonates are often used to give the liquid a 'different' appearance. Aqueous dispersions of an alkali insoluble polymer of styrene, substituted styrene, a co-polymer of styrene with acrylamine, PVC or polyvinylidene chloride are used. All these above materials produce an opacity in the liquid which enhances the appearance. Stability and freeze-thaw studies should, however, be made on all trial formulations.

Liquid formulations can be light sensitive. This can however be overcome either by the addition of ultra-violet absorbers or by packing in opaque bottles.

Liquid formulations allow the incorporation of solvents (solvents are however being incorporated successfully in powders). These solvents need not necessarily be of the water-immiscible type, glycol ethers being examples of excellent fat solvents which are completely water-miscible.

Light-duty liquid detergents can be based on alkyl benzene sulphonic acid, alcohol sulphates non-ionics as such, or sulphated non-ionics, or amphoteric detergents, or combinations of two or more of these materials.

Except when sulphonic acids are used, all the other components are available as concentrated (from 40–100 per cent) neutralised material which needs only to be diluted down to the required activity. With sulphonic acid, however, the manufacturer has the choice of varying his neutralising agent to give his liquid any property he wants, from a low viscosity to a thixotropic gel (Table 10.4).

If non-ionics of the ethylene oxide condensation type are being used in liquid detergents, either alone or in combination, it should be borne in mind that this type of material has a peculiar cloud point, in that it forms a cloudy solution when the temperature is raised and not (as is usually the case) when the temperature is lowered. It is thus important that products should be tested extensively, taking into account the highest possible temperature that the liquid may encounter on storage.

Heavy-duty liquids are coming more and more into use and require special formulation techniques. The problem here is to incorporate into the liquid active matter, sequestering agents, silicates, anti-redeposition and optical brightening agents. To achieve all this and to retain the finished material as a clear liquid with a low cloud point is not easy. If at a moderately high temperature the liquid 'clouds', it is an indication that separation into at least two phases is taking place and each phase will contain different proportions of each of the ingredients.

This problem is best overcome by producing the heavy-duty liquid as a lotion or gel, but this must be formulated so that it will not separate into two phases at the highest ambient temperature to which it may be subjected.

Toilet Preparations

As hair shampoos, shaving creams and bath preparations are all basically detergents, more and more detergent manufacturers are manufacturing either ready-made cosmetic preparations of this type or a concentrate for supply to the cosmetic manufacturer. The alcohol and ether sulphates have proved themselves over the years to be dermatologically safe for this type of preparation.

For hair shampoos the choice depends on the desired physical form of the product. For a cream the sodium slat of one of the fatty alcohol sulphates is recommended. For a liquid an ether sulphate is used or else an ethanolamine salt of a fatty alcohol sulphate. The latter has the disadvantage in that it cannot normally be obtained with as light a colour as the ether sulphate and it also foams less, but this can be enhanced by the use of a foam booster. The pearlescent type of shampoo can use as a base either an ether sulphate or sodium salt of the alcohol sulphate.

In addition to the well-tried sulphated alcohols and ethers, more and more use is being made of the amphoteric detergents as either a portion or all of the active ingredient in shampoos, etc. If they are used as the sole ingredient an additional advantage is that cationic detergents can be incorporated as bactericides.

Perfumes used in shampoos should all be dermatologically tested and in addition preservatives need to be added, as fungi and yeasts grow in detergent solutions. All formulations should be made up to include perfume dyes and preservatives and subjected to extensive storage tests as no hard and fast rule can be laid down for the interaction of the micro constituents. Furthermore, it is best to add a small portion of a chelating agent (0.25 per cent or so) as traces of iron may have an adverse effect on dyes in storage.

Dyes themselves or the constituents of certain dyes have come to be considered health hazards and are now subject to certification for use in toiletries or domestic preparations.

Some basic formulae for shampoos are given below :

Formula 4

Cream shampoo

Sodium fatty alcohol (C_{12-14}) sulphate (50% paste)	25
Coconut monoethanolamide	2
Thickening agent	1
Sodium chloride	2
Chelating agent	0.25
Perfume, dye, preservative	qs
Water	to 100

The thickening agent is predissolved or dispersed in a portion of the water in accordance with the manufacturer's instructions. The sodium chloride is added to increase the viscosity if desired.

Formula 5

Liquid Shampoo

Sodium ether sulphate (C_{12-14}, 2 mol ethylene oxide) (100% basis)	12.5
Thickening agent	qs
Sodium chloride	2
Perfume, dye, preservative	qs
Chelating agent	0.25
Water	to 100

The thickening agent is added as required because hard and fast rules cannot be given as very often the concentrated ether sulphate contains ethyl alcohol or propylene glycol both of which drastically lower the viscosity.

Formula 6

Liquid Shampoo

Triethanolamine fatty alcohol sulphate (40% concentrate)	35
Coconut diethanolamide	3
Chelating agent	0.25
Perfume, dye, preservative	qs
Water	to 100

In place of the triethanolamine salt, the monoethanolamine neutralised fatty alcohol sulphate can be used. This will give a higher viscosity.

A pearlescent effect can be obtained by adding to the paste ethylene glycol stearate, diethylene glycol stearate, polyethylene glycol stearate, magnesium stearate or myristic diethanolamide.

Formula 7

Pearlescent Shampoo

Sodium fatty alcohol sulphate (50% concentrate)		25
Palmitic monoethanolamide		2
Ethylene glycol stearate		1
Chelating agent		0.25
Perfume, dye, preservatives		qs
Water	to	100

The alcohol sulphate, the glycol stearate, the monoethanolamide and the water are charged into the vessel and warmed with gentle stirring until a clear solution free from dispersed particles is obtained. This is then cooled to 50°C and the other ingredients added and passed to packing. When cold the pearlescence will have developed.

Formula 8

Bubble Bath Preparation

Sodium ether sulphate (100% basis)		20
Triethanolamine fatty alcohol sulphate (100% basis)		8
Chelating agent		0.25
Perfume, dye, preservative		qs
Water	to	100

Paste Detergents

In certain parts of the world detergents are used in a finished or semi-finished form as pastes of varying concentrations. Certain detergent raw materials are sold as pastes: for example, the sodium slat of lauryl alcohol sulphate in a concentrated form without any addition of inorganic fillers.

The detergent manufacturer who wishes to manufacture a paste can do so very easily from alkyl benzene sulphonic acid, again either from the 100 per cent or 90 per cent varieties, depending on which is available to him. To produce a snow-white paste from 90 per cent sulphonic acid can only be achieved if the acid is neutralised immediately after sulphonation, and not stored. The 100 per cent sulphonic acid can produce a snow-white paste even though stored for many months.

The manufacture of the paste can be done in a soap crutcher, or in a specially built neutraliser, equipped with a cooling jacket as shown in Fig. 10.8. The neutralising vessel would preferably be built of stainless steel, but if precautions are taken so that the pH of the paste does not fall below 7 during the process of manufacture, a mild steel vessel can be used.

Because sodium alkyl benzene sulphonates are not completely homogeneous in concentrated solution, especially in the presence of electrolytes, various materials must be used to prevent separation into two phases. This can be done by the use of hydrotropes, small proportions of CMC or alkylolamides. However, hydrotropes will tend to produce thinner pastes and this will mean adding inorganic salts to make them stiffer again, depending, of course, on the requirements of the market. In general, tridecyl benzene sulphonate will produce the stiffest paste; dodecyl benzene sulphonate a paste of medium consistency; and the biodegradable linear alkyl benzene sulphonates, the thinnest pastes. Consistency can again be varied by the addition of inorganic salts. In some countries high concentrations of borax are added to produce a stiff paste.

These inorganic salts can be sodium chloride, sodium sulphate, tetra sodium pyrophosphate or sodium silicate solution, or a combination of these. To achieve a stiff paste, it is very rarely necessary to add more than 3 per cent in total. Sodium chloride is the cheapest and gives the greatest stiffening effect, but is corrosive to iron or steel. Sodium sulphate and tetra sodium pyrophosphate suffer from the disadvantage that in cold weather they might crystallise out from the paste into needle-like crystals, while both the tetra sodium pyrophosphate and the sodium silicate raise the pH of the paste, but add to the detergency. The high pH can be regulated by using some of the sulphonic acid to neutralise a portion of the alkalinity of these materials. In particular, to improve the stability of a paste based on SO_3 sulphonated LAS the 1:3.3 ratio sodium silicate is used for partial neutralisation. Colloidal silica is formed in the neutralisation and it serves as a kind of absorbent skeleton to prevent separation. The LAS is thus neutralised by sodium only, the exact amount of caustic soda being dependent on the acid value of the sulphonic acid and also whether partial neutralisation by an inorganic salt has been done. The acid value will vary not only with the average molecular weight of the alkyl benzene (tridecyl benzene has a lower acid value than dodecyl benzene), but also with the method of manufacture (i.e., oleum, sulphuric acid or SO_3) and finally according to the individual manufacturer.

For this reason, although the acid value is usually expressed as the amount of milligrams of caustic potash required to neutralise 1 g of acid, it is better to consider the acid value as the number of grams of caustic soda required to neutralise 100 g of acid. This can readily be calculated from the internationally accepted acid value by the formula :

$$\text{Acid value in g NaOH/100 g acid} = \frac{\text{Acid value in mg KOH / g acid}}{14}$$

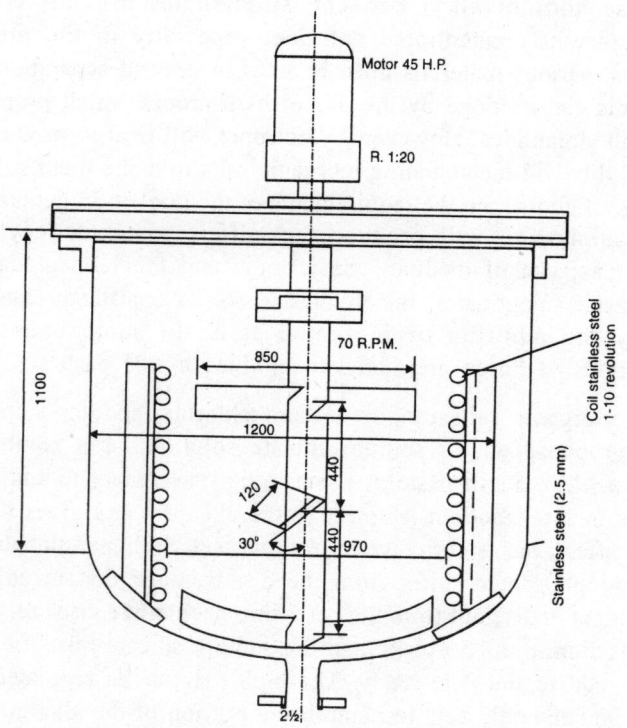

Fig. 10.8. Mixer for liquid and paste-form detergents.

The method of manufacture depends on whether a stainless-steel or mild-steel vessel is being used.

If CMC is to be included in the formulation, it is necessary to dissolve it separately. If a separate mixing vessel with a slow-speed stirrer is available, water to the amount of ten times the weight of the CMC is charged into the container and the CMC poured in slowly with stirring. The stirring is continued until all the lumps of CMC are dispersed, concurrently with the process of the manufacture of the paste. If a mixing vessel is not available, the solution can be made in the same way in a 200 litre drum and the CMC added while stirring with a hand crutcher. This dispersion is best prepared the day before it is required and left to swell and dissolve by itself overnight.

The CMC dispersion, if in contact with iron or steel, will tend to corrode the vessel and also will become discoloured by the rust formed. It is therefore essential to produce this dispersion in a non-corrodible vessel— stainless steel, asbestos cement or steel clad with plastic or rubber. The stirrer itself should also be non-corrodible.

However, some manufacturers are able to achieve rapid wetting, swelling and solution of the CMC in a matter of half an hour by the use of a high speed turbine mixer. The addition of alkali can also help in this dispersion.

If a mild steel neutralising vessel is being used for the manufacture of the paste, all the remaining water called for in the formula (all the water if no CMC is to be added) and the requisite amount of caustic soda solution, with inorganic salts if they are to be used, are run into the vessel. Mixing and cooling is started and the sulphonic acid added at such a rate that the temperature does not rise above 65°C. (If 100 per cent acid is used the heat of reaction is low and the temperature will not rise very high, provided that an efficient cooling coil or jacket is available. If 90 per cent acid is used, the heat of reaction is considerably higher and the rate of addition of the sulphonic acid must be adjusted.)

When about 80 per cent of the sulphonic acid has been taken up and the paste is still distinctly alkaline, the requisite amount of sodium hypochlorite solution is added and mixed in. The balance of the sulphonic acid is now added, taking care that the pH of the paste does not fall below 7. Mixing is continued until the paste is uniform and the CMC solution (if called for) is added, or, if an alkylolamide is to be used, this can now be put in. If a hydrotrope is employed, this can be added with the sulphonic acid.

When a stainless-steel neutralising vessel is being used, the procedure is similar, except that the sodium hypochlorite can be put in at the beginning with the water and caustic soda and no special precautions need be taken about not passing to the acid pH. If an acid pH is reached, the desired pH of 7.5–8 can be achieved by the addition of a litre or two of caustic soda solution.

A slight variation of this paste procedure occurs when a highly concentrated liquid is required. This is done by neutralising the sulphonic acid with an ethanolamine.

In this instance, the requisite water (which depends on the final concentration desired) and the ethanolamine are charged into the vessel. The sulphonic acid is added slowly with intermittent stirring so as not to generate foam. The reaction is instantaneous. The only precautions that need be observed are to avoid the production of foam and to keep the temperature from rising above 60°C.

Calcium Sulphonates

Although efforts of formulators are concentrated in the direction of neutralising the effects of calcium ions by chelation, the calcium salts of alkyl benzene sulphonic acid play an important role in the production of agricultural emulsifiers because of their solubility in the semi-polar solvents used as diluents for insecticides. The calcium ion, being divalent, yields a sulphonate which has the pictorial formula

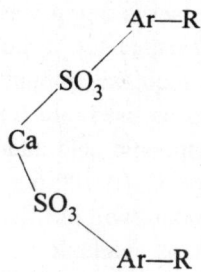

that is, the hydrophilic portion is in the centre of the molecule with two hydrophobic tails. For reasons not readily apparent this allows of spontaneous emulsification, an absolute prerequisite in the field. The sulphonic acid is not necessarily of the same molecular weight as that used in cleaning materials, quite often it can be greater, increasing its oil-solubility. The calcium sulphonate is not the only constituent of these emulsifiers whose formulation is beyond the scope of this volume.

To manufacture calcium sulphonates, good quality SO_3 sulphonated alkyl benzene is essential with an absolute minimum of free sulphuric acid. To attain this completeness of sulphonation is often sacrificed so that (almost) no surplus SO_3 is needed for the sulphonation. As can be understood, the presence of free sulphuric acid will produce insoluble calcium sulphate.

The neutralisation is performed with good quality lime in the presence of a polar solvent as a diluent, and after completion the diluent is distilled off.

SOLID DETERGENTS

Solid detergents are usually manufactured in three forms for the following different applications:

(i) toilet cakes;

(ii) household scrubbing;

(iii) 'one-shot' additions to washing machines.

Detergent Toilet Bars

The manufacture of a detergent toilet bar comparable with soap is fraught with many difficulties, and is the subject of many patent applications throughout the world. Not many of these patents have been successfully exploited commercially for technological or economic reasons, or both. A few toilet detergent bars have been put on the market, but not all of them have captured the sales that the manufacturers had hoped for.

One hundred per cent of detergent active matter is considerably more expensive than 100 per cent soap. This does not matter when detergents in liquid or powder form are being marketed, as 20–25 per cent active matter in a well-made powder can compete very favourably with a 40–60 per cent fatty acid soap powder. To be sold in the same price range as soap a detergent bar cannot have much more than 40 per cent active matter. If only 40 per cent active matter can be used the mass needs to be filled to 100 per cent with relatively inexpensive materials and this causes the problem. The use of soluble inorganic salts is precluded because the cake will be heavy compared to soap and will tend to effloresce. Talc, although not efflorescent, will again make the cake extremely heavy and hard to the feel.

A detergent used as a soap will tend to defeat the skin overmuch and 'superfatting' or retarding agents need to be incorporated into the bar.

One of the big disadvantages to overcome is the behaviour of a detergent tablet when wet compared to a soap tablet. When soap is wetted, water is absorbed over the whole surface of the soap and the viscosity of the water-soap mixture drops dramatically and then increases with further dilution to form the viscous 'middle soap'. Thus when a wet tablet of soap is placed in a soap dish this viscous middle phase prevents water from penetrating into the tablet and the soap remains dry inside, the outside surface gradually drying out by evaporation. This middle phase does not occur with detergents and therefore the water in a wet detergent tablet under the same conditions will penetrate right to the centre of the mass, making it soft and slimy.

For all that, synthetic toilet tablets have one very big advantage in that because of the chemistry of synthetic detergents they do not make a scum or leave a ring in the bath tub. A toilet detergent tablet should therefore not be readily soluble or should not disintegrate, and should have the feel of soap when wet; it should lather freely and when stored should not crack or effloresce. Finally, it should be possible to work the mass in normal soap-milling and stamping machinery.

The bulk of these problems have been overcome by the use of a soap/detergent mixture, but although this improves the physical behaviour of the cake, it does not do away with the disadvantages of soap. Tolerance to hard water can be improved by the addition of lime soap dispersants or even by using them wholly as the synthetic part of the 'combo' soap as it is sometimes called.

There are a few detergent cakes on the market and these are based on sulphated monoglycerides, fatty alcohol sulphates, alkyl isethionates, sulphosuccinates and sulphated fatty acids. Very often the potassium salt is preferred to the sodium because of its better rheological properties. To these active raw materials are added plasticisers such as fatty alcohols, the phosphate and phthalate esters, alkylolamides, particularly coconut or palmitic mixtures, and polymerised ethylene glycols and binders such as starch and talc.

Manufacture is the subject of many patents and trade secrets, but the process consists of using raw material in as nearly an anhydrous form as is possible to maintain, melting it with plasticisers, adding the binders, and then milling, plodding and stamping in the normal soap-making machinery.

General composition for toilet soap :

Surfactant	30–70%
Plasticiser/binder	10–30%
Filler	10–30%
Additives	0–20%
Water	3–10%

A specific formula can be suggested as under :

Coco fatty alcohol sodium sulphate (70–80% AM)	50%
Monoalkanolamide of palmitic/stearic acid derived from either hydrogenated tallow or palm oil methyl ester or fatty acid	20%
White dextrin	30%

The dextrin serves not only as a binder but also as a skin protectant. This formulation contains no insoluble matter.

Household Scrub Bars

Historically, in this field again the first attempts were based on a detergent which had been available as a water solution. Binding the water was done with materials such as bentonite and sodium stearate, i.e., soap produced *in situ*. Because the starting material contained at least as much water as active matter, the final bar or cake contained the same ratio of water to active detergent matter. This limited the amount of active matter that could be incorporated, because the more active matter added, the more water appeared in the final product. Nowadays a bar of this type can be manufactured at concentrations between 20 and 60 per cent, as desired, by the use of alkyl benzene sulphonic acids.

The sulphonic acid is mixed with the dry ingredients in a dough-mixer. A plough-mixer or amalgamator is also suitable provided that the arms and motor are strong enough to withstand the resistance of the material and the neutralising agent is either soda ash or a concentrated caustic soda solution. If caustic soda is used, a percentage of water is automatically added to the material, but the final product can be controlled to have a neutral pH. When soda ash is used (as in the manufacture of powders) a surplus of soda ash needs to be present and it is thus impossible to have a neutral pH. Although a neutral pH is not essential, as it would be for bars meant for toilet use, nevertheless the bar will still come into contact with the hands, and must not be too alkaline. The pH can be lowered by having a large amount of bicarbonate in the formula or by the addition of sodium acid pyrophosphate after the neutralisation has been completed.

After neutralisation, and the addition of any special ingredients, the mass is allowed to age. This ageing process varies with the formulation and further manipulation. If a bar of high active-matter content (60 per cent or so) is being made, the mass should be aged for at least 12 hours. If a low active-matter material (40 per cent) is being manufactured, the mass can be transferred to the next stage immediately. In fact, it is not advisable to allow it to age, as it will set hard and become unworkable. (If by any chance a low active-matter bar has set hard before it can be worked, it is possible to regenerate the plasticity by warming it to 50–60°C in a soap-drying room.)

Low-active concentration bars or tablets are not recommended because they are both inefficient and uneconomical in use. However, in certain of the developing countries it is considered that the unit cost of a bar might be too high if it contains the normal amount of active matter. This is a false consideration because the filling materials as well as the extra manipulations involved all cost money. For all that, bars of 25 per cent active matter or less are being produced.

To make bars of this type filling materials need to be used. Plain water cannot be used alone as this will make the bar too soft. If inorganic salts alone are used the bars become heavy, hard and unattractive in appearance and are also subject to efflorescence on storage. These problems can be partially overcome in various ways.

Paraffin wax at the rate of 2 per cent can be added. The wax needs to be melted separately and poured on the mass while mixing, before the temperature has fallen to 60°C. The addition of the paraffin wax gives a better feel to the tablet and to a certain extent hinders the defatting action of the detergent on the human skin. This skin-protecting action can be enhanced by using a blend of paraffin wax and polyethylene wax. In this instance the polyethylene wax (which has a melting point of the order of 100°C) needs to be dissolved in the molten paraffin wax. Surprisingly enough, this addition of waxes in no way detracts from the detergency of the bar or tablet.

Another method is to use 5 per cent starch and to include in the basic formula a total of 15 per cent water. After the reaction is completed, the starch is added and the mass heated to a minimum of 80°C to solubilise the starch.

Finally, a combination of both the above methods can be used and also on occasion talc is added to improve the feel of the tablets.

After the requisite ageing period, the mass is passed through a soap mill. In certain factories this milling is dispensed with and the aged mass is fed directly into a plodder. However, the appearance and texture of the finished product is definitely inferior if it is not passed through a mill.

For this reason, we stated above that the period of ageing depends on the further processing. If the mass is to be milled, the ageing period can be shorter, as it is desirable to feed the mass into the mill while it is still very plastic. If no milling is to be done, the mass should be less plastic. In some factories the mass is allowed to set hard, then chipped in a soap-chipper, and then passed into a mill or direct into the plodder.

FABRIC SOFTENERS

The fabric softeners have become a necessary adjunct to washing of clothes particularly when machines are being used.

It was becoming apparent that clothes were coming out of the machine 'hard' to the feel. The wool industry had for years used the term 'handle' to test the feel or softness of the fibres and this in turn was a function of the efficient washing or scouring of the wool.

In the same manner as CMC was introduced to produce whiter washing, fabric softeners filled the gap formed by the 'hardness' of the clothes. The reason for the washing coming out hard is a build-up of salts on the fibres and these salts are deposited in such a manner that they are not easily rinsed away. (It should be remembered that no matter how much chelating or sequestering agents are added to the washing powders or liquids, the final rinse in household washing is with tap, i.e., relatively hard, water.)

Certain quaternary ammonium salts (cationic detergents) and also alkylolamides were found to be effective as additives for the softening of clothes. These softeners are adsorbed on to the fibres from the final rinse and the washing comes out softer, probably due to an interfibre lubricating effect.

In addition to softening, the fabric softeners of the quaternary ammonium type also act as bacteriostats if not bactericides. Finally, with the synthetic materials being used nowadays, a very common occurrence is the build-up of electrostatic charges on the fibres. The softener tends to diminish production of this static electricity.

The common cationic detergents in use are di-tallow di-methyl ammonium chloride (or iodide) (I), alkyl di-methyl aryl ammonium chloride (II) or 2-heptadecyl-1-methyl-1-(2'-stearoylamido-ethyl)-imidazolinium methyl sulphate (III) which are pictured below :

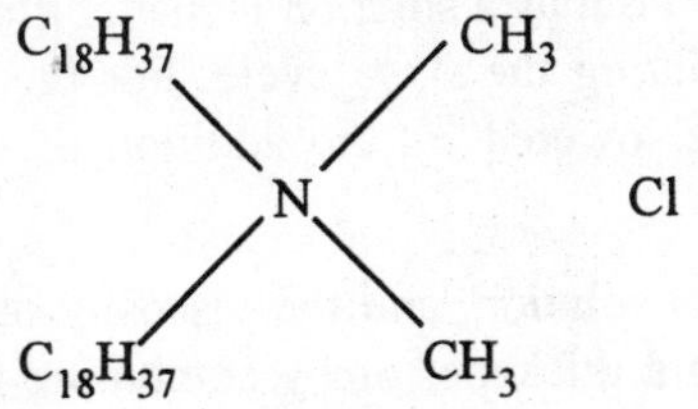

$$C_{18}H_{37} \quad CH_3$$
$$N \quad Cl$$
$$C_{18}H_{37} \quad CH_3$$

I

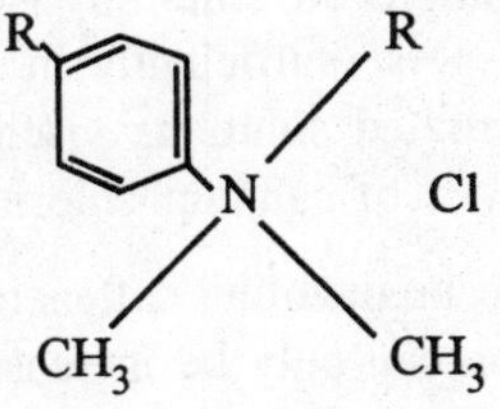

$$R \quad R$$
$$N \quad Cl$$
$$CH_3 \quad CH_3$$

II

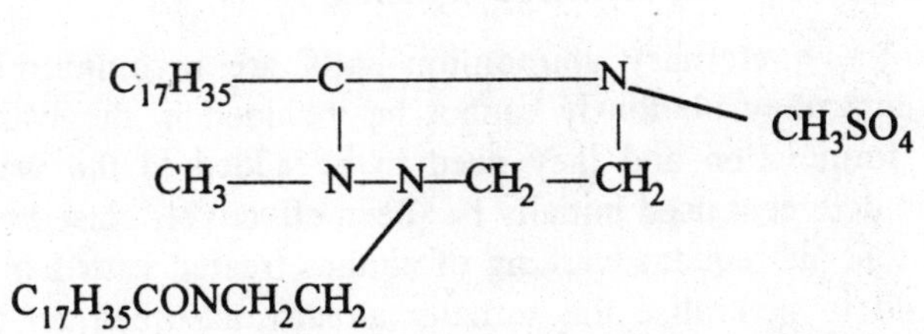

$$C_{17}H_{35}\!-\!\!C\!-\!\!\!-\!\!\!-\!\!N$$
$$CH_3\!-\!\!N\!-\!\!N\!-\!\!CH_2\!-\!\!CH_2 \quad CH_3SO_4$$
$$C_{17}H_{35}CONCH_2CH_2$$

III

It will be observed that III has a quaternary, an ethanolamide and a sulphate group making it an amphoteric. This type has only 70 per cent of the softening efficiency of types I and II.

Viscosity of the softener should of course be constant and here again we come across an anomaly, in that these materials behave differently from other surfactants. If the electrolyte content of the final solution is high the formulation can be almost water-like in its viscosity. If the electrolyte content is very low the solution (dispersion) can be like a jelly. To allow for variations in the hardness of the water supply it is best to use demineralised water for diluting and then control viscosity by addition of electrolyte (sodium chloride or sodium acetate).

These materials are merely dispersed in water to give a convenient concentration so that 0.1 per cent (based on the weight of the clothes) of the softener can be dosed. Thus for a machine taking 5 kg of wash, 5 g of the softener need to be applied. It is therefore advisable to dilute the softener sufficiently so that 5 g (or less for machines taking smaller quantities) of the 100 per cent material can be measured easily.

This concentration is normally in the 5–8 per cent range, but 'concentrated' softeners of double, triple or even greater strengths are available. At the higher concentrations the solubility of the distearyl or ditallow derivatives is poor and the 'dispersion' could separate on storage. To prevent this, recourse to emulsifiers is necessary. Of the emulsifiers suitable for this purpose we mention glycerol monostearate, ethoxylated fatty amines, and fatty alkyl diamines.

Again as the solubility is poor, if too viscous a solution is made, it might not disperse sufficiently in the water during the rinse cycle. For the very concentrated solutions viscosity can be lowered by the addition of small quantities of calcium chloride.

The imidazoline softeners give a thin solution, and the viscosity of this solution can only be increased by using a thickener or by combining them in the ratio of 60:40 or 70:30 with a quaternary softener. This has the added advantage of giving enhanced softening.

As the quaternary ammonium salts are incompatible with anionic detergents, they obviously cannot be included in the washing powder or liquid formulation and they need to be added to the washing when the anionic detergent used initially has been effectively rinsed away. It has been found that subsequent washing of clothes treated with fabric softener does not tend to neutralise the softener already adsorbed on the fibre; rather softener tends to build up with consecutive washes.

However, there are on the market washing powders and liquids which incorporate a softener, that is the softening effect is achieved in the wash cycle. The objection to this mentioned above can be overcome by one of the following methods :

1. Acceptance of the fact that there will be anionic–cationic interaction. This will neutralise a small part of the anionic component (and incidentally reduce foaming) to form the complex (or salt). This complex is still substantive to the fibres, its softening effect will be diminished but will still produce an anti-static function.

2. Use of only non-ionic material as the active matter.

3. Encapsulation of the cationic with a relatively high melting, water-insoluble material (paraffin wax). The softener then becomes dispersed in the water when the temperature in the wash cycle reaches its maximum after the bulk of the washing has been done. There will of course be a competing reaction between the anionic on the one hand and adherence to the fibre on the other.

4. Use of a tertiary amine with two fatty acid groups (di-tallow methyl) for example. These materials are negatively charged at normal wash pH and behave as ordinary bases, although not soluble to any extent in water. This insolubility allows them to adhere to the fibres of the cloth. In subsequent rinse cycles the pH will drop to the neutral range and they become converted to cationic material (they have been called crypto-cationic) and exert the softening effect.

A further tendency has arisen to add optical brightening agents to the softening solution. Care must be taken in the selection of the brightening agents as many of them are incompatible with cationic detergents or unstable in solution. It is best to work in close co-operation with the manufacturers when selecting these dyestuffs.

Fabric softeners for the rinse cycle are always perfumed. The type and quantity of perfume should be chosen with care, preferably with the help of an expert perfumer, as a well-formulated softener can leave the clothes with a pleasant, fresh smell even if they are passed through a drier.

The criteria, apart from the germicidal or bacteriostatic properties, that have been suggested for fabric softeners are :

1. Easy dispersion in a final rinse solution.

2. Substantivity to the fabric.

3. Ability to impart softness and fluffiness (handle).

4. Ability to impart good interfibre lubrication without giving a greasy feel.

5. Ability to impart an anti-static effect.

6. The softener must not adversely affect the colour of the fabric (yellowing or greying) and, optionally, should impart a brightening effect.

7. The fabric must be easily re-wettable after treatment.

ABRASIVE CLEANERS

To this class belong the abrasive hand-cleaners and scouring powders for pots and pans as well as cleaners for tiles and bath-tubs. The vast majority are sold as powders or pastes, but bars and liquids are also available.

Powders, pastes and solids are made in much the same way as the non-abrasive product, with the addition of the abrasive material. In liquids, abrasive matter will always tend to settle out, but, to prevent it from compacting, a gelling agent is incorporated. This increases the viscosity, so that the abrasive can easily be re-dispersed when the liquid is shaken to prevent the abrasive from compacting. The choice of abrasive in the above compounds is one of extreme importance in avoiding undue costs and yet giving the abrasiveness desired without scratching. For example, an abrasive used in a wall-cleaner must be softer than one used on windows, one for aluminium ware softer than one for steelware. Usually the manufacturer employs a compromise type of material. The following materials are listed in order of hardness. Generally, they are used in a size of 200 mesh or finer.

Talc : Hydrous magnesium silicate; very soft; suitable for special scouring agents for tile and porcelain enamel, but not for a general household product.

Diatomaceous earth : Siliceous skeletons of small aquatic plants called diatoms; also called infusorial earth, kieselguhr, tripolite, etc; soft; expensive; not recommended in general household products.

Dolomite and calcite: Natural magnesium carbonate and calcium carbonate.

Marble dust : Calcium carbonate, mild abrasive, commonly used in household cleaners.

Volcanic ash : Sedimentary rock composed of volcanic dust, ash, and cinders; suitable for use in household cleaners. Pumice, also a derivative of volcanic ash, is slightly harder.

Felspar : Mixture of various metal aluminium silicates, chiefly $K_2O.Al_2O.6SiO_2$. Its hardness is just about the equivalent of volcanic ash, and it is usually of a whiter colour, so that it is a very desirable product when available.

Quartz : Crystallised SiO_2. It is harder than pumice and is apt to be abrasive on softer metals such as aluminium and copper. Not recommended where milder abrasives are available.

Sand : Fine grains of disintegrated siliceous rock, chiefly quartz which it resembles in hardness.

In many cases the soap plays a minor role. This fact leads to an interesting development, namely, the use of synthetic detergents to replace soap in the formula. A product containing 5 per cent soap will rapidly lose its sudsing value through the precipitation of the soap by the hard water. This destroys the penetrating action necessary to such cleaners. Synthetic detergents such as alkyl benzene sulphates which are not affected by hard water retain this property under all conditions of use.

Abrasive powders or pastes containing about 5–10 per cent active detergents (the DDBS detergents are specially suitable) have been found by the authors to possess excellent detergent and foaming qualities.

Also specifies cleaners which contain vegetable abrasives. For this purpose sawdust, etc., may be used, and it seems that the fact is taken into consideration that these kinds of abrasive are usually less irritating to the skin than the common mineral abrasives, with the exception perhaps of bentonite or colloidal clay.

Again, it must be emphasized that fineness of particle size of the abrasive material, especially if inorganic, is of the greatest importance; otherwise irritation of the skin may be caused.

The strong defatting action of synthetic detergents and their power of penetrating into the skin may lead to dermatitis. However, the irritant effect of synthetic detergents can be counteracted by adding superfatting agents such as lecithin and lanolin, and also alkanolamides, especially those based on vegetable oils with a high content of double unsaturated fatty acids.

Chapter 11

Formulation and Application of Detergents

INTRODUCTION

An attempt has been made in this chapter to incorporate a wide range of formulations giving various process details. The emphasis has been given on the technoeconomic and functional aspects of the ingredients used in such formulation. The chapter also highlights the various industrial and household applications of detergents.

FOAM

At one time the amount of foam formed by a detergent was considered to be a measure of its effectiveness. It is true that foam is formed when surface tension is low, but lowering of surface tension is not always a criterion of detergency because this is in actual fact related to the lowering of interfacial tension.

In certain detergent operations high foam is a definite requirement, in certain cases it is immaterial whether the detergent foams or not, and in other cases foam can be considered a nuisance if not a prohibition to the use of the detergent for a particular operation.

It is obvious that a hair shampoo, shaving cream (other than the brushless type) and bubble bath preparations need to produce copious foam. Dishwashing compounds which are primarily meant for washing by hand in a sink full of water also need to foam copiously but the reason is not so obvious. As the plates are dipped in the solution and washed oil is freed from the surface of the plate and floats to the top of the solution and in time this fatty layer can become appreciably thick. When the plate is withdrawn from the liquid it will pass through this layer last and part of the oil might become redeposited on the surface. If, however, a thick foam is developed in the liquid by the physical actions of washing, the oil will be trapped in the tremendous surface area of the foam and the amount of oil available for redeposition on the plate in its passage out of the solution is greatly reduced. When the foam becomes saturated with oil it collapses and this is an indication that the solution is no longer suitable for washing. For detergents intended for hand-washing of clothes foam is desirable as a sales-appeal factor.

Commercial and automatic household laundry machines are almost without exception 'foam sensitive'. If the detergent foams unduly the foam overflows on to the floor and also can interfere with the free flow of clothes through the water. In automatic machines the foam can interfere with water level pumps and the proper working of the controls in the machine. A small amount of foam is necessary as this tends to trap dirt particles.

Similarly, dishwashing machines cannot tolerate foam, firstly because again the foam might overflow and also interfere with the pumping of the liquid and secondly, once foam forms, bubbles remain on the dishes and leave spots on drying.

Hard and fast rules cannot be laid down as certain household washing machines (usually those with a propeller) can tolerate foam, and in dishwashing machines using a propeller a small amount of foam is not harmful, but in the case of those working with jets the smallest trace of foam leaves spots on glassware.

The following rough guidelines for foam:

(a) Anionic detergents in general produce voluminous foam; somewhat less is produced in hard water but foam is always increased with an increase in temperature.

(b) Non-ionic detergents foam considerably less than anionics, but this depends on the type of non-ionic. There are 'low-foam' non-ionics which can be used alone for most operations, while other non-ionics still need 'foam control'. Non-ionics usually foam somewhat less in hot water than in cold.

(c) When soap is added to an anionic detergent, foam is depressed.

(d) Non-ionic detergents do not depress foam of anionics; they can even enhance it.

(e) Alkaline builders in general enhance the foaming power of all types of detergents.

We shall later in this chapter include formulations for both powders and liquids with foam control whereby a foam control agent is added to a ready-made powder.

The patent describes a combination of a white mineral oil, paraffin wax or microcrystalline wax and/or glycerol monostearate with microfine precipitated hydrophobic silica in the proportions:

Mineral oil	80 parts
Paraffin wax or microcrystalline wax and/or glycerol monostearate	2–12 parts
Microfine hydrophobic silica	0.3–1 part

The solid waxes are dissolved in the oil with warming and the silica dispersed in it.

To give a non-foaming detergent, 100 parts of finished powder are sprayed with 1 part of this solution/dispersion.

HOUSEHOLD CLEANING

The days of one bar of soap, which served as a hair shampoo, toilet bar, laundry soap, for general washing and (mixed with sand from the garden) for pot-scouring, have long since passed. Now-a-days, materials are manufactured for special purposes, such as heavy-duty laundering (the washing of cotton goods which by their very nature are usually heavily soiled), fine wash (the laundering of delicate fabrics such as silks, nylons, woollens), general purpose, dishwashing, floor-washing, window-cleaning, tile-cleaning, etc.

Heavy-duty Laundering

Detergents should be formulated with the type of clientele in mind. Again washing habits both personal and for clothes differ greatly in the two continents.

Non-ionics perform better than other detergents on fabrics made from synthetic fibres and also under cold water conditions. They also find use as 'rubbing agents' for badly soiled collars.

Cotton goods, which are still the bulk of the household wash, require a moderately high alkalinity. In contact with a solution of a pH of 10 or more, the cellulose fibre swells slightly, allowing the water to penetrate into the fibre and thus loosen adhering dirt.

Household heavy-duty washing powders are generally of two types, for hand washing and for fully automatic washing machines. The hand-washing type of powder requires a copious lather for psychological reasons. Fully automatic washing machines, which have a drum which revolves around a horizontal axis, cannot use a powder with a copious foam. Because of the action of the drum, the foam can spill out of the machine, and also the foam interferes with the action of the automatic floats which adjust the level of the water. However, a small amount of foam is necessary, as some of the dirt is trapped by this foam. For these reasons the powder has to have 'controlled foam'. Semi-automatic machines, i.e., those that have propellers or impellers to do the agitation, or with drums rotating about a vertical axis, can in general, utilize either of the two types of powders.

A formulation for a hand-washing or semi-automatic machine, manufactured by simultaneous absorption and neutralisation, has already been discussed in Chapter 10, Formulae 2 and 3. The addition of sodium perborate is, of course, optional and depends on washing practices in the particular area. We recommend the addition of at least 10 per cent sodium perborate to all laundry powders. In countries where this is not normal practice the addition of sodium perborate might make the powder expensive, but the extra whiteness achieved will compensate for the increased cost.

If the powder is being manufactured by the spray-drying process, the formulation can, if desired, be the same as that given in Formula 2 or 3, but an infinitely better powder can be made by using Formula 1.

Formula 1

Spray-dried heavy-duty household hand-washing powder

Sodium alkyl benzene sulphonate (or equivalent)	25
Sodium tripolyphosphate	25
CMC (66% basis)	2.5
Sodium silicate (1:2.4 ratio)	15
Optical brightening agent	0.2
Sodium sulphate	32.3

and to this powder after spray-drying add sodium perborate at the rate of

Spray-dried powder	90
Sodium perborate	10

Foam Control

The above formula cannot be used in foam-sensitive washing machines. To overcome the foaming problem several methods of foam control are used.

One is to use a low-foaming non-ionic detergent as the active matter. An alkyl phenol condensed with not more than eight molecules of ethylene oxide will not foam unduly, but the physical characteristics (free-flowing, stickiness) both of spray-dried powders and those made by other methods will suffer when using low ethylene oxide content non-ionics and therefore the amount of active matter it is possible to incorporate in the powder is limited.

A typical powder to be made by either spray-drying or spray-mixing is given in formula 2.

Formula 2

Heavy-duty fully automatic washing machine powder

Non-ionic detergent (low-foaming)	10
Sodium tripolyphosphate	30
CMC (66% basis)	1.5
Sodium silicate (1:2.4 ratio)	15
Optical brightening agent	0.2
Sodium sulphate (can be partially replaced by soda ash)	43.3

Another trend is to use a moderately high-foaming non-ionic such as isotridecyl alcohol with 15 molecules of ethylene oxide, tallow alcohol or alkyl phenol both with 10–12 molecules of ethylene oxide. All of these detergents foam quite highly, and to depress the foam a nonyl phenol with $1\frac{1}{2}$ molecules of ethylene oxide is used. The addition of this foam-depressor is critical and if proportions different from the optimum are used the foam is likely to be enhanced. The amount suggested is normally 15 per cent of the total active matter, but this figure should be checked against the particular type of active matter with the exact proportions of builders being used.

Sodium perborate can be added in the same manner as in Formula 9, if the powder is spray-dried. If the powder is manufactured by a spray-mixing process which has arrangements for continuous discharge of the powder, the sodium perborate can be fed continuously on to the conveyor belt which receives the discharged powder. If not, the perborate can be incorporated batch-wise in a powder mixer.

To incorporate appreciable quantities of non-ionics into spray-dried powders, recourse must be made to the 'Pluronics', some of which are available as flakes, prills or flakeable solids. Alternatively all the methods described on Chapter 10 should be considered.

A simpler and cheaper 'controlled foam' powder can be made by the use of a mixture of soap and alkyl benzene sulphonic acid. This will give both lower foam and higher detergency than when either is used alone in comparable proportions. It has been found that the foam is at a minimum when the soap proportion of the total active matter is between 30 and 60 per cent. Outside these limits, foam begins to form. If hard water is prevalent in the area, the detergent/soap ratio should be richer in the synthetic portion. If there is predominantly soft water in the area, more soap can be used if required. A word of caution needs to be given in the use of soap or soap mixtures in automatic washing machines. Some automatic machines have their heating elements protruding directly into the water, while others have the elements protected by a sleeve, which disperses the initial heat over a greater surface area. If the element is itself immersed in the water, it has been found that in areas of moderate to very hard water, a layer of insoluble limesoap can form on the element, decreasing the efficiency of the heat transfer. This can to a certain extent be eliminated by the inclusion of NTA or EDTA in the formula, but this leads to a further complication in that both NTA and EDTA in the presence of oxidising agents (perborate) and corrosive to copper and zinc. Inhibitors have been developed to minimise this corrosion to an acceptable level.

A powder of this type can be made by the absorption and neutralisation process according to Formula 3.

Formula 3

Low-foaming machine powder for soft-water areas

Mix together with warming to 45°C

(a)	LAS (100 per cent)	11.6
	Distilled tallow fatty acid	6.4
	In the powder mixer, mix together:	
	Sodium tripolyphosphate	15.0
(b)	Soda ash, light	55.0
	CMC (66% basis)	2.0

When uniform, add (a) with mixing. When the LAS/tallow fatty acid mixture has been dispersed, add immediately in order

Sodium hypochlorite	2.0
Sodium silicate 40%	8.0

This is mixed for a few minutes until the reaction is seen to have been completed, and then discharged for ageing. Next day the powder is ground and recharged to the mixer in the following proportions :

Milled powder from Formula 6	90.0
Sodium perborate	10.0
Optical brightening agent	0.15

and mixed until uniform and then sent to packing.

If one of the mixers (described in Chapter 10) is used the ageing and grinding can of course be dispensed with.

To manufacture the same formula using ready-made soap powder the formulation is:

Formula 4

*Low-foaming machine powder for soft-water areas using
ready-made soap powder*

Mix together:

Sodium tripolyphosphate	15
Soda ash, light	53.7
CMC (66% basis)	2

add in order :

LAS (100%)	11.4
Sodium hypochlorite	2
Sodium silicate	8

when the reaction is completed add:

High titre soap powder 82 per cent fatty acids 7.9

Thereafter proceed as in Formula 3.

The same powder can of course be made by spray-drying, but a vastly superior one can be made in a spray-drier according to Formula 5. Modern tendencies are to manufacture 'ternary' powders, that is, powders having the active matter made up of three constituents, soap, an anionic and a low-foaming non-ionic detergent, a formula for which is given in Formula 6.

Formulae 5–6

Spray-dried household low-foaming laundry powders

	5	6
Sodium dodecylbenzene sulphonate (100% basis)	15	7
Tallow soap, soda based (100% basis)	10	5
Low-foaming non-ionic detergent	–	7
Sodium tripolyphosphate	15	15
Tetrasodium pyrophosphate	10	10

(optional, can be completely replaced by sodium tripolyphosphate)		
Sodium silicate (anhydrous basis, ratio 1:2.5)	7	7
CMC (66% basis)	2.5	2.5
Optical brightening agent	0.2	0.2
Soda ash	0–20	0–20
Sodium sulphate	to make 100	

Sodium perborate is added at the rate of 10 per cent (or more if desired) after drying.

In place of tallow soap, the soap of behenic acid (a C_{22} saturated acid) is being used. Sodium behenate in the formulation acts mainly as a foam control agent (a very efficient one) and at a maximum concentration of 2 per cent. The synthetic active agents can then be increased to give a more effective wash, even at low temperatures.

Behenic acid has a high melting point (80°C), therefore special precautions need to be applied for its introduction and saponification.

For dry neutralised powders it is best dissolved in the LAS and the liquid non-ionic (if being used) or a solvent with the aid of heat (see Chapter 10 for solvent detergent powders) and the mixture is sprayed hot on to the powders.

For spray-drying it can be dissolved in the LAS, and the mixture kept at a higher than normal temperature in the supply tank. The neutralised paste or slurry, which should also be kept at a higher temperature than normal, will probably have a higher viscosity than normal. Contrary to generally accepted principles, lowering the water content of the slurry slightly will, in this case, also decrease the viscosity. As mentioned in Chapter 10, non-ionics can cause trouble on being spray-dried, pluming and also auto-oxidation.

Pluming can be minimised by the use of the narrow range ethoxylates. Auto-oxidation, which at best can discolour the fines and at worst will discolour the powder, can be inhibited according to a Lever patent by the inclusion of about 1 per cent of charge transfer agent, two examples of which are stannic chloride or tetrachlorobenzoquinone. The most efficient methods of incorporation are of course the methods already described. These methods also lend themselves to the incorporation of enzymes, perborate, etc.

Even in areas where there are no restrictions (total or partial) on the use of phosphates, consideration should be given to the use of zeolites, either the zeolite itself or the zeolite/silicate blend. This will have the added advantage, particularly in dry mixed or agglomerated powders, of surface adsorption. The zeolite can adsorb on its surface larger amounts of liquid surfactants than the normal builders are able to, thus drier, more free-flowing powders with higher active matter can be produced.

Where there are bans or restrictions on the use of phosphate, the STP suggested in the formulations can be replaced by mixtures of zeolites on the one hand and NTA or one of the polycarboxylic acids.

Enzymatic laundry powders for automatic washing machines are now being produced in large quantities. It is difficult to give directions as to the amount of enzyme concentrate to add to powder as the concentration of the 'net' enzyme varies from manufacturer to manufacturer, but they give explicit instructions as to usage. Enzymes for washing machine powders are commonly used together with perborate. The enzyme operates while the temperature of the water is low and it is rapidly inactivated when the temperature reaches 60°C, perborate taking over. Although the manufacturers state that the enzymes are stable to perborate, this stability is only relative and if both enzymes and perborate are included in a powder, it is advisable to increase the enzyme concentration over that normally recommended. Formula 6 with little or no soda ash can serve as a useful base for enzymatic powders.

In all of the above and subsequent formulae which are to be spray-dried, consideration should be given to the addition of 3–5 per cent of sodium toluene sulphonate. This does not serve the purpose that it does in a liquid detergent. In this instance the function of the toluene sulphonate is twofold: it lowers the viscosity of the slurry, allowing a higher solids content, and also promotes the free-flowing characteristics of the finished powder, particularly when linear alkyl benzene sulphonate is used. Also when powders are made by processes other than spray-drying, the inclusion of toluene sulphonate is essential if linear alkyl benzene sulphonate is being used.

Heavy-duty liquid detergents are now making headway for household laundering. The problem here is that the solution for efficient laundering needs a certain amount of alkali, sequestering agent and soil-suspending agent. To incorporate sufficient of these with a detergent into a solution which will stay bright and clear is not easy. As a result, new materials are coming on the market specially for this purpose and new techniques are being employed. Many patents for heavy-duty liquids are daily appearing in the literature and as a result the trend is to move away from conventional materials.

The active ingredient can be chosen from one or more of the alkyl benzene sulphonates, olefin sulphonates, paraffin sulphonates or ethylene oxide condensates. For reasons given below the anionic should preferably not be neutralised with a sodium ion, rather potassium or an ethanolamine. This condition cannot always be observed, particularly if paraffin or olefin sulphonates are bought as the already hydrolysed/neutralised solution.

The linear alkyl benzene sulphonates show better solubility in water than do the branched. For the production of liquids the choice should be in the lower register of molecular weight (C_{10-12} rather than C_{11-13}). Again the high 2-phenyl fraction is more suitable for the production of liquids for reasons of solubility. True the detergency is somewhat lower than for the low 2-phenyl isomers, but the production of liquid detergents does require some compromises and the detergency can be enhanced in different ways.

On this score it should be pointed out that, contrary to expectation, the potassium salt of LAS is not more soluble than the sodium salt; it is less soluble. However, the potassium salts are used on occasion because the inorganic constituents are often potassium salts and to introduce a sodium salt would result in precipitation of the sodium salt of the inorganic compound by double decomposition. This also works the other way in that the potassium ion of the salt could precipitate the LAS. More often than not, however, the LAS is neutralised with an ethanolamine to obviate this problem.

The problem in heavy-duty liquid detergents (HDLD) formulations is in incorporating the builders into the solution in such a way that the liquid has enough of all the necessary ingredients for efficient washing without detracting from the appearance of the product.

The builders in question are sequestering agents, which are usually phosphates; alkalis, which are invariably silicates; anti-soil redeposition agents (CMC) and optical brighteners.

Spray-dried heavy-duty powders invariably contain sodium tripoly-phosphate as the sequestering agent, sometimes with the addition of other sequestering agents. Polyphosphates hydrolyse rapidly in neutral or acid solutions, first to a mixture of ortho- and pyrophosphates, and the pyrophosphates in turn hydrolyse further, albeit at a slower rate, to the simple phosphates. If the pH of the solution is 9 or higher, the hydrolysis of polyphosphates is slow so much so that a solution containing polyphosphate at this pH can be stored for a year at 25°C with no appreciable hydrolysis having taken place.

The solubility of sodium tripolyphosphate in pure water is 15 per cent at room temperature, but this figure will be considerably lower in the presence of other dissolved materials, particularly anionic, due to the common ion effect. The solubility of tetrapotassium pyrophosphate in water is over 60 per cent whereas the corresponding sodium salt is only soluble to the extent of 5 per cent. It is for this reason that tetrapotassium pyrophosphate is often used as a constituent of liquid detergents, however sodium salts need to be excluded almost completely because, due to the common ion effect, tetrasodium pyrophosphate can be precipitated from a solution containing both sodium and potassium ions. Again the use of pyrophosphate is a compromise as it does not have the sequestering and peptizing properties found in the polyphosphates.

Potassium tripolyphosphate, with a solubility in water of 55 per cent, has appeared on the market and this does not suffer from the same defect in that its sodium salt might precipitate due to the increased solubility of its sodium analogue.

Alternatively, phosphates are completely dispensed with and the organic sequestering agents, such as sodium (or potassium) ethylene diamine tetraacetate, or nitrilo triacetate are employed.

The alkali required is obtained from the colloidal silicates. Potassium silicate has been found to be superior to sodium silicate in this respect and silicates act as corrosion inhibitors against the action of phosphates on stainless steels.

However, to get sufficient of all three of the above ingredients into solution is virtually impossible, as the presence of one affects the solubility of the others. For this reason it is necessary to use a hydrotope, i.e., a solvent aid, a product which is itself soluble in the medium and aids in the solution of other products. There are a variety of these, but the most commonly used are potassium (or sodium) xylene, toluene, cumene or ethyl benzene sulphonates. These have to be used in relatively large quantities, of the order of 5–10 per cent of the finished product. They can either be sulphonated (or bought) as such, or else co-sulphated in the correct proportion with the original sulphonate, if the basic detergent is one of the sulphonate types.

Urea, apart from its other uses in the detergent industry, is an efficient hydrotrope, as efficient as the lower alkyl sulphonates. However, it suffers from one disadvantage. If one considers its method of manufacture, the combination of carbon dioxide with ammonia :

$$CO_2 + 2NH_3 \longrightarrow NH_4CONH_2$$
Ammonium carbamate

$$NH_4CO_2NH_2 \longrightarrow NH_2CONH_2 + H_2O$$
Urea

it is conceivable that industrial urea can contain small proportions of unreacted ammonium carbamate. In water solution this can hydrolyse:

$$NH_4CO_2NH_2 + H_2O \longrightarrow (NH_4)_2CO_3$$

to ammonium carbonate. If the solution is alkaline, which it almost invariably is, a smell of ammonia will become apparent, which might be objectionable to some people. Thus if urea is to be considered as the hydrotrope, the purchase specification should stipulate no ammonium carbamate to be present.

Recently potassium salts of lower molecular weight phosphate esters have been developed, specifically as hydrotropes for the production of liquid detergents based on alkyl phenol ethoxylates in admixture with potassium tripolyphosphate. These phosphate esters have a further advantage in that they have detergent or wetting properties in their own right, thus aiding in the washing process, a property which none of the other hydrotropes can claim.

CMC now has to be brought into this solution. It is best incorporated by making a 10 per cent 'solution' separately, when it swells and forms a gel. This swelling and gelling, however, is influenced by ions already in solution and, if present, the CMC in a very short time precipitates out and sinks to the bottom. This problem can be overcome in three ways. One method is to adjust the density of the solution so that it is the same as the density of the precipitated CMC (± 1.37) and the precipitate will therefore not settle. This is achieved by the addition of sodium sulphate or chloride. In another method carboxymethyl cellulose and methyl cellulose are used. Both cellulose compounds tend to precipitate from the solution. By itself, the methyl cellulose would normally rise to the top and the CMC by itself would normally sink to the bottom. By using this pair in equimolar proportions, however, they precipitate, but neither rise nor fall.

With the incorporation of CMC as outlined above, one tends to get an opaque type of suspension. It is felt that once the solution tends to be thick and opaque, it should be made to look like a lotion, and the incorporation of sodium nitrate is said to give a better appearance and easier density control. Alternatively one of the opacifying agents mentioned in Chapter 10 can be used.

If it is decided to dispense with phosphates, one of the organic chelating acids neutralised preferably in this case by monoethanolamine, can be used. Alkalinity can also be obtained by excess monoethanolamine.

In the manufacture of these liquids, very often a sludge of insoluble material settles out. This has been identified as traces of iron contamination in the ingredients or even the water. The addition of 1 per cent triethanolamine, over and above any base needed for neutralisation will eliminate this sediment.

Finally, it is desirable to incorporate an optical brightener in the solution. This is the least of the producer's problems, as optical brighteners are now available which are stable in water solution. The amount is so small that this has no effect on the solubility of the other materials, nor have they any bearing on the solubility of the dye.

Thus, to summarise, the following formulae could be used as heavy-duty liquid detergents.

Formulae 7, 8, 9, 10

Heavy-duty liquid detergents

	7	8	9	10
Alkyl aryl sulphonic acid (ABS)	10	20	9	12
Diethanolamine	3.6	7.2	3.3	4
Non-ionic (100%)	2	—	3	—
PVP (100%)	0.7	—	—	0.7
$K_4P_2O_7$ (100%)	12	12	10	—
Potassium silicate (100%)	4	3	4	—
Monoethanolamine	—	—	—	3
EDTA	—	—	—	5
CMC (100%)	—	1	1	—
Potassium xylene sulphonate or other hydrotope	5	5	4	4
Optical brightening agent	0.1	0.1	0.1	0.1
Water to	100	100	100	100

All the above formulae are of the medium- to high-foaming type and thus unsuitable for fully automatic washing machines.

A formulation for a heavy-duty liquid detergent with a 'controlled foam', using the normal type of CMC is given in Formula 11 :

Formula 11

Heavy-duty liquid detergent with 'controlled foam'

Disperse with gentle mixing:

CMC (66% basis)	2
in water	18

Into a stainless-steel vessel equipped with a slow-speed stirrer charge:

Distilled coconut oil fatty acid	8
Water	25
Caustic potash (40% solution)	5

Stir and warm gently until the solution attains a clear and homogeneous appearance, then add with stirring in order:

Non-ionic detergent	4
Monoethanolamine	1.7
LAS	4
NTA or EDTA (acid form)	1
Hydrotrope	8
Tetrapotassium pyrophosphate or Potassium tripolyphosphate	10
Potassium silicate (40% solution, 1:2.5 mol ratio)	10
Optical brightening agent	0.2

The CMC gel prepared above is now added, mixing continued, and water added to make this solution upto 100.

This will produce a lotion type of liquid. It is not necessary to 'saponify' by boiling, because if all the above directions are adhered to, the coconut fatty acid will be completely neutralised and no free unsaponified oil will be present.

PVP is being used with success in place of the CMC. The PVP is added as a 5 per cent pre-prepared solution, and the water adjusted accordingly. This will produce a clear solution, possibly with a slight haze due to insoluble matter sometimes found in poly- or pyrophosphates.

As mentioned on trace quantities of iron contamination in the raw materials can cause a sludge of ferric hydroxide to separate. This separation is not always immediately visible and an insurance against this is the addition of 1 per cent triethanolamine to chelate the ferric ions. The iron will still be present in the solution, but inactivated.

Gantrez resins as chelating agents are high-molecular-weight polymers with anhydride rings, which on hydrolysis with water form dibasic acids. Thus for a typical molecular weight of 40,000, the hydrolysed resin can have 250 dibasic acid groups per molecule. These need to be neutralised and in fact the maximum chelating action is attained at a pH of 10 minimum. These acid groups can also be esterified and in our particular field the esterification agent they suggest is a non-ionic detergent, which normally has a terminal —OH group available for reaction with a carboxylic acid. If the major portion of the carboxylic groups is esterified, an insoluble mass is obtained, but if a partial ester is formed (1 per cent non-ionic based on the weight of the resin), this serves as a stabilizing agent for non-ionic liquids.

The non-ionic of choice is an alkyl phenol with 15 ethylene oxide units and the procedure is important. A stock solution of the resin is made (medium viscosity). Dissolve in water

Non-ionic	10
Water	890

and raise the temperature to 90°C. Add slowly with stirring

Resin	100

keeping the temperature at 90°C, and continue stirring till a clear solution is obtained. (Caution, a large amount of foam might be formed in this esterification.)

A formulation can give excellent results as a heavy duty liquid is :

Formulation 12

Water	4.0
Gentrez stock solution	10.0
NaOH (45% solution)	2.0
CMC	0.5
Nonyl phenol 9-ethoxylate	3.5
NaOH (45% solution)	29.0
Potassium silicate (1:3 mol ratio, 36%)	51.0
Optical brightener, dye, perfume	qs

The solution is made at 60°C, every ingredient to be dissolved completely before the next one is added.

The first addition of caustic soda is to neutralise the Gantrez resin. The second is to convert the silicate to metasilicate. We have found that a mixture of potassium/sodium metasilicate gives better freeze-thaw characteristics than pure sodium metasilicate as recommended by GAF. This can also be reversed, to use potassium hydroxide and sodium silicate to achieve the same purpose.

An interesting and new development in the liquid heavy-duty household laundering field is a liquid containing a mixture of active chlorine and anionic detergent. It had been generally considered that chlorine (derived from sodium hypochlorite) could not be stable in the presence of considerable amounts of organic matter. However, sodium toluene or xylene sulphonates seem to stabilise sodium hypochlorite, and mixtures of sodium hypochlorite with sodium ether sulphates can now be produced with the chlorine reasonably stable for over three months.

Formula 13

Heavy-duty liquid detergent and bleach

Sodium lauryl ether sulphate (60% concentration)	20
Sodium toluene sulphonate	5
Sodium hypochlorite solution (10% available chlorine)	75

Even better results have been obtained in using one of the range of alkylated diphenyloxide disulphonates of the generic formula

$$SO_3^-X+ \qquad SO_3^-X^+$$

'X' is the sodium ion. This novel surface active material is stable in moderately strong alkali and acid solutions, in the presence of large amounts of inorganic materials and in solutions of oxidising agents.

The high stability of hypochlorite in admixture with di-sulpho-diphenyl type of detergent is due to the inertia against chlorination of the multi-substituted benzene ring, one of the rings being tri-substituted, the other di-substituted.

To achieve maximum stability, it is suggested that the diluted alkylated diphenyloxide disulphonates be heated to 70°C together with 2 per cent of a 12 per cent active chlorine hypochlorite solution, then cooled to 30°C and 20 per cent of a 12 per cent active chlorine hypochlorite solution added. This will give a 2.5 per cent active chlorine solution stable for a reasonable period, when packed in a plastic bottle. For a 6 per cent active detergent solution the formulation could be :

Soft water	65.4
Alkylated diphenyloxide disulphonate	13.0
Sodium hypochlorite, 12% active heat to 70°C, then cool to 30°C	1.6
Add sodium hypochlorite, 12% active	20.0

Free alkalinity should then be adjusted by the addition of caustic soda solution to 0.5–1.0 per cent. This free alkalinity, in addition to stabilising the hypochlorite, reacts with the water hardness lowering the pH for maximum disinfecting and bleaching action of the active chlorine.

Light-duty Household Products

In the field of light-duty detergents, or (as it is sometimes called) fine wash, are the materials for washing delicate fabrics such as wool, nylon, silk, etc. Also, in general, this type of detergent is suitable for household dishwashing by hand as well.

Synthetic detergents first took a hold in the household for this purpose, and liquids consisting only of detergents diluted with water very quickly achieved a large sale and large quantities are still being sold. Now-a-days, liquid detergents are more sophisticated than the plain solutions that were originally sold. They can be based on anionic detergents only, usually dodecyl or tridecyl benzene sulphonate, or can utilize a mixture of the above anionics with a non-ionic, or a sulphated ether. In addition to foam boosters, these liquids can have viscosity increasers and cloud-point depressants. Furthermore, they need not necessarily be transparent liquids, but can be opaque lotions.

The concentration of active matter present in a liquid detergent of this type varies from country to country, but an average figure is 12 per cent for the simpler liquids, but this can go upto 40 per cent for the more sophisticated market.

The cloud point of a detergent is an important factor in its sales appeal, as it is axiomatic that no housewife will buy a liquid which tends to deposit a precipitate or to cloud over on storage. Requirements for the actual cloud point will vary from place to place and no hard-and-fast rules can be given. The cloud point naturally obtained from a particular concentration of active matter depends on the type of material used and the method of neutralisation.

In general, the sodium salt of the alkyl benzene sulphonic acids, at a concentration of 12 per cent active matter, gives a cloud point too high for commercial use. The diethanolamine and triethanolamine salts of the alkyl benzene sulphonic acids give a very low cloud point, lower than is needed generally. Thus for economic reasons, neutralisation is usually done partially with caustic soda and partially with an ethanolamine.

Among the alkyl benzene sulphonates, all other things being equal, the cloud point of the neutralised solution rises in the following order :

(i) Linear dodecyl benzene sulphonate.

(ii) Linear tridecyl benzene sulphonate.

(iii) Branched dodecyl benzene sulphonate.

(iv) Branched tridecyl benzene sulphonate.

The viscosity of the final solution, all other things being again equal, rises in the same order.

In addition to the above factors, the method of sulphonation is important, since on this method depends the amount of free sulphuric acid present. When neutralised, the acid produces inorganic sulphates, both of which raise the cloud point and the viscosity. SO_3 sulphonated material will give the lowest free sulphuric acid present; and normal oleum sulphonation will produce the highest. (The figure is normally as high as 7–8 per cent).

The factors involved in producing an acceptable liquid detergent are therefore dependent on:

(i) The type of alkyl benzene sulphonate.

(ii) The method of manufacture of this sulphonate.

(iii) Whether any other active material is used.

To take an average case, if conventional dodecyl benzene, sulphonated with SO_3, is to be used as the base material for the manufacture of a 12 per cent liquid, a cloud point of below 5°C can be obtained by neutralising half of the active matter with caustic soda and the other half with diethanolamine or triethanolamine. From Table 10.4, Chapter 10, it will be seen that neutralisation with diethanolamine will give a slightly higher viscosity than if neutralisation had been done with triethanolamine. Diethanolamine is a solid, except in very hot climates, and therefore is more difficult to handle. On the other hand, when alkyl benzene sulphonates are wholly or partially neutralised with triethanolamine, which is a liquid, a buffer is formed at a pH of approximately 6. To pass this buffer, it is necessary to use a large excess of base. To preserve a low cloud point, it is not advisable to use too much caustic soda to overcome this buffer, so a fair excess of triethanolamine must therefore be used. The manufacturer must consequently consider the disadvantages involved in handling diethanolamine against the extra cost of triethanolamine.

If in the examples of alkyl benzene sulphonates given above, one of the sulphonates which normally gives a low cloud point were to be used to maintain the same cloud point level, the proportion of caustic soda can be raised and that of the ethanolamine correspondingly reduced.

Furthermore, if an acid sulphonated material is used as the base material, the proportion of ethanolamine needs to be raised considerably (and the caustic soda to be lowered) to maintain the low cloud point.

If it is required to raise the viscosity of the solution, a simple but limited method is the addition of an inorganic salt, like sodium sulphate or sodium chloride. This is limited in that it raises the cloud point. A better and more efficient method of increasing the viscosity is the use of an alkylolamide, preferably a diethanolamide.

Because of the various factors involved in the formulation of a liquid, it is not possible to describe formulations for all possible permutations and combinations.

Formula 14 gives a basic formula to be used as a starting-point. The manufacturer is advised to modify this according to the conditions he requires and the materials available to him.

Formula 14

Light-duty household liquid detergent

LAS (SO_3 sulphonate)	10
Triethanolamine	2
Caustic soda (45% solution)	1.7
Sodium hypochlorite (10% solution)	0.6
Lauric acid diethanolamide	1
Sodium sulphate	1
Water	83.7

Full and detailed instructions on the method of production of this solution are given in Chapter 10.

An opaque lotion type of liquid can be made (taking into account the same factors as mentioned above) from Formula 15.

Formula 15

Lotion-type light-duty liquid detergent

LAS	19.5
Monoethanolamine	4
Lauric acid monoethanolamide	1.5
Sodium hypochlorite (10% solution)	0.6
Sodium sulphate	0.9
Water	73.5

In this case the dodecyl benzene sulphonic acid, the sodium sulphate and the water are mixed in the neutralisation vessel, the sodium hypochlorite added and mixing continued for at least 20 min. The lauric acid monoethanolamide, being solid, is best dissolved in the monoethanolamine with heating. The mixture of the two is then added to the vessel, after the bleaching has if necessary been completed. If an opaque solution is required an opacifier can be added. The product can now be dyed and perfumed.

To manufacture a liquid from an alkyl benzene sulphonate neutralised only with caustic soda (in practice, a concentrated paste of the sodium sulphonate is used as the starting material), a low cloud point can be achieved by the addition of urea. With the sodium sulphonate, however, urea lowers the viscosity considerably. Addition of a fatty acid dialkylolamide will restore the viscosity.

Modern tendencies are to use both stronger solutions and mixtures of active matter. One of the main uses for light-duty liquid detergents is hand dishwashing, and for this particular application foam is important other than from the point of view of sales appeal. Grease released from the dishes floats to the top of the washing solution and forms an oily film which can be redeposited on the plate when it is withdrawn from the sink after having been washed. Due to the enormous surface area in a foam, this grease is held in a thin, easily rinsable film on the bubbles, preventing redeposition. When the foam collapses due to saturation with oil, the solution is considered to be exhausted and methods of test for the efficiency of dishwashing liquids are based on this foam collapse.

Ether sulphates, both of the alcohol and alkyl phenol types are finding more and more use as constituents of dishwashing liquids. If the ammonium salt of the alkyl phenol type is used care must be taken not to raise the pH of the final liquid over 8 or otherwise a smell of ammonia will appear. If a high pH is desired without the ammonia smell, the ammonium ether sulphates can be heated with a stoichiometric amount of caustic soda until the ammonia is distilled off. Alternatively an 'ammoniated' cleaner, which has some sales appeal, can be made at the high pH.

The choice of the type of ether sulphate is rather wide, as the base material can be any one of the alcohols available on the market or any of the alkyl phenols and then again the amount of ethylene oxide can be varied from between 1.7–4 molecules per molecule. Recently viscosity and foam height with different alcohols and different degrees of ethoxylation has been investigated and it is found that the best foamers are a C_{12-14} alcohol with 2 molecules of ethylene oxide. Another important fact in the make-up of the ether sulphate is the amount of polyethylene glycol present. It has also been found that relatively large (of the order of 3 per cent) amounts of polyethylene glycol can act as a solvent or 'thinner' on the final sulphate to reduce the viscosity of the solution.

As the wetting properties of ether sulphates are low they are not recommended as dishwashing agents alone, rather in conjunction with non-ionic or other anionic detergents.

Alcohol sulphates, neutralised with ethanolamines can, of course, be used alone, or with an alkylolamide as a foam booster. It is interesting to note that alkylolamides have no effect on the foaming properties of ether sulphates. Some typical basic formulations are given in formulae 16–20.

Formulae 16–20

Light-duty liquid detergents

	16	17	18	19	20
LAS (SO$_3$ sulphonated)	10	15	20	12	–
Caustic soda (40% solution)	3.2	3.2	3.2	3.2	–
Triethanolamine	–	2.2	4.4	9	–
Sodium alcohol ether sulphate (100% basis)	–	3	3	–	–
Ethanolamine alcohol sulphate (100% basis)	–	–	–	–	24
Coconut diethanolamide*	2	1	1	–	2
Sodium sulphate or chloride	qs	qs	qs	qs	qs
Dye, perfume	qs	qs	qs	qs	qs
Water			—— to make 100 ——		

* The ethanolamide might contain some free ethanolamine, in which case this can be used as a partial neutralising agent for the LAS, reducing the amount of NaOH or triethanolamine.

It will be observed that the first and fourth formulations are relatively cheap, the others more sophisticated.

SO$_3$ sulphonated LAS can be considered the work-horse of liquid detergent formulations. It is good practice, when mixtures of LAS with alcohol or ether sulphates are used, to ensure that the LAS is neutralised completely before the sulphate is introduced into the reactor. This will eliminate hydrolysis of the acid-unstable sulphates. To facilitate formulation the following data show the alkali requirement of a typical LAS.

> 100 kg LAS requires for complete neutralisation :
> 12.8 kg NaOH (100%)
> 18.8 kg KOH (100%)
> 45.5 kg triethanolamine
> 33.6 kg diethanolamine
> 19.7 kg monoethanolamine.

The inorganic salts are included only if it is desired to increase the viscosity. The diethanolamide acts as a viscosity booster but particularly where the ether sulphates are included coconut monoethanolamide will be more effective. To incorporate the monoethanolamide, the solution is merely warmed to 60°C with mixing, when all the monoethanolamide will melt and dissolve and will not be thrown out of solution on cooling.

Household fine-wash detergents are, of course, not limited to liquids. They can be made as spray-dried powders, sometimes only with 20–30 per cent active matter and the balance sodium sulphate as the inert filler, but more often than not they include sodium tripolyphosphate, some silicate, occasionally CMC, and an optical brightening agent, substantive to the fibres for which the powder is being used. The phosphate, of course, helps in the detergency, and the silicate acts in this case as a corrosion inhibitor. Here, the action of the CMC is not as pronounced as it is on cottons, but on woollen fabrics it does aid in preventing redeposition, and as these powders are meant for hand-washing, it does tend to give a protective colloidal action on the skin.

A powder of this type can be manufactured as detailed in Formula 21.

Formula 21

Household fine-wash spray-dried powder

	Minimum	Maximum
Sodium alkyl benzene sulphonate	10	25
Sodium tripolyphosphate	15	25
Sodium silicate (preferably 1:3 ratio) anhydrous	3	5
CMC (100%)	0	1
Lauric acid monoethanolamide	0	2.5
Optical brightening agent	0.2	0.2
Sodium sulphate	Balance	

By virtue of the nature of manufacture, fine-wash powders of this type cannot be made by the absorption methods, as the surplus of soda ash will prove detrimental to the operation of the powder. However, a dry-mixed type of powder can easily be made to Formula 21, using as a base a concentrated powder of about 60 per cent active matter.

In certain parts of the world, neutral pastes of sodium alkyl benzene sulphonate are sold also for fine-wash purposes. These pastes vary in concentration between 20 per cent and 50 per cent active matter. The consistency of the paste varies with the ingredients and the method of manufacture of the alkyl benzene sulphonic acid. Stiffer pastes are obtained with tridecyl benzene sulphonic acid, sulphonated with oleum; and the softest paste is linear dodecyl benzene sulphonic acid, sulphonated with SO_3 gas.

The alkyl benzene sulphonates have limited solubility in water. They, however, can absorb water to form a natural paste. If the amount of water present in above that which can naturally be absorbed, the mass will separate into two phases on standing: a concentrated phase of alkyl benzene sulphonate on top and a weak solution of alkyl benzene sulphonate and inorganic salts at the bottom. The concentration of the sodium alkyl benzene sulphonate in the upper phase varies with the material, but is of the order of 55 per cent active matter. This means that pastes of 55 per cent can be manufactured with no special additions. If lower concentrations are required, it is necessary to add certain ingredients to prevent this separation. Materials that can be used are: hydrotropes, which will make the final paste thinner in consistency, so that stiffening materials have again to be added; alkylolamides, which have the added advantages of foam boosting and skin protection (these pastes are to be used by hand); and CMC. Both the CMC and the alkylolamides tend to increase the viscosity of the mass in such a way that separation cannot take place.

If only CMC is used as the thickening agent, a general rule is to manufacture the paste to a 55 per cent concentration and then to add sufficient of a 10 per cent CMC solution in water to bring the concentration down to the required amount. Sodium sulphate is then added to increase the consistency. The method of manufacture is described in detail in Chapter 10.

A 40 per cent active paste, using both conventional dodecyl benzene and linear dodecyl benzene sulphonic acids, both sulphonated by SO_3 gas, is given in Formula 22.

Formula 22

40 per cent detergent paste

ABS (100%)	40
Caustic soda (45% solution)	11.4
Sodium sulphate	2
Sodium hypochlorite solution	0.6
Water	29.3
{ CMC	1.7 }
{ Water	15 }

This paste may be reduced to 30 per cent by increasing CMC to 2.5 or 3.5 per cent and adding water.

General Purpose Detergents

Powders of this type are the most popular for household use. They are not harshly alkaline, and contain relatively large quantities of both active matter and sodium tripolyphosphate. These achieve the action on cottons without the addition of further alkalis, and the alkalinity naturally present from the phosphate does not harm delicate fabrics.

By virtue of their formulation these powders are truly general-purpose and can be used for virtually every household job. They suffer from the disadvantage, however, of being unsuitable for fully automatic household washing machines, as, in general, they foam too much.

A generally accepted formula for this type of powder to be manufactured by a spray-drier is given in Formula 23.

Formula 23

Spray-dried general-purpose powder

Active matter as alkyl benzene sulphonic acid neutralised with caustic soda	25
Sodium toluene sulphonate*	2.5
Lauric monoethanolamide	3
Sodium tripolyphosphate	30
Sodium silicate (1:2 ratio) anhydrous	10
CMC (100% basis)	2
Optical brightening agent	0.2
Sodium sulphate	27.3

If sodium perborate is to be added, we suggest that 2 per cent of the sodium sulphate be replaced by magnesium sulphate on any anhydrous basis. This magnesium sulphate is added to the slurry prior to spray-drying.

A general-purpose powder can be made by the absorption and mixing process as well, but in this case, due to the limitations of the process, the active matter is limited. The formulation suggested is given in Formula 24.

Formula 24

General-purpose powder

Dodecyl benzene sulphonic acid	18
Sodium tripolyphosphate	25
Sodium silicate (1:2 ratio) 40% solution	5
CMC (100% basis)	5
Sodium bicarbonate	28
Soda ash	20
Optical brightening agent	0.15
Water (or sodium hypochlorite solution)	2

As in Formula 24, the same remarks apply regarding the addition of sodium perborate. If this is to be introduced, it is suggested that the sodium bicarbonate be lowered to 25 per cent and 3 per cent magnesium sulphate crystals added. If sodium hypochlorite is used to bleach the powder and if sodium perborate is to be incorporated.

A general-purpose powder can be manufactured by dry-mixing according to Formula 25.

Formula 25

General-purpose powder

60% alkyl benzene sulphonate powder containing silicate	45
Sodium tripolyphosphate	30
CMC (100% basis)	2
Optical brightening agent	0.2
Sodium sulphate	22.8

Here, if sodium perborate is to be used, no special precautions need to be taken and it can, of course, be added initially with all the other ingredients.

In all the formulations given hitherto, we have included relatively large amounts of sodium tripolyphosphate, this despite the fact that in certain countries there is a complete or partial ban on the use of phosphates. To date no 'plug-in' replacement for this excellent builder has been developed.

Where this is a partial ban, the above formulations to include STP up to the limit allowed and to add a zeolite to make up the difference is suggested. Where there is a total ban, the available alternatives must be considered; increasing the silicate, use of NTA, use of zeolite (the zeolite/silicate cocrystal mentioned might be eminently suitable), or one of the polymers specially mooted for this purpose. The final formulation will probably be a combination of any two or more of the above.

For fully automatic, front-loading household washing machines, foam control needs to be applied. This can be by the use of soap as for heavy duty powders, but infinitely better performance will be attained if the ternary mixture of the anionic/non-ionic/soap is used. With the new processes and mixing equipment for the inclusion of non-ionics into powders, there need be no practical limits to the formulation.

Choice of Non-ionic

As can be appreciated from the descriptions of the possible non-ionic surfactants, the choice is wide and often confusing. Much work has been done on the optimisation of the molecule. It is true that the results cannot be compared directly because of the different substrates, but the results are revealing.

The results can be summarised :

1. From the work on hard-surface cleaning it appears that linear alcohols with a low molecular weight and containing 50 per cent ethylene oxide (average chain length 8.6 carbons with 3.1 moles EO) in a relatively high concentration (5 per cent), are better in cleaning efficiency for grease, wax and particulate soil than the molecules containing C_{10} and higher chain lengths with the appropriate amount of ethylene oxide to maintain the balance.

    This can be explained by the fact that the low-molecular-weight hydrophobe acts as a solvent and this was confirmed by adding a glycol ether when no improvement of soil removal was noted.

2. For low dilutions of the surfactant the optimum chain length was shifted to the C_{8-10} range. Under these conditions of dilution the hydrophobe can no longer act as a solvent and performance was dependent solely on the surface active effect.

3. Comparison of the low carbon number ethoxylated alcohols with a built formulation containing nonyl phenol with $9\frac{1}{2}$ EO units again showed a better performance for the low-molecular-weight alcohol ethoxylates.

4.   By comparing the performance of a linear primary alcohol ethoxylate, a linear secondary alcohol ethoxylate, both of approximately C_{13} average chain lengths, and a branched octyl phenol ethoxylate, each with varying amounts of ethylane oxide and for both cotton and cotton/polyester blends.

5.   The results show that 7–9 EO units on each give optimum performance for sebum, oil and clay removal on the blend but for sebum removal from pure cotton 12–15 EO was superior.

Recent investigations confirms the above in that and thus short-chain non-ionics are better in removing water repellent soils.

The above facts and figures do not take into account the added complications of the presence of anionic detergents and builders but they do indicate a trend.

From the above it appears that any of the commercially available non-ionic types with approximately 8 EO units will give good all round performance. For powders, for technological considerations, the choice of non-ionic has been in a somewhat higher register hitherto.

With the development of systems for incorporation of non-ionics into powders without spray-drying (Chapter 10) the problems associated with making powders using these (relatively) low-molecular-weight materials no longer apply.

The use of this type of non-ionic in liquids will tend to give a low cloud point but this can be overcome by the judicious use of co-solvents and hydrotropes.

The term cloud point can be somewhat confusing. Normally it is the temperature at which a cloud forms on cooling. This demonstrates the lowest temperature at which a liquid can be stored and still remain clear. When applied to non-ionic solutions it indicates the temperature at which the solution becomes cloudy on heating, at this temperature the solution separates into two phases, one of which will be richer in non-ionic than the other.

Concentrated Powders

A new approach to the manufacture of powders is what is called the $\frac{1}{2}$ or $\frac{1}{4}$ cup concentrates, where a minimum of inert filler, if any at all, is used. A patent by Colgate describes the technology for the production of these powders.

In brief the process is to mix STP with water and sodium silicate in a crutcher to allow the STP to hydrate to its hexahydrate, then to add a further quantity of STP and water under conditions that this second addition does not hydrate and to spray-dry the mix. To the spray-dried beads non-ionic detergent is added in a special mixer.

The details are :

Mix together

STP	14.5
Sodium silicate (1:2.4 ratio, 50% solution)	15.2
Deionized water	21.0

Maintain this slurry at 60°C for hydration and then
raise the temperature to approximately 90°C, add

STP	28.3
Deionized water	21.0

At this higher temperature no hydration takes place. This slurry is then spray-dried to a bead containing 10% moisture, with a bulk density of ±0.55 g/ml.

The finished beads are then sprayed, either continuously or batch-wise with a non-ionic, of the alcohol ethoxylate type mixed with minor ingredients; optical brighteners, dye, perfume, etc. The finished powder has a bulk density of 0.68 g/ml and contains :

Base bead	78.0
Non-ionic	19.7
Minor ingredients	2.3

The granules are attractive and dustless and a further claim by the patentors is that they are sufficiently pourable to be packed in a transparent specially designed bottle.

Further embodiments of the patent include adding fillers other than STP to the hydrated STP. No mention of the inclusion of CMC is made but it is envisaged that it could quite easily be added to the slurry prior to spray-drying.

Cold Water Washing

Generally all washing is done at the boil or near it. With the advent of synthetic fibres wash temperatures came down drastically and now in an attempt at energy conservation householders are beginning to use cold water only, for washing of clothes.

The industry is facing and meeting the challenge, and a challenge it is because the highly complicated washing process requires, among other things, energy to break the bond of the dirt to the substrate. This energy can come from lowering of the interfacial tension, the mechanical energy imparted by the motor of the machine and from heat.

For cold washing the thermal energy needs to be replaced, and one method is by increasing the amount of active matter in solution and using more non-ionics which are better for soil removal from synthetics. On studying the parameters involved in cold washing with non-ionics. The conclusions can be summarised as under :

1. Non-polar soils are difficult to remove from polyester/cotton fabrics at low temperatures, thus the inherent detergency needs to be enhanced.

2. Optimum detergency for this system is found with non-ionic mixtures which have cloud points in the range of 15–25°C below the wash temperature.

   This is explained by the fact that a surfactant-rich pseudo-phase separates at temperatures above the cloud point. This can be considered to be globules of concentrated surfactant which are attracted to the soil.

3. Low cloud points were obtained by blending lightly ethoxylated alcohol or unethoxylated alcohol with a normal detergent non-ionic.

A blend of 3 mol ethoxylated alcohol with an 8 mol non-ionic to gives a cloud point close to 0°C enhanced the mineral oil detergency of the 8 mol alcohol ethoxylate by 50 per cent and a blend of a 9 mol non-ionic with decanol, again to give a cloud point close to zero enhanced the detergency of the alcohol-9 ethoxylate by 20 per cent.

As can be seen from the foregoing, possibilities of formulating are varied and later in this chapter we also discuss solvent powder detergents. The actual formulations to be used depend on many factors, mainly the materials easily available and the trends in the area where the powder is to be sold. The above statement also holds good for liquids.

To sum up we give in Tables 11.1 and 11.2 formulations which can be used as starting points for the manufacture of both dry-mixed and spray-dried powders. These formulations do not necessarily parallel those already described in the text, but do give an indication of what is being made.

Table 11.1. Typical formulations for powders produced by dry neutralisation.

	All purpose non-machine	All purpose machine	Light-duty hand	Heavy-duty non-machine	Heavy-duty machine	Solvent powder
LAS	14–15	5	15	12–15	5	5
Dist. fatty acids	–	4	–	–	4	3
Non-ionic	2-3	6	2	-	4	3
STP	30	30	20	15	20	20
Metasilicate ($5H_2O$) or spray-dried disilicate	3	3	–	5–6	5–6	5
Soda ash	15–20	15–20	20 max	to100	to 100	to 100
Sodium bicarbonate	–	–	30	–	–	–
Sodium sulphate	to 100	to 100	to 100	–	–	–
CMC	2	2	–	1.5	1.5	1.5
Optical brightener	0.2	0.2	0.2	0.2	0.2	0.2
Perborate	10	10	–	10	10	–
Enzymes as 300,000 DU	-	0.5	–	0.5	–	–
Perfume	qs	qs	qs	qs	qs	qs
NaOH (30% solution)	2–3	2–3	2–3	2–3	2–3	2–3
Solvent deodorized kerosine	–	–	–	–	–	4

Table 11.2. Typical formulations for powders produced by spray-drying.

	Heavy-duty hand	All-purpose cold water hand	Heavy-duty machine	Light-duty hand
LAS (Na salt)	12	16	6	18
Non-ionic	3	4	4	2
Soap or distilled fatty acids	–	–	6	–
CMC	2	3	2	2
Sodium silicate 1:2.45 ratio (100% basis)	3	3	4	–
Soda ash	10	10	–	–
STP	30	35	30	15
$MgSO_4$	1.5	–	1.5	–
Optical brightener	0.15	0.15	0.15	0.1
Na_2SO_4	22	28	30	60.1
Perborate	15	–	15	–

(*Contd.*)

	Heavy-duty hand	All-purpose cold water hand	Heavy-duty machine	Light-duty hand
Sodium toluene sulphonate	1	1	1	1
Perfume	qs	qs	qs	qs
Enzymes	–	–	0.5–0.8	–
Residual moisture (approx.)	10	5	10	5

Note

1. Part or all of the LAS can be replaced by sulphonated methyl esters.
2. Perfumes for this type of powder should be of low volatility and added after spray drying.
3. All the above formulations lend themselves to production by the combined systems.

Hard-surface Cleaners

A new development in both the household and the institutional cleaning fields in the all-purpose liquid meant specifically for hard-surface cleaning—by hard surfaces are meant those surfaces that cannot be immersed in a bath or basin for cleaning, and the operation needs to be done *in situ*.

These cleaners are usually liquid, alkaline and often contain solvents. Their purpose is to clean grease, mud, and atmospheric grime from walls, doors, glass, tiles, etc. Alkalinity can be derived from alkaline salts, or ammonia (occasionally caustic soda or potash) can be added. The solvent is added because it has been found that oil stains, although theoretically saponifiable by the alkali, cannot always be removed without it. This solvent needs to be soluble in water (unless an emulsion is to be made), reasonably odourless, non-toxic, with a fairly high flash-point and last but not least, a good fat solvent. The glycol ethers, in particular ethylene glycol monobutyl ether or dipropylene glycol methyl ether answer to all the above requirements admirably. Silicates are added both to buffer the alkalinity and to minimise corrosion on metallic surfaces. A typical formulation is :

Formula 26

Hard-surface cleaner

Sodium alkyl benzene sulphonate (100% basis)	15
EDTA sodium salt	4
Ammonia (100% basis) (optional)	5
Butyl cellosolve	7
Sodium silicate (1:2 ratio, 100% basis)	3
Water	to 100

It goes without saying that if ammonia is to be added the solution cannot be used on copper surfaces.

A generalised formulation which is becoming popular in household use is:

Formula 27

Syndet	5–8 wt %
Hydrotrope	6
Alkali	5
Dipropylene glycol methyl ether	4
Pine oil	2
Water	Balance

The syndet can be non-foaming (non-ionic) or medium foaming (LAS) or high foaming (mixture of LAS and ether sulphate). The alkali can be ammonia, trisodium phosphate, tetrasodium pyrophosphate or sodium silicate, or one of the organic amines such as monoethanolamine. If organic alkine salts are not used the hydrotrope can be dispensed with. It is necessary to use softened water for this formulation as otherwise the hardness salts will precipitate or form a haze due to the high pH.

A particular instance of hard-surface cleaning is the oven cleaner. The above formula will theoretically clean ovens, but as grease is normally heavily encrusted on the surfaces of ovens more alkalinity than that provided in this formula is needed. A typical formula could be:

Formula 28

Hard-surface cleaner

Alkyl benzene sulphonate, sodium salt	4
Phosphoric acid (85%)	4.5
Caustic potash (100%) or monoethanolamine	9
Tetrapotassium pyrophosphate	4.5
Ethylene glycol monobutyl ether	6
Isopropyl alcohol	2
Sodium silicate 1:2 ratio (100%)	2
Water	68

It will be noted that the amount of caustic potash suggested is in excess of that required to neutralise the phosphoric acid, the surplus is needed to provide alkalinity. In the above formula the glycol ether is not completely soluble but the isopropyl alcohol acts as a coupling agent. Contrary to general belief, isopropyl alcohol is in itself a good fat solvent.

For certain institutional purposes it might be necessary or desirable to produce an oven cleaner as a paste or viscous liquid so that it will not drain down vertical surfaces. Formulae 34 and 35 can be modified by the addition of either a thickening agent or colloidal silica. If a viscous liquid is to be made it can be conveniently packed as an aerosol. The following formula for an aerosol oven cleaner is suggested :

Formula 29

Aerosol oven cleaner

Veegum-T	1.5
Ammonia solution	6
1, 1, 1-Trichlorethane	18
Water	24
Ethanol	7.5
Tergitol	18
Propellant:	
Dichlorodifluoromethane	17.5
Dichlorotetrafluoroethane	7.5

A low temperature oven cleaner has been described in a patent where mannitol or sorbitol is used for alcoholysis of the fat deposited, with sodium or potassium bicarbonate as the alcoholysis catalyst and salts of low molecular weight organic acids to esterify the alcoholysis product. The formulation is as under:

Formula 30

Sorbitol	2–5
Potassium bicarbonate	0.1–4.0
Eutectic mixture of	
Sodium acetate	
Lithium acetate	1–5
Potassium acetate	
Thickener	qs
Precipitated chalk	20–30
Wetting agent	qs

The eutectic mixture is to lower the melting point of the salts to allow them to react easily. Other salts recommended could be sodium tartrate, Rochelle salt or sodium glycolate.

Machine Dishwashing

Until a few years ago household dishwashing was done by using one of the fine-wash or general-purpose formulations mentioned above. Of recent years the household dishwashing machine has come into popular use and this has requirements of its own. The mechanical cleaning action is done by means of jets of water. These jets are produced by a high-pressure pump or by the whipping action of a fast revolving propeller. In either case it is essential to the operation of the machine that the detergent added be completely non-foaming (and not even with a 'controlled foam' as in household laundry).

Non-ionics, in general, foam considerably less than anionics, but even they do foam slightly, so the amount of detergent is therefore kept to a minimum and the cleaning effect is achieved by the use of alkalis and phosphates. Almost completely non-foaming non-ionics have now been developed by blocking the terminal —OH group. One of the common methods is to add a methyl group to the terminal —OH. If, for example, a non-ionic is to be made by the esterfication of a polyethylene glycol and a fatty acid, instead of the polyethylene glycol a methoxy-polyethylene glycol is used. Another method is to condense to the finished non-ionic detergent a further molecule of butylene oxide. The formulation also depends to a very great extent on the type of water being used. Household water-softeners are only now beginning to appear on the market; so the formula must take into account whether the water being used is soft, moderately hard or hard.

For soft-water areas, soda ash, besides being cheap, can provide a portion of the alkalinity, but in moderately hard and hard water the soda ash will leave a scum of calcium carbonate on crockery and cutlery.

For moderately hard water tetrasodium pyrophosphate can be used to give both detergency and alkalinity, but for hard water sodium tripolyphosphate in combination with strong alkalis must be employed.

Most dishwashing machines have a drying cycle as well as the rinsing cycles, or at least dishes are left hot and the last traces of moisture on them dry very quickly on contact with the air. It is, therefore necessary to provide for the last traces of water to drain from the dishes in a uniform film to avoid water spots, particularly on glassware.

This effect can be achieved by incorporating solvents or an organic chlorine-releasing compound into the powder. Chlorine-releasing materials react with non-ionic detergents and it has been found that these powders work very successfully without any detergent at all. If, however, it is desired to incorporate a detergent, a very small amount of non-ionic detergent can be used. In another method in which non-ionic detergents can be incorporated into these powders in the presence of chlorine releasing agents. It recommends the low-foam modified non-ionic detergents mentioned above. This detergent, at the rate of 1–2 per cent based on the final weight, is pre-mixed with the most alkaline ingredient present (anhydrous metasilicate), the tripolyphosphate added next and then the other ingredients and finally the chlorine-releasing compound.

Highly alkaline materials such as metasilicate can affect the over-glaze on delicate china and cause 'crazing' of the glaze. This can be eliminated by the incorporation of boric acid (not more than 5 per cent), sodium aluminate (2 per cent) or zinc salts in the formula.

Since these powders usually contain large amounts of sodium metasilicate, they are not normally made by spray-drying, as there is a limit to the amount of sodium metasilicate (which is hygroscopic) which can be incorporated into a spray-dried powder. They are produced instead, by a mixing and absorption technique.

To make these powders dustless, granulation techniques are used. The powders and the non-ionic detergent are mixed as described above, without the chlorine-releasing material. On to this mixture a small amount of waterglass is sprayed in a mixer with a revolving or tumbling action, when the waterglass glues the particles together and the revolving action gives them a spherical shape. The chlorine-containing material is then added.

A formula suitable for use in soft-water areas is given in formula 31.

Formula 31

Machine dish-washing powder for soft-water areas

Tetrasodium pyrophosphate	49
Sodium metasilicate anhydrous	25
Soda ash	22.5
Sodium dichloro-iso-cyanurate (60% available chlorine)	1.5–4.5
Non-ionic detergent, e.g., nonyl-phenol 9 mol ethox	2

This formulation is also suitable for dishwashing machines used in catering establishments.

Because of the small amount of detergent involved, it is not necessary to use any special procedure for the incorporation of the active matter. The easiest method is to blend in with the other powders a concentrated detergent powder. Alternatively, any other already prepared detergent powder can be used as the source of the active matter and due allowance must then be made for the other ingredients in this powder.

Where the water is moderately hard the product should be made according to Formula 32.

Formula 32

Machine dish-washing powder for moderately hard-water areas

Tetrasodium pyrophosphates	60
Sodium metasilicate anhydrous	38
Trichloro-iso-cyanuric acid	1–3.5
Non-ionic detergent	1

For very hard-water areas the following formulation will be suitable:

Formula 33

Machine dish-washing powder for hard-water areas

Sodium tripolyphosphate	50
Sodium metasilicate pentahydrate	25
Trisodium phosphate anhydrous	15
Non-ionic detergent	3
Hexylene glycol	2
Isopropyl alcohol	1.5
Water	3.5

The liquids are pre-mixed, and then the powders are charged into the mixer and the liquid sprayed on to the powders while mixing.

For institutional machine dishwashing the tendency is to move away from powders to liquids as these can be dosed either continuously or in accordance with pre-set requirements based on electronic controls.

Liquid for dishwashing machines can be formulated with either a non-ionic component or active chlorine, but not both.

Alkalinity is obtained from caustic alkali (soda or potash) with the addition of the corresponding silicate, together with a condensed phosphate. The same problems of common ion effect occur as described for (heavy duty liquid detergent (HDLD) (Chapter 10)), thus a basic formula could be:

Formula 34

Tetrapotassium pyrophosphate	15
Potassium silicate (1:2 mol ratio, 40% solution)	8
Potassium hydroxide (45%)	10
Soft water	67

Potassium tripolyphosphate, if available, would obviously give vastly improved performance.

To this solution is added a per cent of a non-foaming non-ionic or 2 per cent active chlorine. The chlorine in this instance can be added by direct injection of chlorine gas to form potassium hypochlorite in the solution.

Abrasive-type Cleaners

The most popular household abrasive cleaners are scouring powders. These are usually dry mixes of all of the ingredients. A typical formulation is given in Formula 35.

Formula 35

Household scouring powder

Abrasive	87
Soda ash	5
60% concentrated detergent powder	8

If desired, this can also be manufactured from sulphonic acid by mixing together:

Abrasive (powdered calcite or marble)	85
Soda ash	7

then adding in the mixer:

100% alkyl benzene sulphonic acid (ABS)	5
and water	3

If this scouring powder is to contain active chlorine, as is becoming the fashion now, it is advisable not to use soda ash, as concentrated organic-chlorine-releasing materials are not in general stable in the presence of soda ash, except for the potassium salt of trichloro-iso-cyanuric acid.

Alkalinity can be obtained by anhydrous sodium tripolyphosphate or tetrasodium pyrophosphate. To avoid introducing moisture into the powder, the active matter can best be put in by the use of an already made detergent powder, not necessarily a concentrated one. A chlorine containing powder is best packed in plastic containers.

A suggested formulation is Formula 36 which is a chlorine-containing household scouring powder.

Formula 36

Chlorine-containing detergent scouring powder

Detergent powder containing 20% active matter (without soda ash)	15
Tetrasodium pyrophosphate	5
Abrasive	79
Trichloro-iso-cyanuric acid	1

Where a spray-drier is being operated, Formula 36 is a useful outlet for the cyclone fines produced by the spray-drier.

A scouring liquid can be made according to Formula 37.

Formula 37

Household scouring liquid

Disperse	
Bentonite	5
in water	25
Add 12% active ready-made liquid detergent and dissolve in the liquid:	35
Sodium metasilicate pentahydrate or water-glass and then disperse in this solution:	3
Abrasive	32

Special suspending agents suitable for this type of product are now available. These prevent settling of the abrasives.

Miscellaneous Household Cleaners

Window cleaners of the gentle abrasive type, using whiting, were once popular. These have now been superseded by liquids like those of Formula 38.

Formula 38

Household window-cleaning liquid

Active detergent matter	0.25–0.5
Isopropyl alcohol	15–35
Water	to 100
Dye	as required.

Similarly, for stone or tile floors, although materials such as that given in Formula 1 (Chapter 10) are used, liquid floor cleaners according to Formula 39 are now also being manufactured.

Formula 39

Floor cleaner

Active detergent matter	2–5
Isopropyl alcohol	8–15
Pine oil	1–2
Water	to 100

Commercial Laundering

In commercial laundering, powders are used which do not foam and which are in general more alkaline than household powders. Most commercial laundries use soft water, and we suggest to the detergent manufacturer that if he supplies washing powders to a laundry which does not use soft (or softened) water, he should make every effort to convince his customer that it is essential for him to use it.

Laundry powders can be made both by spray-drying and by absorption and neutralisation. The two formulations below have been used with success in commercial laundries.

Formula 40

Spray-dried industrial laundry powder

Alkyl benzene sulphonic acid	20	} Neutralised with
Distilled tallow fatty acid	15	NaOH to sodium salts
Tetrasodium pyrophosphate	15	
Sodium metasilicate	15	
CMC (100% basis)	2	
Optical brightening agent	0.15	
Soda ash	33	

Formula 41

Industrial laundry powder not spray-dried

Alkyl benzene sulphonic acid	11
Distilled tallow fatty acid	7
Tetrasodium pyrophosphate	10
Sodium metasilicate pentahydrate	7
CMC (100% basis)	1.8
Optical brightening agent	0.1
Soda ash	61.1
Water	2

Directions for the manufacture of Formula 45 are given in Chapter 10. It is obvious that for a given load of washing more of Formula 45 will need to be used than of Formula 40.

Many laundries carry out a pre-wash at a lower temperature. One of the above powders can be added to the pre-wash, but a more effective method is to use for the pre-wash a solvent detergent.

Solvent Detergents

We have already described the manufacture of powders containing solvents in Chapter 10.

It is not too difficult to combine non-ionics with solvents, but anionics are not easy to combine, especially those of the alkyl aryl sulphonate type, as most of these are insoluble, or only slightly soluble, in most non-polar solvents such as kerosene or deodorised kerosene which, of course, are generally the least expensive ones.

Detergents often serve only as emulsifying agents for the solvents, and not so much as detergent or cleaning agents. Emulsifiable insecticide concentrates, for example, are based on a solution of the emulsifying agent within the insecticide solvent mixture, forming a clear stable solution which, on dilution, gives a milky white emulsion with water. Here the emulsifying agent is very often a non-ionic detergent acting as an emulsifier, rather than as a detergent proper. However, we are more concerned with solvent-detergent combinations, in which the detergent acts both as an emulsifying agent for the solvent, and as a detergent in its own right.

In detergent-solvent combinations, the job of the solvent is to dissolve grease and similar oily dirt. The function of the detergent is to act as a penetrating and wetting agent and as an emulsifying agent for carrying off solvent after it has dissolved the oil or grease from the material to be cleaned, but it also keeps solid dirt particles in suspension. In many cleaning operations the detergent, by its surface activity alone, is simply not powerful enough to loosen dirt which is kept strongly attached to the surface by oily or resinous matter. This is very often encountered in metal-cleaning, or during the laundering of oily overalls. On the other hand, it is often impossible for a solvent alone to develop its full dissolving power where the oily matter is covered by solid crusts of insoluble matter. This is the case in the decarbonising operations carried out on the working parts of internal combustion engines. It is the combination of surface activity plus solvent power which makes solvent-detergents so useful in widely different fields of application.

The most easily produced type of solvent-detergent is a combination of non-ionic detergent with solvents. Very often a simple mixing of solvents with detergents is sufficient to obtain a clear, stable product, which generally forms milky emulsions in water. However, not all non-ionics are soluble in any proportion in any solvent. Very often they are only slightly soluble in non-polar solvents of the aliphatic type. Here it is necessary to use so-called 'co-solvents', together with the non-polar aliphatic solvent, to give the desired results.

The subject of solvency is of the greatest importance in working out effective products. By giving concrete examples, it will be made clear how important this type of solvent is in formulating high-grade products.

Generally speaking, the non-ionic detergents are more easily soluble in solvents than most anionic detergents. It is, nevertheless, quite incorrect to assume that they are soluble in all kinds of solvents. Thus, for example, a condensation product of nonyl phenol with 9–10 moles ethylene oxide is readily soluble in chlorinated solvents, xylene, benzene, and in most polar solvents; it is, however, only slightly soluble in kerosene and white spirit and even less so in dearomatised (deodorised) kerosene. To increase the degreasing power of trichlorethylene, it is possible to add a certain percentage of non-ionic, e.g., 3–5 per cent, to the solvent. Furthermore, a stable solution of trichlorethylene may be produced by dissolving 10-15 per cent of non-ionic in trichlorethylene. On dilution of the clear solution with water, a milky-white emulsion will be obtained. In order to prevent corrosion due to free hydrochloric acid, an addition of about 0.5 per cent monoethanolamine to the composition is advisable.

Chlorinated solvents should be used with caution preferably they should be used in closed systems, where the solvent is distilled, condensed and recycled, such as in dry cleaning and closed metal degreasing systems. The low heat of vaporisation of these chlorinated solvents (210 J/g for perchlorethylene) and non-inflammability are distinct advantages.

As already pointed out, most non-ionic detergents are soluble in aromatic solvents and by dissolving 10 per cent in xylene and diluting the clear xylene solution with water, stable milky white emulsions may be obtained which are very useful for metal-degreasing compounds. Aromatic solvents may serve as co-solvents for dissolving non-ionic in aliphatic non-polar solvents such as dearomatised kerosene and white spirit. To obtain clear solution, a proportion of about 30–40 per cent of aromatic solvents is necessary, e.g., 10 parts non-ionic; 30 parts aromatic solvent; 60 parts kerosene.

Formula 42

Detergent-solvent combination	
Nonyl phenol 9 EO	30 parts
Isopropanol	20 parts
Xylene	50 parts

Formula 43

Detergent-solvent combination	
Nonyl phenol 9 EO	30 parts
Methylethylketone	35 parts
Deodorised kerosene	50 parts

Mix the materials in the order given. A clear solution is obtained which becomes blue-white opalescent when diluted with water (hard or soft). Nonylphenol-9-ethox, behaves similarly to octyl phenol ethoxylates.

For the internal degreasing of motors, etc., such combinations with non-ionic detergents and solvents should prove very useful, possibly in combination with flushing oils. Even in the unlikely event of solvent-detergent remaining in the engine, practically no danger of subsequent corrosion should exist, because of the complete volatility of the products of combustion, which are hardly any more corrosive than the combustion products of motor fuels. It would even be possible to formulate an 'internal' decarboniser on the basis of anionic detergents (AB sulphonic acid neutralised with alkanolamine or alkylamine), which does not leave any residues in the engine. Here again, an entire field for further research is open.

Another non-ionic detergent useful for formulating solvent-detergents is alkylolamide. Experiments with this compound (which is a condensation product of ethanolamine with fatty acids) have yielded very efficient solvent-detergents. However, the wetting power of alkylolamide is some-what lower than that of water-soluble alkyl phenol or fatty alcohol ethoxylates.

A special solvent-detergent combination of interest is one which gives a clear solution of kerosene in water according to Formula 44.

Formula 44

Kerosene water solution

Coconut fatty acid diethanolamide	20
Kerosene	20
Water	20

On dilution with soft or hard water, all these products given very stable solvent emulsions of good detergent power.

The alkyl benzene sulphonic acids can also be used as the basis of solvent-detergent combinations.

It is essential to start operations with the unneutralised sulphonic acid. SO_3 sulphonated sulphonic acids, because they contain minimal amounts of inorganic acids, can quite easily be incorporated into these combinations, but oleum sulphonated dodecyl benzene, if treated as described below, can also be turned into an acceptable solvent-detergent.

Formula 45

Solvent-detergent combination

Mix together :

SO_3 produced alkyl benzene sulphonic acid (ABS)	50 parts
Kerosene	50 parts
Aromatic solvent	25 parts

Then slowly add:

Caustic soda solution (38° Be)	17–18 parts

After cooling to about 50°C add:

Isopropanol	10 parts
Pine oil	2 parts

A clear liquid is obtained, which at low temperatures becomes gel-like.

A small amount of water from both the water of solution of the caustic soda and the water produced by neutralisation, will be present. If a completely anhydrous material is required, the lower amines, isopropyl or butyl, can be used for neutralisation. Care should be exercised in their use as they are volatile, toxic and inflammable.

The product is completely soluble in soft water, somewhat turbid in hard water, and is completely soluble in organic solvents. It is almost completely soluble in non-polar paraffinic solvents, such as kerosene, petrol, etc. By diluting the solvent-detergent with solvents, e.g., 1:10 with white spirit or kerosene, a solvent-emulsion concentrate is formed which gives very stable emulsions on dilution with water. This is important for metal-degreasing operations.

It is interesting to note that the wetting power of the ABS detergents is practically unaffected if only aromatic and polar solvents are used, and only slightly affected when non-polar paraffin solvents are present. In all cases, sinking time as measured by the Draves test is not greater than 25s at a concentration of 0.1 per cent active detergent matter in distilled water.

A completely anhydrous solvent-detergent can be manufactured from SO_3 sulphonated LAS according to Formula 46.

Formula 46

Solvent-detergent based on 100 per cent ABS (SO_3 produced)

Kerosene	35
ABS (100% basis)	35
Monoethanolamine	15
Trichlorethylene	13
Pine oil	2

All the ingredients, except for the trichlorethylene, are merely mixed together until uniform. When the liquid has cooled to below 50°C, the trichlorethylene is added and the pH adjusted to 8 by using monoethanolamine. This pH was found to be advisable as the trichlorethylene might release traces of hydrochloric acid and the surplus monoethanolamine will absorb this. Trichlorethylene may be replaced by higher boiling perchlorethylene. Formula 50 is an excellent general-purpose combination and can be used for adding to the pre-wash in laundering, as an additive for dry-cleaning fluids, and for degreasing of engine parts, etc.

One of the most important uses of solvent-detergents is as a dry-cleaning aid. The purpose of the dry-cleaning detergent is a multiple one, to increase the effectiveness of the solvent in dissolving solvent-soluble spots and stains, to aid in the removal of water-soluble stains by dispersing or solubilising a small percentage of water in the solvent, and to increase the dispersion of these water-soluble stains in this water. To achieve this effect generally a combination of non-ionic and anionic detergents is used together with a coupling agent. For example,

Formula 47

Dry-cleaning detergent

Nonyl phenol-9 ethylene oxide	40
Propylene glycol	17
Butyl cellosolve	3
Monoethanolamine	7
Alkyl benzene sulphonic acid	33

(SO$_3$ sulphonated to minimise inorganic salts)

All the ingredients are mixed together and the pH (tested at a dilution of 1:10 in water) is adjusted to between 7 and 8 with small additions of either monoethanolamine or sulphonic acid.

This is added to the dry-cleaning solvent at the rate of 1–1$\frac{1}{2}$ per cent with or without the addition of small quantities of water.

If a readily dispersible product is required, the above formula can be diluted 1:1 with a dry-cleaning solvent (for example, perchlorethylene) and is then used at the rate of 2–3 per cent based on the volume of the solvent in the machine.

Although dodecyl benzene sulphonic acid gives excellent results in the above formulations, even better results are obtainable by replacing the ABS with heavy alkylate-sulphonic acid, which has a higher molecular weight and renders the sulphonate more oil soluble. Of course, with a higher molecular weight, the acid value will obviously be lower. The amount of alkali in the above formulations should, therefore, be correspondingly reduced. (This is best determined by the acid value of the heavy alkylate sulphonic acid.)

CARPET AND UPHOLSTERY CLEANERS

Fabric cleaners of this type differ in their operation from other detergent materials in that it is usually very difficult to rinse the material being cleaned. To overcome this, methods of cleaning have been developed where a solution of the detergent is applied to the carpet by 'shampooing' to form copious foam, or a foam is formed first and the carpet sponged with this foam. In either instance the combination of the detergency of the cleaner and the mechanical energy applied lifts and holds the dirt in the foam.

The foam, having very thin walls and enormous surface area dries relatively quickly into brittle particles of dust and this dust is either vacuum cleaned or brushed away.

Initial formulations for these carpet shampoos were normal light-duty detergents with the addition of tetrasodium pyrophosphate, its function being both to increase detergency and to make the dried residue more brittle. A suitable detergent material which is also in itself fairly brittle when dehydrated is the sodium or magnesium salt of one of the fatty alcohol sulphates. Some formulae also called for the incorporation of a solvent but with the newer fabrics and rubberised bases being used, the solvent should be chosen with care or left out.

These formulations were not, however, the complete answer to the problem. A portion of the active matter became absorbed into the fibre and when dried this left a deposit which tended to attract dirt. Carpets cleaned in this manner became soiled very quickly.

This problem has been overcome by using more crystalline detergents as the base. Such detergents are the half ester of sodium sulphosuccinates used alone or in admixture with fatty alcohol sulphates. The lithium salt of fatty alcohol sulphate for this specific purpose are used. The detergent then absorbed on the fibre is more brittle and when dried will shatter and can easily be brushed away. Another line of attack is to add colloidal silica to the formula to produce the same effect.

TEXTILE DRESSING

Detergents of the anionic type have to a large extent replaced soap in the scouring of wool and cotton. The main factor involved is that anionic detergents do not precipitate their lime and magnesia salts on to the fibres, and thus give a better 'handle' to the finished goods.

As textile processing is a complex process, the detergent manufacturer can only be called upon to supply a concentrated detergent; the various alkalis, phosphates or solvents are added by the textile processer himself. This concentrated detergent is usually manufactured by agreement between the textile scourer and the detergent manufacturer. Fatty alcohol sulphate pastes, neutralised immediately after sulphation, are as often used as are alkyl benzene sulphonate pastes. Depending on the requirements of the textile mill, the alkyl benzene sulphonate can be a concentrated paste neutralised only with caustic soda, or a semi-liquid paste neutralised with ethanolamine. A typical formula can be :

Formula 48

Textile scouring paste

Alkyl benzene sulphonic acid (ABS 100%)	375
Diethanolamine	130
Water	495

The ingredients are mixed together and the final pH adjusted according to the buyer's requirements.

On occasion, especially for degumming, detergents compounded with pine oil are preferred for textile scouring, particularly of wools.

We suggest the following formulation :

Formula 49

Textile degumming detergent paste

Pine oil	250
Alkyl benzene sulphonic acid (ABS 100%)	280
Diethanolamine	97
Water	373

Mercerising

Cotton cloth is treated for anti-shrink properties by dipping in a concentrated (of the order of 12–18 per cent) caustic soda solution. To allow the solution to penetrate well and evenly into the fibres a detergent or rather a wetting agent is necessary. The bulk of detergents are either not soluble or not effective (or both) in strong caustic soda solutions. A commonly used wetting agent for this purpose is the sodium salt of sulphated 2-ethyl hexanol. By itself it is not completely effective but if blended with 10 per cent each of the unsulphated 2-ethyl hexanol and butanol, the material becomes soluble in caustic soda solutions, the solution remains stable and the wetting properties are greatly enhanced. Ethoxylated glycosides (moderate to high foaming) and the low-molecular-weight phosphate esters (low foaming) are also used for this purpose. Another possibility, which has not yet been exploited commercially, is SO_3 sulphonated olefins.

FOOD AND DAIRY INDUSTRIES

As with any food industry, dairies require spotless cleanliness to prevent spoilage, and sterilisation against bacterial contamination. The tolerance for residual bacteria is, however, lower for dairies than for other food industries, as in this case bacteria can cause the milk to turn sour. A further complication is the presence of high amounts of lime salts in the milk.

Many plants use strong alkalis only for cleaning, and sterilisation is achieved by steam. These materials have their limitations, as can be seen from Table 11.3, detailing the properties of various alkalis for dairies.

It will be noticed that no one material supplies all the requirements. Chlorinated trisodium phosphate adds a sterilising action. For this reason it is often desirable to combine several of these alkalis to produce a blended material which will provide all the requirements. The addition of a detergent greatly enhances the performance of these alkalis, but in certain processes foaming will interfere with the operation. Even low-foaming non-ionics foam slightly, and the use of detergents is thus sometimes excluded. Formulae 50–52 give some typical formulations for food- and dairy-cleaning alkaline detergents.

Formula 50–52

Food and dairy alkaline detergent cleaner

	Non-foaming	Medium foam	High foam
	50	51	52
Trisodium phosphate	15	–	10
Sodium carbonate	10	39	35
Sodium metasilicate pentahydrate	40	20	20
Tetrasodium pyrophosphate	–	40	–
Sodium tripolyphosphate	35	–	30
Non-ionic detergent	–	1	–
Concentrated anionic detergent powder	–	–	5

Table 11.3. Alkalis for dairy cleaners.

Agent	Wetting power	Solution of heat-deposited milk solids	Emulsification of fats	Buffering power	Rinse-ability	Water-softening properties	Corrosive action
Caustic soda	Poor	Very good	Poor	Poor	Poor	Good	Strong
Carbonates	Poor	Good	Fair	Poor	Fair	Good	Weak
Trisodium phosphate	Good	Poor	Fair	Good	Good	Good	Weak
Orthosilicate	Good	Good	Very good	Very good	Fair	Fair	Strong
Metasilicate	Good	Good	Good	Fair	Fair	Fair	Weak
Sesquisilicate	Good	Good	Fair	Good	Fair	Fair	Medium
Chlorinated trisodium phosphate	Good	Fair	Fair	Good	Good	Good	Medium

To produce a powder that simultaneously sterilises and cleans, the metasilicates must be of the anhydrous variety, and soda ash and non-ionic detergents should not be used. Subject to the above conditions, sufficient organic-chlorine-bearing materials can be added to the above formulation to give, say, 3 per cent active chlorine in the finished powder.

For bottle-washing in the food industry the pH required is very high. Sodium gluconate should be added for its sequestering action.

Formula 52

Bottle-washing compound

Caustic soda flakes	55
Sodium metasilicate	20
Soda ash	17
Sodium gluconate	5
60% anionic detergent concentrate	3

In addition to the above alkaline cleaning materials, in the dairy industry it is necessary as a routine to give the pipelines, etc., an acid wash to dissolve milkstone deposited on the walls.

Very often concentrated hydrochloric acid is used. This is very corrosive and although the dairy plant is invariably stainless steel it is advisable to add an inhibitor. Nitric acid is also used on occasion. Although nitric acid is both a strong acid and on oxidising agent, sight must not be lost of the fact that the original stainless steels were developed to withstand this acid, so no inhibitor is necessary. Both these acids are unpleasant to handle and environmentally unsound, therefore, to overcome these difficulties less corrosive materials have been developed.

Combinations of lactic acid with acid-resistant detergents have found application. So, also, have combinations of phosphoric, citric, and tartaric acid with acid-resistant detergents. The most recent development in this important field is the application of sulphamic acid. This has the structural formula :

$$H_2N-\overset{\displaystyle O}{\underset{\displaystyle O}{\overset{\|}{\underset{\|}{S}}}}-OH$$

It is the half amide of sulphuric acid.

Molecular weight   97.09

Specific gravity   2.03

The following advantages of sulphamic acid are enumerated.

1. Its crystalline, non-volatile nature, which makes it easy, safe and economical to handle, and eliminates the evolution of objectionable fumes when solutions are needed.

2. Its strong acid character gives the 'bite' necessary for the removal of deposits.

3. All salts of sulphamic acid (e.g., calcium, magnesium, iron) are readily soluble in water. Hence, adequate dissolving and thorough rinsing of scale is assured. Further, it is expected that less sulphamic acid is needed than other commonly used acids.

4. In spite of its strong acid character, dilute sulphamic acid solutions are not unduly corrosive to dairy equipment.

5. It is simpler and less hazardous in handling.

The following five general steps are recommended for cleaning heat-transfer equipment and vats, using sulphamic acid:

1. Thoroughly rinse or flush the equipment with warm water until the water is no longer milky.

2. Prepare an adequate quantity of sulphamic acid solution of 0.2–2 per cent concentration (1.6–16 kg/m^3), heat the solution to 65–70°C, and circulate for 10–30 minutes. Because of the many variables in equipment, composition of deposits and preferred cleaning practices,

precise recommendations suitable for all cleaning applications cannot be given. The above ranges of concentration, temperature and time have developed from experience and should cover most situations.

3. Rinse thoroughly with warm water.

4. Flush the equipment with a cleaning and neutralising solution consisting of 15–85 g of trisodium phosphate or other alkaline-type detergent compound for each gallon of water. The alkaline treatment should consist of approximately the same concentration, temperature and time as used with the sulphamic acid cleaning solution.

5. Rinse the equipment again with warm water.

In the case of some stubborn or fat-containing deposits which tend to be water-repellent, it is sometimes advantageous to use a small amount of a wetting agent along with the sulphamic acid, so as to enhance the rate at which it penetrates and attacks the material to be moved. Alkyl benzene sulphonic acid is suitable for this purpose.

Since uninhibited sulphamic acid is rather corrosive to ordinary steel and cast iron, it is wise to provide a separate pump for circulating solutions if existing pumps have ferrous metal parts in contact with the fluid.

On conducting experiments with 100 per cent dodecyl benzene sulphonic acid to clean milkstone. The calcium salts of DDBS are dispersible in water and therefore the acid should be suitable. However, the 100 per cent acid dissolves with difficulty in water in low concentrations. It was found that the addition of 10 per cent xylene sulphonic acid to the 100 per cent DDBS rendered the mixture very easily soluble. A 5 per cent solution of this mixture dissolves the milkstone readily and the surplus is easily rinsed away.

DETERGENT SANITISERS

The halogens have long been known to be effective germicides, and chlorinated trisodium phosphate is used pretty effectively to clean and disinfect milking equipment.

There also has been a tendency in the past years to use bromine as such for disinfection but it has no real advantage over chlorine and is much less pleasant to handle.

Iodine is now gaining considerably in this field, particularly combinations of iodine with non-ionic surfactants.

These products called iodophors with a 1–3 per cent iodine content are active against bacteria at a concentration of 0.012–0.025 g/litre and it has been found that maximum activity is obtained in an acid environment.

The iodophors are prepared by mixing nonylphenol adducts containing at least 8 mol ethylene oxide or ethoxylated propylene glycols with 20–25 per cent iodine and heating this mixture to 50–60°C. Care should be taken with ventilation as iodine sublimes, even at room temperatures. Under these conditions 75 per cent of the iodine combines chemically with the nonylphenol ethoxylate.

Typical formulae 53, 54 are given below.

Formula 53 and Formula 54

	53	54
	%	%
20% available iodine in nonylphenol ethoxylate	8.75	8.75
Phosphoric acid (75%)	8.00	0.60
Nonylphenol + 30 mol ethylene oxide	5.00	5.00
Water	78.25	85.65

METAL CLEANERS

The cleaning of metals is done with either acid or alkaline materials. The acid treatment is meant for the removal of rust and other products of corrosion, and for the solution of 'scale'. This is a very wide term and includes both the layer of insoluble heavy metal salts precipitated on the walls and tubes of steam boilers, and also the layer of oxides (as opposed to rust) formed on steel surfaces under certain conditions of heating. The alkaline treatment is for the removal of grease, oil, paint and foreign matter, and is also necessary as a preliminary treatment prior to pickling (acid treatment) to remove the film of oil which at best will hinder the pickling acid in its work and at worst will only allow this acid to work in patches, producing an uneven finish.

Scale removal by pickling calls for cleaning compounds (which are added to help the acid to penetrate and give a uniform finish) that will resist the strong acids used in pickling processes. Generally, detergents without any special additions of builders are used in these cases. However, it is of great advantage to use detergents in conjunction with 'inhibitors' which prevent (or retard) the attack of the pickling acid on the metal itself. The detergent has not much inhibiting effect, but its wetting action is apt to prevent pitting to a certain extent—a very important reason for using detergents in the pickling bath. The most suitable detergents are the alkyl benzene sulphonates, but certain petroleum sulphonates or heavy alkylate sulphonates may also be used with advantage. In these instances it is not necessary to neutralise the sulphonic acids, merely to add the material to the bath. If, however, a ready-neutralised material is all that is available, this can be used. These detergents need to be highly acid-resistant and to stand up to severe conditions, such as a 5 per cent sulphuric acid solution at 70°C. The amount of detergent to be added to the bath is on the average between 0.1 and 0.2 per cent, calculated as active detergent matter.

As inhibitors, phenylthiourea and thiourea are very effective. They are solid substances which can be compounded with powdered detergents to increase the rapidity of the pickling effect (in this case a neutralised concentrated detergent powder is used), and to prevent excessive corrosion of the metal proper. The amount of corrosion inhibitor required is 0.01 to 0.05 per cent, based on the sulphuric acid content of the pickling bath. If the concentration of the active detergent in the bath is of the order of 0.1 per cent, it follows that the compounding of an effective pickling bath is not very difficult. An example would be: a concentrated detergent powder containing 40 per cent active dodecyl benzene sulphonate is mixed with 10 per cent of its weight of phenylthiourea or thiourea. This mixture is added to the pickling bath in the proportion of 0.25–0.5 per cent. An alkyl aryl sulphonate of the alkyl naphthalene type was effective in concentrations of 0.1 per cent, both as an inhibitor and a wetting agent.

The following inhibitors were also found to be effective in sulphuric acid pickling baths:

Butyl sulphide, o-tolylthiourea, p-tolylthiourea, butyl disulphide, amyl mercaptan, ethyl selenide, propyl sulphide, butyl-methyl-sulphide, butyl mercaptan, p-thiocresol, iso-butyl-mercaptan, trianyl amine, m-thiocresol, trihexylamine, ethyl sulphide, phenyl morpholine, ethyl mercaptan, formaldehyde, methyl sulphide, 2-thionaphthol, crotonaldehyde. (This is only a selection from a list of more than a hundred compounds listed.)

In the past types of inhibitors especially suitable for acid cleaners, such as cleaners for automobile cooling systems, etc., have been developed, such as alkyl-substituted thioureas, diethyl- and dibutylthiourea, on to the market. These are suitable for use with 10 per cent hydrochloric acid solutions in concentrations as low as 0.05 per cent. Other types of acid inhibitors, are based on rosin amine derivatives, namely polyoxyethylated dehydroabeitylamines. These inhibitors were found to be effective in the range of 0.05-0.2 per cent in hydrochloric acid.

The effect of inhibitors in conjunction with surface-active agents is of great importance when removing water scale from steel pipes with hyrdrochloric acid. Again it was found that thiourea and phenylthiourea are very effective inhibitors whose effect is strongly enhanced by the use of detergents. A further line of research is the use of surface active agents in conjunction with corrosion inhibitors, such as sodium chromate, in brine solutions, calcium chloride refrigeration solutions and in automobile radiators.

For automobile cooling systems, acid mixtures are often used to remove hard and adherent scale. The effect of these mixtures is often greatly increased by the addition of detergents, which help in the penetration and removal of grease and oil which may have entered the cooling system and which prevent the acid cleaners from exerting their full effect on the scale. Mixtures of the solid acids, such as oxalic acid or sulphamic acid, and acid salts such as sodium bisulphate, may be improved by the addition of DDB sulphonate powders. The following formula is a suggestion for such a product:

Formula 55

Acid cleaner for water-cooling systems

Oxalic acid	80
Sodium bisulphate	10
Concentrated detergent 40 per cent active	10

The concentrated detergent can be either dodecyl benzene sulphonate or an acid-resistant petroleum sulphonate.

Aluminium is often cleaned and brightened by acid solutions containing hydrofluoric acid. One of the few detergents found to be compatible with this highly corrosive acid is an amphoteric of the dicarboxyethylated derivative of cocoimidazoline. A formulation suggested found to be effective is:

Formula 56

Miranol conc.	5
Glycol ether	6
Phosphoric acid (85%)	38
Hydrofluoric acid (70%)	8
EDTA	1
Water	42

Although this is called a cleaner, it actually removes the oxides on the surface thus giving a brightening effect. This solution should be used after an alkaline wash to remove the surface film of dirt and oil. Aluminium is attacked by alkali, but silicates effectively inhibit this attack.

Besides the acid-type metal cleaners described above, metals are cleaned in strongly alkaline solutions to remove grease, oil and foreign matter prior to plating, painting, enamelling or other protective treatments used nowadays.

The method of removing this foreign matter determines the composition of the alkaline cleaner. It is obvious that if a spray method is to be used the cleaning solution cannot foam. The principal methods of alkaline metal treatment are soaking, spraying and electrolytic, or combinations of two of these processes. In the soaking process, the cleaning effect is obtained by the high concentration of the detergent, and also a possible circulation of the solution by a pump. The spray process adds mechanical energy from the jet to aid in the removal of the dirt. The electrolytic process produces the cleanest surface because of the scrubbing action of the gases evolved and the attraction of the electrically charged dirt particles by the electrodes.

In addition to the types of cleaning process used, the type of metal to be cleaned influences the choice of formulation of the detergent. No detergent can do the universal job of cleaning every surface, and various formulations for different types of metals and also the different cleaning processes are given in Formulae 57–71.

In these formulae, mention is made of sodium resinate, which is saponified tall oil, obtained as a by-product in the paper industry. This soap is relatively cheap and in addition has good emulsification properties for oil. As its detergency is poor, it is invariably used in conjunction with a detergent. We suggest that the same emulsification can be achieved by the use of a sulphonated heavy alkylate (see Chapter 10). The formulae also call for alkyl aryl sodium sulphonate. This is meant to be a concentrated powder containing 40 per cent sodium dodecyl benzene sulphonate. If another concentration is available the proportion should be varied accordingly.

MISCELLANEOUS CLEANERS

Lavatory Cleaner

The universal ingredient of lavatory cleaners is sodium bisulphate ($NaHSO_4$, technically called nitre cake). The most popular lavatory cleaner is based on this compound. Caking of sodium bisulphate can easily be overcome by adding to the ground salt some 0.5–1 per cent pine oil or a mixture of pine oil and kerosene. This also largely prevents corrosion of the metal

Formula Nos 57–71.

Alkaline metal cleaning compounds

	Aluminium		Copper			Copper-plate	Iron and steel			Magnesium			Zinc	
	Soak	Spray	Soak	Spray	Electro-lytic	Electro-lytic	Soak	Spray	Electro-lytic	Soak	Spray	Soak	Spray	Electro-lytic
Builders	Composition of cleaners, % by weight													
Sodium hydroxide, ground	–	–	20	15	15	55	20	15	15	20	20	–	15	15
Sodium carbonate, dense	–	–	18	–	–	8	18	29	8½	18	29	–	–	–
Sodium bicarbonate	21	24	–	34	34	–	–	–	–	–	–	–	35	34
Sodium tripolyphosphate	30	30	–	–	10	–	20	20	10	20	20	90	10	10
Tetrasodium pyrophosphate	–	–	20	10	–	10	–	–	–	–	–	–	–	–
Sodium metasilicate, anhydrous	45	45	30	40	40	25	30	30	25	30	30	–	40	40
Surface-active (wetting) agents)														
Sodium resinate	–	–	5	–	–	–	5	–	–	1	–	5	–	–
Alkyl aryl sodium sulphonate	3	–	5	–	–	–	5	–	–	5	–	5	–	–
Alkyl aryl polyether alcohol	–	–	2	–	–	1	2	–	–	2	–	–	–	–
Nonionic high in ethylene oxide	1	1	–	1	1	1	–	1	½	–	1	–	–	–
	Other conditions													
Operating temperature of solution °F	70	70	80	75	70	80	95	75	80	95	75	80	75	80
Concentration of cleaner, kg/m³ H_2O	25	6	50	6	50	50	50	6	50	50	6	25	6	37

containers in which these products may be marketed. (Recently, high density polythene containers have been adopted in many European countries. These are completely unaffected by lavatory-cleaning compounds.)

The cleaning effect can be improved by adding to the sodium bisulphate not only the 0.5–1.0 per cent pine oil, but also approximately 1 per cent alkyl benzene sulphonic acid.

In the last few years liquid toilet bowl cleaners have appeared in market. These are packed in specially designed plastic bottles so that they can be dispensed on to the inner, invisible, rim of the toilet pan. These liquids are acid, either mineral or organic acids being the acidic component, contain relatively large amounts of surfactants, are dyed and perfumed and have medium viscosity. A typical representative formulation is given as under :

Formula 72

Non-ionic detergent	11.0
Hydrochloric acid (33%)	10.0
Phosphoric acid (85%)	10.0
Perfume	0.2
Water	68.8

Due to the acidity of the solution, the non-ionics need to be chosen with care and are usually a blend of alkyl phenol ethoxylates of both short and long chains. The perfume needs to be matched to the formulation to render it stable under acid conditions.

A viscous lavatory cleaner with both sanitizing and deodorizing effects may be produced by using LAS (SO_3 sulphonated) and pine oil as its two components :

Formula 73

LAS (acid form)	100
Pine oil	50–70
Water	150

The LAS and the pine oil are mixed and the water added immediately to prevent reaction of the sulphonic acid with the pine oil.

The water and/or the pine oil are varied to produce the desired viscosity. An oily liquid with lower viscosity but faster dispersibility in water is obtained by adding 10 to 20 parts isopropanol.

Hand Cleansers

The manufacture of hand cleanser is similar to that of other abrasive cleaners. Thus a hand soap in paste form may be produced according to Formula 74 :

Formula 74

Hand cleanser

40% sodium ABS paste	130 parts
Bentonite	150 parts
Fine sand	200 parts
Sodium carbonate	10 parts

Or it can be manufactured by using as the base a detergent paste according to Formula 75.

Formula 75

Detergent hand cleanser

40-50% sodium alkyl benzene sulphonate paste	100 parts
Bentonite	30 parts
Abrasive powder	200 parts
Sodium carbonate	10 parts

The solid materials are incorporated into the soft soap or paste. Some water may need to be added to regulate consistency.

A hand cleanser in powder form may be produced with Formula 76.

Formula 76

Hand cleanser in powder form

Pure powdered soap or concentrated detergent powder	26 parts
Abrasive material	70 parts
Borax	4 parts

Some of the abrasive material may be replaced by vegetable abrasives, sawdust, etc.

A special hand-cleansing compound containing approximately 75 per cent borax and 25 per cent of dry soap was found by the laboratories to possess desirable fungicidal properties, and yet to be so mild in its action on the skin as to reduce any tendency to dermatitis. The borax is in finely granulated form, so that, when mixed with dry, powdered soap, gumming or caking is avoided. The product is described as a good cleanser, and will effectively, and without the use of waste, remove dirt from hands. The borax is not merely a diluent for the soap. Its hardness of two makes it even softer than chalk; it is readily soluble, and its abrasive action is only temporary, as the sharp edges of the grains become blunted almost immediately. It has detergent and water-softening properties of its own and it is a mild alkaline salt possessing the characteristic properties of imparting to a soap solution a pH value lower than that of soap alone.

Waterless Hand Cleansers

This type of hand cleanser is specially suitable for motorists, for removing oil, grease and grime after changing tyres or doing repairs while on the road. We confine ourselves to giving some formulae which we have worked out and found suitable for the purpose. The formula 77 especially useful to motorists for removing oil, paint, tar, etc., from the hands.

Formula 77

Waterless hand cleanser

Stearic acid and/or 100% alkyl benzene sulphonic acid	25
Lanolin and/or lecithin	15
are dissolved in :	
Deodorized kerosene	350

The mixture is kept at a temperature of about 70°C.

To this solution the following mixture, having the same temperature of 70°C, is added

Triethanolamine	25
Water	95

Stir constantly until the mixture is cold

3.5 parts of pine oil is recommended as a cheap perfume and disinfectant.

Formula 78

Waterless hand cleanser

White mineral oil	40.5
Oleic acid	10.5
Non-ionic detergent	6.0
Propylene glycol	5.0
Triethanolamine	2.6
Morpholine	1.0
Water	34.4

Formula 79

Waterless hand cleanser

Deodorized kerosene	42.8
Lanoline	0.9
Oleic acid	5.9
Cetyl alcohol	0.4
Triethanolamine	2.9
Propylene glycol	2.7
Sodium lauryl alcohol sulphate	1.4
Water	43.0

It is suggested not to add abrasives to this type of waterless hand cleanser, because they cannot be completely wiped off the hands after use and are likely to choke the pores and thereby irritate the skin. If, for very heavy-duty waterless cleansers, an abrasive cannot be dispensed with, only the softest kinds, such as whiting, bole, kaolin, etc., should be used, and in the instructions for use it should be recommended to wash the hands with soap and water as soon after the application of the cleaner as possible. Neither halogenated solvents nor aromatic solvents should be used.

Chapter 12

Perfuming of Soaps and Detergents

INTRODUCTION

The terms 'fragrance' and 'perfume' are synonymous and are used interchangeably throughout the perfumery and cosmetics industries. This will also be so throughout this chapter. Fragrance (or perfume) is a blend of two or more materials characterised by having olfactive properties, and is incorporated into a preparation with the intention of imparting specific odours characteristics to that preparation. Very occasionally, the fragrance may consist of one raw material only.

The most important function of the perfume is usually to impart a pleasant and suitable odour to the product into which it is to be incorporated. It is the job of the creative and evaluation teams to define the terms 'pleasant' and 'suitable' for each particular brief that is received. The perfume may also be required to perform other roles in the product, which may take priority over, or be additional to the primary role of imparting odour.

Most soap products have a mild but detectable odour, which is derived from the raw materials in the formulation. Although this 'base' odour is usually inoffensive, it can detract from the overall aesthetic character of the product and should at least be 'masked'. This can be effectively achieved by the use of a suitable fragrance. Very pungent bases, such as permanent-wave lotions, are almost impossible to mask, even with the strongest of perfumes.

The role of the perfume in these products is therefore to produce as pleasant an odour as possible, and this can often be attained by harmonising the fragrance with the base odour.

Fragrances can be designed to give subliminal support to a particular product. For example, a fragrance may appear to make a shampoo wash the hair more cleanly, a perfumed night cream may seem more nutritional to the skin, or a perfumed body spray may appear to impart all-over freshness, all-day. In other words, the perfume may have the role of apparently enhancing product performance.

Some fragrance materials have antimicrobial activity, particularly against bacteria. Both bacteriostatic and bacteriocidal properties are evident. Consequently, fragrance blends often take on the antimicrobial activity of some of the individual components. Fragrances that have spicy and herbal notes usually exhibit some antibacterial activity owing to the presence of phenols such as eugenol (from clove oil), and thymol and carvacrol (from thyme oil). Consequently, the fragrance may also have the role of assisting the preservative system and even in extreme cases, of replacing the preservative system altogether. Similarly, the fragrance may assist a bacteriocide/bacteriostat in its effect on the skin's microflora (reducing body odour) or may be required to carry this attribute of the finished product on its own.

TYPES OF PERFUMES

The cost of soap is determined largely by the perfume. A perfect sample should be strong without being harsh and should retain its fragrance right up to the last thin wafer in use. The stability in the presence of alkali of the aromatic substances in any compound is of importance, as also any change in colour which they may effect in the soap after manufacture. Successful soap perfuming therefore requires a knowledge of the durability of each individual raw material when in contact with the soap and any of its constituent impurities. One or 2 per cent of compounded perfume, spread through a large and non-volatile mass has to yield an odour suggestive of the finer alcoholic perfumes in which the solvent plays an important part in determining the delicacy of the finished odour. While alcohol develops the finer ingredients of a perfume, soap on the contrary modifies the odour to such an extent as to often make it unrecognisable in the absence of correct blending. Raw materials listed in the odour classification under basic notes play an important part in solving this problem, and the durability of a perfume in soap largely depends upon their correct selection. Natural oils and resins are often the key, and if price will allow they should be used liberally. Some makers rely upon a very small range of raw materials, and in consequence their products lack variety. Many of the newer synthetics have proved of great value in soap perfumery, but in all cases, before adoption, experiment is necessary. For this purpose a miniature mill and plodder are used. To save expense, however, many chemists use an ordinary mincing machine, and after passing the perfumed ribbons through two or three times, pound them into a mass in a mortar, and stand aside under ordinary commercial conditions for three months. At the end of this time the durability of any perfume compound can be well judged. To facilitate choice a list of substances with their properties are discussed. The stability and keeping qualities are indicated as follows—*very good, A; good, B; weak, C; poor, D.* These are all that are necessary for the standard lines of soap, but may be added to when the creation of a new bouquet is desired:

Ajowan oil will replace thyme oil. It is stronger and imparts a pleasing freshness (*A*).

Almond oil, bitter, useful in acacia and fancy compounds, generally replaced by benzaldehyde, free from chlorine.(*C*).

Ambergris tincture, not often used excepting in very high-class soaps, generally replaced by labdanum (*B*).

Amyl cinnamic aldehyde, useful in jasmin soaps and very persistent.(*B*).

Amyl salicylate, very good in trefle compounds and quite strong and lasting in use.(*A*).

Aniseed oil, very powerful. Use traces only.(*B*).

Anisic aldehyde, excellent in May blossom and acacia compounds. About 10 per cent sufficient.(*A*).

Atlas cedarwood oil, excellent in acacia, mimosa, santal, and violet. Very strong and lasting. Resinoid also good in fern and chypre.(*A*).

Basil oil, indispensable for mignonette, but requires to be well blended(*B*).

Bay oil, sometimes used in place of clove oil.(*B*).

Benzaldehyde, cheap and useful in almond soaps, traces in fancy compounds yield pleasing results.(*C*).

Benzoin, used as a resinoid and also as a tincture made with industrial methylated spirit, inclined to darken soaps.(*B*).

Benzyl acetate, 5 or 10 per cent will give a refreshing and sweet odour to most soaps, useful in jasmin and orange blossom when blended with petitgrain oil.(*B*).

Benzyl alcohol, weak and faintly balsamic, useful as a diluent.(*C*).

Benzyl benzoate, a good solvent for musk xylene.(*A*).

Benzylidene acetone, useful in the sweet-pea type, blends well with bromstyrole for cheap bouquets. Owing to its irritating effect on the skin, should be used with care.(*B*).

Bergamot oil, very good, often replaced by lemon oil.(*A*).

Beta naphthol ethers, very powerful, but require to be well blended, otherwise coarse. Use in moderation.(*A*).

Bois de rose oil, indispensable in all flower compounds, much improves Bourbon geranium.(*A*).

Bornyl acetate, the base of all pine perfumes, useful in lavender compounds.(*A*).

Bromstyrole, very powerful and stable, good hyacinth base.(*A*).

Cananga oil, useful in violet and santalwood, not very strong and lacks body.(*D*).

Caraway oil, an indispensable constituent in brown windsor. Rather fleeting.(*B*).

Cassia oil, base for all brown windsors, good blender in bouquets, and should not be overlooked in some rose compounds. Causes darkening.(*A*).

Castoreum tincture, excellent and cheap substitute for musk.(*A*).

Cedarwood oil, strong and persistent, good basis for violets.(*A*).

Cinnamic alcohol, weak but good hyacinth base.(*C*).

Cinnamic aldehyde, occasionally used to replace cassia oil.(*A*).

Cinnamon leaf oil, good substitute for clove oil, darkens in soap.(*C*).

Citral, sometimes used as a substitute for lemon oil.(*B*).

Citronella oils. Both Java and Ceylon very largely employed in all forms of cheap soap compounds, Ceylon frequently preferred and not always on account of its lower cost—very strong.(*A*).

Citronellol, useful in rose compounds.(*B*).

Citronellyl esters, useful as modifiers and blenders.(*C*).

Civet tincture, made with industrial spirit, often replaced by synthetic civets based on phenylacetic acid and skatole.(*A*).

Clove oil, much used for imparting a pleasant sweetness to compounds and indispensable in carnation, darkens rapidly.(*A*).

Copaiba resin (balsam), good fixative.(*A*).

Coumarin, very good in all compounds, not forgetting lavender. Gives a yellowish tinge in time.(*A*).

Para-cresol methyl ether, very strong and stable. Use only about 1 per cent.(*A*).

Diphenyl methane and oxide, very strong and stable, substitutes for geranium oil.(*A*).

Ethyl cinnamate, very powerful and balsamic.(*B*).

Eucalyptus oil, traces useful in fern compounds, much used in medicated soaps.(*A*).

Eugenol, good in carnation, but darkens less than clove.(*B*).

Fennel oil, very useful, requires moderation.(*C*).

Geraniol, stable in rose and other compounds.(*A*).

Geranium oils, both Algerian and Bourbon indispensable, latter rather coarse, equal parts of each give best results, used in rose and all bouquets.(*A*).

Geranyl acetate, excellent in rose and lavender.(*C*).

Ginger-grass oil, very good, but requires to be well blended.(*B*).

Guaiac-wood oil, an excellent base for all compounds.(*A*).

Heliotropin, only permissible in coloured soaps, odour good, and indispensable in heliotropes.(*A*).

Hydroxy-citronellal. Use the residues—quite stable and persistent.(*C*).

Ionone, unpurified 100 per cent or residues will yield good results, about 20 per cent sufficient in violets, less in bouquets.(*B*).

Iso-butyl esters, good modifiers and blenders, rather weak excepting the phenylacetate and salicylate.(*B*).

Iso-eugenol, excellent in carnation but darkens rapidly, colours the soaps containing it.(*B*).

Labdanum, very persistent in small quantities, will replace ambergris, good in lavender.(*A*).

Lavender oil, useless alone, must be strengthened and well blended, oakmoss excellent for the purpose, also rose-mary, thyme, and borneol.(*B*).

Lavender oil, spike, powerful and very good in lavenders, ferns, etc.(*A*).

Lemon-grass oil, basis of all verbenas. Modify with bois de rose or palmarosa—turns the soap yellowish.(*C*).

Lemon oil, inclined to darken slightly. Ten per cent will make a marked difference to many compounds, indispensable in Cologne soaps, but must be well fixed to retard oxidation of terpenes.(*C*).

Linalol. Better use bois de rose oil.(*A*).

Linalyl acetate, good, but better to use bergamot oil.(*B*).

Methyl acetophenone, must be used in small quantities and requires well blending, useful in acacias and mimosas.(*A*).

Methyl anthranilate, very cheap and strong, but inclined to darken.(*A*).

Methyl benzoate, very strong and stable.(*A*).

Methyl cinnamate, better than ethyl ester, small quantities give an amber note.(*A*).

Methyl heptine carbonate, traces sometimes used in violets, sharp at first but softens after a few days.(*C*).

Methyl salicylate, used in chypre and fern soaps.(*B*).

Mirbane oil. Don't use it.(*B*).

Musk ambrette, about 2 per cent is excellent, more gives a sickly sweetness. Residues are good and cheaper.(*A*).

Musk tincture, seldom employed.(*B*).

Musk xylene, very useful, but make sure it is all dissolved, otherwise the soap will be spotty—turns yellowish in time.(*A*).

Nutmeg oil, used occasionally in lavender compounds.(*D*).

Oakmoss, excellent in lavender and fern, much used in fancy bouquets, very persistent. The green resin is good enough for all purposes.(*A*).

Orange oil, excellent sweetener but must be well fixed, inclined to darken.(*C*).

Orris oil concrete, much used in violets, but often replaced with the oleo-resin or resin extracted from the spent rhizomes. Industrial tinctures of this material useful.(*B*).

Palmarosa oil, good and stable, useful in rose and similar compounds. (*A*).

Patchouli oil, very good and stable, more than 10 per cent becomes unpleasant, much less generally enough, useless in rancid bases.(*A*).

Pepper oil, traces employed in fancy bouquets and carnation compounds.(*B*).

Peppermint oil, traces are useful for developing other odours.(*A*).

Peru balsam, imparts warmth, darkens in time.(*A*).

Petitgrain oil, good in many compounds, but turns yellowish in time, often used for predominating note in glycerine and cucumber and Cologne soaps. Very durable.(*B*).

Phenylacetic aldehyde, not very valuable in soap, better use bromstyrole. (*D*).

Phenylethyl alcohol, useful in all flower compounds.(*A*).

Phenylethyl esters, good blenders, rather weak.(*B*).

Rhodinol, excellent in rose and other compounds, usually too expensive. Use Bourbon Geranium.(*A*).

Rosemary oil, good and stable, imparts freshness, useful in many compounds, especially lavender.(*A*).

Safrole, much used in cheap soaps.(*A*).

Sandalwood oil, excellent blender in violets and roses, requires to be developed in santal soaps.(*A*).

Sassafras oil, indispensable in chypre.(*A*).

Spearmint, powerful at first but evanescent.(*C*).

Storax, the best all-round resinous fixative.(*A*).

Terpineol, often the only perfume in cheap soaps, excels rather as a basis on which to build fanciful odours. Use 10 per cent or thereabouts — very stable and strong.(*A*).

Terpinyl esters, fair blenders, but not so stable.(*B*).

Thyme oil, excellent for imparting freshness to all kinds of soap compounds, antiseptic value good.(*A*).

Tolu balsam, good fixative, but inclined to darken.(*B*).

Vanillin, traces useful in fancy bouquets, turns the soap yellowish.(*B*).

Vetivert oil, small quantities are most persistent, good in violets.(*A*).

Flower oils are only used in the highest priced soaps. They yield excellent results. A cheaper jasmin is made at Grasse by distilling the flowers with cedarwood oil. This product is of considerable value. Mimosa absolute should not be overlooked since it is fairly cheap.

MATCHING A PERFUME

Matching a perfume of known composition requires experiment and is often necessary when a toilet soap is included in any series of cosmetic products. As will be readily understood it would be impossible, from a pecuniary point of view, to use a first-class compound in soaps. From another point of view it would be equally unsatisfactory, since some of the ingredients might undergo decomposition, cause discolouration, or generally upset the finished odour to such an extent as to make it unrecognisable. A formula therefore requires very careful examination and analysis. Those ingredients which are known to be stable and not too dear are allowed to remain although the quantities may have to be adjusted. The others are eliminated but replaced with raw materials having similar odours. Other substances may have to be added to diffuse the perfume. The most important point is to choose the right combination of basic notes and add just sufficient to regulate the volatility of the perfume. There are, of course, many essential oils having distinctive odours which at the same time act in this way as for instance, vetivert and santal in violets; patchouli and guaiac-wood in roses. These are often more stable than synthetics but, nevertheless, gum-resins in liberal quantity are imperative constituents. The soap perfumer's stand-by is styrax, but this is by no means the only one of value; almost all the resinous substances are excellent and many useful variations can be made with them. It should be noted that it is useless to judge the odour of a soap compound by the usual methods. The perfume must be incorporated in the soap by milling and the finished tablet left on one side for three months. The experiments leading up to this stage may be carried out with a mincing machine.

In order to indicate the lines upon which matching may be conducted an example of violet will be examined. A good quality compound for general use in perfumery will be approximately as follows :

Formula 1

500	Ionone alpha (1)	500
80	Orris oil, concrete (2)	80
50	Heliotropin (3)	50
150	Bergamot oil (4)	150
30	Ylang-ylang oil, Manila (5)	30
40	Sandalwood oil (6)	40
30	Violet leaf absolute (7)	30
25	Cassie absolute (8)	25
50	Jasmin absolute (9)	50
1	Aldehyde C_{12} (10)	1
25	Rose otto, Bulgarian (11)	25
20	Benzoin R. (12).	20
1001		1001

This formula would be adjusted as follows :

Replace 1 with ionone 100 per cent for soaps and reduce the quantity: Replace 2 with orris oleo-resin and reduce, only when costs compel: Retain 3 and 4: Replace 5 with cananga oil and increase: Increase 6: Replace 7 with methyl heptine carbonate and reduce: Omit 8: Replace 9 with benzyl acetate or inexpensive jasmin compound: Omit 10: Replace 11 with Algerian geranium oil and increase: Retain 12: To strengthen and diffuse add cedarwood, clove, and vetivert oils, terpineol and musk xylene or residues.

The formula for the violet soap compound having a similar odour to the above would therefore read (Formula 1a)

Formula 1(a)

Ionone, 100 per cent for soaps	200
Orris oleo-resin	80
Heliotropin	50
Bergamot oil	150
Cananga oil	50
Sandalwood oil	60
Methyl heptine carbonate	1
Benzyl acetate	50
Algerian geranium oil	40
Benzoin R.	20
Cedarwood oil	130
Clove oil	20
Vetivert oil	30
Musk residues	50
Terpineol	70
	1001

In making violet soaps some manufacturers add palm millings which emit a violet-like odour. Others add powdered orris-roots, but the use of the oleo resin dispenses with the necessity.

Formulation of Perfumes in Soap Compounds

Two examples of each of the principal toilet soap perfumes will now be given. The first and better quality one being based where possible largely upon essential oils, and the second and cheaper more particularly upon synthetics. They are all workable and are capable of endless modification to suit individual tastes. By applying the principles outlined above it will be quite easy to further cheapen the compound as desired. With a view to assisting the experimenter in the choice of raw materials for odours of specific type, the more important components are enumerated in accordance with the classification of odours (Formula 2)

Formula 2

Acacia soap (Based on essential oils)

Rosewood	100
Methyl acetophenone	20
Petitgrain para	200
Terpineol	50
Bergamot	50
Methyl cinnamate	20
Cananga	40
Anisic aldehyde	200
Methyl anthranilate	50
Clove	20
Geranium, African	50
Methyl naphthyl ketone	50
Santal	50
Tolu balsam	30
Vetivert	40
Musk xylene	30
	1000

Formula 3

Acacia soap (Based upon synthetics)

Rosewood	50
Petitgrain para	100
Bromstyrole	20
Terpineol	100
Geraniol, Java	100
Cedarwood	100
Methyl cinnamate	20
Cananga	40
Anisic aldehyde	200
Ionone	50
Methyl anthranilate	50
Clove	20
Methyl naphthyl ketone	50
Styrax resin	70
Musk xylene	30
	1000

Formula 4

Almond soap

Benzaldehyde	400
Rosewood	150
Clove	100
Geranium, African	50
Orris resin	70
Peru balsam	100
Santal	100
Musk xylene	30
	1000

Formula 5

Amber soap

Rosewood	150
Phenyl ethyl alcohol	50
Bergamot	150
Amyl salicylate	100
Geranium, Bourbon	40
Castoreum absolute	10
Coumarin	100
Labdanum resin	200
Musk xylene	30
Oakmoss	20
Phenyl acetic acid	50
Resin (styrax)	100
	1000

Formula 6

Cedarwood

Santal	150
Bromstyrole	10
Cananga	100
Terpineol	100
Methyl ionone	150
Cedarwood	250
Geranium, Bourbon	100
Vetivert	30
Orris resin	60
Heliotropin	50
	1000

Formula 7
Carnation soap

Amyl salicylate	100
Geraniol, Java	100
Eugenol	450
Geranium, African	50
Cassia	30
Orris resin	50
Iso-eugenol	70
Musk xylene	30
Pepper	20
Peru balsam	100
	1000

Formula 8
Chypre soap

Bergamot	400
Methyl ionone	100
Geranium, African	100
Civet absolute	5
Orris resin	50
Castoreum absolute	5
Coumarin	50
Musk xylene	30
Oakmoss	50
Patchouli	30
Santal	80
Peru balsam	100
	1000

Formula 9
Cologne soap

Petitgrain para	200
Lavender	50
Bergamot	200
Lemon	150
Lemongrass	50
Methyl anthranilate	40
Rosemary	60
Geranium, African	100
Citral	100
Methyl naphthal ketone	50
	1000

Formula 10
Heliotrope soap

Benzaldehyde	10
Benzyl acetate	40
Bromstyrole	10
Phenyl ethyl alcohol	90
Geraniol, Java	100
Sweet orange	50
Heliotropin	300
Anisic aldehyde	50
Ionone	50
Benzoin resin	100
Musk xylene	50
Peru balsam	100
Vanillin	50
	1000

Formula 11
Herb

Thyme, red	100
Rosemary	100
Lavender, spike	200
Fennel	50
Spearmint	50
Geranium, bourbon	300
Cassia	50
Clove	100
Camomile	50
	1000

Formula 12

Herb compound	500
Citronella, Java	100
Terpineol	300
Rosewood	100
	1000

Formula 13

Bergamot	200
Rosewood	200
Geranium, Algerian	100
Miel compound	100

Amyl cinnamic aldehyde	50
Civet absolute	10
Peru balsam	60
Musk xylene	30
Phenylethyl alcohol	150
Terpineol	100
	1000

Formula 14
Honey

Citronella	450
Lemon	200
Cinnamon leaf	100
Bay	50
Iso-butyl phenylacetate	20
Phenylethyl alcohol	100
Phenylacetic acid	30
Musk residues	50
	1000

Formula 15
Hyacinth soap

Benzyl acetate	200
Rosewood	200
Bromstyrole	50
Phenyl ethyl alcohol	100
Methyl cinnamate	30
Cananga	50
Ionone	50
Clove	70
Geranium	50
Galbanum resin	50
Amyl cinnamic aldehyde	50
Styrax resin	100
	1000

Formula 16
Jasmin soap

Benzyl acetate	200
Rosewood	100
Petitgrain para	80
Phenyl ethyl alcohol	100
Bergamot	100

Amyl salicylate	70
Sweet orange	25
Cananga	100
Clove	50
Civet absolute	3
Orris resin	50
Undecalactone	2
Amyl cinnamic aldehyde	40
Liquidambar resin	50
Musk xylene	30
	1000

Formula 17
Jasmin *(Another formula)*

Benzyl acetate	200
Rosewood	200
Terpineol	100
Petitgrain para	150
Amyl salicylate	50
Cananga	50
Ionone	50
Methyl anthranilate	50
Clove	20
Amyl cinnamic aldehyde	50
Musk xylene	30
Resin (styrax)	50
	1000

Formula 18
Lavender soap

Bergamot	70
Geraniol, Java	50
White thyme	20
Spike lavender	100
Lavender	130
Rosemary	100
Clove	10
Orris resin	30
Benzoin resin	70
Coumarin	50
Musk xylene	30
Oakmoss	30
Patchouli	10
Santal	30
	1000

Formula 19
Lily soap

Benzyl acetate	50
Rosewood	200
Terpineol	200
Citronellol	200
Bergamot	100
Cananga	50
Heliotropin	50
Ionone	30
Hydroxy citronellal	100
Cyclamen aldehyde	10
Amyl cinnamic aldehyde	20
Coumarin	10
Styrax resin	80
	1000

Formula 20
Musk

Musk ambrette	20
Musk ketone	30
Vetivert	40
Santal	100
Methyl ionone	200
Liquidambar resin	70
Bergamot	100
Geranium, Bourbon	200
Musk extract, 3 per cent	100
Civet absolute	5
Castoreum absolute	5
Terpineol	100
Cassia oil	30
	1000

Formula 21
Palm and olive oils

Palm compound	500
Terpineol	200
Lavender, spike	300
	1000

Formula 22
Frose soap

Citronellol	200
Phenyl ethyl alcohol	200
Geraniol palmarosa	200
Verbena	20
Ionone	50
Clove	10
Geranium, African	150
Orris resin	50
Musk xylene	30
Santal	50
Trichlor phenyl methyl carbinyl acetate	30
Vetivert	10
	1000

Formula 23
Santal soap

Rosewood	100
Geraniol, Java	100
Clove	30
Geranium, African	50
Orris resin	50
Coumarin	70
Musk zylene	30
Patchouli	20
Peru balsam	100
Santal	400
Vetivert	50
	1000

Formula 24
Violet soap

Rosewood	100
Bergamot	150
Methyl heptine carbonate	10
Cananga	40
Methyl Ionone	300

Mimosa absolute	20
Anisic aldehyde	30
Ionone	200
Orris resin	50
Iso-eugenol	10
Musk xylene	30
Styrax resin	30
Vetivert	30
	1000

Antiseptic and Medicated Soaps

Antiseptic and medicated soaps are made up with all kinds of medicaments, such as iodoform, thymol, phenol, betanaphthol, sulphur, etc. Many of the essential oils used in perfuming the soaps are highly antiseptic.

Perfuming Boxes

When packing wrapped soap it is advisable to spray the boxes with perfume—an example of which is given :

Formula 25

Violet soap

Bergamot oil	30
Geranium oil, Algerian	40
Rose, synthetic	30
Jasmin, synthetic	10
Tuberose, synthetic	20
Oakmoss	5
Benzyl cinnamate	20
Coumarin	20
Musk ambrette	10
Tincture of civet, 3 per cent	100
Tincture of benzoin, 10 per cent	715
	1000

Chapter 13

Environmental and Safety Aspects

INTRODUCTION

The performance chemicals occupy a vital place in modern chemical industry. These are used for human comfort, cleanliness and for industrial surface-active applications. Amongst these chemicals, surfactants can be ranked the first both by volume and value. The main types of chemical surfactant products, - anionic, -cationic and -nonionic, find applications in household products, personal care and industrial uses, such as detergents, paints and coatings.

ECOLOGICAL EFFECTS

Very recently, control of the environment and the effects of chemicals on the ecology have become prominent and well-publicised subjects. As far as the detergent industry is concerned, the control of foam and syndets in water systems as an indication of pollution and its effect on the ecology was the first step in adequate control.

Biodegradable Detergents

Soap use is historical and it is exceptional to find reference to water contaminated by it. The fact that water hardness reduces soap to a nonfoaming state is one reason why it has not been considered a major pollutant, and the further fact that it is almost without exception a straight-chain compound is another more pertinent reason. The latter characteristic is important to river water and sewage disposal as was evident when replacement of soap by syndets became more universal. Sewage plants began to display beds of foam at their effluent ends, some of them covering entire streams for several hundred feet. Then too, foaming of drinking water from ground-water supplies where septic fields were in use was a further indication of pollution, both by syndets and viable organisms. Presence of foam was a sure indication that biodegradation had not occurred.

It became evident that something must be done about "hard" syndets. "Hard" because they were not straight-chain compounds and resisted sewage treatment and river-water degradation. "Soft" or easily degraded surfactants were to become a necessity. The surfactant chosen for study because of its wide usage was alkyl-benzene sulphonate. The alkyl portion of the compound was the focus of attention: its source may be chlorinated kerosene or a suitable olefin. Extensive research demonstrated that a tertiary carbon atom in the alkyl chain was the point at which biodegradation ceased. The organisms were unable to utilise the chain any further when such an atom was encountered, whereas with straight-chain compounds such as soap no difficulty was found. The answer to this was a straight-chain kerosene or olefin, and suitable synthesis control to prevent chain rearrangement, or more suitable use of a preferred straight-chain olefin in the synthesis.

The straight-chain product finally used was termed LAS, linear alkylbenzene sulphonate (as contrasted with branched chain ABS). The manufacturers by a mass effort were successful, at considerable cost, in meeting the 1965 target date. Before LAS, sewage treatment plants were successful in degrading ABS only to 60 to 70 per cent; afterwards a 90 per cent effectiveness was achieved. Today any surfactant to be used successfully must be biodegradable, as shown in Fig. 13.1. A third generation of agents is beginning to draw attention, and includes nonionics and nonionic sulphates.

Detergent Additives

Phosphorus compounds

Immediately following the development of biodegradable syndets to control pollution by surfactants, the overall pollution of rivers and lakes became the next focus of attention. Lakes in particular, as recipients of inflow from agricultural lands and either the effluent from inadequate sewage treatment plants or raw sewage were becoming so tainted that in effect it might be said the lakes were dying. The apparent effect was a change in the aquatic life with more frequent and extensive algal blooms. These blooms reduced the oxygen content of the water, killing much of the aquatic life, and in effect stagnating the water body by depositing organic debris which consumed still more oxygen. Reasons for this change were attributed principally to high phosphorus contributions from run-off and sewage effluent, and secondarily to nitrogen and carbon dioxide. Ordinarily this process, called eutrophication, would take thousands of years, but it has been greatly accelerated by overnourishment.

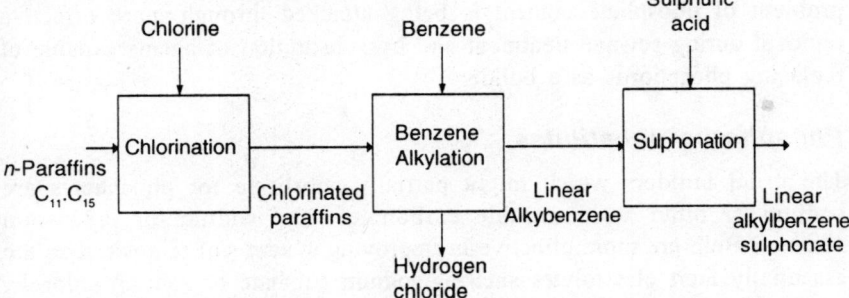

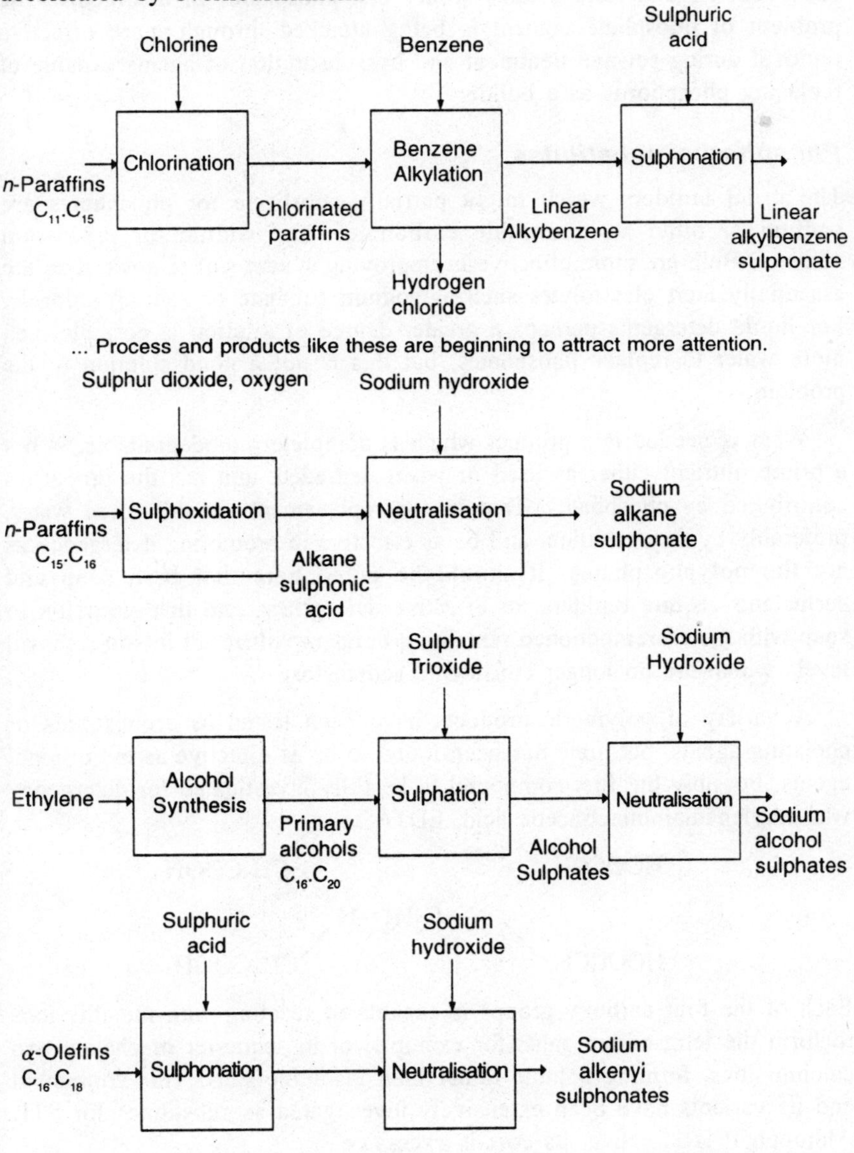

... Process and products like these are beginning to attract more attention.

Fig. 13.1. Flow diagrams for manufacture of some biodegradable surface-active agents.

Since the phosphorus content of sewage is not greatly reduced by treatment, syndets which contain from 5 to 50 per cent phosphates became the immediate target for control. The pros and cons continue, while congressional and state actions would ban phosphates in detergents. The problem of phosphate control is being attacked through more effective removal during sewage treatment and by substitution of agents capable of replacing phosphorus as a builder.

Phosphorus substitutes

The usual builders which might partially substitute for phosphates are sodium or other water-soluble carbonates, and sodium or potassium silicates. Both are more effective in improving syndet soil removal than are essentially inert electrolytes such as sodium sulphate or sodium chloride. For liquid detergents perhaps a greater degree of dilution is possible, i.e., more water to replace phosphates, but this is not a good solution to the problem.

What is needed is a product which is completely biodegradable, is not a prime nutrient either as used or when degraded, and has the properties contributed by phosphates. That is, the replacement should soften water, preferably by sequestration, and be as effective in promoting detergency as are the polyphosphates. It should be noted here that both soap and surfactants require builders for effective detergency, and that reversion to soap with the aforementioned remaining builders will result in soil removal levels which are no longer considered satisfactory.

A variety of polymeric products have been tested as sequestrants or chelating agents, but none has been found to be as effective as the original agents. Possibly the first compound to be fully investigated for detergency was ethylenediaminetetracetic acid, EDTA :

$$\text{HOOCCH}_2 \diagdown \qquad\qquad \diagup \text{CH}_2\text{COOH}$$
$$\qquad\qquad \diagup\text{N}-\text{C}_2\text{H}_4-\text{N}\diagdown$$
$$\text{HOOCCH}_2 \diagup \qquad\qquad \diagdown \text{CH}_2\text{COOH}$$

Each of the four carboxy groups is capable of reacting with metallic ions to form the tetrasodium salts, for example, or to sequester or chelate two calcium ions, forming soluble rather than insoluble salts. This compound and its variants have been extensively investigated as substitutes for STP. Although it is effective, its cost is excessive.

Chelating agents, organic phosphonates, polyelectrolytes, and sequestering agents were reviewed in some detail by McCutcheon.

Another compound which received attention as a substitute for phosphates is also a chelating or sequestering agent. Nitrolotriacetate, NTA, when used in adequate amounts proved essentially as effective as EDTA but at lower cost.

$$N \begin{array}{l} \diagup CH_2COOH \\ - CH_2COOH \\ \diagdown CH_2COOH \end{array}$$

NTA was the compound which the detergent manufacturers started to use to replace STP. Its use was terminated, however, when it was disclosed that birth defects in rats and mice occurred when they ingested NTA complexed cadmium or methyl mercury, chemicals which are potentially present in ground waters. This left no suitable candidate for replacement, and for the time being the only replacements are the conventional alkaline builders and water.

LAUNDRY WASTES

One of the most common liquid wastes discharged is laundry waste. As a rule the waste is discharged into municipal sewers. Larger volumes of soapy wastes from power laundries, scouring establishments and plants employing soap solutions may require pre-treatment before discharge into municipal sewers or must be treated separately.

Soapy solutions have high oxygen demands; pure soaps require almost twice their weight in oxygen for satisfaction of the 5-day BOD and commercial varieties of soap about 1½ times.

Table 13.1. Average analyses of commercial and domestic laundry wastes.

	Commercial	Domestic
pH	10.3	8.1
Total alkalinity (ppm)	511	678
Total solids (ppm)	2,114	3,314
Volatile solids (ppm)	1,538	2,515
BOD, 5-day (ppm)	1,860	3,813
Oxygen consumed (ppm)	868	1,045
Grease (ppm)	554	1,406

The volume of laundry waste normally is 5 to 10 per cent of the average daily flow of sewage, but the waste is from 10 to 20 times stronger.

analyses of domestic and commercial laundry waste are shown in Table 13.1, but laundry practice varies and both stronger and weaker wastes are frequently observed.

Separate treatments for clarification by means of chemicals have included lime, alumino-ferric, calcium chloride, iron salts, alum and salt (sea) water. Treatment with alum and acid has been claimed to be less costly than ferric sulphate-lime or acid-lime treatment, but purification results comparable to acid-alum treatment with ferric sulphate and lime at about equal cost have been reported. It appears that adjustment of pH to 6.4 to 6.6 with acid prior to coagulant addition produces optimum purification at lowest cost. The use of carbon dioxide to lower alkalinity and the pH value prior to coagulation has produced excellent results.

The oxidation of laundry waste can be successfully accomplished on trickling filters, handling rates of application of 1.5 mgad despite grease loadings up to 490 pounds per acre foot per day, pH values of 11.0 and alkalinities of 1,000 ppm in the raw waste.

Activated sludge oxidation of laundry waste is feasible, but longer periods of aeration than commonly used for sewage are required.

One difficulty seldom mentioned, but frequently encountered in treating the waste with chemicals, is settling. The release of carbon dioxide on treatment causes partial flotation of the sludge. Experiments on treatment with flotation by vacuation produced better results than settling. Treatment of waste waters after equalisation reduces the chemical demand and aids in clarification. Pre-aeration or flocculation prior to settling aids removal of impurities. Laundry waste treatment can be done as under :

(1) To remove about 75 per cent of oxygen consumed, solids, and grease, laundry waste can be treated most economically by acidification with H_2SO_4, CO_2, or SO_2, followed by coagulation with alum or ferric sulphate. It can also be coagulated partly and completely by many salts and acids and by lime, but in most cases such treatment is too costly.

(2) Laundry waste can be purified to a high degree by means of conventional and probably high-rate trickling filter, or by the activated sludge process if long aeration periods are used.

(3) The sludge obtained can be dried on sand beds directly or very probably digested anaerobically, or filter pressed. Final disposition of undigested sludge can be accomplished by a soap recovery process or by incineration.

(4) Waste treated by chemical coagulation can be further purified by passage through a biological filter or by the activated sludge process.

(5) Sewage containing any percentage of laundry waste can be treated on a biological filter of adequate capacity and correctly designed for this purpose.

(6) The activated sludge process employing normal aeration periods can handle sewage containing laundry waste in concentrations approaching 20 per cent.

ENVIRONMENTAL PROBLEMS

General

As in other industries, some problems of an environmental nature can arise in, and in the environs of a factory, due to noise, gaseous and liquid effluents, and solid wastes. This section is concerned, however, with two problems—lack of adequate biodegradability and eutrophication, which can occur when waste waters containing certain detergents are discharged from the homes and factories in which they have been used. The two problems are quite distinct.

SYNTHETIC SURFACTANT WASTES

In the past, the increased use of synthetic surfactants, especially detergents reached to the point of creating problems in municipal sewage plants due to excessive foaming and inability to reduce the organic content of the sewage effluent in the developed countries. These compounds which are not degradable are called bio-hard. The need for biodegradable surfactants has become very prominent due to the emerging lobbies of environmental activists. The biodegradable types can be oxidised to simple end-products such as methane is known as biologically soft syndets. This has presented the industry with both challenges and opportunities. The molecular structure of chains is important in determining whether a compound is biologically soft or not. As the straight chain-normal paraffin structure gives a soft compound and a branched or iso-paraffin structure resists biodegradation, the biological-persistent branched alkyl benzene chains are being replaced by linear chains made by new technologies.

Biodegradability

When waste soapy water is discharged into a sewage system, or directly into a river or lake, no real problems normally arise. Excess soap is precipitated by the hardness in the receiving water, and it is also readily broken down by micro-organisms such as those which play a major role in sewage treatment. Soap is, therefore, described as biodegradable. This is not the case with certain non-soapy surfactants. Problems arose with foam on small rivers due to the discharge of waste liquors from textile factories. NSD products had serious adverse effects on sewage treatment plants and on the rivers, etc. into which the effluents, treated or untreated, were discharged. Even water from wells near places where these effluents were discharged was sometimes found to foam. At that time the trouble was traced to alkyl benzene sulphonates with highly branched alkyl chains, in particular those derived from propylene tetramer (the so-called PT alkyl benzenes, or alkylates). These surfactants do not break down sufficiently readily in sewage treatment plants and they are described as biologically 'hard', or as not biodegradable. The micro-organisms attack side chains two carbon atoms at a time and the process is stopped if a tertiary carbon (one to which only alkyl groups and no hydrogen atoms are attached) is reached. This led to the linear alkyl benzene sulphonates, often called LABS or LAS, in which the side chains are derived from suitable, substantially straight-chain, hydrocarbons. In fact, this was a reversion to improved forms of the 'keryl' benzenes which were used before the PT type was developed and came to be the dominant anionic surfactant. LABS surfactants are accepted as adequately biodegradable, although they are not broken down as readily and completely as soaps and other surfactants derived from fats, or synthesized to contain a completely unbranched chain with an even number of carbon atoms and no benzene ring.

Nonionic surfactants can also lack adequate biodegradability, but problems of analysis of low concentrations in effluent waters make regulation more difficult than for anionics. The alkyl phenol ethylene oxide types, once widely used, are being, or have been, phased out by voluntary agreement. There is still some argument, but they are probably not sufficiently degradable to be acceptable. The octyl and nonyl side chains in these compounds are highly branched. Other nonionics based on straight-chain, or relatively straight-chain, alcohols are currently acceptable, but there is some uncertainty regarding the fate of the polygylcol chains. Surfactants with long polyethylene oxide, and/or polypropylene oxide, chains are particularly suspect.

The biodegradability of cationic surfactants has been little discussed, but it is probably not important, because of the mutual precipitation when most anionic and cationic surfactants are mixed.

Eutrophication

It is a natural process for lakes and ponds to silt up gradually. When little nutrient is present, and hence there is little plant life, the condition is called oligotrophic. When rivers and rains bring in nutrients, algae and other plants grow and the condition becomes eutrophic. 'Cultural' eutrophication can occur if human activities, usually the discharge of large quantities of sewage into relatively small lakes, or into lakes from which little, or no, water is discharged to the sea, result in the presence in the water of excessive quantities of nutrients. This abundance of nutrients leads to a massive explosion of the population of algae, with production of oxygen. At the end of the growing season most of the algae die and bacteria feed on the dead plant material. This process requires oxygen, that is the destruction of the dead algae imposes a biological oxygen demand (BOD), similar to that caused by the discharge of milk, or any other destructible organic matter, including normal sewage. In extreme cases, this can eventually turn the lake waters from an aerobic to an anaerobic condition (one lacking dissolved oxygen), which kills fish and other animals. Under anaerobic conditions, anaerobic bacteria flourish, and may produce toxins and bad odours.

LINEAR ALKYL BENZENE (LAB)

Environmental Issues

LAB sulphonate (LAS) emerged in detergent market as an alternative to sodium dodecyl benzene sulphonate (DDBS) which has been phased out due to problems of persistent foam in rivers and sewages. In view of growing concern about impact of LAS on environment, several studies have been undertaken on biodegradability and possible hazards of this product. Some inferences drawn are as follows :

(i)   LAS is mineralised in the environment to form CO_2, water and inorganic sulphates.

(ii) Its biodegradability as per OECD method is over 98%.

(iii) The half life of the product in natural water is in the range of 4 to 9 days.

Though LAS is biodegradable, efforts are being made to develop detergents with more positive biodegradability profile. These include n-paraffin sulphonates, alpha-sulphomethyl ester of detergent alcohols and alkyl polyglycosides.

METHODS OF SAFETY EVALUATION OF SYNTHETIC DETERGENTS

Detergents are used almost universally on regular basis. Their safety aspect has been the subject of numerous investigations in the past. The committee responsible for the formulation of this standard felt the need for specifying test methods which can evaluate the safety aspects of synthetic detergents used by the consumers. These test methods have been specified in two parts of this standard. Part 1 of this standard prescribes the test methods for irritant potential of synthetic detergents. These tests are designed to investigate the irritancy potential of detergent products which, during use, involve repeated immersion of hands/forearms and skin contact with the detergent for extended periods of time. The test for skin sensitisation is being covered in this second part.

Detergent products which come in contact with skin should be free from skin sensitising/skin allergy producing components. Such components even when present in minute levels (ppm — part per million concentration) in the formulation, would cause itching, redness, swelling and blister formation on the skin. The onset of such episodes is usually a delayed response from the time of first contact with the product which makes it difficult to identify the incriminating product/ingredient, even by an experienced medical expert. The detection involves cumbersome procedure of patch testing on skin with all suspect materials on people who are susceptible to such skin allergy. The intractable nature of this skin condition may cause unnecessary hardships to the consumers. Hence, it is essential that manufacturers of detergent products make sure that their products are free from such hazard, by subjecting their products/ingredients to a predictive skin sensitisation test using a sensitive animal model.

The allergic skin reaction described above is not caused by toxic or irritant properties of the substance (which need to be tested separately) but is due to the increased sensitivity and harmful immunological response to these substances caused by repeated contact. Identification of such allergenic ingredients/products is possible through the guinea pig maximisation test (GPMT). This standard describes the principle and method of test.

Presently, the facilities for these test methods are limited. However, there is scope and imperative need for rapidly expanding and providing these facilities.

These tests are designed to investigate the irritancy potential of detergent products which, during use, involve repeated immersion of hands/ forearms and skin contact with the detergent for extended periods of time. The test for allergenicity (skin sensitisation) will be covered in the subsequent part of the standard, when the test procedure and other details have been standardised. Till such time, if it is reported that a particular detergent is producing very frequent allergic reactions, the matter should be referred to a competent dermatological laboratory.

Presently, the facilities for these test methods are limited. These facilities are available at organisations like the Haffkins Institute, Mumbai; Industrial Toxicological Research Centre, Lucknow; and the All India Institute of Medical Sciences, New Delhi. However, more and more organisations are acquiring these facilities.

This standard (Part 1) covers skin irritation test and repeated insult test for determination of irritant potential of synthetic detergents. For the purpose of this standard the following definitions shall apply (also Ref. BIS-7597-1974).

Allergic reactions — Symptoms/signs caused by exposure of an allergic individual to the corresponding substance.

Dermatitis— Inflammation of the skin.

Erythema — Redness of the skin due to dilatation of the blood vessels.

Hypersensitivity — A state where an individual is capable of developing an allergic reaction due to some external agent.

Occlusion area — An area of skin which is cut off from the environment.

Repeated insult irritant — An agent which causes an irritant reaction only after repeated applications to the surface of the skin.

Sensitisation — A process in which an individual develops the capability of reacting in an allergic (abnormal) manner to a particular substance.

Skin Irritation Test

The test shall be carried out with synthetic detergent powders and bars (concentration 1.0 percent m/m). The duration of test shall be 1 hour daily for four days. The test shall be carried out on six Albino guinea pigs of body mass 250–275 g. These guinea pigs shall be on a standard diet and shall be housed in individual cages. The whole trunk of the guinea pigs shall be subjected to test.

Procedure

After clipping the hair from the whole trunk, the animal is immersed up to the neck in 1.0 per cent (m/m) of fabric washing powder bar solution (37°C) taken in a 500 ml beaker. The animal is restrained by the help of fibre board piece enveloped in a polyethylene sheet with a central opening to hold the head up comfortably. After the requisite immersion period, the animals are washed with water and dried with a towel followed by hair blower (air dryer). This procedure is repeated on 4 successive days.

Body mass of the animals is recorded daily before immersion during the four-day immersion period. Mortality, if any, is also noted. The animals are also observed macroscopically during this period. Reactions are scored on the next day after the fourth immersion on the basis of intensity and extent (overall area affected) per animal. The responses in general are in the nature of erythema, oedema, scaliness, cracking, leathery skin and epidermal breakdown. The scoring pattern would be: No reaction = 0, very slight = 0.25, slight = 0.5, distinct = 1.0, well developed = 2.0, severe = 3.0. Mean scores of less than 1.5 would be considered to be an acceptable level of irritancy and values higher than that would be considered to have an unacceptable level of irritancy potential.

Microscopic assessment of the skin may also be carried out by collecting biopsy samples from the skin for further confirmation of macroscopic evaluation.

Repeated insult test

Ten human volunteers are required to dip their hands in 1 per cent mass by mass solution of detergent formulation for half an hour daily for five days. The temperature of the detergent solution shall be room temperature (27 ± 2°C). The hands of the human volunteers shall be examined after dipping for half-an-hour for any adverse reaction. Itching, cracking, redness and swelling shall be considered as adverse reactions and samples in such cases shall be declared as failing in this test. Mild scaling without accompanying symptoms shall not be considered as a failure in this test.

Method of Test for Skin Sensitisation Potential of Synthetic Detergents (Guinea Pig Maximisation Test)

This standard prescribes the guinea pig maximisation test for determination of the skin sensitisation potential of synthetic detergents/ingredients.

If the product fails in the skin irritation test : [see IS 11601 (Part 1) 1986 'Methods of safety evaluation of synthetic detergents : Part I- Method of test for irritant potential of synthetic detergents'], it is not necessary to carry out this test.

This test needs to be carried out whenever an ingredient or formulation undergoes a change. Ingredients incorporated need to be evaluated, if their source of supply, processing methods or chemical composition undergo a change.

Principle

Guinea pigs are treated with detergent products/ingredients intradermally and topically in such a way as to induce a state of sensitisation to any allergenic material present. In order to enhance the sensitising process Freund's Complete Adjuvant (FCA) is also injected along with the substances. The sensitising component termed 'hapten' forms a conjugate with the skin protein which triggers T-lymphocytes of the area possibly through the mediation of Langerhans cells. The triggered lympocytes enter the draining lymph nodes and proliferate and get transformed into memory cells and effector cells. After a fortnight when the animals are topically challenged on the flank region with the same detergent product/ingredient, its sensitising potential is revealed by an inflammatory response (redness, swelling) at the challenged site, while no such response occurs if the substance is not a sensitiser. The responses of the experimental animals are compared with those obtained in control animals.

Outline of the Method

Initially using four guinea pigs a slightly irritant concentration of the detergent product/ingredient is determined by intradermal administration at concentrations ranging from 0.1 to 10 per cent (say 0.1, 0.5, 1, 2, 5, 10, etc.).

Another four guinea pigs are similarly tested by using topical application to identify concentrations which are slightly irritant and which are non-irritant.

Main Test

The main test comprises three phases, namely :

(a) Induction,

(b) Boosting, and

(c) Challenge.

Induction

Twenty-eight guinea pigs are used. Ten termed 'test animals' are subjected to intradermal injections of slightly irritant concentrations (as determined in above) of the test substance and Freund's Complete Adjuvant (FCA) for enhancing immunological reactivity of the animals. Another 10 animals, termed positive controls are treated with a well known skin sensitiser, for example, dinitrochlorobenzene (DNCB) or dinitrofluorobenzene (DNFB) or potassium dichromate (PD) or picric acid (PA) in a manner similar to the experimental animals. Four animals, called 'treated controls' are given injection of FCA only, that is, the test substance is not injected. Four animals, called 'untreated controls' are not given any treatment.

Boosting

One week after the induction treatment, the area of injection site is allowed to have continuous 48 hours contact with the test substance in the case of test animals and with the DNCB (or DNFB or PD or PA) in the case of positive controls at a slightly irritant concentration (tropical) as determined above.

Challenge

Two weeks after the boosting, that is, three weeks from the initial intradermal treatment, all the test and positive control animals are challenged topically, with the test substance and DNCB or PD or PA, as the case may be, on the flanks by the highest non-irritant concentration (as determined above) allowing a 24-hour continuous contact of substance by occluded patch.

The skin reaction at the challenge sites are examined 24 and 48 hours after removal of patch. The eight control animals will show no reaction or very slight reaction. Any animal showing a reaction greater than that observed in the treated and untreated control animals is deemed to show a positive reaction. If there is no reaction in the positive control group the test needs to be repeated.

Details of test method

General

Albino guinea pigs of either sex bred from a stock which is disease-free as well as showing positive sensitisation response with any of the well known skin sensitisers, such as dinitrochlorobenzene (DNCB) of dinitrofluorobenzene (DNFB), should be used. The latter procedure is necessary so as not to miss any weak contact sensitisers because of lack of sensitiveness of the animals.

Housing and Feeding

Guinea pigs shall be housed individually in airy clean cages having approximately 30 cm length, 25 cm breadth and 20 cm height. The cages shall be located in well ventilated room with controlled temperature (22°C + 2), relative humidity (55 + 5 per cent) and lighting.

The animals shall be well looked after and given nutritious feed and water *ad libitum*. Good house-keeping and normal hygienic practices shall be maintained in the animal house.

The method consists of 2 parts, namely :

(a) preliminary irritancy test, and

(b) main sensitisation test.

Preliminary Irritancy Test

Intradermal irritancy test. This is carried out to find out the suitable intradermal injection concentration which should be used for sensitising the animals. Four guinea pigs of the same sex and weighing around 300 g each are injected intradermally on the flank clipped and shaved of hair, with 0.1 ml each of 0.25, 0.5 and 1 per cent of detergent product in distilled water using a sterilised tuberculin or disposable syringe fixed with 26 gauge needle. The injections are made at least one centimetre apart. The animals are then put in individual cages and fed *ad libitum*. 24 hours later, reactions are examined for size in millimetre (length and breadth) or erythema (redness) and oedema (swelling). The concentration which produces a slightly irritant reaction (7 mm × 7 mm erythema with oedema–mean from the 4 animals) is selected as the intradermal injection concentration. Usually for detergent products a concentration around 0.5 per cent shall be found suitable for the main sensitisation procedure. For the different ingredients as well as for the positive control substance one has to try different concentrations, say 0.25, 0.5, 1, 2, 5, 10, or more in the above manner to arrive at the suitable slightly irritant concentration. For water insoluble material, either refined groundnut oil or pharmaceutical grade paraffin oil or propylene glycol may be used for solubilising or making a fine dispersion of the material.

Topical irritancy test. Eight millimetre diameter chromatography paper discs are saturated with a range of concentrations, namely, 1, 5, 10 and 20 per cent detergent product/ingredient (as well as positive control substance) in distilled water/suitable solvents and held in place under 10 mm diameter aluminium discs. These are placed at least one cm apart on the clipped and shaved flank of four guinea pigs (weighing about 400 g each). The patch test discs containing the paper are held in position with surgical adhesive plaster tapes and cloth bandage. The animals are then put in individual cages, fed and looked after. Patches are removed 24 hours after application. The treatment sites are examined 24 hours and 48 hours after the removal of the patches. The resulting reactions are scored for irritation on a 0–3 scale. For boosting the sensitisation response through topical application a concentration giving a slightly irritant (score 1) reaction is selected. The highest concentration which cause no visible reaction (score 0) is selected for challenge treatment (final treatment to determine whether the product/ingredient is a sensitiser). For detergent products these concentrations would be around 10 and 1 per cent respectively for boosting and topical challenge. For different ingredients a range of concentrations is to be tried in the above mentioned manner using suitable solvents.

The main sensitisation test

20 guinea pigs weighing about 300 g each are selected from the stock.

Sensitisation treatment — induction

This consists of 2 stages, intradermal injection followed one week later by topical application.

(a) *Intradermal injection* — The hair is clipped from a 2 × 4 cm area of skin on the dorsal shoulder region and 3 pairs of intradermal injections are made within the clipped area using a sterilised tuberculin or disposable syringe fixed with a 26 gauge needle as follows :

    (i)   Two 0.1 ml injections of 50 per cent FCA in distilled water in sites marked '1';

    (ii) Two 0.1 ml injections in sites marked '2' of detergent product/ingredient/positive control substance at the concentration selected for sensitisation from the intradermal irritancy test, usually about 0.5 per cent concentration for detergent product); and

    (iii) Two 0.1 ml injections in sites marked '3' of detergent product/ingredient/positive control substance in distilled water/suitable solvent mixed with 1:1 FCA, such that the final concentration of test substance injected is the same as that in (ii) above.

    Injections (i) and (ii) are given close to each other while injection (iii) is given slightly away from (ii).

(b) *Topical treatment* — boosting—One week after the injection, the same 20 mm × 40 mm area is clipped and shaved. A 20 mm × 40 mm chromatography paper is saturated with detergent product at selected concentration (usually 10 per cent) or with the test ingredient/positive control substance as determined and placed over the shaved site.

This is covered by 40 mm × 60 mm piece of thin polyethylene. The paper saturated with test substance covered by polyethylene is held in place for 48 hours by surgical adhesive plaster tape and cloth bandage.

Topical challenge

Fourteen days after the boosting treatment, the guinea pigs which are treated with test material and positive control substances are respectively, challenged (final treatment to determined the sensitisation) with test material and positive control chemical on the clipped and shaved flank by occluded patch. For each animal, an 8 mm diameter chromatography paper disc is saturated with the detergent product/ingredient/positive control chemical at the selected challenge concentration (highest topically non-irritant concentration) and placed in a 10 mm diameter aluminium patch test disc. This is then applied on one flank and held in position using surgical adhesive plaster tape and cloth bandage for 24 hours. The challenged site is examined for inflammatory response — redness (erythema), swelling (oedema) 24 and 48 hours after removal of the patch using the scoring system. One week after the first challenge is made on the opposite flank exactly in the same manner as the first. A third challenge may be made one week later on the opposite flank.

Treated and Untreated controls

Treated controls. At the same time as the main test animals are selected, four guinea pigs of the same weight range are selected as treated controls. They are given mock sensitisation treatment at the same time and in the same way as for the main test animals except that the test substance is omitted from the intradermal injection induction and topical application boosting. They are treated exactly the same way as the test animal at every challenge with the test material/positive control chemical.

Untreated controls

Four animals which did not receive any treatment previously and are of the same weight range as the main test animals are challenged exactly in the same manner as the test and 'treated controls' animals.

Identification of sensitisation response

Skin reactions resulting from treatment can be scored using the following scale :

0 = no reaction

0.5 = very faint erythema,

1 = faint erythema,

2 = moderate erythema, and

3 = marked erythema with or without oedema.

Thus a reaction in a test animal is considered to be positive response if it is significantly greater than the response on treated and untreated control animals. If the positive control animals do not show any expected response, then the whole test needs to be repeated. When there is no reaction on treated and untreated control animals, a reaction score of 1 or more in any of the test animals is considered to be a positive sensitisation response.

Detergent products/ingredients producing a positive sensitisation response pose a risk to the consumer.

Chapter 14

Testing of Soaps and Detergents

INTRODUCTION

For a practical and realistic evaluation of quality of soap and detergents, performance test constitute the ideal yardstick. Thus the need of the testing is to prove performance that is material has to pass through various specifications in a process. Physico-chemical analysis alone is not adequate. This chapter highlights current practices followed during testing from time to time to improve their precision and accuracy.

TEST METHODS

Few methods for measurement of physico-chemical properties of surface active agents have been adopted by various industries. Since there are upwards of a thousand available surface active agents whose claims of activity widely overlap, for choice of the best agent for a given purpose it is necessary that all their properties be recognised and selection made accordingly. Certainly if an agent is not stable to hard water is resembles soap, and must possess other properties sufficiently different from soap to make it competitive. Screening tests can save much effort and can increase the general knowledge of utility when properly used. The tests described have been used successfully for years and cover the general properties of synthetics which make them industrially useful.

An over simplified statement of the detergent process indicates that it is the removal of soil (matter out of place) from a surface in some medium, generally through the application of mechanical force, and most frequently in the presence of a substance which may lower the adherence of the soil to the surface, then maintain the soil in suspension so that it can be rinsed or flushed away. When it is considered that the medium in which the cleansing can take place may be aqueous, nonaqueous, or mixed, that the force applied may be of almost any kind, and that the surfaces and soils to be separated are legion, then the real complexity of the detergency picture becomes apparent. Since some surfaces may be satisfactorily cleaned in one or more media using a wide variety of cleaning agents on any of them, a choice must be made which will prove economically satisfactory. When all these factors are recognised it may not appear so unusual that no accord has been reached as to recognised testing methods.

The most effective method, and the most reasonable one, is to perform the laboratory screening tests, meanwhile building up a fund of data to be correlated with practice, progressing as soon as feasible to a semi- or full-scale trial under practical conditions of operation. Regardless of verbal expression to the contrary, this is the procedure most generally followed.

STATISTICAL PRESENTATION OF DATA

Knowledge of the reliability of results obtained in any test is necessary. Frequently this is accomplished simply by comparison with standard samples. Certainly the use of samples of known characteristics for comparison purposes is highly desirable but this method lacks precision. In the presentation of data the ASTM Manual on Quality Control of Materials is a useful reference.

In the operation of a test method it should be realised that the error of the average obtained is inversely proportional to the square root of the number of observations made. This simply means that in order to improve upon the information obtained, the method must either be improved to reduce variables which are recognised as resulting in lack of control, or the number of replicate samples must be increased. However this can be carried to extremes when it is recalled that the average of four observations reduces the error of the average only to half that of the single observation. Likewise that the average of 16 samples or observations has 1/4 the error of the single observation. It quickly becomes apparent that for minimisation of effort that the test method itself should be more rigidly controlled if this is possible.

The minimum data which should be presented in recording experimental results is the mean ($\overline{X}$), the standard deviation (σ), the number of observations (n) and sometimes the range. From these data it is a relatively simple matter to compare results and certainly offers much over no statistical data of any kind, as a means for ascertaining the reliability of the experimental work.

Having developed the experimental data, the next question that will arise is whether samples which have been tested actually differ. It is suggested that a rough approximation of the differences between two mean values can be obtained by applying the twice standard error criterion to the data. In other words, if the difference between two mean values is equal to, or greater than twice the larger standard error, the chances are roughly 95 out of 100 that the two samples are different. This is a useful and reliable method when more than six replicate samples have been used in the observation of the mean value. However, for less than this number of replications the approximation becomes even more rough. The significance level between the difference of two mean values can be ascertained more accurately by several methods (Ref. Table I in the ASTM Manual) which presents the factors for calculating 90, 95, and 99% confidence limits for averages which includes the number of observations from 4 up to 25 and greater. These values were computed from Fisher's table of "t". Application of this method will indicate for these several levels of confidence whether the differences between observed mean values are real.

Still a further method is use of Student's t for the comparison of two means. The calculations used are given below :

$$t = \frac{\overline{X} - \overline{X}'}{\sigma} \sqrt{\frac{n_1 \times n_2}{n_1 + n_2}}$$

where

$$\sigma^2 = \frac{\left[\sum(X'^2) - \frac{(\sum(X'))^2}{n_2} + \sum(X^2) - \frac{(\sum(X))^2}{n_1} \right]}{n_1 + n_2 - 2}$$

Although these calculations may look rather formidable they actually are relatively simple and straight forward. Having calculated σ^2 it is possible to enter student's table of t for the number of degrees of freedom observed, to determine the significance of the difference between two means.

In analytical work it is generally assumed that the reproducibility of data should lie within the 99.25% range. In other words, the chances of being in error should be extremely small. However, in tests such as those described here, where control of many factors is very difficult, the significance level may be considerably broader. Whether this is a 95 or 90% level of significance depends upon the number of samples used, the reproducibility of the method, and the judgment of the operator. Obviously the significance level should be as high as possible, without, as previously suggested, increasing the number of observations too extensively.

Radioactive Tracer Technic

As a new medium, this technic offers possibilities for the investigation of the quantitative characteristics of a system not readily susceptible to other methods. It also makes possible the estimation of extremely minute amounts of materials. Other uses for the radioactive tracer technic lie in the calibration of other methods of examination and in some instances can be used in certain industrial applications for estimation of quality or quantity.

There are two general classifications in the detergent industry to which this method has been applied. One is in estimating the surface active characteristics of agents and the other has been applied to metal cleaning processes. In the former investigation, synthetic agents were synthesized containing a radioactive isotope and the orientation of these at solution-air interfaces measured. The adsorption or relative solubility of ordinarily insoluble salts of surface active agents was measured using a radioactive calcium salt as the insolubilising agent. In the metal cleaning application tagged hydrophobic materials such as stearic acid were used in the soiling medium and the ease of removal from metal surfaces was measured.

Health hazards in the use of radioactive materials must always be considered and it was this that ruled against the use of graphitised carbon 14 as a constituent of a graphite soil for fabric.

In general the radioactive tracer technique would appear to be a most useful tool where the sensitivity of a method is to be determined or for which the usual methods of investigation fail. It currently appears that for the detergent industry, the radioactive tracer technic will find its greatest use in research rather than in regular control operations.

Water Hardness

The water hardness values chosen were for both soft and hard areas in order that detergents might be evaluated for use under the extremes of operating conditions ordinarily encountered. The water itself is comprised of Ca and Mg salts in the ratio of 60 to 40. There are those who object correctly that many synthetic detergents are more affected by the Ca in the water than by the Mg ions, hence that the ratio of Ca in the water should be increased. Others will criticise the range of water hardnesses chosen. The obvious answer to both criticisms is that a really effective agent will be only mildly affected by either type of water hardness, and that at any rate a full investigation of the characteristics of a material will include concentration-water hardness curves for a range up to 1,000 ppm. The present method has successfully served the purpose of screening potential detergents :

2000 ppm. water

Ratio of Ca: Mg of 60:40 is 29.6 g. $MgSO_4 \cdot 7H_2O$, 26.4 g. $CaCl_2.2H_2O$

Dissolve separately in distilled water and make up to 15 litres. Dilution to hardness desired :

$$\frac{No.\, of\, ml.\, wanted \times hardness}{2000} = (A)\, ml.\, of\, 2000\, ppm.\, water\, to\, use$$

Make (A) up to volume wanted with distilled water.

SCREENING TESTS

With so many available agents it is desirable to use screening tests to minimise the effort involved in running laborious detergency evaluation, since there are frequently other factors involved which may aid in controlling their application. Frequently the conditions which either precede or follow a deterging or cleansing operation, or the chemicals with which they may come into contact may control the type of agent which may be used.

It frequently is necessary that the agent to be used in a plant must not only be an effective detergent, but must perform several other additional functions. For example, it may be necessary that the detergent be an excellent wetting agent, must have high effectiveness for promoting emulsion formation, or if not completely rinsed from the fabric may beneficially aid in rewetting that fabric satisfactorily.

The tests which follow are designed to differentiate between the characteristics of surface active agents so that suitable selection may be made. These tests can frequently be initiated prior to wash testing to save much time and effort. The discussion which accompanies each test will further suggest some of the applications for these or other similar materials and reasons for choice.

Acid Stability

A surface-active agent chemically unaffected by acid has many potential uses. Soap is not useful under these conditions since it is decomposed, forming inactive, and many times obnoxious, oily or waxy fatty acids. In the textile industry, the acid-stable agents are used among other purposes for acid-dyeing operations and in carbonising baths. The metal trades utilise them in pickling and electroplating baths, and they may even be used in the acidisation of oil wells.

Chemical stability may vary markedly among the available agents. Esters may be stable to slight acid concentrations for some time, but increase in acid concentration, temperature, or time of exposure may induce decomposition. Sulphonic acid types of agents generally are quite stable to acids, but sulphonic acid separation may be mistaken for decomposition.

One possible method for evaluating these agents is to add a known excess of acid to a given solution of the agent, boil under reflux, then back titrate, thus quantitatively measuring any change. This method, while quantitative, does not provide acid concentrations of the degree many times encountered in industry. Consequently the qualitative method later described has given adequate results.

The other agents may therefore be used in acid media of considerable strength, but a further elimination would be required by evaluation under conditions approaching actual use as closely as possible. Combinations of tests such as acid stability, surface-tension measurements in acid media, and stability to metallic ions under acid conditions may prove adequate means for comparison of several otherwise almost identical materials.

Acid stability test method

Apparatus : 250 ml wide-mouth extraction flasks, Allihn condensers, four-bulb, Porous plate fragments.

Procedure : Transfer 100 ml of a solution containing 1.0 g. of agent to a 250 ml. wide-mouth extraction flask and add a piece of porous plate to prevent bumping. Attach condenser.

Bring to the boil and record appearance.

Add 0.1 per cent (1 ml. of 10 per cent) sulphuric acid and boil 15 min. Record appearance. Lack of stability is exhibited by turbidity, separation of an oil, and loss of lathering power.

If no change occurs or the change is minor, add acid to provide 1 per cent acidity (0.5 ml. of concentrated sulphuric acid) through the condenser. Boil for 15 min. and again record appearance.

If no change occurs, or the change is minor, add acid to provide 3 per cent acidity (1.0 ml. of concentrated sulphuric) and boil a further 15 min. period.

If the product is still stable add sufficient acid for 10 per cent concentration (6.5 ml. concentrated sulphuric). Again boil 15 min. If no change occurs record. this fact and discontinue test. At this acid concentration there may be a separation of oil which may perhaps be sulphonic acid insoluble at the conditions prevailing. Dilute with an equal volume of water and agitate vigorously to determine whether foaming power has been destroyed. If separated sulphonic acid has resulted this test indicates its stability but relative insolubility.

Record as "stable," "partially decomposed," or "decomposed" at the acid concentrations indicated.

Alkali Stability and Solubility in Caustic Soda Solutions

Although soap is chemically stable to alkali, it may be salted out and loses its effectiveness whenever a sufficiently high caustic alkali concentration is reached. Consequently, there is a definite place for an alkali-soluble surface-active agent. Perhaps the outstanding use for an agent of this type is in mercerisation, but surface-active agents are also used in the manufacture of viscose yarn. There are many other processes where lower alkalinity values are used in which fully soluble, active agents may be effective. It is obvious, however, that such agents must not be subject to decomposition, so that a suitable test may serve to screen out those agents most likely to successful application under the foregoing conditions of usage.

A second test of a more quantitative nature is used to determine the approximate solubility of the wetting agent in sodium hydroxide solutions, while saponification determinations may be made if the materials are of saponifiable nature and such quantitative data may prove necessary.

Few materials are fully soluble in 25 per cent caustic soda solutions, but in many cases such an agent is most desirable, so that an adequate screening test or tests may materially simplify practical testing.

The low surface tension of the more dilute solutions of surface-active agent in dilute alkali are to be expected, but what actually is desired is surface tension lowering even in concentrated alkali, as shown by Areskap 100. However, lowered effectiveness of even this compound may be expected at caustic soda concentrations in excess of 21.6 per cent, or Areskap concentrations greater than 0.8 per cent at 21.6 per cent NaOH solution.

Test method—alkali stability

Apparatus : 250 ml. wide-mouth extraction flasks, Allihn condensers, four-bulb, Porous plate fragments.

Procedure : Dissolve 1.0 g. of sample in 74 ml. of water. Add 25.0 g. of sodium hydroxide (c.p.) and a small piece of porous plate and swirl to dissolve the sodium hydroxide. Record appearance of system after sodium hydroxide is dissolved.

Boil under reflux for 15 min. and again record appearance which most generally will be a salting-out effect.

Cool the contents of the flask. Decant the solution through a fluted filter.

Transfer the insoluble agent from the paper to a beaker containing 25 ml. of distilled water. Add 3 drops of methyl orange indicator solution, stir and titrate with dilute acid to a faintly acid end point.

Stir thoroughly, heat to boil, permit to cool to room temperature, and note separation of oil. If oil separates under these conditions, and material has proven unstable to dilute acid, record the material as being "unstable." If the insoluble material completely dissolves in acidified solution and shows no separation, consider the product as "stable."

Test method—solubility

Apparatus : 1-litre Erlenmeyer flask, Dispensing burette.

Reagents : Solution sodium hydroxide, c.p., 30 g. per 100 g. of solution.

Procedure : Dissolve 1.0 g. agent in 9.0 g. distilled water, add 90 g. of 30 per cent sodium hydroxide solution and transfer to a 1-litre Erlenmeyer flask. Rinse the beaker in which the solution was prepared into the flask with small increments of distilled water. Continue the addition of distilled water either until the solution becomes clear and free from undissolved particles, or when 900 ml. of water have been added.

The surface tension of the resultant solution is measured at 25 ± 1 C.

Calculations :

$$\text{Percentage Agent} = \frac{1.0}{\text{wt. water (g.)} + 100 \text{ g.}} \times 100$$

$$\text{Percentage NaOH} = \frac{27.0}{\text{wt. water (g.)} + 100 \text{ g.}} \times 100$$

Stability toward Metallic Ions

Water solubility of the metallic salts of surface-active agents will greatly influence their ultimate application. First, these agents must be stable to the calcium and magnesium ions of hard water, since one of the factors unfavourable to soap is its ready formation of insoluble, sticky lime and magnesia soaps. In textile processing many operations require the use of acids, as for example, in dyeing, and metal from equipment may be dissolved, so that in this case the agent must not only be a good wetting agent but must be chemically stable both to acid and metallic salts incidentally formed. In carbonising wool a wetting agent may find considerable usage, but must be stable to acid or to a combination of acid and aluminium salts. For scouring pulled wool, a detergent is required which is stable to lime salts. Other agents may be useful in metal salt solutions as in electroplating, pickling or soldering fluxes. Although the foregoing is but a partial list of uses where metallic salts are encountered, it should be evident that knowledge of solubility of the metallic salts of the surface-active agent should be available.

There are two general methods for obtaining these solubility data: One is to prepare the salts by metathesis or other process and determine their solubility; the other is to add the metallic ion in question to a solution of the agent and determine the point at which a given maximum degree of turbidity is obtained, where lathering ceases, or both. There are advantages to both methods. The preparation of the pure compound permits accurate measurement of values, but experience with metallic soaps, for example, has shown that the solubility characteristics are entirely different after drying than they are in solution or dispersion, prior to drying. Furthermore, attempted preparation of a salt will present an entirely different problem from one in which the metallic ion represents a small portion of the system. Casual contamination in many cases will give vastly different results from an attempt to measure the solubility of the salt in water, since the pH value of the solution will have profound effect upon degree of solubility.

The specific metallic ions are chosen for this test for the following reasons:

(a) Calcium and magnesium because they are responsible for water hardness.

(b) Copper and nickel because of possible contamination from processing equipment.

(c) Aluminium and iron salts because they readily form insoluble salts and any agent stable to them should prove useful for many purposes.

(d) Barium was chosen because sulphates and even sulphonates are precipitated or become turbid in its presence.

(e) Heavy metal salts because they are represented by lead contamination.

(f) Zinc salts since they are somewhat amphoteric in nature and are present in certain processing baths.

(g) Other salts or varying concentrations at varying pH values may advantageously be evaluated, but their applications are more specific.

Test method

Sample : 1.0 per cent solution in distilled water.

Metallic Salt Solutions: 1.0 per cent solutions (dry weight basis) of the following c.p. salts are used :

Salt	Weight for 1 per cent (dry basis) g. per 100 ml.
$CaCl_2.2H_2O$	1.327
$MgSO_4.7H_2O$	2.048
$CuSO_4.5H_2O$	1.564
$Al_2(SO_4)_3.18H_2O$	1.948
$BaCl_2.2H_2O$	1.173
$Fe_2(SO_4)_3.xH_2O$	—
$Pb(NO_3)_2$	1.000
$ZnSO_4.7H_2O$	1.781
$Ni(NO_3)_2.6H_2O$	1.591

Dissolve the required amount of salt in water, cool if necessary, and make up to volume.

Equipment : 50 ml. Erlenmeyer flasks, Mohr measuring pipettes.

Procedure : Transfer 10 ml. volumes of 1 per cent surface-active agent solution to each of nine Erlenmeyer flasks.

Add solution of metallic salt dropwise from a Mohr measuring pipette to the solution of surface-active agent until the solution becomes turbid or precipitation is noted. Record the volume of metallic salt solution used, and heat the flask containing the mixture to the boil. If the solution becomes clear, add additional metallic salt: (*a*) Until 10 ml. of metallic salt solution has been added. (*b*) Terminate the test when the addition of metallic salt causes no further precipitation or when lather no longer is formed.

The total volume of metallic salt solution used to attain the end point is recorded.

Calculation : Metallic ion stability = volume metallic ion solution × 10.

Surface and Interfacial Tension Values and Spreading Coefficient

Potential detergent and wetting action may be ascertained by measurement of surface and interfacial tension. In general, the greater the ability to reduce surface tension and to lower the interfacial tension between oil and water, for example, the more useful the composition may be for producing desired effects under actual conditions of usage. These measurements have a practical application since dilution effect and quality may thus be indicated. The activity of certain baths, as for example electroplating baths, in which a surface-active "anti-pitting" agent has been used, may be controlled by use of these methods.

Surface-tension measurements generally refer to the liquid-air interface, while interfacial tensions refer to interfaces such as liquid-liquid or solid-liquid. Since these interfaces are so interrelated, it is difficult to discuss them separately, but an attempt will be made to do so as a means for simplification.

Surface tension

(a) Capillary height or pressure.

(b) Drop weight.

(c) Maximum bubble pressure.

(d) Capillary waves.

(e) Vibrating jet.

(f) Drop height.

(g) Drop-shape.

(h) Maximum pressure in drops.

(i) Method of Sentis (capillary tube hanging drop and measurement of meniseus height in tube above drop).

(j) Ring method.

Interfacial tension

Wetting of liquid by a liquid, or solid by a liquid may be estimated by several methods. The solid-liquid interface may be measured by several techniques which require estimation of the angle of contact :

(a) Pressure of displacement.

(b) Horizontal plate.

(c) Vertical bar.

(d) Deposited film.

Liquid-liquid interfaces may be determined by several methods :

(a) Stalagmometer.

(b) Drop weight (Millard)

(c) Ring

(d) Ripple.

The facility with which a solution may wet another surface, whether solid or liquid, may have considerable influence upon subsequent operations. Cleaning or dyeing a fibre, wetting a metal surface or foliage, and formation of emulsions all are dependent upon optimum interfacial tension properties.

The lower the surface and interfacial tensions of a system, the better the chance of forming an emulsion or wetting a given surface. Cupples has applied these data to foliage wetting by utilising the measurement of surface tension against air and against mineral oil. The spreading coefficient (SC) of a system may be calculated as follows :

SC = surface tension oil − (surface tension solution
                                   + interfacial tension)   (1)

The more positive the value obtained, the greater the spreading value of a system. Consequently, the possible utility of the wetting agent for insecticidal sprays may be predicted as an emulsifier for mineral or other oil, or as a means for wetting oiled soiled surfaces.

In general, the du Nouy instrument may most satisfactorily be used for liquid-liquid systems. One instrument may furthermore be used for the determination of both surface and interfacial tensions. For control purposes, the dial reading after calibration may be adequate, but if highest accuracy is desired, ring corrections have been developed.

Surface Active Agents

Since surface activity is increased by this addition, the term "builder" for added sodium sulphate is warranted, and this effect may be attributed at least in part to orientation of the surface-active agent at the interfaces involved.

The generally high surface activity of these agents upon dilution indicates their potential utility for the wide variety of uses open to surface-active agents. Those materials with most highly positive spreading coefficients are potentially useful as inseetidical spreading agents and are most likely to be good emulsifying agents.

Test Method

Apparatus : Either the precision (surface tension) or interfacial tensiometers may be used.

Calibration :

(a) Level the tensiometer.

(b) Adjust the torsion wire until it is taut.

(c) Adjust vernier of scale to zero.

(d) Adjust platinum ring.

(e) Place piece of paper on ring and adjust until index level of arm is opposite reference line of mirror (thus compensating for weight of paper).

(f) Place weights totalling 600 mg. on paper platform.

(g) Turn adjusting head (controlling dial reading) until the index level of the arm is opposite the reference line of mirror. Record dial reading to 0.10 division.

(h) Calculate scale reading as follows : Weight added to ring (m) = 0.600 g. Mean circumference of ring (L) = 4.00 cm. Acceleration of gravity (G) = 980.3 cm. per sq. sec. (at Chicago) :

$$Y = \frac{MG}{2L} = \frac{0.600 \times 980.3}{2 \times 4.00} = 73.52 \text{ dynes per cm.}$$

For the interfacial tensiometer substitute the pertinent values in the above equation.

(i) If the recorded value is greater than the calculated value of h, shorten the torsion arm, or lengthen it if the recorded value is less.

(j) Repeat the calibration procedure, readjusting to the zero position after each change of torsion arm length until the dial reading corresponds to the calculated value.

(k) Adjust the dial reading to zero, remove the weights and the paper platform and readjust so that the torsion arm indicator is in alignment with the reference line of the mirror.

Precautions :

(a) The wire of the ring should lie in one plane, be horizontal, and round.

(b) The vessel holding the liquid should be large enough so that curvature of the free surface shall have no effect upon the shape of the liquid column raised by the ring.

(c) The liquid surface should be free from motion.

(d) The ring should be only very slowly raised.

(e) There should be no evaporation or coating of the surface.

Procedure : Prepare 1.0 per cent solutions of the agents in distilled water. Retain a portion of dilution. Determine surface tension by averaging four measurements. The reading is taken at the point where the film breaks, while every effort is made to maintain the torsion arm indicator exactly opposite the reference line of the mirror. The interfacial tension measurement is made by immersing the ring in the aqueous layer, then carefully flowing a sufficient depth of layer of paraffin oil (Nujol) over the aqueous surface that the inverted V of the ring shall at all times during the measurement be immersed in the oil. The measurement is carefully made, since extremely low interfacial tensions are hard to discern, and in some instances no more than one measurement may be made because of preferential wetting of the ring by the oil. The results are averaged and recorded.

The ring is cleaned between determinations by immersion in turn in carbon tetrachloride and dilute hydrochloric acid solution followed by flaming.

Similar readings are obtained for 0.25 per cent solutions of the agents, and 0.0625 per cent solutions.

Corrections for surface tension measurements may be made by reference to calibration curves or to the data developed by Harkins and Jordan (4) or Zuidema and Waters (7) for interfacial measurements.

Lime Soap Dispersion

The synthetic surface-active agents will themselves form calcium and magnesium salts of varying degrees of solubility, and highly effective compounds will exhibit exceptional tolerance for these ions. A possible use for these agents is in combination with soda soap as a dispersant for subsequently formed insoluble soaps. This lime soap dispersing action has been put to use in the war effort since both the Army and Navy use certain synthetics in combination with soap in their all-purpose or sea-water detergents. A further possibility is that these synthetics may be used to cleanse lime soap-containing garments, since these soaps may readily be removed either by solubilising them or removing them by peptisation or emulsification. In such cases high washing temperatures will have important effect upon degree of removal by the synthetic. Two possible conditions may thus be encountered : (1) presence of large amounts of insoluble soap; (2) presence of relatively light films of insoluble soap. Laboratory test conditions should be chosen which will duplicate the most severe conditions which will be encountered, so that relatively large amounts of lime soap should be present in the test medium.

Nonlathering agents may also be useful for lime soap dispersion (phosphates) but in cases where soap is actually regenerated, by exchange of ions to form a soluble calcium or magnesium complex, a lesser amount of lime soap should be used or formed in situ, and the solution should be titrated with the regenerating agent to permanent lather.

Test Method

Reagents : 2.3 per cent sodium oleate (dry weight). Either pure sodium oleate or oleate flake soap.

0.34 per cent $CaCl_2$ solution.
0.2 N HCl.
Methyl orange-xylene cyanole indicator solution (Eastman No. 2216).
10 per cent volume solution of auxiliary agent under investigation.

Procedure :

(a) Add 1-ml. increments of auxiliary or dispersing agent up to 8 ml. to separate 20 ml. volumes of 2.3 per cent soap solution in 125 ml. flasks. The solutions are then shaken and to them are added 25 ml. of 0.34 per cent calcium chloride solution. Heat for 15 min. on a water bath, then suction filter through a prepared No. 4 Whatman paper on a Gooch crucible. The flask and crucible are then washed with three 10 ml. portions of distilled water. If complete filtration is impossible with 8 ml. of auxiliary agent, this fact is recorded without further examination.

(b) The filtered solutions are titrated with 0.2 N HCl to the xylol-cyanole end point which is gray in colour. A blank for each agent comprising 20 ml. of soap solution, 25 ml. of calcium chloride solution, and the volume of auxiliary agent required, are titrated without filtration. When the filtered test sample has a titration value practically equal to that of the blank, the volume of auxiliary agent required to produce complete dispersion is recorded.

(c) The results are expressed as centigrams of auxiliary agent required to disperse 45.5 mg. of calcium oleate formed. One millilitre of auxiliary agent is equal to a dispersion number of 10.

Calculation : Dispersion number = ml. auxiliary agent × 10.

Practical Wetting Tests

Since the usual surface and interfacial tension measurements limit the systems under investigation to liquid-air, liquid-liquid, or liquid-solid, more practical systems may prove useful in evaluation. In these systems capillarity may be a sufficient criterion of suitability, and the height to which the solution rises in a strip of fabric or paper may provide adequate comparisons. In these cases the interfaces involved may be liquid-solid-air-oil, but other materials may further complicate the picture.

Generally Draves-Clarkson skein test and the canvas disk method are used. The canvas disk test can readily be used under conditions not suited to the use of skeins.

Regardless of what test surface is used, standardisation of the test is necessary. With skein preparation wetting times will vary for a given weight of skein with humidity, spool, tension on the skein winder, and age of yarn or variation from lot to lot. The canvas used may age to give progressively slower wetting times as the finishing agents become progressively more water repellent. In fact, whenever *any fibre, yarn* or fabric is used, steps should be taken to standardise against a stable wetting agent under a given set of conditions, and this standardisation made periodically, to save embarrassment. The Draves-Clarkson and canvas disk tests have in their favour simplicity of operation, minimum of equipment, and adequate reproducibility. Both methods are subject to variation in the standard fabric or yarn used in the determination, either as a result of aging or actual variation in the character of the manufactured article. They furthermore have been most widely used, and evaluation data on either basis are available from the literature. Either method provides data which, under careful control, can be used for the evaluation of comparative wetting efficiency, but their values are not directly comparable.

Test method : Draves-Clarkson test

Equipment : 500 ml. graduate.

Hook or sinker, 1.50 g. in weight (of No. 12B. and S. gage copper: with ¼-in. hood opening

Anchor, 20 g. in weight.

Skeins : A 5.0 g. skein of two-ply (70/2 combed peeler yarn) is prepared such that an 18-in. loop can be formed. The yarn may best be standardised against a standard sample of wetting agent, adjusting the weight of skein to yield a specific wetting time.

Wetting Test :

(a) Prepare 500 ml. of 1.0 per cent test solution.

(b) Transfer 250 ml. of 1.0 per cent solution to a 600 ml. beaker, add 250 ml. distilled water and adjust the temperature to 25 ± 1 C.

(c) Transfer the temperature-adjusted 0.50 per cent solution to a 500 ml. graduate.

(d) Attach the hool or sinker to the 18-in. loop, hold the skein in the middle and cut the end opposite the hook.

(e) Transfer the hook, anchor, and skein to the graduate of solution and dip the hook into the solution until the surface is just touched, then drop the skein and start the stopwatch simultaneously.

(f) The sinking time is attained when the skein sinks, and the hook just reaches the level of the anchor, at which time the stopwatch again is snapped.

(g) Repeat steps d, e, and f until five replicate skeins have been tested, using the same solution throughout.

(h) In a manner similar to step c, prepare an 0.25, 0.125, 0.0625 or further dilution until a wetting time in one dilution of + 180 sec. is attained for two replicate skeins. The test is then terminated.

(i) The mean values for each dilution shall then be reported.

Note : This test is applicable to acid, alkaline, or saline conditions.

Canvas disk test method

Equipment : Vernon duck.
Steel die, 1 in. in diameter.
Carbon filter tube of 75 mm. stem and 38 mm. inside top diameter.
600 ml. Griffin beaker with spout.

Disks: The disks may best be cut using a 1-in. steel die. Cut the disks on a smooth, hardwood surface. The disks for use must be flat and free from rough edges and ravelings. Cut the disks from the fabric area bounded by 1 in. within either side of the selvedge.

Wetting Test :

(a) Transfer 500 ml. of test solution at 25 ± 1 C. to a 600 ml. beaker.

(b) Drop the flat canvas disk into the funnel and quickly invert and immerse in the test solution.

(c) Start the timer as soon as the disk enters the solution. The individual test is considered as terminated when the disk first begins to sink.

(d) Four replicate tests shall be made for each dilution tested.

(e) The average results shall be reported as seconds wetting time for each of the concentrations tested, which shall be 0.5, 0.25, 0.125, 0.0625 or further dilution until the disks give wetting times greater than 180 sec., at which point the test shall be terminated.

Note : This test is applicable to acid, alkaline, or saline conditions.

The test equipment should be so aligned that a disk held horizontally by the filter tube should be 1 in. below the surface of the liquid.

Rewetting Test

It frequently is desirable to know whether the agent in question is suitable for rewetting purposes since this may be an important factor in its choice for subsequent processing.

Briefly, this test consists of applying the solution of rewetting agent by means of dipping in a bath and running through squeeze rolls (a pair of regular hand wringer rolls will prove satisfactory). Good impregnation is obtained by running two ends (i.e. opposite end of tape fed through wringer first) at a sufficiently high roll pressure to give a 100% weight pick-up. The test strips are then either air or oven force-dried and allowed to condition over night. The next morning the rewetting times are run in much the same manner as wetting times. Though pretested, the tape has been found to vary from lot to lot, and drying conditions have been found to alter the rewetting times considerably so that it is necessary to run a standard agent for comparison with the experimental sample.

Test Method

Equipment : Stopwatch, 500 ml. graduate, Hook, 1.00 g., Anchor, 40 g., Distance hook to anchor, 20 mm., Tape, length 9 in., width 1¼ in, Laboratory padder or wringer rolls.

Impregnation : A 21 in. length of tape (sufficient to allow for shrinkage and trimming) is required for duplicate determinations at each concentration of surface active agent.

An initial bath concentration of 0.0625% is suggested, though this may be adjusted to double or half this concentration, depending upon effectiveness in speed of wetting.

The bath temperature is adjusted to 70°C.

The tape is run through the padder or wringer twice, reversing ends at 200 pounds pressure (20 lb. pressure/inch face). Adjustment of add-on or solution pick-up is made for 100% weight increase of tape.

The tape is oven-dried 10 minutes at 105 C. or dried overnight and conditioned overnight in either case at 70°F. and 65% R.H.

Rewetting time :

(a) The 500 ml. graduate is filled with water at the required temperature. (A fresh portion of water is used for each test to avoid contamination by surfactant).

(b) The hook is inserted through the tape about ¼ in. from one end on the centre line.

(c) The tape is held by the free end so that the anchor and bottom hook are just submerged in the liquid in the graduate.

(d) The tape is dropped and the stopwatch is started when the anchor strikes the bottom of the graduate.

(e) When the hook and tape start to sink at a constant rate the sinking time (rewetting time) is reached and the watch is stopped.

Lather Values

The relative lathering action of surface-active agents is obviously important with detergents and wetting agents used largely for their foaming action. Soap, as the time-recognised detergent, foams profusely in soft water or even in hard water if a sufficient amount is present to overcome both water hardness and the insoluble soap formed. As a consequence, foam is now considered a requisite to detergent action. Many detergent adjuncts, which are detergents in their own right, produce no foam. These adjuvants are the alkaline salts such as the carbonates, silicates, borates, and phosphates. The fact that these materials did not foam disproved the theory that foam was useful as a measure of detergency, but the volume of foam formed is still a very valuable property of surface-active materials. Ability to foam at extreme dilution or under adverse conditions may be an important factor in application. Wetting agents have been used in the wash water of conditioning systems to remove foreign substances, but excessive foam would be a liability rather than an asset.

Many methods for measuring foaming characteristics of compositions have been devised. From the consideration of equipment involved they may be classified as follows :

(a) Bubble cylinder.

(b) Manual manipulation of closed container.

(c) Capillary under constant pressure.

(d) Solution drop, open cylinder.

(e) Mechanical agitation in closed vessel.

(f) Single bubble.

The subject of foams is a complicated one, but for many purposes a measurement of foam height will suffice as a means for preliminary comparison. The method often used in the absence of other equipment is to shake a volume of solution in a stoppered graduate in a standard manner, and measure the foam formed. This method provides only a rough estimate, since a variation of ± 15 per cent is about the best that may be expected under carefully controlled conditions.

The most satisfactory method for the evaluation of foaming agents is that developed by Ross and Miles. This test is easy to make, the equipment made according to ASTM specifications is commercially available and relatively inexpensive, outside variables are reduced to a minimum, and reproducibility is good. Figs. 12.1 and 12.2 give the necessary specifications which must be checked against the equipment as supplied. Careful evaluation will require tests over a range of concentrations and temperatures.

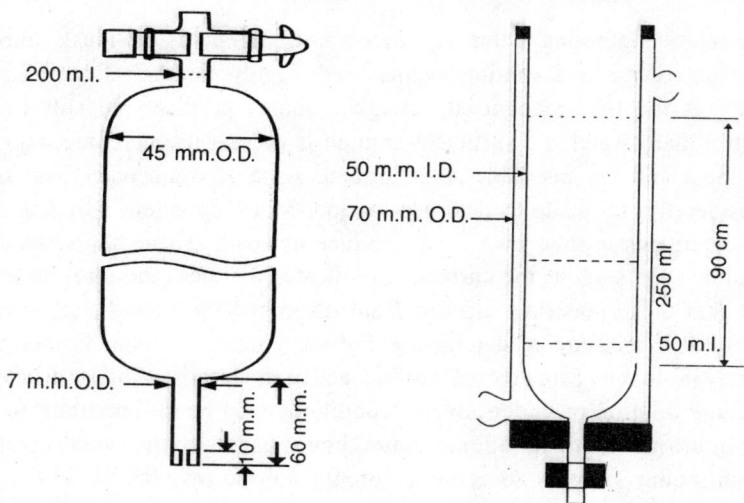

Fig. 14.1. Ross-Miles lather test foam pipette.

Fig. 14.2. Ross-Miles lather test foam receiver.

The anionic synthetic detergents will form salts of varying degrees of solubility and these will foam to greater or lesser extent. Consequently, hard water (calcium or magnesium ions) will have more or less effect upon the lather of the sodium salts. In some cases, the calcium and magnesium ions may depress foam formation, in others foam may be increased, while in still others no change will be observed.

Test method : Ross-Miles pour foam test

Apparatus: Level cylinder so that drops fall vertically into centre of liquid at bottom of cylinder.

Solution Preparation :

(a) Dissolve sample in distilled water at double the concentration desired by bringing to the boil then cooling to 25°C.

(b) Make back to weight with distilled water for loss by evaporation.

(c) Dilute to correct concentration by using water at 25°C of double the hardness or strength finally required.

Procedure :

(a) Start circulation pump on column to maintain temperature at 25 ± 0.2°C.

(b) Rinse cylinder walls with distilled water, drain 10 min. and close stopcock.

(c) Age solution for 10 min. in constant temperature both at 25 ± 0.2°C.

(d) Wet down walls of cylinder with 50 ml. of test solution by washing walls in circular motion with 50 ml. pipette of solution.

(e) Immediately transfer 200 ml. of test solution to tipped pipette. Insert in holder in position vertical to base and open stopcock.

(f) Record later height in centimeters, at once, and in 5 min.

(g) Drain cylinder, rinse with distilled water, and permit to drain for 10 min.

(h) Repeat steps of through g for next sample making one determination for each sample.

Solubility—Organic Solvents

In spite of the presence of polar groups, many compounds are soluble in nonpolar solvents. Their possible applications under these conditions are in paints, dry cleaning solvents, insecticides, printing inks, and as self-emulsifying agents.

Since the surface active agent generally is added to produce some given effect, it is not necessary that solubility be determined under conditions of greatest accuracy: Small amounts of agent are generally sufficient or must be maintained. It is therefore felt that the accuracy of the method used is entirely adequate.

Only the pure materials are evaluated for this purpose, since even traces of inorganic salts or of water will reduce actual or apparent solubility.

Other solvents may be used in this test when further information is desired, although chlorinated hydrocarbons, a polar solvent, and a hydrocarbon are represented.

Test Method

Sample Weight : 0.50 g.
Equipment : 16-oz. glass stoppered bottles dispensing burettes.

Temperature : 25 ± 3°C.

Solvents : Transfer 0.50 g. of sample to the glass-stoppered bottle. Add 1.0 ml. of solvent, shake thoroughly, and determine whether solution has taken place. Continue adding increments of solvent, either until solution is effected, or until 450 ml. of solvent has been added.

Calculation :

$$\text{Percentage solubility} = \frac{0.50}{\text{vol. solvent} \times \text{sp. gr.}} \times 100$$

COTTON WASHING

Through the process of forming fibre into yarn and into fabric, and prior to conversion into garments, rigorous cleaning schedules are followed which require considerable amounts of detergents. When of sufficiently high tensile strength characteristics, those fibres and fabrics which can withstand the cleaning process most satisfactorily will be chosen for the hardest wear. Cotton still retains its historic leadership even though the synthetic fibres are gaining much favour. Indeed several of the synthetic fibres are simply cellulose which has been treated to provide fibres differing from the original cotton in luster, evenness of character, and offering opportunities for styling that cotton does not provide. Cotton as well as many of the fabrics made from these fibres can be washed repeatedly, wear well, and retain their dimensional stability. These are the fibres which are washed most often and their cleansing represents the large outlet for cleansing agents.

The controversy as to whether an applied soil should simulate a "real" soil or be a synthetic one continues. Practically speaking, the soil and its method of application and the laboratory machine used in washing are minor factors so long as they remain constant, provided that the data so obtained can be correlated with practice. Advancement from the laboratory to the practical scale for proof of utility is a "must" and should proceed as soon as laboratory data are convincing.

Water Hardness

In general, most detergents become less effective when water hardness increases. Certain of the anionic type synthetics are improved in detergency in the presence of moderate amounts of calcium and magnesium salts but excesses render the surface active agent totally ineffective or at least reduce effectiveness in cleansing action. Calcium is the ion which causes most difficulty. Nonionics are generally only slightly affected by water hardness.

Exactly what ratio of calcium to magnesium to use and what shall be the water hardness for detergent testing are factors which must be decided. Many investigators have used a ratio of calcium to magnesium of 60 to 40 while others have increased the ratio or have excluded magnesium entirely. Regardless of the ratio used it is desirable that the detergency test be made over a range of water hardnesses, and if this is done, and the range is sufficiently extended, the effect of having used only calcium as the hardening ion will have been equivalated.

A standardised water hardness is desirable and one which can be made in a stock solution for dilution purposes should be used. For general screening purposes hardness levels of 50 and 300 ppm. have been used successfully. First and most important, they demonstrate the comparative effectiveness of a detergent when subjected to hard water usage; and second they represent levels of practical significance. A 50 ppm. level probably represents the maximum (with exceptions) for the large population areas, while the 300 ppm. hardness is one frequently approached in hard water areas.

Because of test discrepancy, suspected inaccuracy in preparation of standard water, or necessity for analysis of unknown water, it is valuable to have several test methods available. Both of the following have been used successfully : Langelier method or the Versenate method. The former is a variation of the soap titration method, and the latter a colorimetric titration procedure. The APHA soap titration method is still useful, but in inexperienced hands can prove misleading and takes longer to perform than either of the foregoing methods.

Machines

When a decision has been reached as to the manner in which mechanical action shall be applied in the washing operation, the machine should then be designed so that all of the other factors involved in the washing operation can be controlled. It should be obvious that mechanical action is applied in the washing operation and depending upon the fibre and fabric, and the state of manufacture, the application of mechanical "elbow grease" may be varied.

A very short review of some of the operations in washing (scouring or cleansing) may indicate why some of the types of washing machines have been developed. In the manufacture of textile fibres there are many points where cleansing must be performed before subsequent processing can be properly controlled. In the cotton mill raw cotton is mechanically cleaned of burrs and extraneous matter and formed into yarn which is then woven into a fabric. Before this fabric can be further processed it must be kiered, which is simply removal of the cotton wax and lignin-like materials inherent to the cotton; then it is bleached and further processed. Following the dyeing operation it generally is necessary to "soap off" or scour for final finishing. Washing of fabrics in the home may take many forms. It may be simple hand washing or it may be accomplished with any one of several different types of washing machines, either manually or automatically operated. Here the detergent is selected with a view to providing the greatest soil removal for the amount of mechanical energy applied to the fabric in scrubbing and removing the closely adherent soil.

Many laboratory machines have been designed which provide control of operating conditions for the evaluation of detergents, most of them varying in the method of applying force to the cleaning operation. Generally the evaluation of a washing machine, as contrasted with detergent testing, will be accomplished on a full scale basis.

Standard Cotton Soil Fabric Preparation

The requirements for a standard soiled fabric are :

The fabric should be uniform.

The soiling operation should provide the same initial amount of soil, and the same removal value for any given detergent under identical conditions of cleansing.

The soil removal values should be reproducible, preferably of analytical accuracy.

The effect of aging should be minimised.

The method for determination of soil removal or retention should be reproducible, preferably quickly accomplished, and closely correlated with actual practice.

The process of preparing a standardised soil fabric may be subdivided as follows: fabric, desizing of fabric, soil, soiling machine, wash test machine, wash test method, storage, evaluation of washed swatches, soiled fabric testing, statistical data.

Fabric choice

Indian Head fabric is a plain weave sheeting, has a thread count of 48 × 54, is available in several widths, and has a weight of 5 ounces to the square yard. To minimise the desizing problem, purchase is made on specification requiring from the manufacturer a scoured, bleached, dried, but unmercerized or otherwise finished grade. This meant that chemical change or deposition of chemicals liable to change the surface reactivity of the fabric were eliminated since a mercerized or chemically treated surface is much more reactive than one not so treated.

Desizing

For purposes of uniformity all sizing should be removed. Variation in fabric characteristics resulting from small scale operation led to desizing in a 25 × 25 in. pony wash wheel.

Soil

A test solid generally will comprise a binding agent and a material which permits quantitative estimation of removal and which in itself may be a natural soil. Since soot is a common soiling agent and certainly lends itself to reflectance measurement against a white fabric, it is not unusual that carbon in some form should be used. Carbon will vary widely in composition depending upon its method of manufacture. Some may be very finely divided, others coarse, and the particle size distribution may vary widely. Some carbons may be predominantly oil-dispersible and others water-dispersible, but apparently satisfactory results have been obtained using either type.

Oildag concentrate has been used satisfactorily for at least 12 years and is comprised of colloidal graphite and apparently an oil-soluble dispersing agent, dispersed in mineral oil. This forms relatively stable dispersions in nonaqueous solvents. So long as the code number which appears on the Oildag package is specified, satisfactory reproducibility by the manufacturer has been noted. The binding agent for the graphite may be chosen from a wide variety of agents, but preferred is a vegetable oil such as Wesson oil, though Mazola oil can be substituted for it. Others have used lard, tallow, or other natural oils. Yet others have used mineral oils because they are not readily susceptible to atmospheric oxidation. Wesson oil was chosen because when it was properly aged or cured on the fabric, it made the graphite much more difficult to remove than did mineral oil, giving a satisfactory spread in soil removal. The difficulty in the use of an oxidizable oil however is that the degree of oxidation must be controlled and the storage conditions maintained at an optimum to prevent further oxidation.

The medium to be used for dispersion of the carbon and oil is obviously important. In some systems graphite or carbon is deposited upon the fabric in water, loose carbon removed, and the soiled fabric either used as such or the dried fabric impregnated additionally with an oil. Water is a safe ?nd inexpensive medium for soil application, but the fabric must be treated to remove wrinkles and loose carbon. Choice of a chlorinated solvent may be preferable to a naphtha solvent wherever fire hazard is involved, though the latter is less expensive and relatively nontoxic.

A technical grade of carbon tetrachloride has been used and while there is no evidence that inhibitors added to the carbon tetrachloride may affect oxidation, this is a possibility which has not been explored.

The soiling solution (or suspension) used has the following composition :

    5.0 grams Wesson oil
    10.0 grams Oildag concentrate (Acheson Colloids Corp.)
    12.0 litres carbon tetrachloride (Dowelene, Dow Chemical Co.)

The amounts of Wesson oil and Oildag are weighed to the accuracy indicated, and the carbon tetrachloride measured from a two-litre graduate. Care is taken to disperse the Wesson oil and Oildag thoroughly during dilution.

Soiling machine

The soiling operation is important, and to insure reproducibility of application it is general practice to use a machine of some sort. Generally a chlorinated solvent or hydrocarbon is used as a means for uniformly applying them to the fabric. The machine was developed from an initially hand operated heated tube to the presence equipment. Since moisture as dew deposited on the fabric (as a result of solvent volatilisation) has such a marked effect upon soil characteristics, the machine was designed to minimise the effect of condensed moisture by raising the dewpoint.

Soil storage

The ideal soiled test fabric should be completely stable toward ageing or storage. The soil described continues to age over a considerable period of time, but to minimise this effect it was found that ordinary mechanical refrigeration during storage of the soiled cloth in a dry atmosphere will prevent change in soil removal characteristics over a period of a month. All of the usual methods for inhibiting oxidation such as vacuum, absence of light, inert atmosphere, desiccator, etc. (and other than the use of an inhibitor) were evaluated in arriving at the most desirable storage conditions noted above.

Evaluation of washed swatches

Since the soil is designed to permit a spread of results from original soil to original white (which seldom if even is attained), and since this is what the eye sees, regardless of actual weight percentage soil distribution, a photometric method for soil removal evaluation is indicated. Other investigators prefer the measurement of soil removed from the fabric and held in suspension in the solution, essentially disregarding the fabric as a means for evaluation of removal. This latter method utilises the photometer, reading transmittance against a curve for removal, but possible turbidity of detergent in solution interferes, and somewhat narrows the utility of this method.

Standardisation of Solid Fabric Testing

Since every load of samples includes at least one standard detergent product, for many years it was considered sufficient simply to use the soiled fabric as it was cured or artificially aged and without preliminary evaluation. Apparently this is the general method followed by a large proportion of the investigators reporting detergency data.

In an attempt to improve reproducibility each roll of soiled fabric is tested immediately after preparation and without further treatment. There are variations from roll to roll as evidenced through the use of the standard detergent sample. Having ascertained the degree of soil removal immediately after soiling, and knowing the degree of retentivity of soil desired, the freshly produced fabric is then baked in a circulatory oven in festooned form for periods of time which experience indicates will provide adequate resistance to removal. After baking, the soiled fabric is again tested under standard conditions to be certain that specified control has been attained.

COTTON WASH TEST METHODS

The wash test method chosen will depend to a considerable extent upon the soiled test fabric and the washing machine available. Depending upon the mechanical energy expended in the washing operation, a soiled fabric may be of single or multiple wash characteristics.

Frequently, a soiled fabric designed for one type of operation will be found useful for others, and the influence of the washing machine will be minimised.

Having decided that the fabric may prove satisfactory with the available washing machine, a decision will be made as to whether a single or multiple washing operation shall be followed. Home laundering may encompass pre-wash and wash operations; a single wash; and single or multiple rinse procedures. Power laundries follow a multiple suds, multiple rinse procedure. Which procedure is chosen will be arbitrary, since the laboratory machine used frequently cannot be compared exactly with a full scale washer.

Next the duration of the wash period must be decided. It should be long enough that stabilised conditions of soil removal can be attained, but not so long that unnatural redeposition of soil can occur. For laboratory work where many samples must be screened under a variety of concentrations and water hardnesses, the shorter the washing period (within reason) the greater will be the work output.

Whether the rinse shall be by machine or by hand, and whether the test pieces shall be wrung out by hand or by using wringer rolls will depend upon the degree of stability of soil removal. Certainly, if the rinse removes considerably further amounts of soil, this should be done under controlled conditions and not haphazardly, and the rinse will be by machine. However, if the soil removed in the rinse is negligible, a simple hand-rinsing is indicated to save time and effort. The same argument is applicable to hand vs. roll wringing of swatches, but in either case the rinse water should be as well removed as possible to prevent possible migration of soil or retained discoloration.

Drying of the washed swatches is important. Migration frequently may occur when the test pieces are laid out flat or hung on a line and permitted to dry overnight. Forced hot air drying in either vertical or horizontal position is therefore indicated.

Deter-Meter Test Method

Dynamics. General testing, 120 cycles/minute. 50 min. amplitude. No. 2 impact surface spacing. 5 minute wash time.

Concentrations. At three chosen levels.

Temperature. 120, 140 160°F.

Replicate Tests. Three, using fresh swatches in each. Where a relatively large number of concentrations or temperatures are involved two swatches at each point are suggested.

Sensitiveness of Test Method. Ability is claimed to distinguish between concentrations which differ as little as 0.01% with a single swatch at each concentration.

Conventional Washer Method

The water capacity of machine with an 8-pound load (or other recommended load) is determined for calculation of detergent requirement. The load is weighed into the machine and should consist of normally soiled, fast-coloured or white garments.

Test Cloth Preparation. Each load contains, in addition to the soiled garments, five 5 × 6 inch swatches of standard soiled fabric marked for identification.

Charging the Washer.

(a) The machine is filled to the predetermined level with water at 120°F.

(b) The detergent is added with the agitator in motion to permit rapid solution.

(c) The 8-pound weighed load is added.

(d) The load is washed for 10 minutes.

(e) The garments are then wrung into a stationary tub of 9 gallons of water of the same hardness as the wash, at 100 ± 5°F.

(f) A hand rinse is administered in the water of step (e) by thoroughly moving the garments up and down in the water.

(g) The rinsed garments are wrung into the machine filled with 17 gallons of fresh water of the hardness in use at 100 ± 5°F.

(h) A machine rinse is made for 5 minutes.

(i) The garments are wrung from the rinse.

Drying the Washed Swatches. The soil swatches are removed and pressed dry at a low iron temperature in a flat work ironer.

Practical Wash Tests

For complete assurance that any developed detergent composition is satisfactory for the purpose designed, it is necessary that the product at least be pilot-plant tested. Satisfactory screening on a full scale with home washer and on a pilot scale with a pony washer for commercial laundering has been the rule.

Home laundering

If after completion of the small scale bench testing of the detergent an apparently satisfactory product has been developed, it is then reasonable that the sample be evaluated under conditions which at least approximate those for large scale use. For the home washing machine this means that a load of fabric of about a maximum of 9 lbs. dry weight (dependent upon machine capacity) will be required. Obviously a mechanical set-up for the supply of water of the proper temperature and hardness is necessary. Quite satisfactory results have been obtained with several types of home washing machines.

Sea Water Detergency Tests

A sea water detergent should be soluble in this medium, and should exert reasonable soil removal characteristics. Any detergent which can be operated successfully in this medium can probably be used in the hardest of waters. This led to a combination of tallow soap, as a binder, and alkyl benzene sulphonate as the active detergent ingredient.

Detergency methods

These methods can be made to exactly duplicate fresh water tests with the possible exception that a final fresh (distilled) water rinse may be used. In laboratory as well as power laundry trials this rinse will remove some of the stiffness from the fabric without markedly increasing soil removal: for garments it is feasible aboard ship and greatly improves the wearer's personal comfort.

A *synthetic sea water* used and successfully applied to evaluation either in laboratory machines or the 25-pound capacity pony wash wheel follows:

$$1.60 \text{ g. } CaCl_2.2H_2O$$
$$11.00 \text{ g. } MgCl_2.6H_2O$$
$$4.00 \text{ g. } Na_2SO_4 \text{ anhydrous}$$
$$25.00 \text{ g. } NaCl$$

Dissolve in 500-700 ml. of distilled water then make up to 1 litre.

Concentration. For laboratory evaluation it may be desirable to cover a range of sample concentrations such as 0.10, 0.20, and 0.40%. For power laundry work the several materials used successfully have been used at near normal power laundry concentrations of near 0.1%.

Redeposition

Soap is relatively effective in preventing redeposition of soil once it has been removed from the fabric, but this property is not a "built-in" attribute to most synthetic detergents. It is generally understood that the synthetic detergents are less colloidal in character than soaps, and consequently lack certain properties of which this is one. Certain soaps may likewise be lacking in these redeposition-preventing characteristics, but these are not the soaps generally used in commercial work.

Even though the individual factor or factors involved in preventing redeposition have not been isolated, it is possible to measure the effectiveness of detergent compositions in preventing this occurrence. It has long been recognised that some detergent type materials may remove soil from a fabric and then redeposit it in such a form that the washed fabric is darker than the original before immersion in the test solution. It is this redeposition of soil particles under conditions of high soil, adverse temperature conditions, low solution volume and low detergent concentration that promotes graying of fabrics.

It has been recognised that synthetic detergents were deficient in preventing redeposition of soil and developed sodium carboxymethyl cellulose to remedy this shortcoming. This is one of the very few materials for use with synthetic detergents which prevent redeposition of loosened soil. So extensive now is the use of CMC that almost without exception, there are no synthetic detergents on the market which do not contain this compound in greater or lesser proportion.

General methods

The evaluation of redeposition in washing compositions, and of agents potentially useful for this purpose, can be accomplished by two general methods. In one case a sample of soiled fabric for the wash test is accompanied by a swatch of white fabric, and the net detergency or the soil redeposition measured photometrically after a proper washing period. The second and preferred method is to add a known amount of soiling agent to the detergent solution, add a swatch of white fabric, then determine the reflectance of the swatch after washing. The second method permits better control of amount of soil used than the first, and is generally under better statistical control.

Redeposition test procedure

Launder-Ometer. In duplicate swatches per sample.

Time of Wash. 20 minutes.

Temperature. 140 ± 2°F.

Water. 300 ppm. hardness unless otherwise noted.

Fabric. Indian Head Permanent Finish as received. Cut into swatches 6 × 5 inches and mark for identification.

Standard Value. This is obtained by making 6 or 8 readings on the Hunter Reflectometer on swatches taken at random from those to be used in test. This standard value is repeated for each separate test.

Solution Preparation. Make 200 ml. of 0.2% concentration as follows: Weigh 0.4 g. of sample and dissolve in 100 ml. of water of desired hardness. After sample is dissolved, add 2 ml. of 10% Aquadag suspension and 98 ml. of water of same water hardness. If excessive heating is required to dissolve sample, enough water should be added to maintain total volume at 200 ml.

Preheating. The solutions are preheated on a hot plate to 60°C. They are then transferred to jars in the preheating tray on Launder-Ometer, each jar to contain 100 ml. of solution and 10 rubber balls.

Washing. After all solutions have been transferred to the preheating tray, the test swatches are transferred to them. The jars are at once tightly capped, tested for leaks and placed in the Launder-Ometer.

Blotting. At the end of the 20-minutes wash period, the jars are removed from the Launder-Ometer and the contents poured into a strainer. Each duplicate set of swatches is removed before another set is poured into the strainer. The swatches *without being squeezed*, are placed between filter papers (or paper towels). The papers are then placed between glass plates and a 5 or 6 lb, weight applied for one minute.

Rinsing. While the wash is running, the rinse is prepared. Water of the same hardness is heated on the hot plate to 60°C and 100 ml. is transferred to each jar. 10 rubber balls are included in the rinse. After all swatches have been blotted they are placed in the jars for rinsing. The jars are tightly capped and tested for leaks as before, and then placed in the Launder-Ometer and run for 5 minutes. At the end of the rinse period the swatches are removed, as before and blotted.

Drying. The swatches are then either air or preferably oven dried, and the reflectance measured on the Hunter Reflectometer.

Record of Values. Standardise the Hunter instrument and record the reflectance values for each swatch and average them.

Redeposition Index

$$\frac{\text{Reflectance after treatment}}{\text{Reflectance of whites watches}} \times 100 = \text{Redeposition index}$$

Standard Sample. A standard comparison sample of known quality is desirable, and should be included with each wash load.

WOOL WASHING

Detergents and their evaluation, or the evaluation of cleansing methods are important throughout the processing of wool, from the scouring of raw wool, back washing, wash-off after dyeing or fulling to the cleansing of soiled garments after the wool has been fabricated and worn. This may be simplified to: raw wool scouring, in-process scouring, and finished fabric cleansing.

Raw Wool Scouring

The first requirement in the treatment of raw wool is mechanical removal of the dirt and extraneous matter by a simple opening and beating or picking operation. Following this, the raw wool is generally carried through what is known as a train of scouring bowls which vary in number. Since wool is sensitive to mechanical action in the scouring operation, the agitation involved in the wool scouring train is mild and accomplished by rakes which simply pull the mass of raw wool through each of the bowls between a nip, sometimes rinsed between bowls, passing into the next bowl in the train. In general these bowls consist of a steep bowl where the greatest portion of the dirt is removed and where some of the fatty acids are saponified (by soda ash, a desirable sodium bicarbonate content being built up). The second bowl consists of a detergent-builder mixture, either soap and soda ash (or modified soda) or synthetic detergent and soda ash. The next bowl may either be a final rinse bowl in simpler operations, or is a second soap-detergent bath of lower concentration, followed by one or two water rinses. Temperatures of scour and rinse are carefully controlled and seldom exceed 130°F.

Laboratory evaluation of raw wool scouring is difficult unless a satisfactory sample of raw wool of the type expected to be used in the plant is available. To satisfy this requirement a quantitative method for the evaluation of wool-scouring detergents has been developed by using small skeins of wool-in-the-grease (in this case carpet yarn), the use of a laboratory bowl (beaker) scouring train and determination of the point at which the bath became exhausted or lost its effectiveness in the removal of the grease (as determined by solvent extraction of the skeins and determination of residual grease content). The method is reasonably reproducible from laboratory to laboratory and has the decided advantage that it closely approximates plant scouring conditions using wool which might actually be encountered. This method has been used successfully in the evaluation of surface active agents for this purpose, and detergents tested in the laboratory by this method have been taken directly to the mill and have met successful acceptance.

Continuous scouring of raw grease wool

This specification covers a method for the laboratory evaluation of continuous wool scouring. It is adaptable to the comparison of detergents, builders, and other chemical factors in the scouring operation; and also to the study of mechanical and other control variables, such as temperature, the number of bowls, immersion time, etc.

A standard apparatus and set of operating conditions are specified for purposes of calibration. Beyond that, conditions and apparatus may be varied in accordance with the desires of the operator to study any particular system which may exist in his own plant.

The criterion for scouring effectiveness is the extent of grease removal. Extraction is done by agitating the scoured and dried wool specimens in a piston and cylinder system while immersed in carbon tetrachloride. An aliquot sample of the carbon tetrachloride is then evaporated in a tared aluminium cup on an infrared dryer to determine the grease residue

In-Process Scouring

The Detergency Comparator for the evaluation of detergents on woollen and worsted materials is used.

The Detergency Comparator is constructed of stainless steel and has a thermostatically controlled, electrically-heated outer water bath, and the solution may be circulated by means of an electric pump to further insure close temperature control. Stepped pulleys are used for speed selection and it has Hyear-covered nip rolls (3 inch diameter), driven take-off rolls, baffles to prevent suds from cross-contaminating the detergent solutions under test and suds boxes with piping arrangements to permit the expressed liquor to be returned to the main washing tank either by cascading or by-passing through the pipe drains or to be drained off into the waste line. The adjustable pot eyes help to prevent the formation of hard creases in the knitted tubing samples. The rinsing process is controlled using a reservoir fitted with calipered openings and a valve for close control.

The Detergency Comparator is said to be the only device that will compare detergents under the conditions of high concentration and low bath ratio similar to those obtained in good plant practice.

1. **Soiled Suspension.** The composition recommended follows : commercial grease oil 50% (by wt.); refined mineral oil 50% (by wt.).

   To one gallon of the oil blend was added ½ pound of Bear Brand Lamp Black (Monsanto). This was mixed thoroughly before use.

2. **Soiled Fabric Preparation.** The mixture of oil and lamp black was applied to worsted top sliver during the first drawing operation so as to deposit 5.0 to 5.5% oil on the stock. The worsted stock was then drawn and spun to 2/40's count and 1.75 inch wide knitted tubing was prepared (2 yards long weighing 40 grams).

3. **AATCC Detergency Comparator Test Method.** This method of test is applicable to the evaluation of detergents on woollen and worsted material. It provides an ideal method for classification without conducting costly, time-consuming plant runs.

Finished Fabric Scouring

In comparison with cotton, much less effort has been spent on the standardisation of a test method for wool. The soil or soils vary, but in general are similar to those used for cotton. The fabric chosen should be scoured, in the grey, neutral, and as free from adsorbed soap or synthetic detergent as possible. The laboratory washing machines can be the same as those used for cotton, the wash temperature being adjusted to a level to prevent excessive felting.

Wash test method

Soiled Fabric Preparation. The fabric is Botany Worsted Mills, Style 404, prescoured.

The proportions for the soiling ingredients are :

Edible tallow	2.0	g.
Nujol	6.0	g.
Grinders	0.125	g.
Lamp black		

These ingredients are mixed in the Waring Blender then passed through a homogeniser. To 8.125 g. of the homogenised mixture add 4 litres carbon tetrachloride and mix thoroughly.

Using the soiling machine a 7 inch width of fabric 22 feet long, is passed six times through the soiling bath and machine until a reflectance of 28.32% is obtained. This soil is baked one hour at 90°C.

WASHING PROCEDURES FOR OTHER FIBRES

The increasing number of new fibres and mixtures of them with cotton or wool may change some of the practical washing procedures mainly as to temperatures used, but laboratory wash test procedures described in Section 12.3 and 12.4 are applicable.

Dry Cleaning

Dry cleaning establishments are as numerous as power laundries and indeed both are frequently maintained.

The effect of the moisture regain and the relative humidity of the air in the cleaning operation, and of added moisture in certain instances has marked effect upon both soil removal and soil redeposition. It has been found that when the relative humidity is lower than 40% that 1.5-3.0% water, added on the weight of the fabric, will reduce soil redeposition. It has been investigated that of a variety of detergents tested, certain ones were more susceptible to the addition of moisture or to relative humidity conditions than others. Thus it was found that both Stoddard solvent and tri-chloroethylene were beneficiated in preventing redeposition, when the garments were preconditioned, or when water was added to the cleaning solution.

Evaluation can consist either of redeposition or soil removal, or both. In any event, the effect of preconditioning the fabric or of adding water to the cleaning medium must be considered.

Shampoos

Since the materials which may be used in the production of shampoos are legion, it might be expected that fairly well defined methods would exist for their evaluation. The reverse is true, however, since the cleansing of the scalp is a rather difficult process to control. Generally the evaluation of shampoos has been performed either through experienced operators in beauty parlors or by the panel technique.

One of the important factors in a shampoo is the abundance and stability of the lather formed, and this phase of the evaluation can be performed in the laboratory and separately from the actual cleansing operation. A recent innovation for detergency evaluation has been the use of a variation of the raw would scouring technique.

Detergency method

Fabric. Wool yarn "in-the-grease" is used for this test. A small skein about 4 inches from end to end when stretched is prepared from a 20 foot length of yarn and will weigh between 4.5 and 5 grams.

Detergent Solution. A sample of detergent weighing 0.50 g. (100% basis) is dissolved in 200 ml. of water (hardness selected). The solution is adjusted to 38°C.

Procedure

(1) Accurately weight the skein of wool yarn.

(2) Transfer to a 500 ml. Erlenmeyer flask containing the temperature-adjusted detergent solution and stopper.

(3) Slowly invert flask back and forth 50 times per minute for 4 minutes.

(4) Remove skein and either hand or wringer-squeeze to remove excess solution.

(5) Drain and rinse flask, add 100 ml. of water of hardness in use at 38°C and add the squeezed skein of yarn.

(6) Rinse by shaking for 2 minutes in same manner as before.

(7) Remove wool, squeeze as before and repeat rinse as above for 1 minute.

(8) Remove wool, squeeze to remove water and dry in an oven at 45°C.

(9) During drying of scoured yarn accurately weigh two 400 ml. extraction flasks and a second skein of wool yarn.

(10) Place washed sample and unwashed control in separate extraction thimbles and reflux each with 200 ml. of petroleum ether to remove residual grease.

(11) Remove solvent in hood over a steam bath and dry to constant weight. (this generally will take one hour).

Calculation. For the *control* :

$$\frac{(\text{wt. flask} + \text{contents}) - \text{wt. empty flask}}{\text{wt. of wool sample}} \times 100$$

$$= \% \text{ grease content}$$

For the *scoured wool* sample :

$$\frac{\text{grease content of control - grease content scoured sample}}{\text{grease content of control}} \times 100$$

$$= \% \text{ grease removal by detergent.}$$

Lather - Shampoo

This method is a tentative standard for the American Society for Testing Materials, D-1173, Method of Test for Foaming Properties of Surface-Active Agents.

Use of soiling agents in combination with the foaming agents under test is not convenient with the Ross-Miles test equipment because it is so difficult to clean. Consequently other methods for estimation of foam volume and stability in the presence of soils must be devised. Moderately successful results have been obtained using a simple stirrer agitation method in a graduated, temperature controlled vessel, to which may be added successive increments of a chosen soil such as lanolin, mineral or vegetable oil.

HARD SURFACE CLEANING

Hard surfaces as contrasted with soft, or textile surfaces, exhibit several differences which strongly affect detergency operations. In general, hard surfaces have a comparatively restricted surface area, though still of considerable dimensions when closely examined.

Metal Cleaning

A metal surface offers a number of conveniences in the estimation of soil removal. Though it may seem perfectly flat, when observed microscopically may contain a tremendous number of pores or hills. Many of the metals are chemically reactive and either oxidize rapidly, presenting an adsorptive surface, or can react preferentially to provide chemisorbed surfaces as with adsorbed corrosion inhibitors, etc. Another factor which may have influenced the extensive investigation in this field is the tremendous tonnages of metal processed.

Performance tests

The main performance test is effectiveness of soil removal under actual (or simulated) conditions of use. However, there are many individual factors which can be measured and when properly used can be made excellent supplements to detergency testing. The more generally investigated performance tests are :

> pH and titration values
> Colloidal properties
> Water softening
> Conductivity
> Solvent action
> Surface and interfacial tension measurements
> Emulsification
> Rinsability
> Stability under conditions of usage.

A very short review will indicate the variety of metal cleaning processes:

> Wiping (by hand or machine)
> Brushing (by hand or machine)
> Sand blasting
> Tumbling
> Hydraulic spraying
> Steam gun spraying
> Dipping in still tank

> Machine washing
> Electrolytic cleaning (anodie or cathodie)
> Continuous acid or alkaline dip
> Vapour degreasing
> Solvent degreasing.

To this may be added:

Emulsion degreasing : (1) Organic solvent containing oil soluble emulsifying agent. (2) Organic solvent containing oil-soluble emulsifying agent mixed with kerosene and solvent to form emulsion.

Diphase cleaning

The main general types of cleaning, however, may be reduced to the following :

> Soak tank cleaning
> Mechanical tank or spray cleaning
> Electrolytic cleaning (anodie, cathodie, alternate current)
> Solvent or vapour degreasing
> Emulsion degreasing
> Diphase cleaning.

Surfaces

Cleaning of metals represents the entire gamut of pure metals and alloys. All must be cleaned to a greater or lesser extent for either processing or for use as a finished product. The chemical reactivity of the surfaces to form either oxide coatings or reactions with components of corrosion inhibiting compositions or cleaning compounds, may greatly influence the effectiveness of the compositions used and of the finished product. Lard oil which has a stronger bond of attachment for metal than mineral oil, is readily removed by the addition of surface active agents in contrast with mineral oil which is not. Presence or absence of oxide film is an important factor in the ease with which oil may be removed, since this probably depends upon the bond between the oil and the metal surface. The effect of freshly pickled surfaces in increasing the attraction of the oil for the metal is an important one and is most marked with mineral oils. In contrast, fatty oils containing considerable amounts of free fatty acid are more strongly adsorbed by the oxide films. Condition of the metal as it is to be worked in the plant will be the controlling factor in metal cleaning operations.

Soils

The variety of soils encountered in the cleaning of metals is numerous. Generally, however, they are comprised of fatty or mineral oils, debris from previous processing, corrosion inhibitors, drawing compounds and the like. These oils act as entraining agents for solids which may be oxides or simply debris picked up in the processing operation.

Machines for removal

Probably the simplest device which is used in the metal cleaning industry is the so-called soak tank method of cleaning. This may be manually or mechanically operated and in either event is strongly dependent upon the temperature of the bath and the degree of agitation (with steam, air or pump) to provide the desired amount of mechanical action.

Laboratory machines for duplicating the results obtained in plant machines are simple and are frequently quite representative of what can be done in the plant, particularly when samples of the actual soiled surface can be cleaned with the material expected to be used. Generally the simple soak bath is used in laboratory evaluation but spraying operations may be set up with spray nozzles and pumps which simulate plant application. A device which agitates the article to be cleaned in the soak bath has been suggested. Whichever piece of machinery is used for laboratory evaluation is not as important as control of the surface prior to, and subsequent to, the cleaning operation and full recognition of the desired state of cleanliness is important for maximum economy.

Sample

The size and shape of the panel to be soiled and cleaned can vary at will. Disks one inch in diameter to panels 6 × 6 inches or more have been used. The shape of the panel can likewise vary, some being perforated at two points on one edge so that when held at an angle they drain from a corner, while still another was shaped like a flat spearhead so that drainage can be made to take place from the tip.

Estimation of soil removal

In general, either a metal surface is clean or it is not, and any degree between these two extremes may be considered unsatisfactory. In effect, this is a "go," "no go" type procedure. However, for the estimation of the effectiveness of cleaning agents in the laboratory it is desirable that satisfactory quantitative methods be available. Historically the water-break method is the one which has provide most satisfactory. It can be made in the plant as well as in the laboratory, and is indicative of general cleanliness. In either case, maximum advantage is obtained by an acid dip prior to evaluation of the degree of soil removal. An obvious method of attack is to submit the cleaned sample to *solvent extraction* and to weigh the oil or soil which is removed. In cases where almost complete cleanliness has been obtained, sensitivity of this test is obviously dependent upon technic and a sufficiently sensitive weighing device. The *radioactive trace technic* has been used in the evaluation of metal cleaning compositions and likewise for the evaluation of the several methods available to determine the cleanliness of cleaned metal surfaces. The *copper dip method* has been used for the estimation of the cleanliness of surfaces which are to be electroplated. The adherence and the brightness of the copper coating which results from dipping the cleaned ferrous metal in an acid copper sulphate solution is indicative of the quality of electroplate to be expected. The *fluorescent dye method* depends upon the inclusion of an ultraviolet sensitive fluorescing material combined with the oil used as a soiling agent. This has certain advantages for pictorial evaluation of data but is not entirely satisfactory as a quantitative method for estimation.

Dishwashing

The cleansing of dishes has been investigated with one or more objectives which are: deposition of hard water film, soil removal, bacteriological cleanliness, and washing machine design and efficiency.

Frequently several of these factors can be combined and a clear over-all conception of the effectiveness of the detergent solution or of the machine design can be obtained. Mahlmann and his co-workers in a number of different publications. It has been found that for bacteriological cleanliness a temperature of not less than 170°F. in the wash and rinse will generally provide the desired bactericidal action. Consequently the temperature of wash and of the abundance of the supply of wash water to a mechanical washing machine is most important.

Prior to the introduction of polyphosphates there was considerable difficulty in the deposition of an insoluble hard water film upon glassware which was most unsightly and could potentially harbor bacteria and the soil particles. It has been found that the inclusion of polyphosphates in the washing composition is most beneficial in reducing hard water precipitate and deposition. The essentials of a test for dish washing evaluation are comprised of the following :

Soil

In evaluating the removal of these soils it generally is necessary that a tracer compound of one sort or another be added with the expectation that it will be removed in the same ratio as the soil itself and thus can be used as a device to measure the soil retained on the surface. Agents which have been used for this purpose are lamp black, India Ink, a combination of printing ink and linseed oil, a test organism suspension, and also ferric chloride.

The most recent thinking on this general subject is that the method for deposition of the soil is important to the extent that the viscosity of the soil medium should be adjusted and the soil applied in some given and reproducible manner.

Measurement of soil retention (or removal)

A number of different photometers, colorimeters and glossimeters are available on the market and it seems hardly necessary that a special design of equipment now be developed. Obviously if the desire is to ascertain bacteriological cleanliness, a properly standardised plating technique will be used and several of these have been described.

Machines for washing

Most of the small scale laboratory devices for evaluation of washing efficiency have been based on the use of microscope slides and either an agitating, rocking or rotating device in which these slides are held while immersed and moved by mechanical action in the bath to provide soil removal.

MISCELLANEOUS TESTS

There are several tests which may prove useful in the further evaluation of compounds to be market tested. Most synthetic detergents are deficient in the ability to suspend soil once it is removed from a fabric surface. The redeposition method will indicate the comparative ability of a composition to prevent this graying, effect. A compound almost universally used for this purpose is carboxymethylcellulose, and it might be thought that its isolation from the detergent composition and its analysis would prove the most effective method for estimation. Such is not the case even though methods of analysis exist, and the redeposition characteristics will be fully as revealing as an analytical procedure. Furthermore, though there are a few other compounds which possess this same power to prevent redeposition, analysis means little when measure of comparative effectiveness is desired.

Another useful addition to the detergent field, and used both in soaps and synthetic detergents, is the fluorescent dye. These are also called brightening or "bleaching" agents. The best criterion of effectiveness again is a direct comparison, under use conditions of either competitive detergents or competitive dyes. While not actually increasing detergency or soil removal, the effect is as if increased cleanliness had occurred and is especially evident in actual practice over a sequence of washings.

Determination of pH can be very helpful, but unless properly executed, can be most misleading. This determination should be a routine operation.

A product, before introduction to the market, should be as free from possibility of harm to equipment as possible. Corrosion of metal, glass or vitreous enamel surfaces can cause many complaints and even result in rejection of a product. The methods chosen cover aluminium and vitreous enamel surfaces.

Brightening Agents

These compounds are colourless to the eye but are excited by ultraviolet light, fluorescing blue. They have been termed "bleaching" agents because they are so effective in converting daylight ultraviolet to visual, making white fabric appear bleached or brightened in comparison with one not so treated. Brightening is a better term since colours are also increased in intensity as well.

The most generally used dyes are chosen for their substantivity to cellulosic fabrics, though some less generally used are available which exhaust onto wool rather than cellulose. Depending upon chemical structure, these dyes will fluoresce as a reddish blue or preferably as a brilliant, almost true blue colour.

Choice of brightening agent will depend upon the hue of the dye, the degree of substantivity or exhaustion from the bath, stability toward excess untraviolet illumination, stability toward bleaching compositions, buildup of colour and cost. A reddish blue fluorescence may not have the intensity of a dye giving a deep blue. Unless the dye is evenly adsorbed in a short time, and essentially quantitatively, it may fail under adverse conditions of operation or in the small amounts used. Some dyes will "sunburn" when exposed continuously to ultraviolet, producing an undesirable visual colour. Other dyes may be oxidised to undesirable visual "off" colours by the common bleaching agents. Early experience with these agents in repeated washings resulted in "off" shades of pink built up by adsorption on the fabric. The evidence of the effectiveness of well selected brightening agents is so clear that a detergent manufacturer can hardly afford not to use them, especially as they are used in such minor percentages.

The evaluation of these dyes can be complicated or relatively simple, depending upon the size of the detergent operation, and upon the accuracy and sensitivity of the method chosen. Certainly a dye manufacturer can afford a complicated evaluation method if it will either control his operation or permit development of better, more effective agents. The detergent manufacturer is interested in producing maximum effectiveness at minimum cost of dye and of process control and the method which follows has proved quite adequate.

Evaluation

Machine. Launder-Ometer, or Terg-O-Tometer, set at 345° rotation and 117 half-cycles per minute.

Fabric. Desized Indian Head or white terry cloth towelling, 8 × 9 inch size.

Temperature. 120°F.

Detergent concentration. 0.25%

Volume of solution. 1000 ml. Terg-O-Tometer. 200 ml. Launder-Ometer.

Time of wash. 10 minutes.

Time of Rinse. 5 minutes.

Procedure

(1) Wash for 10 minutes.
(2) Remove swatches, run through hand wringer.
(3) Rinse for 5 minutes.
(4) Remove swatches, run through hand wringer.
(5) Oven dry.

Number of washes. Selection of fabric swatch size will make it possible to remove a portion after every wash or alternatively after the fifth and tenth was periods.

These same tests can be performed in varying water hardnesses with various fabrics, and in the presence of commercial bleaching agents in the amounts recommended by the bleach manufacturer. If temperature of wash water or rinse is important, tests can be made to cover a range to indicate effectiveness of the agent.

pH Determination

The pH values of soaps, alkalies, and detergents are valuable indications useful in application. It should be recalled that soaps, as "neutral" salts of fatty acids, have pH values much higher than organic sulphates or sulphonates. Should soap pH values fall in a range lower than those for the normal salts, difficulty in the way of reduced detergency or actual deposition of insoluble fatty acids may be expected. The surface active sulphate or sulphonates may be finished off at pH values over a considerable range, but they generally closely approach pH 7 as the normal salts.

"Building" of soaps or synthetic detergents with alkaline salts frequently is accomplished to conform to some predetermined pH level dependent upon the sensitivity of the product or equipment with which they are to be used. Most built soap or synthetics to come into contact with the hands are maintained at levels below pH 10.5 since otherwise they may cause excessive skin irritation. This buffering ability to maintain a given level is most readily ascertained using a combination of pH and titration methods.

Colorimetric methods for pH determination can be most useful, but for precise estimation the electrometric method is to be preferred. Simplified apparatus is available for electrometric measurement, largely with the glass electrode which may be operated from line current or from batteries and may be secured for versatility of measurement in different media.

The first item to recall in making pH measurements is that distilled water unless freshly boiled may absorb much carbon dioxide from the air and will then have a quite acid value. Freshly boiled distilled water is indicated for the measurement of pH values for detergent solutions since the synthetic detergents are frequently poorly buffered. The fact is that very many of the neutral or sodium sulphate-built detergents will simply assume the pH of the water used in the test. Attempts to obtain check pH values between laboratories may frequently fail unless this precaution of using freshly boiled distilled water is observed.

The second item of importance is to standardise the pH meter with appropriate standard buffer solutions using the proper glass electrode for the pH range and temperature used.

The American Society for Testing Materials Specification E-70 is a tentative method for Determination of the pH of aqueous Solutions of Soaps and Detergents with the glass electrode and gives specific data for the proper use of this equipment.

Proposed method of test for pH of aqueous solutions of soaps and detergents

Scope. This method covers procedures for the preparation of aqeuous solutions of soaps and detergents and for the determination of their pH.

Sampling. The material shall be sampled in accordance with the Standard Methods of Sampling and Chemical Analysis of Soaps and Soap

Products, or the Standard Methods of Sampling and Chemical Analysis of Alkaline Detergents, whichever is applicable.

Reagent, Distilled Water. Distilled water shall be boiled thoroughly, or purged with CO_2-free air, to remove CO_2 and shall be protected with soda-lime or soda-asbestos (Ascarite) while cooling and in storage. The pH of this water shall be between 6.2 and 7.2 at 25°C. The residue on evaporation, when heated at 105°C for 1 hr., shall not be more than 0.5 mg. per litre.

Preparation of solutions. Weigh 1 ± 0.001 g. of the soap or detergent and transfer to a 1-liter volumetric flask. Partially fill the flask with distilled water and agitate until the sample is completely dissolved. Adjust the temperature of the solution and the distilled water to 25 ± 0.5°C. and fill to the calibration mark with distilled water. Stopper the flask, mix thoroughly, and allow the solution to stand at a temperature of 25°C. for 2 hours prior to measuring the pH.

Determination of pH. Measure the pH of the solution as directed in the Tentative Method for Determination of the pH of Aqueous Solutions with the Glass Electrode (A.S.T.M. Designation: E-70). Use the "low sodium-error" glass electrodes for solutions having pH values higher than 9. Use type I or type II meters.

Corrosion

Detergent compositions are designed for many purposes, but one thing they should *not* do, is to cause corrosion or remove protective coatings. Dishwashing compositions should not remove decorative decals or overglazes, nor should they cause corrosion of silverware or of the machines in which they may be used. Washing machine detergents should not cause corrosion of metal parts of the machine, nor remove vitreous enamel or damage organic finishes. The machine designer and the metallurgist can do much to eliminate complaint, but because detergent compositions contain alkaline builder salts and are generally very effective electrolytes, they must be suitably inhibited to minimise attack.

Aluminium is a metal commonly found in machines and is rather easily attacked by either acid or alkaline media, but can be protected through the use of silicates in alkaline compositions. Chrome or nickel plating, when available and well deposited, prove quite satisfactory but when skimped are quickly removed. Substitute platings frequently are less resistant and may even cause electrolytic corrosion with rapid removal and complaint. This latter difficulty can be alleviated by sound metallurgy, but in any event, the less corrosion caused by the detergent, the better the washer.

Polyphosphates have been shown to remove vitreous enamel under conditions of high temperature and high concentration, but even this corrosion can be controlled. In this same category would be removal of ceramic overglaze from chinaware. Complaints from organic finishes originate from wear, brittleness, crazing or impact, with exposure of base metal to electrolytes; however under normal conditions they are unaffected by detergents.

Corrosion and corrosion inhibition studies can be carried out using the following two methods as applied to metal or to ceramic surfaces.

Glossary

The glossary of technical terms will help the processors, consumers and others connected with this trade in correct interpretation of the terms.

Acid oil. It is normally produced by acidulation of soap-stock and contains neutral oil, fatty acids and different levels of other lipids but having free fatty acids more than 35 per cent by mass.

Active matter. In a composition, the whole of the specific constituents responsible for the activity specified.

Adsorption layer of surface active agent. In the case of surface active substances in solution, a layer stretching more or less across an interface and the thickness of which is determined by the fact that at any random point in that layer, the concentration of the adsorbed product is greater than that in each of the contiguous phases.

Advancing wetting angle. The introduction of a solid surface at slow and constant speed into a liquid phase gives rise to the formation of a contact angle which may depend upon the nature of the surface and the speed of entry. This angle is called *advancing wetting angle*.

Advancing wetting tension. Wetting tension corresponding to the formation of an advancing wetting angle.

After-treating agent for prints. Product covered by the definition corresponding to the dye-fixing agent.

Amide formation. Chemical reaction giving rise to the formation of amides by the action of ammonia or primary or secondary amines on acids, their halides or their esters.

Amphiphilic product. Product comprising in its molecule, at the same time, one or more hydrophilic groups and one or more lipophilic groups. (*Note* — Surface active agents are amphiphilic products.)

Ampholytic surface active agent. A surface active agent having two or more functional groups which, depending on the conditions of the medium, can be ionized in an aqueous solution and give to the compound the characteristics of an anionic or a cationic surface active agent. This ionic behaviour is similar to that of amphoteric compounds in the broadest sense.

Ancillary (for detergents). A complementary component of a detergent which imparts added properties unrelated to the washing action as such. *Examples*: Fluorescent whitening agents, corrosion inhibitors, anti-electrostatic agents, colouring matter, perfumes, bactericides. Ancillaries are usually present in small quantities.

Anionic surface active agent. A surface active agent which has one or more functional groups and ionises in aqueous solution to produce negatively charged organic ions responsible for the surface activity.

Anti-foaming agent (anti-foamer). A substance which prevents the formation of a foam or considerably reduces foam persistence.

Anti-redeposition agent. A complementary component of a detergent, usually organic, which imparts to the latter the property of preventing redeposition. *Example* : Carboxymethyl cellulose.

Anti-foaming agent for the textile industry. Product which prevents the formation of foam or which considerably reduces its stability. In the textile industry, it is used particularly in sizing, finishing and dye baths, in printing pastes, etc. *Note* – These products include, among other things, certain surface active agents and preparations comprising them, for example: those based on oils, phosphoric acid esters, and alcohols of high molecular mass.

Anti-redepositing power. The ability of a substance to prevent insoluble particles from redepositing on the washed surface and, possibly, to maintain the particles in suspension.

Anti-static agent. Product which, when applied to a textile article during or after processing, makes it possible to eliminate the disadvantages due to phenomena of static electricity. *Note* – These products are generally surface active agents, for example: alkylsulphonates, alkylphosphates, alkylamines and their derivatives, and also the ethoxylation products of fatty acids, fatty alcohols, fatty amines, fatty amides, alkylphenols, and quarternary ammonium salts.

Apparent density. Mass of unit apparent volume.

Apparent volume. Volume determined by the exterior limits of quantity of substance, under the experimental test conditions. The volume includes possible bubbles, pores and interstices.

Aqueous emulsion (symbol L-H: oil in water). An emulsion in which the continuous phase is aqueous. *Note* – L from the Greek *lipos*: H from the Greek *hudor*.

Artificial soil. Soil of selected composition, prepared for detergency tests.

Assistant for desizing and for removal of printing thickeners. Product which accelerates the removal of print thickeners as well as the desizing insofar as starch products are concerned, possibly in combination with linseed oil or other sizes based on linseed oil. To this end, appropriate surface active agents, such as wetting agents or detergents can be added to agents based on enzymes. Similarly, the removal of linseed oil or sizes based on linseed oil is facilitated by the use of surface active agents as indicated above in association with solvents and/or oxidising agents.

Auto-oxidation. Chemical reaction involving the unaided coupling, either fast or slow, of molecular oxygen to an organic or inorganic compound.

Biodegradability. Aptitude of an organic substance to undergo biodegradation.

Biodegradable surface active agent. Surface active agent which is susceptible to biodegradation causing a loss of its surface active characteristics.

Biodegradation. Molecular degradation of an organic substance resulting from the complex actions of living organisms, ordinarily in an aqueous medium.

Bleaching assistance. Product which makes it possible to speed up the bleaching operation and to make it more even in effect. *Note* – These products are generally surface active agents, mainly wetting agents which are stable in bleaching baths.

Booster (for detergents). A complementary component of a detergent, usually organic, which strengthens certain characteristic properties of the essential constituents. *Examples* : Certain alkanolamides and amine oxides.

Brightening agent. Product intended to improve or restore the purity of the colour; it may also play a part in giving the textile articles the desired finish from the point of view of feel and sheen. *Note* – These products are generally those mentioned under the heading of processing aid.

Bubble. A volume of gas enclosed by a thin film of liquid.

Builder (for detergents). A complementary component of a detergent, usually inorganic, which, with reference to the washing action, enhances the characteristic properties of the essential constituents. *Examples* : Polyphosphates, carbonates, silicates.

Capillary activity. The property of a substance in solution to decrease interfacial tension, associated with an augmentation of the concentration of the substance at the interface.

Carbonising assistant. Product facilitating and speeding up the penetration of carbonising agents (acids or generators of acids) into the vegetable impurities in wool, thus promoting their destruction during subsequent heat treatment. *Note*– These products are wetting agents with sufficient stability in the presence of acids.

Cationic surface active agent. Surface active agent which has one or more functional groups and ionizes in aqueous solution to produce positively charged organic ions responsible for the surface activity.

Chelate. A metallic complex in which the metal ion is suppressed by chelation.

Chelating agent. A substance having a molecular structure embodying several electron-donor groups which render it capable of combining with metallic ions by chelation.

Chelating power. The ability of certain bodies to complex cations to form a so-called ring structure.

Chelation. The formation of complexes in which the suppressed metal ion is held in a so-called ring structure, with one or more molecules having, sometimes, several electron-donor groups.

Chemical bleaching agent. A product which, by chemical action, usually oxidising or reducing, acting under controlled conditions on textile or other materials, converts substances which affect adversely the white appearance of the material, into substances of less intense colouration.

Coacervate (coacervated phase). A concentrated phase of a system which has undergone coacervation.

Coacervate system. The whole of the phases of a system which has undergone coacervation.

Complexing power. The ability of certain bodies to combine with cations which then lose their ionic character.

Coalescence. The joining together of droplets of an emulsion.

Coacervation. The separation into colloidal phases containing the same constituents in equilibrium but in different proportions.

Cloud point. In the case of certain non-ionic surface active agents, the temperature above which their aqueous solutions become heterogeneous with the formation of two liquid phases.

Complexing. The transforming of a metal cation into a new ion by the action of molecules containing at least one electron-donor group.

Coning oil. Product intended to make yarns suitable for winding and for subsequent textile operations, such as knitting, by making the yarns more flexible and slippery. *Note*–These products are oils, or oils which are emulsifiable in water, which can be prepared with the aid of surface active agents, such as oil soluble polyglycol esters or ethers.

Contact angle. In a plane perpendicular to the line of separation formed by three phases, the tangents to the two curves formed when the plane cuts one of the three phases are considered. The angle formed by these tangents is called the *contact angle* of this phase in relation to the other two.

Critical concentration for micelle formation. Due to the formation of micelles aqueous solutions of surface active agents exhibit a more of less abrupt change in their physical properties over a narrow concentration range. The concentration of surface active agent at which the concentration of micelles suddenly becomes appreciable is referred to as the critical concentration for micelle formation. The physical characteristics that are changed abruptly are conductivity, surface tension, osmotic pressure, interfacial tension, detergency, etc.

Cutting oil. A lubricating compound, emulsifiable or not, which facilitates the work of machine tools and the dissipation of heat produced. *Note* – It may contain an additive imparting anti-corrosive properties.

Dehydration. Physical operation resulting in the removal of all or part of the water bound to a product; *or* chemical reaction resulting in the removal of one or more molecules of water.

Detergency (detergence). The process by which soil is dislodged and brought into a state of solution or dispersion. In its usual sense, detergency has the effect of cleaning surfaces. It is the result of the action of several physico-chemical phenomena.

Detergent. A product specially formulated to promote the development of detergency.

Detergent for the textile industry. Product used in the textile industry to eliminate fats and soiling matter on textiles during manufacture and finishing. Its composition and/or its formulation meet the requirements for the various stages of the work, for example: for scouring raw wool, yarns or pieces and for the back-washing of dyed and printed fabrics, etc. *Note*–These products are surface active agents or compounds comprising them, such as soaps, alkylsulphates, alkylsulphonates, fatty acid condensates, alkylaryl-sulphonates, polyglycol esters and ethers.

Detergent oil. A lubricating compound, generally based on mineral oils and surface active agents, which facilitates the suspension or resuspension of solid particles emanating from the operation of an internal combustion engine.

Dilatancy. Under isothermal and reversible conditions, increase without hysteresis of the apparent viscosity under a shearing load.

Discharging agent. Product which, when added to a printing paste of discharge, makes it possible to discharge colour satisfactorily in the case of a dye which is difficult to discharge. *Note* –These products are mainly based on derivatives of quarternary ammonium and ethoxylated amines.

Disperse phase. The discontinuous phase of a dispersion.

Dispersing agent. A substance capable of promoting the formation of a dispersion.

Dispersing power. The extent of the ability of a product to bring about the formation of a dispersion.

Dispersion medium. The continuous phase of a dispersion.

Distilled fatty acids. These are technical grade of fatty acids which are subsequently distilled having specific free fatty acids and colour.

Dispersion. A system consisting of several phases, of which one is continuous and at least one other is finely dispersed.

Detersion (cleaning). The act of bringing the phenomenon of detergency into effect.

Dye-fixing agent. Product intended to improve the fastness of dyes from certain points of view. To increase the rubbing fastness, use is made of detergents for the textile industry which eliminate the loose dye. To increase the wet fastness, use is made of products which, together with the dye, form stable compounds which do not dissolve easily. *Note* – In the latter case, these products include cationic substances, such as amines and amine derivatives, for example, quarternary ammonium salts and ethoxylated amines.

Emulsion. A heterogeneous system made by dispersing small globules of one liquid in another liquid which forms a continuous phase.

Emulsifying agent (emulsifier). A substance which permits or facilitates the formation of an emulsion.

Emulsification. The action causing the formation of an emulsion.

Emulsifiable liquid. A liquid suitable for constituting the disperse phase of an emulsion.

Emulsifying agent for the textile industry. Product which makes possible or facilitates the formation of an emulsion. In the textile industry, it is generally used in the preparation of batching, brightening and/or preparing agents, coning oils, etc., in order to obtain a special effect. *Note*–These products are surface active agents or preparations comprising them such as soaps, alkylsulphates, alkylsulphonates, fatty acid condensates, alkylaryl-sulphonates, polyglycol esters and ethers, esters of fatty acids and polyhydroxyl compounds.

Emulsifying liquid. A liquid suitable for constituting the continuous phase of an emulsion.

Emulsifying power. The ability of a substance to facilitate the formation of an emulsion.

Emulsion persistence. The ability of an emulsion to persist.

Endophilicity. Constitutional property which denotes the tendency of the whole or a part of a molecule to penetrate into, or remain in a phase. It is characterised, with regard to the functional groups in the molecule, by the fact that such groups give rise to decrease in the variation of the chemical potential when the molecules of the product pass from a gaseous ideal state to the phase under consideration.

Esterification. In the particular case of surface active agents, chemical reaction giving rise to an ester derived from an acid and an alcohol, enol or phenol with the elimination of water.

Ethoxylation. In the particular case of surface active agents, chemical reaction leading to the addition of one or more molecules of ethylene oxide at a liable hydrogen compound.

Exophilicity. Constitutional property which denotes the tendency of the whole or a part of a molecule to pass out of, or not to penetrate into, a phase. It is characterised, with regard to the functional groups in the molecule, by the fact that such groups give rise to increase in the variation of the chemical potential when the molecules of the product pass from a gaseous ideal state to the phase under consideration.

Fibre humectant. Product intended to control the desired humidity of the yarns, to maintain it throughout the subsequent textile operations, and possibly to increase the strength of the yarns. *Note* –These products are generally solutions of wetting agents with hygroscopic agents and/or preserving agents added.

Fibre protecting agent. Product used for preserving fibres, particularly animal fibres, during the operations of bleaching, dyeing and stripping. *Note* – These products are based, for example, on degraded proteins, fatty acid and protein condensates, ammonium alkylsulphates and alkylsulphonates, salts of ligno-sulphonic acids.

Filler (for detergents). An organic or inorganic product, usually inert, employed to produce the desired type or presentation and/or concentration. *Examples* : Sodium sulphate, water.

Film. A thin layer of matter, homogeneous or not.

Finishing assistant. Product added to finishing compounds to impart fluidity, body and/or stability and to alter the finishing in the desired way. *Note* — These products include, among others, sulphated oils and greases, and products mentioned under the heading of processing aid.

Flocculate (floc). Matter which has undergone flocculation.

Foam. A mass of gas cells separated by thin films of liquid and formed by the juxtaposition of bubbles, giving a gas dispersed in a liquid.

Foam booster. A product which increases foaming power.

Foam drainage. The return to the liquid phase of the excess of liquid entrained by bubbles during foaming.

Foam persistence. The ability of a foam to persist.

Foam stabiliser. A product which increases the stability of foam. *Note* – According to the conditions of test or use, or according to the nature of the foaming product, the effect of stabilisation can also result in the formation of a greater volume of foam as well as lead to a greater persistence of the foam produced.

Foaming. The action causing formation of foam.

Foaming agent (foamer). A substance which, when introduced into a liquid, confers on it an ability to form foam.

Foaming power. The ability to produce foam.

Free energy of adhesion. The work required to achieve, in an isothermal, isobaric and reversible manner, a separation at the interface between two phases (liquid/solid) with the formation of a new free liquid surface of the same dimensions as the initial interface, is reflected in an increase in the free energy of the system. This energy is called *free energy of adhesion*. It is the sum of the free energy of wetting and the free surface energy. It is expressed in ergs.

Free energy of wetting. The work obtained when a surface is wetted in an isothermal, isobaric and reversible manner without changing the size of the free liquid surface, is reflected in a diminution in the free energy of the system. This part of the free energy is called *free energy of wetting*. It is expressed in ergs.

Free interfacial energy (liquid-liquid). The energy which manifests itself in the work required to increase or form the interface separating two liquid phases, in an isothermal and reversible manner. The free interfacial energy is expressed in joules.

Free surface energy. The energy which manifests itself in the work required to increase or form a surface in a liquid in an isothermal and reversible manner. The free surface energy is expressed in joules. The joule (symbol : J) is the SI unit; usually this energy is expressed in ergs, the erg (symbol : erg) being the CGS unit. $1J = 10^7$ ergs.

Fulling assistant. Product intended to facilitate the formation of felt during the fulling operation. *Note*–Generally, the fibres are made more slippery by surface active agents or preparations containing them, such as soaps, alkylsulphates and fatty acid condensates, possibly in conjunction with mineral or organic swelling agents.

Hydrolysis. Reaction of splitting by water. In the particular case of surface active agents, hydrolysis is more particularly the reverse action to esterification or amide formation and is characterised by the formation of an acid and an alcohol, enol or phenol or of ammonia or an amine. Hydrolysis of fats gives rise to fatty acids and glycerol; that of soap, to fatty acids and a base.

Hydrophilic group. Molecular group having an endophilic behaviour in relation to water.

Hydrophilic-lipophilic balance. The relative power of the polar group or groups and of the non-polar part conditions the affinities of the molecule for water and for organic solvents of low polarity respectively. The relation between these affinities represents the hydrophilic-lipophilic balance of the compound. (*Note* — This definition relates only to emulsifying agents.)

Hydrophobic group. Molecular group having an exophilic behaviour in relation to water.

Hydrophily. Endophily in relation to water.

Hydrophoby. Exophily in relation to water.

Hydrotropy. The increase in solubility of a substance which is only slightly soluble in water by the addition of a third substance. This third substance is called a 'hydrotrope' or a 'hydrotropic agent'.

Interfacial tension. The force per unit length arising from the free interfacial energy. It is numerically equal to the free interfacial energy per unit of interface and is expressed in dynes per centimetre.

Kier boiling assistant. Product intended to increase the effectiveness and speed of the processing, under pressure or otherwise, of textile materials or articles made of natural or regenerated cellulose fibres, either alone or mixed with similar fibres or with synthetic fibres, with alkaline lye, water, salt solutions or acid solutions. It is applied, for example, in the boiling of grey cotton (under pressure or otherwise), the scouring of linen, the rendering of cotton articles absorbent to water by the continuous process, etc. *Note* – These are generally special wetting products, often mixed with solvents.

Krafft point. The temperature (in practice, narrow range of temperature) at which the solubility of an ionic surface active agent reaches the critical value for micelle formation. Above this temperature, the solubility curve shows a rapid increase. In the soap industry, 'Krafft Point' is the temperature below which a transparent soap solution becomes cloudy.

Levelling agent. Product designed to promote the even dyeing of textiles. *Note* – These products are surface active agents or preparations comprising them, such as sulphated oils, esters and amides of fatty acids, fatty acid condensates, esters of fatty acids, fatty acid condensates, alkylsulphates, alkylarylsulphonates, alkyl and alkylaryl polyglycolethers, polyglycol esters of fatty acids, amine derivatives. Agents with the properties of protective colloids, such as fatty acids and protein condensates can also be used.

Lipophilicity. Endophilicity in relation to a non-gaseous non-polar organic phase.

Lipophilic group. Molecular group having an endophilic behaviour in relation to a non-gaseous organic phase.

Lipophibicity. Exophilicity in relation to a non-gaseous non-polar organic phase.

Lyophilic group. A molecular group which possesses endophilic behaviour with respect to a liquid phase.

Lyophily. A predominant tendency to endophily of a substance dispersed in a medium.

Lyophobic group. A molecular group which possesses exophilic behaviour with respect to a liquid phase.

Lyotropy. The increase in solubility of a substance which is only slightly soluble in a solvent by the addition of a third substance. This third substance is called a 'lyotrope' or a 'lyotropic agent'.

Mercerizing assistant. Product used to improve the wetting power of mercerising lyes and thus to speed up their uniform penetration into the fibres. *Note* —These products are wetting agents which are stable in highly concentrated lyes; they are based both on a component which is effective as a surface active agent and as an emulsifier in lyes (alkylsulphates of low molecular mass, highly sulphated oils, cresols, xylenols) and anti-foaming and wetting substance, itself insoluble in lye, but made soluble by hydrotropy (for example, butyl glycol, ethoxylated amines, etc.).

Micelle. The essential feature of surface active solutions is the existence of colloidal size particles called micelles, formed by the spontaneous association of the ultimate molecules and/or ions of the solute. The micelles exist in thermodynamically stable equilibrium with these simpler ions and molecules. Micelles may be laminar, cylindrical or spherical in shape.

Monomolecular layer (Monolayer). Adsorption layer which, under determined conditions of concentration, is limited to unit molecular thickness of a surface active agent.

Non-biodegradable surface active agent. Surface active agent which resists biodegradation.

Non-ionic surface active agent. A surface active agent which does not produce ions in an aqueous solution. The solubility in water of non-ionic surface active agents is due to the presence in the molecules of functional groups which have a strong affinity for water.

Non-polar group. The organic part of the molecule, in which the distribution of electrons does not cause a considerable electrical dipole moment. Such a group conditions the affinity for organic solvents of low polarity and consequently the lipophilic character of the molecule.

Oil emulsion (Symbol H-L: water in oil). An emulsion in which the continuous phase is a liquid insoluble in water. *Note* –H from the Greek *hudor*: L from the Greek *lipos*.

Peptisation. The formation of a stable dispersion from flocs or aggregates.

Peptising agent (peptiser). A substance capable of promoting peptisation.

Phosphation. In the particular case of surface active agents, chemical reaction giving rise to the formation of phosphoric esters.

Phosphonation. Chemical reaction or sequence of chemical reactions, leading to the introduction into a molecule of one or more phosphonic radicals by direct carbon/phosphorus linkage.

Polar group. A functional group, in which the distribution of electrons tends to give a considerable electrical dipole moment to the molecule. Such a group conditions the affinity for markedly polar surfaces, the affinity for water in particular and the hydrophilic character of the molecule.

Polar-non-polar structure. The structure of molecule which has at least one polar group and a large non-polar group. Such a structure conditions the hydrophilic and lipophilic character of the molecule.

Processed fixed oils. Vegetable oils, animal fats and marine oils processed individually or in blends to remove their objectionable impurities or even chemically modify their composition to make them suitable for the manufacture of soap.

Processing aid. Product intended, in general, to make a textile material better suited to undergo a subsequent operation, for example: spinning, winding, knitting, etc. *Note–* These products are generally surface active agents or preparations of the latter with oils and greases. The surface active agents used include, for example: sulphated oils and greases, alkylsulphates, esters and amides of fatty acids, fatty amine condensates, and also the ethoxylation products of fatty acids and fatty alcohols, fatty amides or fatty amine condensates.

Propoxylation. In the particular case of surface active agents, chemical reaction leading to the addition of one or more molecules of propylene oxide at a liable hydrogen compound.

Protective colloid. Substance which, within a certain concentration range and when acting as a lyophilic colloid, retards or prevents the aggregation of particles of a lyophobic dispersion.

Pure fatty acids. These are distilled and fractionated fatty acids, for example, lauric acid, stearic acid, etc.

Receding wetting angle. The withdrawal of a solid surface at slow and constant speed from a liquid phase gives rise to the formation of a contact angle which may depend upon the nature of the surface and the speed of withdrawal. This angle is called the *receding wetting angle*.

Receding wetting tension. Wetting tension corresponding to the formation of a receding wetting angle.

Reduction inhibitor agent. Product lessening the reducing effect of foreign matter on dyes, and consequently combating the destruction of the latter. *Note*–The products include, for example, preparations based on buffer substances and oxidising substances with surface active agents, such as degraded proteins, fatty acid and protein condensates, ammonium salts of alkylsulphates and alkylsulphonates.

Rheopexy (anti-thixotropy). Under isothermal and reversible conditions, increase with hysteresis of the apparent viscosity under a shearing load.

Reversible hydrolysis. The action of water on ions from a dissolved salt, which assumes a state of equilibrium in which exist both ions and molecules of acid or base capable of forming the salt. The molecules of acid or base can revert to the ionic state when the conditions in the medium change. Reversible hydrolylsis is particularly noticeable in the case of salts of weak organic acids or weak organic bases, possessing appreciably hydrophobic radicals.

Saponification. A chemical reaction permitting the separation of an ester into its constituent parts, acid and alcohol or possibly phenol, by the action of a base, with the formation of a salt from the acid. Saponification of fats produces soap.

Sedimentation. In a liquid medium, the accumulation of particles in dispersion under the effect of gravity or centrifugal force.

Semi-acid oil. It is partially acidulated soap-stock and eliminates most of the water content of soap-stock thus giving transport economy but at the same time should not be highly acidic. It consists of soap, fatty acids, neutral oil and water.

Sequestering agent (sequestrant). A substance having functional characteristics which make it capable of suppressing metallic ions and ensuring that they remain in solution in the medium.

Sequestering power. The ability of certain substances to keep cations in solution, in a more or less labile condition, so that the reactions of the cations are then, for the most part, masked.

Sequestration. The masking of metallic ions dissolved in a medium, the ions being normally liable to form precipitates in the presence of certain reagents, particularly surface active agents. The masking is accomplished by the formation of complexes which remain in solution in the medium.

Sizing assistant. Product which, when added to sizing compounds, makes the warp yarns more flexible and slippery for the subsequent operation of weaving. *Note–* These products may, for example, be sulphated or emulsified waxes and greases, possibly with the addition of wetting agents.

Soap. An alkaline salt (inorganic or organic) of a fatty acid, or mixture of fatty acids, containing at least eight carbon atoms. This anionic surface active agent exhibits the phenomenon of reversible hydrolysis by the action of water. Because of this fact, water soluble soaps or true sopas, exhibit characteristic properties. Their reaction is usually alkaline.

Soap-stock. It is soap containing material not completely freed of unsaponified material or other impurities. It is normally produced in the removal of free fatty acids from triglycerides during neutralisation.

Sodium alkyl aryl sulphonate, technical. It is the sodium salt of alkyl aryl sulphonic acid containing atleast 40 per cent of active ingredient.

Sodium Oleostearate, technical, stabilised. It is sodium salt of fatty acids comprising chiefly oleic and stearic acids with added preservatives and necessary electrolytes. It may be built or unbuilt but in a form not suitable for direct domestic use.

Softening agent. Product used to make the processed textile more flexible, and consequently to obtain a given feel. It is also used as an additive in sizing and finishing baths, etc. *Note –* These products are generally surface active agents or preparations based on greases and oils with suitable emulsifying agent. Brightening agents may also be used.

Soil. The undesirable deposit on the surface and/or within the substrate, which changes some characteristics of appearance or feel of clean surfaces.

Solubilising and/or dispersing agent for dyestuffs. Product promoting the solubilisation and/or the aqueous dispersion of dyes and consequently improving their dyeing properties (efficiency, penetration). *Note –* These products are surface active agents, with or without the addition of solvents, for example esters and amides of sulphated fatty acids, condensates, alkylarylsulphonates, polyglycol esters and ethers, derivatives of aliphatic amines.

Solubilising and/or dispersing agent for pigments. Product covered by the definition corresponding to the solubilising and/or dispersing agent for dyestuffs.

Solubilising power. The extent of the ability of a dissolved surface active agent to confer on certain bodies, of low solubility in the pure solvent, an apparent solubility by micelle formation.

Spent bleaching earth. It is produced during bleaching of oils and fats by the process of adsorption bleaching and contains fatty matter, soap, bleaching earths, and other impurities.

Spinning bath additive. Product used, among other things, for clarifying the spinning bath and preventing the nozzles from becoming clogged. *Note* –These products are generally surface active agents or preparations comprising them, for example: sulphated oils, alkylsulphonates, fatty acid condensates, oxyethyl alkylamines, quaternary ammonium derivatives.

Spinning oil. Product which, applied to the fibres in the course of their preparation for spinning, makes them more slippery and more flexible and possibly gives them other surface qualities (for example, cohesion) with a view to the operations of combing and spinning. Spinning oils are also called 'batching oils', 'tearing oils', etc. Depending on the purpose for which they are used, spinning oils may also possess the properties of wetting agents and fulling assistant and other secondary properties, for example, that of promoting the loosening of hard or stem fibres.

Spinning solution additive. Product added during the preparation of the spinning solution, for the purpose of improving the suitability of the solution for spinning and possibly to alter the quality of the filaments.

Spotting agent. Product intended to improve local stains on textile articles. A distinction is drawn between 'dry' and 'wet' spotting agents, depending on whether they act in a solvent or an aqueous medium. *Note* –These products are mainly preparations based on solvents and surface active agents with emulsifying and detergent properties, such as amine soaps, alkylsulphates, alkylsulphonates, fatty acid condensates, alkylarylsulphonates, polyglycol esters and ethers.

Spreading ability. The property of a liquid, particularly of a solution of surface active agents, enabling a drop of this liquid to cover spontaneously another liquid or solid surface.

Spreading tension (spreading coefficient). The tendency of a liquid to spread on a solid surface is expressed by the difference between the wetting tension and the surface tension of the liquid. This value is called *spreading tension.* It is expressed in dynes per centimetre. When the spreading tension is positive the liquid spreads spontaneously on the solid surface.

Stripping agent, partial or total. A partial stripping agent is intended to lighten a dye which is too dark. It acts by eliminating part of the dye. Levelling agents are suitable for this operation. A total stripping agent, or dye removing agent, is used to eliminate the dye from a dyed fabric. Reducing agents are generally used for this, in conjunction with levelling agents.

Sulphation. Chemical reaction giving rise to the formation of a sulphuric ester. In practice, a mono-sulphuric ester is formed.

Sulphite addition. Sulphonation brought about by the reaction of sulphur dioxide or, in more general terms, its derivatives (sulphites, bisulphites) with an electrophilic group.

Sulphonation. Chemical reaction leading to the introduction of a sulphonyl group into a molecule by direct carbon sulphur linkage.

Surface active agent. A chemical compound which, when dissolved or dispersed in a liquid, is preferentially adsorbed at an interface, giving rise to a number of physico-chemical or chemical properties of practical interest. The molecule of the compound includes at least one group with an affinity for markedly polar surfaces, ensuring in most cases solubilisation in water, and a group which has little affinity for water.

Surface active derivative. Product endowed with surface activity.

Surface activity. All the properties particular to the surface active agents in solution and operating at the interfaces.

Surface phenomena. Phenomena the effects of which (mechanical, electrical, optical, etc.) become apparent at the surface separating two phases (liquid-gas, liquid-solid, liquid-liquid, or gas-solid).

Surface tension. The force per unit length resulting from the free surface energy. It is numerically equal to the free surface energy per unit of surface and is expressed in dynes per centimetre. *Note* — The dyne per centimetre (dyn/cm) is the unit in CGS system. The SI unit is newton per metre (N/m): $1 \text{ N/M} = 10^3 \text{ dyn/cm}$.

Suspending power. In the case of solutions of surface active agents, the ability of certain substances to maintain in suspension particles insoluble in the solution. *Note* – The suspending power can vary very considerably depending upon the nature of these particles.

Structural viscosity. Under isothermal and reversible conditions, reduction without hysteresis of apparent viscosity under a shearing load.

Synergy (synergistic effect). A mixture, in given proportions of two surface active agents or other chemicals, within certain limits of their respective concentration exhibits a given effectiveness for a given concentration of the mixture in the medium in which the measurement is carried out. There is synergy when this concentration is lower than that which would result from the linear combination, in the same proportions, of the concentrations which, for each constituent considered separately, would be necessary to achieve the same effectiveness.

Temperature of clarification. In the case of certain non-ionic surface active agents exhibiting a cloud temperature, the temperature at which the mixing of the two liquid phases becomes homogeneous on cooling. Note — The temperature of clarification is often determined as 'Cloud Point'.

Thixotropy. Under isothermal and reversible conditions, reduction with hysteresis of the apparent viscosity under a shearing load.

Undistilled fatty acids. Produced by splitting of triglycerides and have free fatty acids, minimum of 90 per cent by mass.

Unsaponifiable matter. The whole of the constituents soluble in fatty matter and insoluble in water, which cannot be modified by reaction of saponification producing a salt. *Note* –In practice, and for analytical determination, the whole of the product present in the substance analysed which, after saponification of the latter with an alkaline hydroxide and extraction by a specified solvent, remains non-volatile and the defined condition of test.

Unsaponified matter. A saponifiable substance which remains after a saponification reaction.

Unsulphatable matter. A constituent which is not capable of undergoing a sulphation reaction.

Unsulphated matter. A sulphatable substance which remains after a sulphation reaction and/or is converted into an unsulphatable product during this reaction.

Unsulphonatable matter. A constituent which is not capable of undergoing a sulphonation reaction.

Unsulphonated matter. A sulphonatable substance which remains after a sulphonation reaction and/or is converted into an unsulphonatable product during this reaction.

Washing power. Degree of aptitude of a surface active agent or of a detergent to promote detergency.

Wetting hysteresis. Under certain conditions, the work required to effect the introduction at slow and constant speed of a part of a solid physical surface into a liquid phase differs from that corresponding to the comparable withdrawal of the same part of the surface from the liquid phase. The difference between these two quantities, related to unit surface area, represents a hysteresis termed *wetting hysteresis*. It is expressed in ergs and is numerically equal to the difference between the advancing and receding wetting tensions.

Wetting tension. The force per unit length resulting from the free energy of wetting. It is numerically equal to the free energy of wetting per unit of surface. It is expressed in dynes per centimetre.

Wetting agent. A substance which, when introduced into a liquid, increases its wetting tendency.

Wettability. The ability of a surface to become wetted.

Wetting. In the special case of a surface active agent in solution, action corresponding to bringing into effect the properties of wetting tendency and wettability.

Wetting tendency. The tendency of a liquid to spread over a surface. A decrease in the contact angle between the solution and the surface is shown by an increase in wetting. A zero contact angle corresponds to spontaneous spreading.

Wetting power. The ability to wet.

Wetting agent and dyeing oil. Products increasing the wetting power of the dye bath for the textile fabric. The dyeing oil generally confers on the article dyed an additional brightening effect. *Note* –These products are surface active agents or preparations comprising them, such as alkylsulphates, alkylsulphonates, alkylarylsulphonates, fatty acid condensates and sulphated oils.

BIS SPECIFICATIONS

Essential oils sampling and tests 326–1971

Soap Products

Laundry soap 285–1974

Laundry soap powder 2887–1974

Liquid soap 4199–1974

Shaving soap 5784–1970

Soft soap 7532–1974

Toilet soap 2888–1974

Glycerine 1796–1977

Linear Alkyl Benzene **286– 1978**

Methods of sampling and test for soaps 286–1978

Water for general laboratory use 1070–1977

Methods of sampling of petroleum
and its products 1447–1960

Methods of test for petroleum and its
products, colour by Lovibond tintometer 1448–1960

Methods of test for petroleum and its
products, colour by Saybolt chromometer 1448–1960

Methods of test for petroleum and
its products, Distillation 1448–1960

Methods of test for petroleum and its
products, Doctor test 1448–1960

Methods of test for petroleum and
its products, Flash point 1448–1970

Methods of test for petroleum and its
products, Density and relative density 1448–1972

Methods of test for petroleum and its
products, Bromine number by electrometer 1448–1983
titration method

Specification for alkyl benzene 8401–1977
sulphuric acid (acid slurry)

APPENDICES

APPENDIX-1

USEFUL INFORMATION

Data on Air

Average dry analysis by volume

N_2	78.03%
A	0.94
O_2	20.99
	99.96%
CO_2	0.03
H_2, Ne, He, Kr, Xe	0.01
	100.00%

Values to Use in Combustion Calculations. In combustion calculations, the 0.04% of CO_2, H_2, and rare gases may be ignored. Furthermore, the argon may be lumped with the nitrogen; this is referred to as atmospheric nitrogen.

	% by volume	% by weight	Molecular weight
Atmospheric nitrogen	79.00	76.80	28.16
Oxygen	21.00	23.20	32.00
	100.00	100.00	

Molecular weight   =   28.97

Physical Constants

The Gas-law constant R

Numerical value	Units
1.987	g-cal/(g-mole) (K°)
1.987	Btu/(lb-mole (R°)

82.06	(cm^3) (atm)/(g-mole) (K°)
0.08205	(liter) (atm)/(g-mole (K°)
10.731	(ft^3) (lb)/(in.)^2 (lb-mole (R°)
0.7302	$(ft)^3$ (atm)/(lb-mole) (R°)

1 faraday = 96,493.1 (abs. coulomb) / (g-equivalent)

Avogadro constant

= 6.02380×10^{23} atoms per gram-atom or molecules per gram-mole.

Density

1 g-mole of an ideal gas at 0°C, 760 mm Hg = 22.4140 litres

= 22,414.6 cc

1 lb-mole of an ideal gas at 0°C, 760 mm Hg = 359.05 cu ft

Density of dry air at 0°C and 760 mm Hg = 1.2929 g per litre

= 0.080711 lb per cu ft

1 gram per cc = 62.43 lb per cu ft

1 gram per cc = 8.345 lb per U.S. gal

Length

1 in.	2.540 cm
1 micron	10^{-6} metre
1 Angstrom	10^{-10} metre

Mass

1 lb (avoirdupois)	16 oz
1 lb (avoirdupois)	7000 grains
1 lb (avoirdupois)	453.6 grams
1 ton (short)	2000 lb (Av.)
1 ton (long)	2240 lb (Av.)
1 gram	15.43 grains
1 kilogram	2.2046 lb (Av.)

Mathematical Constants

e	2.7183
π	3.1416

ln N	2.303 log N

Power

1 kw	56.87 Btu per min
1 kw	1.341 hp
1 hp	550 ft-lb per sec
1 watt	44.25 ft-lb per min
1 watt	14.34 g-cal per min

Pressure

1 psi	2.036 in. Hg at 0°C
1 psi	2.311 ft water at 70°F
1 atm	14.696 psi
1 atm	760 mm Hg at 0°C
1 atm	29.921 in. Hg at 0°C

Temperature Scales

Degrees Fahrenheit = 1.8 (degrees centigrade) + 32
Degrees Kelvin = degrees centigrade + 273.16
Degrees Rankine = degrees Fahrenheit + 459.69

Volume

1 cu in.	16.39 cc
1 litre	61.03 cu in.
1 litre	1000.028 cc
1 cu ft	28.32 litres
1 cu metre	1.308 cu yd
1 cu metre	1000 litres
1 U.S. gal	4 qt
1 U.S. gal	3.785 litres
1 U.S. gal	231 cu in.
1 British gal	277.42 cu in.
1 British gal	1.20094 U.S. gal
1 cu ft	7.481 U.S. gal
1 liter	1.057 U.S. qt
1 U.S. fluid oz	29.57 cc

Conversion Factors

To convert from	To	Multiply by
Acres	sq. ft.	43,560
Acres	square metres	4,046.87
Acres	square miles	0.0015625
Acres	sq. yd.	4,840
Ampere-hours (abs.)	coulombs (abs)	3,600.0
Angstrom units	in.	3.937×10^{-9}
Angstrom units	micromicrons	100
Angstrom units	microns	1×10^{-4}
Angstrom units	millimicrons	0.1
Atmospheres	bars	1.0133
Atmospheres	cm. of mercury at 0°C.	76
Atmospheres	dynes/cm.^{2}	1.0133×10^6
Atmospheres	ft. of water at 39.1°F.	33.899
Atmospheres	grams/sq. cm.	1,033.3
Atmospheres	in. of mercury at 32°F.	29,921
Atmospheres	lb./sq. ft.	2,116.32
Atmospheres	lb./sq. in.	14,696
Barrels, cement	lb. of cement	376
Barrels, oil	gal. (U.S.)	42
Barrels	cu. metres	0.11924
Barrels	gal.	31.5
Bars	atmospheres	0.98692
Bars	lb./sq. in.	14.504
Board feet	cu. ft.	1/12
Boiler horsepower	B.t.u./hr.	33,479
Boiler horsepower	kw.	9.803
B.t.u.	calories (gram)	252
B.t.u.	ft-lb	777.9
B.t.u.	hp-hr.	3.929×10^{-4}
B.t.u.	joules	1,055
B.t.u.	liter-atmospheres	10.41
B.t.u.	lb. carbon to CO_2	6.88×10^{-5}
B.t.u.	lb water evaporated from and at 212°F.	0.001036
B.t.u. (mean)	calories, gram (mean)	251.98
B.t.u. (mean)	cu.ft-atmospheres	3676

B.t.u. (mean)	ft-lb.	777.98
B.t.u. (mean)	hp-hr.	3.9292×10^{-4}
B.t.u. (mean)	joules (abs.)	1,054.8
B.t.u. (mean)	kw-hr.	2.930×10^{-4}
B.t.u. (mean)	liter-atmospheres	10.409
B.t.u. (mean)	watt-hours	0.2930
B.t.u. (mean)/minute	hp.	0.023575
B.t.u. (mean)/pound/°F.	calories,gram/gram/°C	1
B.t.u. (mean)/square foot/minute	kw./sq.ft.	0.17580
B.t.u. (mean)/square foot/second for a temp. gradient of 1°F. per in.	calories, gram (15°C.)/ sq.cm. per sec. for a temp. gradient of 1°C. per cm.	1.2404
B.t.u. (60°F.) / °F	calories, gram/°C	453.59
Bushels	cu. ft.	1.2444
Calories, gram	B.t.u.	3.968×10^{-3}
Calories, gram (mean)	B.t.u. (mean)	3.9685×10^{-3}
Calories, gram	ft-lb.	3.087
Calories, gram (mean)	ft.-lb.	3.0874
Calories, gram	joules	4.185
Calories, gram (mean)	joules (abs.)	4.186
Calories, gram	liter-atmospheres	4.130×10^{-2}
Calories, gram (mean)	liter-atmospheres	4.1311×10^{-2}
Calories, kg. (mean)	ft-lb.	3,087.4
Calories, kg. (mean)	hp-hr.	0.0015593
Calories, kg. (mean)	kw-hr.	0.0011628
Calories, kg. (mean)/sec.	kw	4.186
Carats (metric)	grams	0.2
Centimetres	angstrom units	1×10^{8}
Centimetres	ft.	0.032808
Centimetres	in.	0.393700
Centimetres	microns	10,000
Centimetres of mercury at 0°C.	atmospheres	0.013158
Centimetres of mercury at 0°C.	ft. of water at 39.1°F.	0.44604
Centimetres of mercury at 0°C.	lb./sq.ft.	27.845
Centimetres of mercury at 0°C.	lb./sq.in.	0.19337
Centimetres per second	feet/min.	1.9685
Circular mils	sq. cm.	5.0671×10^{-6}
Circular mils	sq. in.	7.854×10^{-7}
Circular mils	square mils	0.78540

Cords	cu.ft.	128
Cubic centimetres	cu. ft.	3.5315×10^{-5}
Cubic centimetres	gal.	2.6417×10^{-4}
Cubic centimetres	oz.	0.033814
Cubic centimetres	qt.	0.0010567
Cubic feet	bu.	0.80357
Cubic feet	cu.cm.	28,317
Cubic feet	cubic meters	0.028317
Cubic feet	cu.yd.	0.037037
Cubic feet	gal.	7.481
Cubic feet	liters	28.316
Cubic feet-atmospheres	ft-lb.	2116.3
Cubic feet-atmospheres	liter-atmospheres	28.316
Cubic feet of water (60°F)	lb.	62.37
Cubic feet per minute	cu.cm./sec.	472.0
Cubic feet per minute	gal./sec.	0.1247
Degrees per second	radians/sec.	0.017453
Drams (apothecaries or troy)	grams	3.8879
Drams (avoirdupois)	grams	1.77185
Fathoms	feet	6
Feet	cm.	30.4801
Feet per minute	cm./sec.	0.508001
Feet per minute	miles/hr.	0.011364
Foot-poundals	B.t.u. (mean)	3.9951×10^{-5}
Foot-poundals	joules	0.04214
Foot-poundals	liter-atmospheres	4.1588×10^{-4}
Foot-pounds	B.t.u. (mean)	0.0012854
Foot-pounds	calories, gram (mean)	0.32389
Foot-pounds	foot-poundals	32.174
Foot-pounds	hp-hr.	5.0505×10^{-7}
Foot-pounds	kw-hr.	3.7662×10^{-7}
Foot-pounds	liter-atmospheres	0.013381
Foot-pounds per second	hp.	0.0018182
Foot-pounds per second	kw.	0.0013558
Furlongs	miles	0.125
Gallons	barrels (liquid)	0.031746
Gallons	cu.cm.	3,785.4
Gallons	cu.feet	0.13368

Gallons	gal.	0.83268
Gallons	liters	3.78533
Gallons	oz. (U.S. fluid)	128
Gallons per minute	cu.ft./hr.	8.0208
Gallons per minute	cu.ft.sec.	0.002228
Grains	grams	0.064799
Grams	drams (avoirdupois)	0.564383
Grams	drams (troy)	0.257206
Grams	pounds (avoirdupois)	0.0022046
Grams	pounds (troy)	0.002679
Grams per cubic centimeter	lb./cu.ft.	62.43
Grams per cubic centimeter	lb./gal. (U.S)	8.3454
Grams per liter	grains/gal. (U.S)	58.417
Grams per liter	lb./cu.ft.	0.062427
Grams per square centimeter	lb./sq.ft.	2.04817
Grams per square centimeter	lb./sq.in.	0.014223
Horsepower	B.t.u. (mean)/min.	42.418
Horsepower	B.t.u. (mean)/hr.	2,545.08
Horsepower	ft-lb./min.	33,000
Horsepower	ft-lb./sec.	550
Horsepower	kw. (g = 980.665)	0.74570
Horsepower	lb. carbon to CO_2/hr.	0.175
Horsepower	lb. evaporated per hr. at 212°F.	2.64
Joules (abs.)	B.t.u. (mean)	9.480×10^{-4}
Joules (abs.)	calories, grams (mean)	0.23889
Joules (abs.)	cu. ft. atmospheres	3.485×10^{-4}
Joules (abs.)	ft-lb.	0.73756
Joules (abs.)	kw-hr.	2.7778×10^{-7}
Joules (abs.)	liter-atmospheres	0.009869
Kilograms per square centimeter	lb./sq.in.	14.223
Kilowatt-hours	B.t.u. (mean)	3,413.0
Kilowatt-hours	ft-lb./hr.	2.6552×10^{6}
Kilowatt-hours	hp-hr.	1.3410
Knots (per hour)	miles/hr.	1.15155
Liter-atmospheres	cu.ft.-atmospheres	0.035316
Liter-atmospheres	ft-lb.	74.735
Liters	cu.ft.	0.35316
Liters	gal. (U.S)	0.26418

Microns	angstrom units	1×10^4
Microns	cm.	1×10^{-4}
Miles (nautical)	ft.	6,080.2
Miles (nautical)	miles (U.S. statute)	1.1516
Miles	ft.	5,280
Miles per hour	cm./sec.	44.7041
Miles per hour	ft./sec.	1.4667
Mils	cm.	0.00254
Mils	in.	0.001
Pounds (avoirdupois)	grams	453.59
Pounds per square foot	grams/sq. cm.	0.48824
Pounds per square inch	kg./sq. cm.	0.070307
Pounds per cubic foot	grams/cu. cm.	0.016018
Pounds per square foot	atmospheres	4.7252×10^{-4}
Pounds per square inch	atmospheres	0.068046
Slugs	lb.	32.174
Square centimetres	sq. ft.	0.0010764
Square feet	sq. cm.	929.0341
Square inches	sq. cm.	6.4516
Tons refrigeration	B.t.u./hr.	12,000

SPECIAL CONVERSION FACTORS

h = Heat-transfer Coefficient

P.c.u./(hr.) (sq.ft.) (°C)	B.t.u./(hr.)(sq.ft.) (°F)	1
Kg-cal./(hr.) (sq.m.) (°C.)	B.t.u./(hr.)(sq.ft.) (°F)	0.2048
G-cal./(sec.) (sq.cm.) (°C.)	B.t.u./(hr.)(sq.ft.) (°F)	7,380
Watts/(sq.cm.) (°C.)	B.t.u./(hr.)(sq.ft.) (°F)	1,760
Watts/(sq.in.) (°F)	B.t.u./(hr.)(sq.ft.) (°F)	490

μ = Viscosity

Centipoises	g./(sec.) (cm.) or poise	0.01
Centipoises	lb./(sec.) (ft.)	0.000672
Centipoises	lb./(hr.) (ft.)	2.42
Centipoises	kg./ (hr.) (m.)	3.60

k = Thermal conductivity

G-cal./(sec.) (sq.cm.) (cm.) (°C.)	B.t.u./ (hr.) (sq.ft.) (in.) (°F)	2,903.0
Watts/(sq.cm.) (cm.) (°C.)	B.t.u./ (hr.) (sq.ft.) (in.) (°F)	694.0
G-cal./(hr.) (sq.cm.) (cm.) (°C.)	B.t.u./ (hr.) (sq.ft.) (in.) (°F)	0.8064

Values for the gas constant R

1543 ft-lb./(°K.) (g.-mole)
0.08206 liter-atmosphere/(°K.) (g.-mole)
8.315 joules/(°K) (g.-mole)
1.987 cal./(°K.) (g.-mole)

APPENDIX-II

MANUFACTURERS OF RAW MATERIALS

HYDROCHLORIC ACID

Atul Products Ltd., Dist. Bulsar, Atul-396020, (Gujarat).

Ballarpur Industries Ltd., Thapar House, 124, Janpath, New Delhi-110001.

DCM Chemical Works Ltd., Najafgarh Road, New Delhi-110015.

Hindustan Heavy Chemicals Ltd., Globe Building, 7-E, Lindsay Street, Calcutta-700 016.

J.K. Chemicals Ltd., J.K. Building, Ballard Estate, Fort, Mumbai-400 038.

Jayshree Chemicals Ltd., 14, Netaji Subash Road, Calcutta - 700 001.

Tata Chemicals Ltd., Bombay House, Homi Mody Station, Fort, Mumbai-400 023.

SULPHURIC ACID

Adarsh Chemicals & Fertilisers Ltd., Udhana-394210, Surat.

Alkali & Chemical Corporation of India Ltd., 34, Chowringhee, Calcutta-700 016.

Andhra Sugar Ltd., Venkatarayapuram, Tamuka-534211.

Baroda Rayon Corporation Ltd., Udhana-394 210, Surat.

Century Rayon & Spinning & Mfg. Co. Ltd., 159, Church Gate, Reclamation, Mumbai-400 020.

DCM Chemical Works Ltd., Najafgarh Road, New Delhi-110 015.

Dharmsi Morarji Chemical Co. Ltd., 317/21, Dr. D. N. Road, Fort, Mumbai-400 001.

Gujarat State Fertiliser Ltd., P.O. Fertiliser Nagar, Vadodara-390 750.

Gwalior Rayon Silk Mfg. (Wvg.) Co. Ltd., P.O. Birlagram, Nagda-456 331.

Hindustan Organic Chemicals Ltd., P.O. Rasayani, District Kolaba, Maharashtra.

Hindustan Heavy Chemicals Ltd., Globe Building, 7-E, Lindsay Street, Calcutta-700 016.

Hindustan Lever Ltd., Lever House, 65/166, Backway Reclamation, Mumbai-400 020.

Indian Rayon Corporation Ltd., Industry House, Ist Floor, 159, Churchgate Reclamation, Fort, Mumbai-400 020.

Indian Dyestuff Industries Ltd., Mafatlal Centre, Nariman Point, Mumbai-400 001.

Jayshree Chemicals & Fertilisers, Industry House, 10, Comac Street, Calcutta-700 017.

J.K. Chemicals Ltd., J.K. Building, N. Morarjee Marg, Ballard Estate, Fort, Mumbai-400 038.

National Rayon Corporation, Eros Theatre Building, IIT Road, Mumbai-400 020.

BORAX & BORIC ACID

Borax Morarji Ltd., Rajmahal, 84, Veer Nariman Road, Church Gate, Mumbai-400 064.

Southern Borax Ltd., 34, Barnby Road, Kilpark, Chennai-600 010.

CALCIUM CARBONATE

Aditya Chemicals Ltd., 407, Red House, 49–50, Nehru Place, New Delhi-110 019.

Carbonate (India) Ltd., Hong Kong House, 31, Dalhousie Square, Calcutta-700 001.

Gulshan Sugar & Chemicals Ltd., 9th Km. Janpath Road, Muzaffarnagar (U.P).

Studies Chemicals Ltd., Nevile House, Graham Road, Ballard Estate, Mumbai.

CAUSTIC SODA

Atul Products Ltd., District Bulsar, Gujarat.

Ballarpur Industries Ltd., Thapar House, 124, Janpath, New Delhi - 110 001.

Gwalior Rayon Silk Mfg Co. Ltd., (Chemical Division), P.O. Birlagram, Nagda, Madhya Pradesh.

Indian Explosives Ltd., 34, Chowringhee Road, Calcutta.

J. K. Chemicals Ltd., J. K. Building, Dougal Road, Ballard Estate, Fort, Mumbai-400 001.

Modi Alkalis & Chemicals Ltd., 18, Community Centre, New Friends Colony, New Delhi - 110064.

National Rayon Corporation Ltd., Homi Modi Street, Ewart House, Fort, Mumbai-400 023.

Punjab Alkalis & Chemicals Ltd., SCO 125-127, Sector 17-B, Chandigarh-160 017.

Saurashtra Chemicals Ltd., P.O. Porbander, 2, Birla Nagar, Gujarat.

Shri Ram Food & Fertilisers Ltd., Najafgarh Road, New Delhi - 110 015.

Shri Ram Vinyl & Chemicals Ltd., Shriram Nagar, Kota-324 004.

CAUSTIC POTASH

Andhra Sugar Ltd., Venkatarayapuram, Tamuka-534 211.

Atul Products Ltd., Hospital Premises, Near Ashoka Mills Ltd., Naroda Road, Ahmedabad-380 002.

Standard Mills Co. Ltd., Mafatlal Centre, Nariman Point, Mumbai-400 021.

HYDRATED LIME

Gulshan Sugars & Chemicals Ltd., 45-B, New Mandi, Muzaffar Nagar (U.P).

Mayur Chemicals, 1205, Delhi Road, Rattan Nagar, Meerut (U.P).

Mettur Chemicals & Industrial Corporation Ltd., Mettur Dam - 636 402, District Salem, Chennai.

HYDROGEN PEROXIDE AND SODIUM PERBORATE

National Peroxide Ltd., Nevile House, J.N. Heredia Marg, Ballard Estate, Mumbai-400 038.

SODA ASH AND SODIUM BICARBONATE

Dharangadha Chemical Works Ltd., Dharangadha, Gujarat.

Saurashtra Chemicals Works Ltd., Birla Nagar, Gwalior, Madhya Pradesh.

Tata Chemicals Ltd., Bombay House, 24, Homi Modi Street, Mumbai.

BLEACHING POWDER

Kanoria Chemical & Industries Ltd., 16-A Brabourne Road, Calcutta - 700 001.

Mettur Chemical & Industrial Corporation Ltd., Salem District, Mettur Dam-636 402.

Modi Alkalis & Chemicals, 18, Community Centre, New Friends Colony, New Delhi-110064.

Shri Ram Food & Fertilisers Ltd., Shivaji Marg, Najafgarh Road, New Delhi-110015.

PERFUMERY CHEMICALS

Asian Chemical Works, 124-126, Shamaldas Gandhi, Princes Street, Mumbai-400 002.

Bengal Chemicals & Pharmaceuticals Ltd., 6, Ganesh Chandra Avenue, Calcutta.

Bush Boakes Allen (I) Ltd., 1-5-7, Sells Street, St. Thomas Mount, Chennai - 600 016.

Calcutta Chemical Co. Ltd., 35, Pandita Road, Calcutta.

Camphor & Allied Products Ltd., Jahangir Building, 13, Mahatma Gandhi Road, Mumbai-400 001.

East India Sandal Wood Oil Factory, Mysore, Karnataka.

Govt. Sandal Wood Oil Factory, Mysore, Karnataka.

Hindustan Lever Ltd., Lever House, 165-166, Backbay Reclamation, Mumbai-400 020.

Industrial Perfumes Ltd., Hay Bunder Road, Near Tank Road, P.O., Sweri, Mumbai.

Karnataka Soaps & Detergents Ltd., 5, Crescent Road, High Ground, Bangalore - 560 001.

Meghalaya Essential Oils, Norgrim Hills, Shillong (Assam).

Naarden (I) Ltd. Saki Vihar Road, Saki Naka, Mumbai-400 072.

New Sandal Wood Oils Co., Kannauj (U.P).

Sh. Kelkar & Co. Ltd., LBS Marg, Mulund, Mumbai-400 080.

ACETIC ACID

Andhra Sugar Ltd., Venkatarapuram, Tamuka - 534 211.

Gujchem Distilleries (I) Ltd., National Chamber, Ashram Road, Ahmedabad-9.

Indian Organic Chemicals Ltd., New Excelisor Building, Wallace Street, Mumbai-400 001.

Sirsilk Ltd., Lingapur House, 3-6-237, Himayat Nagar Road, Hyderabad-560 020.

Somaiya Organic (I) Ltd., Narang House, 34, Shivaji Marg, Mumbai-400 039.

Vam Organic Chemicals Ltd., Skyline House, 85, Nehru Place,
New Delhi - 110019.

ACETIC ANHYDRIDE

Andhra Sugar Ltd., Venkatarapuram, Tamuka - 534 211.

Indian Organic Chemicals Ltd., New Excelisor Building, Wallace Street,
Mumbai-400 001.

Sirsilk Ltd., Lingapur House, 3-6-237, Himayat Nagar Road,
Hyderabad-560 020.

Vam Organic Chemicals Ltd., Skyline House, 85, Nehru Place,
New Delhi - 110019.

ACETONE

Sirsilk Ltd., Lingapur House, 3-6-237, Himayat Nagar Road,
Hyderabad-560 020.

Herdillia Chemicals Ltd., Air India Building, Nariman Point, Mumbai-400 021.

National Organic Chemical Industrial Ltd. (NOCIL), Mafatlal Centre,
Nariman Point, Mumbai-400 021.

BENZENE

Indian Petrochemicals Corporation Ltd. (IPCL), P.O. Petrochemicals, District
Vadodara, Gujarat.

National Organic Chemical Industries Ltd. (NOCIL), Mafatlal Centre, Nariman
Point, Mumbai-400 021.

Steel Authority of India Ltd., Ispat Bhavan, Lodhi Road,
New Delhi - 110 001.

BUTYL ACETATE

Indian Organic Chemicals Ltd., New Excelsior Building, Wallace Street,
Mumbai-400 020.

Somaiya Organics (I) Ltd., Narang House, 34, Shivaji Marg, Mumbai-400 039.

CARBOXY METHYL CELLULOSE

Ashok Cellulose Ltd., Sharda Chambers, 33, Marine Lines, Mumbai - 400 020.

Cellulose Products of India Ltd., P.O. Kathwada, Maize Products, Ahmedabad.

Gujchem Distillers (I) Ltd., National Chambers, Ashram Road, Ahmedabad.

Indian Organic Chemicals Ltd., New Excelsior Building, Wallace Street, Mumbai-400 020.

ZINC OXIDE

Associated Pigments Ltd., 14, Netaji Subash Road, Calcutta - 700 001.

Murarka Paints & Varnish Works Pvt. Ltd., 4 - E, BBD Bagh, 29, Stephens House, Calcutta - 700 001.

Pigments & Chemicals Industries Pvt. Ltd., 32, Armenian Street, Calcutta - 700 001.

Waldies Ltd., Gilliander House, Netaji Subash Road, Calcutta - 700 071.

LABORATORY CHEMICALS

Glaxo Laboratories (I) Ltd., Dr. Annie Besant Road, Worli, Mumbai - 400 025.

Merck (I) Ltd., Shiv Sagar Estate, 'A' Block, Worli, Mumbai-400 025.

Ranbaxy Laboratories Ltd., Nehru Place, New Delhi - 110 019.

S D Fine Chemicals Ltd., 315 - 317, TVI Estate, 248, Worli Road, Mumbai - 400 025.

LABORATORY AND SCIENTIFIC GLASSWARE

Borosil Glass Works Ltd., 44, Khanna Construction House, Abdul Gaffar Road, Worli, Mumbai-400 018.

Seralcella Glass Works Ltd., Konnagar Railway Station, District Hooghly, West Bengal.

BARIUM AND SALTS

Barium Chemicals Ltd., P.O. Ramavaran, R S Bhahadra Chellum Road, District Khammam, Andhra Pradesh.

Chawla Chemical Industries Ltd., Jaji Adam Mansion, 6, Homoji Street, Mumbai-430 001.

Jyoti Chemicals, Mustafia Building, Sir P M Road, Mumbai - 400 001.

Travancore Chemical & Mfg. Co. Ltd., Kalamasseri, Alwaye, Kerala.

ACTIVATED CARBON

Anil Starch Products Ltd., Anil Road, Ahmedabad- 380 025.

ACTIVATED EARTH

Shri Ram Food & Fertiliser, Shivaji Marg, Najafgarh Road, New Delhi-110 015.

Indian Earth Ltd., III, Maharishi Karve Road, Mumbai-400 020.

POTTASSIUM META BISULPHIDE

Liberty Chemical Works Ltd., Nagardas Road, Mogra West, Andheri (E), Mumbai-400 069.

Kesar Sugar Works Ltd., Insurance Building, 7, Jamshedji Tata Road, Churchgate, Reclamation, Mumbai-400 001.

TRI-SODIUM PHOSPHATE

Indian Rare Earths Ltd., Udyog Mandal, P.O. Alwaye, Kerala.

Star Chemicals, 21, Noble Chambers, Parsee Bazar Street, Fort, Mumbai.

FOAMING AGENTS

Hico Products Ltd., Megal Lane, Mahim, Mumbai - 400 016.

Modern Chemical Works, 22, Janki Niwas, N. C. Kelkar Road, Dadar, Mumbai-400 028.

OPTICAL WHITENING AGENTS

Amar Dye-chem Ltd., Rang Udyan, Sitladevi Temple Road, Mahim, Mumbai - 400 016.

Atul Products Ltd., District Bulsar, Atul - 396 020, Gujarat.

Suhrid Geigy Ltd., Express Building, E - Road, Church Gate, Mumbai - 400 020.

APPENDIX III

MANUFACTURERS OF PLANT AND MACHINERY

SOAP AND SYNTHETIC DETERGENT MACHINERY

Anup Engineering Ltd., Anil Starch Premises, Anil Road, Ahmedabad - 380 002.

Ashwamedh Machines Pvt. Ltd., E - 64, MIDC Shiroli-416 122, District Kolhapur, Maharashtra.

Bellestra (India) Ltd., Tulsiani Chambers, 212, Nariman Point, Mumbai.

Hindustan Lever Ltd., Hindustan Lever House, 165–166, Backbay Reclamation, Mumbai.

Hemant Engg. Industries, Plot No. 135-B, 2-CD, Govt. Industrial Estate, Kandivi-West, Mumbai-400 067.

Jaswindra Mechanicals, Near Esis Hospital, Ulhasnagar - 421 002, Maharashtra.

Maharashtra Engineers & Founders, 20, Gorupdep X Lane, 3, Tank Road, Mumbai - 400 033.

Munish & Company Pvt. Ltd., 3-A, Nahur Road, Mulund (W), Mumbai.

Precision Machinists, Plot No. 36-D, 2-CD, Govt. Industrial Estate, Kandivi (W), Mumbai-400 067.

EFFLUENT TREATMENT PLANTS

Batliboi & Co. Ltd., V. B. Gandhi Marg, Fort, Mumbai - 400 023.

Thermax Ltd., 9, Community Centre, Basant Lok, New Delhi - 110 057.

Voltas Ltd., Manekji Wadia Building, 127, Mahatma Gandhi Road, Mumbai - 400 023.

INDUSTRIAL FILTERATION EQUIPMENTS

Anup Engineering Ltd., Anil Starchys Premises, Anil Road, Ahmedabad.

Hein Lehman (I) Ltd., 16, Hare Street, Calcutta - 700 001.

John Fowler (I) Ltd., Marshala Building, Shoorjit Vallabdas Marg, Mumbai.

Vulcan Laval Ltd., Mustafa Building, Sir P M Road, Mumbai-400 001.

SPRAY DRYING EQUIPMENT

Larsen & Toubro Ltd., L&T House, Ballard Estate, Mumbai-400 038.

Macneil & Magar Ltd., Machinos Mackenzie Building, Ballard Estate, Mumbai - 400 038.

AGITATORS, MIXERS & DRYERS

APV Equipment Co. Ltd., Kalwa Hamilton House, 3, Graham Road, Ballard Estate, Mumbai - 400 038.

Larsen & Toubro Ltd., L&T House, Ballard Estate, Mumbai - 400 038.

New Standard Engg. Co. Ltd., NSE Estate, Goregaon East, Mumbai - 400 002

Sudershan Chemical Industries Ltd., 102, Wellesley Road, Pune - 411 001.

EVAPORATORS & CONDENSERS

APV Equipment Co. Ltd., Kalwa Hamilton House, 3, Graham Road, Ballard Estate, Mumbai - 400 038.

Bharat Heavy Plate & Vessels Ltd., Vishakapatnam - 530 102

Larsen & Toubro Ltd., L&T House, Ballard Estate, Mumbai - 400 038.

Saurashtra Engg. Corpn. Ltd., Khatani I. Estate, Mumbai - 400 070.

Test Steel Ltd., Navdeep Opp. Akashwani, Ashram Road, Ahmedabad.

HEAT EXCHANGERS & COOLERS

APV Equipment Co. Ltd., Kalwa Hamilton House, 3, Graham Road, Ballard Estate, Mumbai - 400 038

Bharat Heavy Plate & Vessels Ltd., Vishakapatnam - 530 102.

Hindustan Development Corporation Ltd., Hindu Family Building, 27, Sir R N Mukerjee Road, Calcutta - 700 039.

KCP Ltd., Rama Krishna Building, 39, Mount Road, Chennai - 600 017.

Larsen & Toubro Ltd., L&T House, Billard Estate, Mumbai - 400 038.

Richardson & Gruddas Ltd., Byculla Iron Works, Sir J. J. Road, Mumbai - 400 008.

Texmaco Ltd., 506, Surya Kiran Building, 19, K. G. Marg, New Delhi - 110 001.

Vulcan Laval Ltd., Mustafa Building, 7-A, Sir P M Road, Mumbai - 400 001.

Walchand Industries Ltd., Construction House, Walchand Hirachand Marg, Fort, Mumbai.

PRESSURE VESSEL, REACTION VESSELS, AUTOCLAVE, CHEMICAL STORAGE TANKS, CHLORINE TANK CONTAINERS

Anup Engineering Ltd., Anil Starch's Premises, Anil Road, Ahmedabad Gujarat.

APV Equipment Co. Ltd., Kalwa Hamilton House, 3, Graham Road, Ballard Estate, Mumbai - 400 038.

Bharat Heavy Electricals Ltd., 18–20, K.G. Marg, H. T. House, New Delhi - 110 001.

Fertilisers & Chemicals Travancore Ltd., Udyog Mandal, Palluruthy, Cochin - 682 006.

Hindustan Galvanizing & Engg. Co. Ltd., 96, Garden Reach Road, Calcutta - 700 023.

KCP Ltd., Rama Krishna Building, 39, Mount Road, Chennai - 600 017.

National Rayon Corporation Ltd., Eros Theatre Building, 1, Jamshedji Tata Road, Mumbai - 400 020.

Richardsan & Grudas Ltd., Byculla Iron Works, Sir J. J. Road, Mumbai - 400 008.

Sudershan Chemical Industries Ltd., 102, Welesley Road, Pune - 411 001.

Texmaco Ltd., 506, Surya Kiran Building, 19, K. G. Marg, New Delhi-110 001.

WATER TREATMENT PLANT

Bharat Process & Mechanical Engg. Ltd., Chartered Bank Building, Calcutta.

Ion Exchange (I) Ltd., Tiecicon House, Dr. E. Moses Road, Mahalaxmi, Mumbai - 400 011.

INDUSTRIAL PROCESS CONTROL INSTRUMENTS

Bestobell India Ltd., 31, Camac Street, Calcutta - 700 016.

Instrumentation Ltd., Jhalwar Road, Kota-324 005.

Industrial Electrodes & Gauges Ltd., 15, Ganesh Chandra Avenue, Calcutta - 700 001.

National Instruments Ltd., 1/1, Raja Subodh Chandra Mulick Road, Calcutta - 700 032.

Taylor Instruments Co. Ltd., 14, Mathura Road, Faridabad (Haryana).

Toshniwal Industries Ltd., Industrial Estate, Makhupura, Ajmer - 305 002.

Siemens (I) Ltd., 134–A, Dr. Annie Besant Road, Worli, Mumbai.